M 2257515

AHDB LIBRARY
☎ 02476 478839
ACC.NO. 2246 PO NO 803166
PRICE: £ 60.72 DATE: 3/12/14
CLASS.NO. 344.4 MEU

D1423150

EU Food Law Handbook

Also available in the 'European Institute for Food Law series':

Fed up with the right to food?
The Netherlands' policies and practices regarding the human right to adequate food
edited by: Otto Hospes and Bernd van der Meulen
ISBN 978-90-8686-107-1
www.WageningenAcademic.com/righttofood

Reconciling food law to competitiveness
Report on the regulatory environment of the European food and dairy sector
Bernd van der Meulen
ISBN 978-90-8686-098-2
www.WageningenAcademic.com/reconciling

Governing food security
Law, politics and the right to food
edited by: Otto Hospes, Irene Hadiprayitno
ISBN: 978-90-8686-157-6; e-book ISBN: 978-90-8686-713-4
www.WageningenAcademic.com/EIFL-05

Private food law
Governing food chains through contract law, self-regulation, private standards, audits and certification schemes
edited by: Bernd van der Meulen
ISBN: 978-90-8686-176-7; e-book ISBN: 978-90-8686-730-1
www.WageningenAcademic.com/EIFL-06

Regulating food law
Risk analysis and the precautionary principle as general principles of EU food law
Anna Szajkowska
ISBN: 978-90-8686-194-1; e-book ISBN: 978-90-8686-750-9
www.WageningenAcademic.com/EIFL-07

The enforceability of the human right to adequate food
A comparative study
Bart F.W. Wernaart
ISBN: 978-90-8686-239-9; e-book ISBN: 978-90-8686-791-2
www.WageningenAcademic.com/EIFL-08

EU Food Law Handbook

edited by:
Bernd van der Meulen

EAN: 9789086862467
ISBN: 978-90-8686-246-7

ISSN 1871-3483

First published, 2014

© Wageningen Academic Publishers
The Netherlands, 2014

This work is subject to copyright. All rights are reserved, whether the whole or part of the material is concerned. Nothing from this publication may be translated, reproduced, stored in a computerised system or published in any form or in any manner, including electronic, mechanical, reprographic or photographic, without prior written permission from the publisher, Wageningen Academic Publishers, P.O. Box 220, 6700 AE Wageningen, the Netherlands, www.WageningenAcademic.com copyright@WageningenAcademic.com

While the authors and publisher have used their best efforts in preparing this book, neither the authors nor the publisher can accept any legal responsibility for any errors or omissions that may be made. No warranty is made, express or implied, with respect to the accuracy or completeness of the material contained in this book.

Dedicated to Menno van der Velde

Foreword

The political climate throughout the European Union at the end of the 1990s was heavily influenced by the outbreak of several food scares in a number of Member States, the most serious being Bovine Spongiform Encephalopathy (BSE) and dioxin. Public confidence in food and in politicians collapsed and governments fell. The BSE epidemic in the UK first reported in 1986 quickly spread throughout the country and at its height 1,000 new cases were being reported each week. It subsequently spread to other European Union countries including Germany, Spain, France, and also Switzerland. Finally, after years of denial that BSE could be transmitted to humans, an admission by the UK authorities in 1996 that there was a possible connection between BSE and Creutzfeldt-Jakob Disease (CJD), the human equivalent, caused widespread public alarm. The resulting loss of consumer confidence caused sales of beef to plummet.

At that time in the EU, food law was concerned with facilitating the free movement of food and food products within the internal market and with the implementation of the Common Agricultural Policy. Its focus was more on the producer than on the consumer.

It was clear that the EU framework for food safety was inadequate. Following the enactment of the Maastricht Treaty (1992) and the Treaty of Amsterdam (1997), which extended the EU competence in health, a secure legal base now existed for legislative change to provide an integrated approach to food safety throughout the food chain.

The first step was the establishment of a new Directorate General in 1999 with responsibility for health and consumer protection, DG Sanco. Within six months food safety law in the EU found its voice with the publication of the White Paper on Food Safety in January 2000.

The White Paper laid down policy objectives which were well received by the Council of Ministers and the European Parliament. It emphasised that food safety policy must be based on a comprehensive integrated approach. It placed the primary responsibility for food safety on the food and feed manufacturers, the farmers and food producers. Responding to the realisation that contaminated animal feed was the primary cause of the BSE and dioxin crises, a 'farm to table' approach was advocated covering all sectors of the food chain, including feed production. The White Paper noted that a successful food policy requires the traceability of feed and food and their ingredients. This, together with the establishment of the Rapid Alert System for Food (RASF), would enable adequate procedures to be put in place for the immediate withdrawal of feed and food from the market where a risk to consumer health was identified.

Risk analysis became the foundation on which food safety policy was to be based. 'The EU must base its food policy on the application of the three components of risk analysis: risk assessment (scientific advice and information analysis); risk management (regulation and control); and risk communication'. Independent scientific advice would be provided by a European Food Authority to be established by legislation.

The White Paper identified the need for the enactment by the Council and European Parliament, where necessary, of 84 new pieces of legislation. The most important of these was the General Food Law Regulation 178/2002 (GFL), which gave legal effect to the fundamental principles set out in the White Paper. The GFL established the European Food Safety Authority (EFSA) and ensured that its scientific opinions would serve as the scientific basis for the drafting and adoption of European Community law in this domain. In recognition of the uncertainty that sometimes surrounds scientific knowledge, it also provided for dispute resolution mechanisms designed to facilitate the achievement of consensus, or failing that, to identify and publish the precise reasons for any divergence of opinion. This process is required to be undertaken in a fully transparent manner in the knowledge that consumer confidence cannot be mandated; it can only be earned. The GFL also clearly recognised the role of the Precautionary Principle in providing a valuable tool for the risk manager in appropriate circumstances.

The officials in DG Sanco were charged with carrying out the Herculean task of drafting all these legal instruments, 90% of which were enacted by the end of the Prodi Commission in November 2004. Hardly without exception, every one of the monthly parliamentary sessions in Strasbourg throughout the period 1999-2004 saw a new piece of legislation become law.

The subsequent implementation of all this new legislation has been so well executed by the Commission, with the advice of EFSA, and by the competent authorities in Member States, that the confidence of the European consumer has been regained and retained. There is now widespread acceptance that the occurrence of any food safety issue will be effectively dealt with by the application of a set of rules and procedures designed for the protection of the consumer.

As the safety dimension has become more settled, attention has been given to the science associated with food quality and the maximising of the nutritional benefits of food. Coupled with that there is now a recognition of a requirement to make sure that food companies provide accurate information thus ensuring that the public is not misled. The Nutrition and Health Claims Regulation (NHCR) and the Food Information Regulation provide important contributions to the achievement of that objective.

Of all the laws enacted over the past fifteen years, the most controversial are those associated with GMOs and the NHCR. It is a curious phenomenon that those voices who tend to criticise EFSA in its scientific assessment of GMOs tend also to be those who most strongly defend the scientific assessment of health claims – and *vice versa*. This supports the widespread opinion that EFSA is judicious in its decision making, relying on the faithful adherence to the scientific method by its scientists in the formation of their opinions.

The EU is the biggest importer and exporter of food in the world today and as food is one of the most highly regulated internationally traded products, every care must be taken to ensure that the application by the EU of its food safety regulations must be credible and officials exercising their decision-making powers must act with integrity. Above all, decisions must be free of the taints of commercial opportunism and trade protectionism. Undoubtedly disagreements will occur from time to time, but every effort must be made to resolve such differences in the international fora established for this purpose, such as the *Codex Alimentarius* Commission, with full regard to internationally accepted norms like those found in the provisions of the Sanitary and Phytosanitary (SPS) Agreement.

The publication of the first edition of the *European Food Law Handbook* in 2008 added greatly to the knowledge and understanding of EU Community law. As much new law has been promulgated over the past six years, this new volume has been eagerly awaited, and not just by legal practitioners. A clear understanding of the law protects the interests of the consumer, provides a stable platform for food companies and promotes the transparent application of the law for the benefit of international trade and economic prosperity. The authors of this second edition admirably achieve these objectives.

David Byrne
EU Commissioner for Health and Consumer Protection (1999-2004)

Preface

Bernd van der Meulen

New, continuing and celebrating

This book appears almost to the day *ten years* after the publication of 'Food Safety Law in the European Union: an introduction'. That book was the first in which we set out our approach to EU food law. In 2008 this book was succeeded by the European Food Law Handbook. As witnessed by the similarity in title, the current book in its turn can be regarded as a successor to the book from 2008. However, it is as much an entirely new book as it is an update. The previous two books were the joint effort of two authors – Menno van der Velde and myself. The current book is a multi-author book. New authors have joined the team from all over the EU. They have updated, rewritten and replaced existing chapters and added new ones. As a consequence, each chapter is considered a separate publication indicating its own author(s). Previously existing chapters feature the original author as last author and the new author as first author(s). The sequence in which authors are indicated therefore does not imply any form of ranking. The team of authors includes legal scholars, scientists and practitioners, who have their base in Ireland, Denmark, Portugal, Italy, Germany, Belgium, the Netherlands, Switzerland, Macedonia and South Africa. I'm proud and grateful to have them on the team.

Since the publication of the first book in 2004, the European Institute for Food Series at Wageningen Academic Publishers has grown to include research books, conference proceedings of the European Food Law Association (EFLA) and PhD dissertations, on a wide area of food law related topics such as the human right to food, food security, food law and agriculture, the impact of food law on the competitiveness of the sector, risk analysis, food governance and private food law.

Objectives

The aim of this book is fourfold:

- to provide a thorough understanding of food law to both lawyers and non-lawyers who take an interest in the field of food and its regulatory framework;
- to assign the functional field of food law its proper place within European law;
- to uncover the system of EU food law;
- to provide a framework for further research in this exciting area.

Scope

This book is about food and food safety in European Union law. Food safety is a topic that attracts a considerable amount of interest from all kinds of different scientific angles. In the mid-1990s food safety scares brought food safety to the top of the European policy agenda. There it has remained ever since. On the one hand this is due to the persistence of food safety problems, on the other hand to the fact that an atmosphere of crisis has been mobilised and sustained to achieve policy goals. Food safety is not a stable world. In a hostile environment, it is under constant attack from microorganisms, chemicals, human ignorance, criminal intent and many other adversaries.

This state of affairs contributed to the flourishing of technical sciences concerned with food safety: food microbiology and food toxicology to name but a few. All those who assist in dealing with food safety problems do so within the framework of an applicable set of rules. To get a grip on the situation, this set of rules has been adapted and expanded. We can call the collection of rules that are concerned with food 'food law'. The aim of this book is to present this framework. We encounter such rules at different levels: the global level, the regional level and the national level. This book focuses on the European Union.

We address the subject matter from the assumption that food law is not a world of its own. There is no watershed between rules that concern food and other rules. For this reason, we present food law within its context. This context comprises law at national, international, European and business level, but also history and science.

Presentation and references

The reference list at the end of this book provides for all EU legislation the source in the *Official Journal* (OJ) and for EU case law in the *European Court Reports* (ECR). This was formerly considered the proper way of referring to EU law sources. As all official EU documents are disclosed in online databases we think that this way of referencing will soon be considered outdated. To help readers find their way in the current digital legal environment, an annex to this book explains how law sources can be found in the official databases.

We have applied several techniques to make food law accessible. The text is straightforward. Quotes from legal sources are presented in Law text boxes to make them stand out. Diagrams have been used to represent the subject matter in graphic form. The table of contents, indexes, and list of abbreviations give entries from different sides.

Language

The EU has 24 official languages. Among these is British English. As much as possible we have followed British spelling and grammar. Apart from possible errors – for which we apologise – other forms of English will mainly be found in quotes and in the names of international organisations.

The EU as we know it today has resulted from a development over time. Each stage of development seems to have been marked by the use of different language. At different moments, different labels applied such as the European Economic Community (EEC); the European Communities (EC), the European Community (EC), the European Union (EU), the European Court of Justice (ECJ) and the Court of Justice of the European Union (CJEU); also different concepts were used such as common market, single market and internal market. Even though it may sometimes seem a little anachronistic, in this book we use the current vocabulary also when referring to situations in the past when different names applied. Only in quotes or when the change is the topic being discussed, are the old expressions used.

Dedication

This book is dedicated to Menno van der Velde. Due to his health and retirement Menno laid down his contributions to the discovery of food law. From the very beginning Menno was part of the development of the thinking set out in this book and of the shaping of food law as an academic discipline in its own right. I am grateful for all that Menno has contributed.

Bernd van der Meulen
Professor of Law and Governance at Wageningen University (www.law.wur.nl); director of the European Institute for Food Law (www.food-law.nl).

Comments are welcome at: Bernd.vanderMeulen@wur.nl

Table of contents

Abbreviations

AB	Appellate Body (WTO)
ADI	Acceptable Daily Intake
ADNS	Animal Disease Notification System
ADOI	Annual Declaration of Interests
AFC	Panel on additives, flavourings, processing aids and materials in contact with food (EFSA)
AGRI	Agriculture and Rural Development Committee of the European Parliament
AHAW	Panel on animal health and welfare (EFSA)
ALARA	As low as reasonably achievable
ALOP	Appropriate level of protection
ANS	Panel on food additives and nutrient sources added to food (EFSA)
ARD	Agricultural and Rural Development Department of the World Bank
ARfD	Acute Reference Dose
Art.	Article
AO	Appellations of Origin
AOC	Appellations d'origine contrôlées
AVMSD	Audiovisual Media Services Directive
BADGE	Bisphenol A diglycidyl ether
BCPs	Border Control Posts
BEUC	Bureau Européen des Unions de Consommateurs – the European Consumer Organisation
BFDGE	Bisphenol F diglycidyl ether
BIOHAZ	Panel on biological hazards (EFSA)
BIPs	EU Border Inspections Posts
BIRPI	United International Bureaux for Protection of Intellectual Property
BRC	British Retail Consortium
BSE	Bovine spongiform encephalopathy
BSS	Basic Safety Standards
Bt	*Bacillus thuringiensis*
B-to-B (B2B)	Business to business
B-to-C (B2C)	Business to consumer
bw	Body weight
C	Celsius
CAC	Codex Alimentarius Commission
CAP	Common Agricultural Policy
CCD	Colony Collapse Disorder
CCP	Critical Control Point
CED	Common Entry Document
CEF	Panel on food contact materials, enzymes, flavourings, and processing aids (EFSA)

CEN	Comité Européen de Normalisation (European Committee for Standardization)
CENELEC	Comité Européen de Normalisation Electrotechnique (European Committee for Electrotechnical Standardization)
CFI	Court of First Instance (EU)
CFSP	Common Foreign and Security Policy of the European Union
cfu/g	Colony forming units/gram
CGIAR	Consultative Group on International Agricultural Research
CHED	Common Health Entry Document
CIES	Comité International d'Entreprise à Succursales (the Food Business Forum)
CIF	Costs, Insurance, Freight
CINDI	Countrywide Integrated Noncommunicable Disease Intervention Programme (WHO)
CISG	United Nations Convention on Contracts for the International Sale of Goods
CJEU	Court of Justice of the European Union
cm	centimetre
CMO	Common Market Organisation (Regulation 1308/2013)
CN	Combined Nomenclature (Customs tariff numbers)
COE	Council of Europe
CONTAM	Panel on contaminants in the food chain (EFSA)
COOL	Country of origin labelling
COREPER	Comité des Représentants Permanents (Committee of Permanent Representatives) (EU)
CPs	Control Points
CPVO	Community Plant Variety Office
CSA	Comité Special d'Agriculture (Special Agriculture Committee)
CVED	Common Veterinary Entry Document
DCFR	Draft common frame of reference
DDT	Dichloro-diphenyl-trichloroethane
DES	Diethylstilbestrol
DG	Directorate General of the European Commission's civil service
DG Agri	EC Directorate General Agriculture and Rural Development
DG Entr	EC Directorate General Enterprise and Industry
DG Sanco	EC Directorate General Health and Consumer Protection
DM	Deutsche Mark
DNA	Deoxyribonucleic acid
DOI	Declarations of Interests
DPEs	Designated Points of Entry
DPIs	Designated Points of Import
DSB	Dispute Settlement Body (WTO)
DSU	Dispute Settlement Understanding (WTO)
EC	European Community

ECE	United Nations Economic Commission for Europe
ECHR	European Court of Human Rights (COE)
ECLI	European Case Law Identifier
Ecofin	Economic and Financial Affairs (EU)
ECOSOC	Economic and Social Council (UN)
ECJ	European Court of Justice (EU) (now CJEU)
ECR	European Court Reports
ECSC	European Coal and Steel Community
ECU	European Currency Unit (now EURO)
EDI	Estimated daily intake
EEA	European Economic Area
EEC	European Economic Community
EFFL	European Food & Feed Law Review
EFLA	European Food Law Association
EFSA	European Food Safety Authority
EHEC	Enterohemorrhagic *Escherichia coli*
EMA	European Medicines Agency
ENVI	Environment, Public Health and Consumer Policy Committee of the European Parliament
EP	European Parliament
EPC	European Patent Convention
EPIC	European prospective investigation into cancer and nutrition
EPO	European Patent Office
EU	European Union
Euratom	European Atomic Energy Community
EUREP	Euro-Retailer Produce Working Group
EUROPHYT	European Union Notification System for Plant Health Interceptions
FAO	Food and Agriculture Organisation (UN)
FCD	Fédération des entreprises du Commerce et de la Distribution (Federation of Enterprises of Trade and Distribution)
FCM	Food contact materials
FCS	Food categorization system (CAC)
FDA	Food and Drug Administration (USA)
FDE	FoodDrink Europe (formerly CIAA)
FeBO(s)	Feed business operator(s)
FEEDAP	Panel on additives and products or substances used in animal feed (EFSA)
FFV	Fresh fruit and vegetables
FIAP	Food improvement agents package
FIC	Food Information to Consumers Regulation (Regulation 1169/2011)
FMI	Food Marketing Institute
FOB	Free On Board
FP	Framework programme
FSG	Food for Specific Groups

FSMS	Food Safety Management System
FSO	Food Safety Objective
FVO	Food and Veterinary Office (EU)
g	grams
GAP	Good Agricultural Practice
GATS	General Agreement on Trade in Services (WTO)
GATT	General Agreement on Tariffs and Trade (WTO)
GDA	Guideline Daily Amounts
GFL	General Food Law; Regulation 178/2002
GFSI	Global Food Safety Initiative
GIs	Geographical Indications
GM-food	Genetically Modified Organisms used for human consumption
GMO	Genetically Modified Organism
GMO	Panel on Genetically Modified Organisms (EFSA)
GMP	Good Manufacturing Practice
GTZ	Gesellschaft für Technische Zusammenarbeit (German Agency for Technical Cooperation)
HACCP	Hazard Analysis and Critical Control Points
HCH	Hexachlorocyclohexane
HDE	Hauptverband des Deutschen Einzelhandels (Central Association of German Retail Trade)
HFSS	High Fat, Salt and Sugar
HR	Highest Residue level
ICC	International Chamber of Commerce
ICESCR	International Covenant on Economic, Social and Cultural Rights (UN)
ICJ	International Court of Justice (UN)
ICRP	International Committee on Radiation Protection
ICTSD	International Centre for Trade and Sustainable Development
ICUMSA	International Commission for Uniform Methods of Sugar Analysis
IEC	International Electrotechnical Commission
IFAD	International Fund for Agricultural Development (UN)
IFIC	International Food Information Council
IFS	International Food Standard
INFOSAN	International Food Safety Authorities Network (WHO)
IP	Intellectual Property
IPC	International Patent Classification
IPPC	International Plant Protection Convention
IPRs	Intellectual Property Rights
ISBN	International Standard Book Number
ISO	International Organization of Standardization
ISSN	International Standard Serial Number
JECFA	Joint FAO/WHO Expert Committee on Food Additives
kcal	Kilocalorie
kg	Kilogram

kJ	Kilojoule
L	Lot
LCB	Licensed Certification Body
LOD	Limit of detection
MB	Management board (EFSA))
MCPDs	Monochloropropanediols
mg	Milligram
µg	Microgram
ml	Millilitre
mm	Millimetre
MOAH	Mineral oil aromatic hydrocarbons
MOH	Mineral oil hydrocarbons
MOSH	Mineral oil saturated hydrocarbons
MRL	Maximum Residue Limit (or Level)
MS	Member State(s)
MSM	Mechanically separated meat
NACMCF	National Advisory Committee on Microbiological Criteria for Foods (USA)
NDA	Panel on dietetic products, nutrition and allergies (EFSA)
NFR	Novel foods regulation (Regulation 258/97)
NGO	Non-Governmental Organisation
NHCR	Nutrition and Health Claims Regulation (Regulation 1924/2006)
NOAEL	No observed adverse effects level
NOGE	Novolac glycidyl ether
NS	Non-numerical Acceptable Daily Intake 'Not Specified'
NVLR	Nederlandse Vereniging voor Levensmiddelenrecht (Dutch Food Law Association)
OECD	Organization for Economic Cooperation and Development
OHIM	Office for Harmonization in the Internal Market
OIE	International Office of Epizootics
OIML	Organisation internationale de métrologie légale (International Organization of Legal Metrology)
OJ	Official Journal of the European Union
OJ C	Official Journal Communication Series
OJ L	Official Journal Legislation Series
OLFs	Other legitimate factors
PARNUTS	Foods for particular nutritional uses
PAHs	Polycyclic aromatic hydrocarbons
PBRs	Plant Breeders' Rights
PC	Performance Criterion
PCA	Permanent Court of Arbitration
PCBs	Polychlorinated biphenyls
PCT	Patent Cooperation Treaty
PDO	Protected Designation of Origin

PGI	Protected Geographical Indication
pH	Measure of acidity or basicity
PLH	Panel on Plant health (EFSA)
PM	*Pro memoria*, i.e. to be decided
PO	Performance Objective
PoEs	Point of Entry
ppb	Parts per billion
PPR	Panel on plant protection products and their residues (EFSA)
PRAC	Procédure de réglementation avec contrôle (Regulatory procedure with scrutiny)
PVP	Plant Variety Protection
QS	Quantum Satis (the required quantity and no more)
QUID	QUantitative Ingredients Declaration
RCF	Regenerated cellulose films
RDA	Recommended daily allowance
R&D	Research and development
RASFF	Rapid Alert System for Food and Feed (EU)
REACH	Registration, Evaluation, Authorisation and Restriction of Chemical substances (Regulation 1907/2006)
SARS	Severe acute respiratory syndrome
SCF	Scientific Committee on Food (EU)
SCFCAH	Standing Committee on the Food Chain and Animal Health (EU)
SCVPH	Scientific Committee on Veterinary measures relating to Public Health
SDOI	Specific Declaration of Interests
SG	Secretary-General
SML	Specific migration limits
SPS	Agreement on the Application of Sanitary and Phytosanitary Measures (WTO)
SQF	Safe Quality Food
SQFI	SQF Institute
STEC	Shiga-toxin producing *E. coli*
STMR	Supervised Trials Median Residue
TBR	Trade Barriers Regulation (Regulation 3286/94)
TBT	Agreement on Technical Barriers to Trade (WTO)
TEC	EC Treaty
TEU	Treaty on European Union (Lisbon)
TFEU	Treaty on the Functioning of the European Union (Lisbon)
TPRM	Trade Policy Review Mechanism (WTO)
TRACES	Trade Controls and Expert System
TRIPs	Agreement on Trade Related Aspects of Intellectual Property Rights (WTO)
TSE	Transmissible spongiform encephalopathy
TSG	Traditional Speciality Guaranteed
UK	United Kingdom

UN	United Nations
UNCITRAL	United Nations Commission on International Trade Law
UNECE	United Nations Economic Commission for Europe
UNICEF	United Nations Children's Fund (previously United Nations International Children's Emergency Fund)
UNGA	United Nations General Assembly
UNTS	United Nations Treaty Series
UPOV	Union internationale pour la protection des obtentions végétales (International Union for the Protection of New Varieties of Plants)
US / USA	United States of America
USDA	US Department of Agriculture
vs.	versus (against)
vCJD	Variant Creutzfeldt-Jakob disease
VTEC	Verotoxicogenic E. coli
WFP	World Food Programme (UN)
WHO	World Health Organisation (a United Nations Specialised Agency)
WIPO	World Intellectual Property Organization
WTO	World Trade Organisation
ZLR	Zeitschrift für das gesamte Lebensmittelrecht (Periodical for all food law)

1. Introduction

Bernd van der Meulen and Menno van der Velde

1.1 The food sector and its law

The food sector is the largest production sector in the European Union. It is also heavily regulated. According to the EU's directorate general for Enterprise and Industry, the food industry is in the top three of most regulated industrial sectors in the EU.[1] The motives for regulating the food sector have developed over time. Initially the focus was on creating a level playing field for a well-functioning internal market. Currently the emphasis is on protecting consumer interests, primarily life and health, but also the possibility to make informed choices and to be protected from misleading practices.

European food law has an impact on the interests of many stakeholders: consumers, officials and civil servants at the European and national level in the Member States of the EU, candidate members and neighbouring countries, businesses in the EU or those trading with EU partners, legal and regulatory affairs managers, advisors, consultants and many more.

1.2 Multi-layered food law

Eating and drinking are necessities of life. From the beginning of time the gathering, production and distribution of food has been central to many human relations. And in all societies rules and regulations concerning this primary human activity have developed. As those relations have become more complex, stretched across borders and ecosystems, organisation has become increasingly necessary. Food law is now formulated and used at many levels: from the international community to several smaller parts of the world. The European Union plays a key role, as it provides the legal framework and substance for rules applied on a continental scale to societies that belong to the most developed in the world, coming from very differing economic and political traditions, and has had to develop a new infrastructure to accommodate the style and standard of the Union's food law. At the national level there is enough left to work on and assist the implementation of EU law into the national law. Even the private sector participates in regulation.

1.3 This book

This book gives an account of European Union (EU) food law. After 2000 an overhaul of European food law was undertaken. The blueprint was laid down in the White Paper on Food Safety (2000). The general concepts and principles have

[1] For details, see Chapter 8.

been enacted in Regulation 178/2002, known as the General Food Law. In 2003 there followed a package of regulations regarding genetically modified foods, in 2004 a package regarding food hygiene and enforcement, in 2005 legislation on packaging[2] and allergen labelling, in 2006 a regulation on nutrition and health claims, in 2008 a package of regulations on food additives and other so-called food improvement agents, and in 2011 a regulation on food information to consumers. At the moment of writing these lines, proposals are being looked at for novel foods and official controls.

This large-scale reformulation of food law in the EU is driven by several concerns, two of which stand out: the intimate connection of food law with the internal market, and the lessons learned from the major food scares of the 1990s. The innovated body of European food law is at the centre of this book. It is described and analysed. It is situated in the midst of those elements of European Union law that determine the conditions and restrictions for food law in the Union and in the Union's Member States. Food law on the level of the European Union is flanked by many legal rules from the wider international plane, from the national legal systems of the Member States and from private regulation by the food sector. This book presents just a few excursions into these flanks. Our main attention is focussed on the food law of the EU itself, whilst the excursions serve to fit EU food law to these surroundings, or to underline its specific characteristics.

The development of European Union food law is traced over the four decades of its history since the end of the transition period used to create the common market. Knowledge of certain aspects of this development is necessary to understand the present situation of the internal market, harmonisation of national legislation, and food law. Readers will learn how law is used as an instrument to achieve food safety and to deal with food safety problems.

1.4 Overview

The book is subdivided into three parts. The first part we call 'prerequisites'. It provides the background information that readers need to be able to understand the analysis of food law. Food law is a field where law and science meet. It is practiced just as often by professionals with a background in food sciences as by professionals with a background in law. For the benefit of those readers who do not have a background in law, we provide a short introduction to legal thinking and legal method in general (Chapter 2), to international law (Chapter 3, in particular Section 3.2) and to the law and institutions of the European Union (Chapters 4, 5 and 6). The external context of EU food law – international food law – is discussed in Chapter 3. Against this background the turbulent development of food law in

[2] For details, see Chapter 19.

Europe from the beginnings of the European Community to the release of the White Paper on Food Safety in 2000 is described in Chapter 7.

The second part of the book – the systematic analysis of food law – forms the heart of the book. It sets out the analytical thinking underlying this book (Chapter 8) and goes on to apply this thinking in eight chapters. These address the general principles (Chapter 9), rules on the authorisation of products (Chapter 10), the chemical and physical safety of products (Chapter 11), the biological safety of products (Chapter 12), the processes to prevent and deal with food safety problems (Chapter 13), and communication from businesses to consumers (Chapter 14). These chapters largely take the perspective of the legislature (Chapter 9) and businesses (Chapters 10-14). Chapter 15 takes the perspective of public authorities and discusses their powers of enforcement and incident management. The final chapter of part II (Chapter 16) focuses on the position of the consumer in EU food law.

The third part of the book is labelled 'Selected topics'. It provides a collection of topics subjected to further analysis, in particular: special foods – formerly known as PARNUTS (Chapter 17), import requirements (Chapter 18), food contact materials (Chapter 19), nutrition policy (Chapter 20), animal feed (Chapter 21), intellectual property (Chapter 22) and private food law (Chapter 23). A short conclusion is drawn in Chapter 24.

Writing about law is reciting, describing, comparing and analysing the law. In order to separate the original instrument of law from description, analysis and comments, the legal texts including articles from treaties, legislation, or case law appear in Law text boxes. In this way the reader is invited to look at the original and to develop personal views on the description and analysis made by the authors. To further stimulate recourse to the sources, references to the official documents are often made, especially to readily available sources on the World Wide Web. The account is supported by diagrams that give a more or less graphical image of these basics and provide some additional information in a nutshell.

1. Prerequisites

2. Introduction to law

Bernd van der Meulen and Menno van der Velde

Food law is a specialisation in law with many interdisciplinary characteristics. Since not all readers will be familiar with legal science, its vocabulary and methods, this chapter provides a brief introduction to the science of law. Readers with a background in law may want to skip it.

2.1 Introduction to legal science

The word 'law' refers to limits set on various forms of human behaviour. Legal science does not concern itself with so-called descriptive laws, such as those found in natural sciences and economics, which describe how people or natural phenomena usually behave. Legal science is concerned with prescriptive laws, which are rules governing how people ought to behave. The rule that food shall not be placed on the market if it is unsafe,[3] does not say that this does not happen but that it should not happen. It imposes a duty on those who put food on the market. Likewise, there are rules prescribing what is to be done if – despite this rule – unsafe food does appear on the market. Law in this sense is intended to prevent and resolve conflict by organising and describing the rights and duties in our society. Those rights and duties are overseen and interpreted by a system of courts, which is backed by the enforcement powers of police services.[4]

To a very large extent law is national. It is closely linked to the political structure of a given society. Political debates in parliaments, for example, often result in legislation. Obviously, this leads to different laws from one country to the next, even where such laws are intended to regulate or define the same fields of activity.

[3] Article 14(1) General Food Law. Discussed repeatedly in this book.

[4] Other means to promote the observance of the law, such as arbitration or mediation, are ultimately based on the public court system, for persons participating in these voluntary proceedings to apply the law know that ultimately the public power will back up the decisions reached in the voluntary proceedings, unless they are blatantly unjust (in which case the public court system can provide a solution itself).

Diagram 2.1. Law as a multifunctional word.

Law, a word about a word

Readers may have noted that already in the few lines above we have used the word 'law' with two different meanings, one referring to a system or academic discipline and one to a text. Other languages have different words for these two meanings. Law as a system is called *ius* in Latin, *le droit* in French, *das Recht* in German, *het recht* in Dutch. The law (the article is rarely used when referring to the system) as a text that forms part of legislation also known as 'act' or 'enactment' is *lex* in Latin, *la loi* in French, *das Gesetz* in German and *de wet* in Dutch.

These languages, however, may suffer from another language difficulty. Where English has separate words for law (as a system) and right; the expressions ius, droit, Recht and recht cover both meanings.

Most national legal systems in the world can be grouped into one of two basic systems: the common law system, used in England and most of the former British colonies, including the USA, India and Australia; and the civil law system, which is used across most of continental Europe and in its former colonies like parts of Latin America. English law is called common law because it was common to the whole of England. This uniformity was achieved at a very early stage. It began soon after the Norman Conquest of England in 1066, when King and Court travelled around the country hearing grievances. Because the same people sat in different courts, the same understanding of the law applied whichever court made the decision.

The common law system is based on the principle of deciding cases by reference to previous judicial decisions (known as 'precedent'), rather than by reference to written statutes drafted by legislative bodies. English law has evolved from the 12th century onwards in this way, through a body of reported cases, which present specific problems out of which general points of law are extracted. Formulation of the law is bottom-up from a specific event to a general principle. Judicial decisions accumulate around a particular kind of dispute and general rules or precedents emerge. These precedents are binding on other courts at the same or a lower level in the hierarchy. The same decision must result from each situation in which the material or relevant facts are the same. The law evolves when opinions change as to which facts are relevant, and when novel situations arise. The role of the legislator is limited to filling in gaps in the common law or to correct it.

The civil[5] law system is used by most of continental Europe and parts of Latin America. The main body of the law is written down in statutes or codes in a more or less logical and organised (codified) way across all the subject areas. Many

[5] The word is derived from Latin. It referred to the law for the Roman citizens. Roman law has strongly influenced the development of legal thinking and scholarship in continental Europe.

modern codifications find their origins in the Napoleonic (civil[6] and criminal) codes of the 18th century.

In such systems, precedents are not normally recognised as binding on the courts, although court rulings often have a strong influence through the quality of their reasoning and the authority of the courts that gave them. The civil law system results in a top-down system of codified law books, which are based upon broad principles and then broken into legal topics similar to those of the common law countries.

Generally speaking, civil law approaches are more cautious than common law approaches.[7] For almost a millennium, common law has developed in reaction to problems that took place and were dealt with in terms of liability. In civil law on the other hand, legislatures make it their business to foresee and prevent societal problems. In other words, civil law lays down the rules before the game is played, while common law invents the rules while the game is in progress.

In practice, the differences between common law and civil law systems are not as big as they used to be. In common law countries the amount of statutory legislation is rapidly increasing and in civil law countries the courts take it upon themselves to fill in the gaps that have been left open by the legislators and to give authoritative interpretation to the written enactments. Changes in interpretation may have effects similar to changes in legislation. In both systems legislation and court rulings (case law) are the two most important sources of law. It is mainly the way of thinking and of analysing the law that is different today.

This book reflects the civil law tradition in that it takes legislation and its underlying structure as a frame of analysis.[8]

2.2 Sources of law

Rules do not just 'exist', they must originate from a body that is accepted by doctrine and the courts as having rule-making powers. These are the so-called sources of law.

Although differing in relative weight, in both common law and civil law systems statutory legislation ('written law') and case law[9] are the most important primary sources of law, with textbooks, journal articles, encyclopaedias, indices and digests

[6] In the sense of 'private law', see Section 2.7.

[7] J.C. Coffee, Jr., The Rise of Dispersed Ownership: The Roles of Law & the State in the Separation of Ownership and Control, 2001, Yale Law Journal 111: 62-63.

[8] The abundant use of footnotes for references of a detailed nature and for additional information is typical for legal literature.

[9] We avoid the word 'jurisprudence'. It may be used to refer to case law, but also to legal science.

making up a body of secondary sources. Primary sources of law are those sources where the rules are found. Other primary sources of law are international treaties and so-called unwritten law (like legal custom and general principles of law). Secondary sources provide tools that may help to understand the primary sources. They themselves do not give law, only opinions about law.

Diagram 2.2. Sources of law.

Written law	International treaties Legislation
Unwritten law	Case law Customary law Principles of law

Somewhere in between primary and secondary sources of law are policy documents holding interpretations of the law by the authorities responsible for implementing the law. These documents do not hold 'law' in themselves, but have more weight than scholarly writings. They are often referred to as 'soft law'.[10]

European Union law resembles the civil law system more than the common law system. Legislation (treaties, regulations and directives) is the major source of European Union law. However, the importance of the case law of the European Court of Justice can hardly be overestimated.

The sources of international law are listed in the Statute of the International Court of Justice, presented in Law text box 2.1. The Statute is an integral part of the Charter of the United Nations, the world's largest international organisation.[11] Therefore, the statement of Article 38 carries some authority.

2.3 Legal analysis

The core of legal craftsmanship is to solve cases by applying legal rules. From a legal perspective, a case is solved once the rights and obligations of the respective parties have been identified.

2.3.1 Legal relation

The elementary particle in any legal analysis, whether in scientific research, legal design, application of the law *ex ante* to decide on a course of action or *ex post* to sort out the consequences of actions from the past, is the individual legal

[10] On soft law see for example: L..J. Senden, Soft Law in European Community Law, Hart Publishing, Oxford, UK, 2004. It is beyond the scope of this book to discuss the concept of soft law.
[11] As from the accession of South Sudan in 2011, the UN has 193 members.

Law text box 2.1. Sources of public international law, Article 38 Statute of the International Court of Justice.

1. The Court, whose function is to decide in accordance with international law such disputes as are submitted to it, shall apply:
   a. international conventions, whether general or particular, establishing rules expressly recognised by the contesting states;
   b. international custom, as evidence of a general practice accepted as law;
   c. the general principles of law recognised by civilised nations;
   d. subject to the provisions of Article 59, judicial decisions and the teachings of the most highly qualified publicists of the various nations, as subsidiary means for the determination of rules of law.
2. This provision shall not prejudice the power of the Court to decide a case ex aequo et bono, if the parties agree thereto.

relation. This is the relation between stakeholders (any legal entity such as; people, corporations, states, institutions) framed in terms of rights and obligations. This is true when a given relation is subject to scrutiny, but also in legislative drafting. In the latter case, the general nature of the law notwithstanding, the legislator needs to conceptualise what relations governed by this law should look like.

In performing legal analysis, the analyser has to connect different levels (or scales). For the sake of simplicity, the person performing this analysis is referred to in this essay as 'the lawyer'. Legal analysis is, however, by no means the exclusive domain of legal practitioners. According to an old Roman maxim still applied today, each citizen is deemed to know the law.[12] Applying and invoking the law is thus relevant to all, and – up to a point – thus also analysing it. The concept of 'lawyer' as used in this essay therefore potentially includes all; and in particular researchers in food law, food regulatory affairs managers and students.

Let us look at a simple example taken from food law: a food business operator has the intention to make a delivery of milk to a dairy processor. It transpires that the milk is contaminated with dioxins. Question: Would the food business operator be allowed to deliver the milk?

To answer this question, the 'lawyer' would need to find an applicable rule. From the body of all the existing laws he calls upon Article 14(1) of Regulation 178/2002: food shall not be placed on the market if it is unsafe.

[12] '*Nemo censetur ignorare legem*' or '*Nemo ius ignorare censetur*'. This means that the law can be applied to all, in principle without allowing for recourse to ignorance of its content as a justification for non-compliance.

To find this one particular rule among all the existing rules requires a certain skill and some knowledge regarding this legal landscape. From the information regarding the case 'milk contaminated with dioxins to be delivered to a processor' the lawyer makes an intellectual journey to all the existing rules to single out one that may be applicable, which s/he now has to apply to the case. Regarding this application, at least two questions present themselves: does milk contaminated with dioxins qualify as food which is unsafe and, if so, does delivering to the processor qualify as placing it on the market? Behind these questions further questions may emerge such as: does milk qualify as food? To answer these questions, further travels within the totality of rules may be needed to find more rules that may apply and further information regarding the case may be needed such as: what level of dioxins is present in the milk?[13]

2.3.2 ORC grid

The example shows that even in the simplest of cases, at least three levels have to be included in the analysis: (1) the level of the case; (2) the level of the rule; and (3) the level of all the rules. The latter can be called the legal system or the legal order. At the level of the legal order the rule(s) must be found, at the rule level the rule(s) must be interpreted[14] and at the case level, the rule(s) must be applied.

A basic level of legal analysis knowledge must be acquired of the legal order/system such that the lawyer knows what kind of rules can be found where.[15] Along with this knowledge, the skill must be developed to actually find the required rule and apply it to the case. Diagram 2.3 shows for each of the three, abbreviated as ORC) a student's tasks and required skills along with the data used and methods applied.

Diagram 2.3. ORC grid – simplified legal method.

Level/aspect	Task	Data	Method	Skills
Legal **O**rder (Macro)	Find the rule	Literature/ sources of law	Systematisation	Knowledge
Rule (Meso)	Interpret the rule	Sources of law	If-then-syntax; qualify facts as antecedent and conclude in terms of consequent	Selection
Case (Micro)	Apply the rule	Sources of rights and obligations		Application

[13] For a more detailed discussion of legal analysis connecting different levels of abstraction, see: P. Wahlgren, Legal reasoning. a jurisprudential model, 2000, Scandinavian Studies in Law 40: 199-282. Wahlgren subdivides the legal reasoning process into six interrelated steps: identification, law-search, interpretation, evaluation, formulation and learning.

[14] That is, in order to be understood they must be given meaning.

[15] For example, searching a database of rules for 'milk' and 'dioxin' would not have yielded the lawyer in the example Article 14(1) of Regulation (EC) 178/2002.

The knowledge of the legal order needed to find the rule is acquired from literature that systematically presents the legal order at issue. As discussed in Section 2.2, within the legal order, sources of law are recognised in which the rules can be found.

2.3.3 If-then syntax

These rules must be given meaning in the context of the case at issue. Rules can be analysed on the basis of syllogism: i.e. of an if-then syntax. If the conditions the rule sets ('the antecedent') are fulfilled, then the rule states the consequences ('consequent'). In our example, if 'milk contaminated with dioxins' qualifies as 'unsafe food' (which it does at certain levels of contamination), then the rule states: 'it shall not be placed on the market'. If delivering to a processor qualifies as 'placing on the market' (which it does), this means that the milk may not be delivered.[16]

The antecedent can be seen as the independent variable, the consequent as the dependent variable. Fulfilment of the antecedent is decisive for applicability of the rule, the consequent for its operation.

The objective of the remainder of this chapter is to give the reader insight into legal systems in general. The objective of the remainder of this book is to give the reader insight into food law in particular. It is up to the reader to use these insights to find the rules needed to solve issues and to apply these rules for this purpose.

2.4 Branches of law

In legal doctrine different branches of law are distinguished. Most novices think first of criminal law, but there is more to law than just crimes and punishment. A first distinction is between public and private law. Private law is concerned with relations between parties who have no public power or do not exercise it.[17] There are different kinds of private parties with different legal status. Among them are natural persons, associations of persons without legal personality, and various corporate legal persons. National private law can be subdivided in several ways; one of them is an important division in the law of persons, law of property, contract law and non-contractual liability law (or tort law). Public law covers relations in which public authorities are involved as such: they are using their public powers. National public law can be subdivided into constitutional law, administrative law

[16] For a more detailed discussion of the if-then syntax, see: P. Wahlgren, Legal reasoning. a jurisprudential model, 2000, Scandinavian Studies in Law 40: 199-282. For a more detailed discussion of our approach to legal method, see: B.M.J. van der Meulen, Governance in Law. Charting legal intuition, in: A.L.B. Colombi Ciacchi *et al.* Law & Governance. Beyond the Public-Private Law Divide? Eleven international publishing, The Hague, 2013, pp. 275-309.

[17] The State, for example, is exercising public power when it collects taxes, but usually not when it is buying products.

and criminal law. Diagram 2.4 presents an overview of the branches of law that are most relevant for food law.

2.5 Constitutional law

Constitutional law is concerned with the recognition of human rights and the organisational structure of the state and its government. Most countries have

Diagram 2.4. An overview of the branches of law that are relevant for food law. The most relevant branches have a lighter shade.

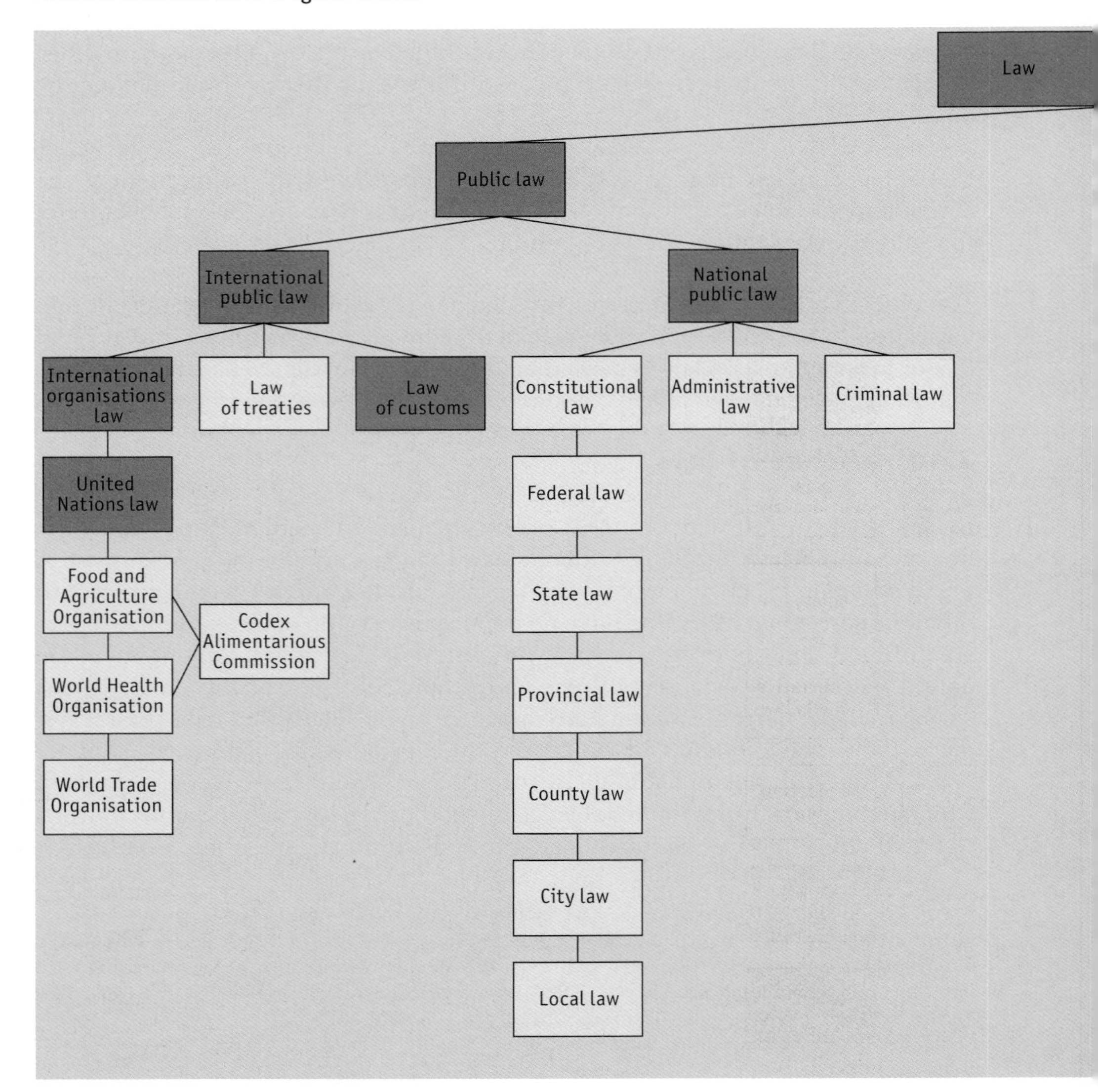

a written constitution defining the legislator, the executive, the powers of the courts, etc. For the European Union the founding treaties fulfil the function of a constitution.[18] Countries that do not have a written constitution[19] still have parts of their customary law and legislation that deal with constitutional issues and define one way or another the legislator and the other essential organs of their state. Many constitutions create a balance of power between the organs or institutions of the state. By dividing the powers of the state over several organs

[18] See Chapters 4 and 5.
[19] Like, for example, the United Kingdom.

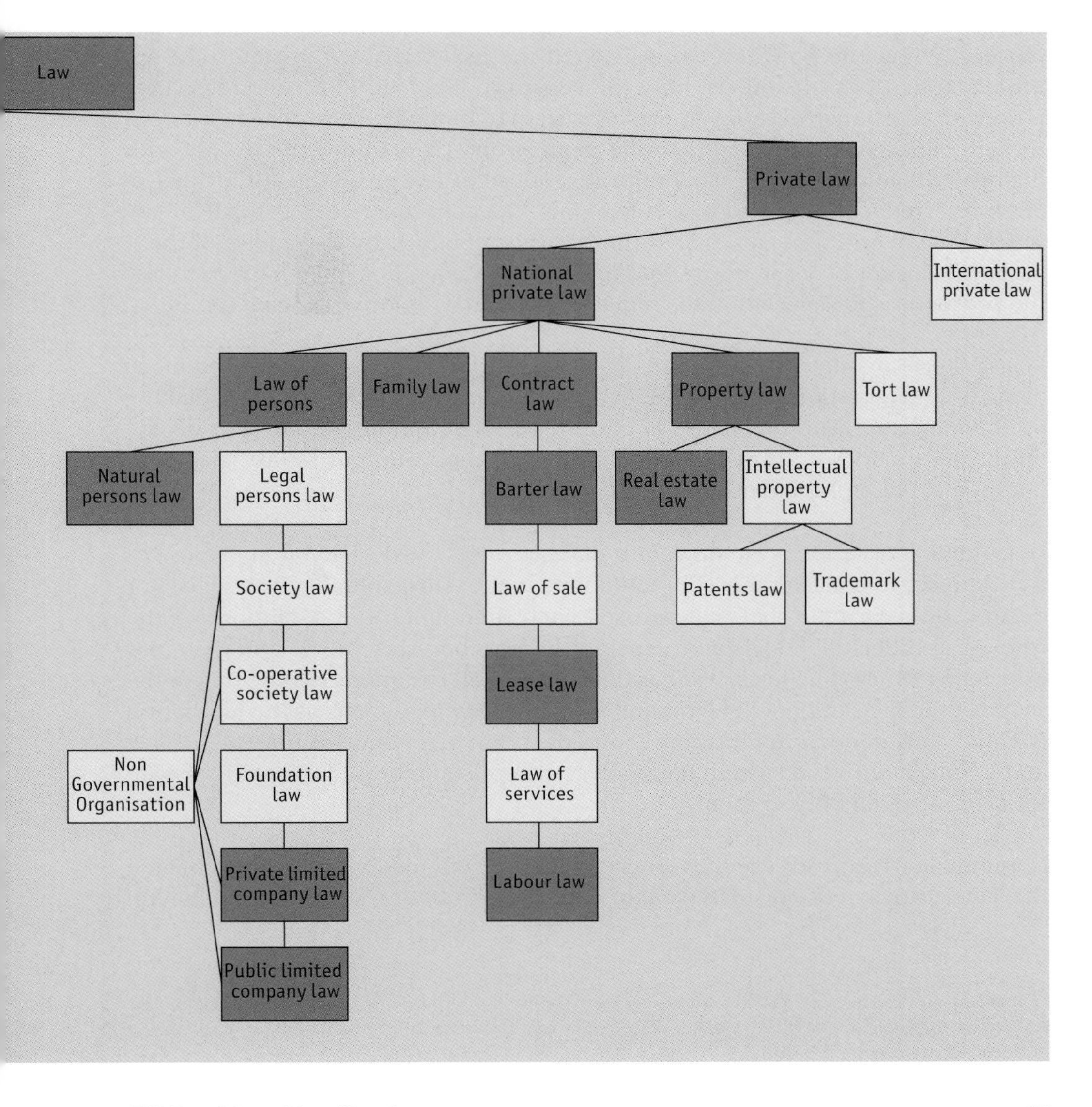

they aim to protect the freedom of their citizens. Constitutions also differentiate several basic functions of the state, such as maintenance of public order and peace, rulemaking, execution of policies, impartial appraisal of conflicts about facts and law, and settlement of those conflicts by an authoritative judgement. All constitutions differentiate the functions of legislator, executive and judiciary, but differ in how they attribute these functions to different organs. The question 'who creates rules on food' is thus a question of constitutional law, and may be answered differently in different national systems.

In most countries and in the European Union (EU) public authorities – especially the executive branch – only have those powers that have been explicitly conferred on them by the legislator. This means that only the legislator can create new powers. When the powers involved touch the essentials of the state that are protected by the constitution, only the constitutional legislator can create new powers or re-arrange existing powers. A system of limited powers that guarantees the protection of human rights to the population and provides the people with instruments to defend themselves against abuse of power by the authorities, is often referred to as applying 'the rule of law'.[20] A consequence of the rule of law is that the government who makes the law is itself bound by these laws. If these authorities want to act in a way that the law does not allow, they have to follow the procedures for legislation to change the law first, before they can act.

Legislation is an important means of organising the law and expressing the opinions of society. Legislation is the name of a specific procedure for decision making by the legislator. The procedure is often used for different purposes: to decide on the state's budget, to make rules for general application or to take a solemn decision about a particular issue.

In general, legislation is equated with making rules, the decision that prescribes a certain set of consequences for situations that occur generally. The legislator decides on a formula that will be used each time when the conditions of the rule are fulfilled. The legislator does not know when this will be the case, nor how many times. The legislator has decided that all situations that belong to the scope of the legislation will be treated the same way for as long as the legislation is valid. This general application is one of the cornerstones of the rule of law and a breach with the system of privilege where some or all have their private legislation, given only to them.

Common law countries use the terms Bill and Act. Bill indicates a proposal which the legislature can decide to make into binding law, the Act. The Acts that contain

[20] The German expression '*Rechtsstaat*' is also well known. On the rule of law see: B.M. Hager, The Rule of Law: A Lexicon for Policy Makers, 2nd ed., Mansfield Center for Pacific Affairs, 2000, available at: http://tinyurl.com/pdr5ozq.

general prescripts, rules, are often called statutes. The word legislation is used primarily to indicate the process of making these statutes.

Within the confines of this book the word legislation has the meaning that is common to the civil law traditions: the rules made by the legislature.

2.6 Administrative law

Administrative law deals with the exercise of public powers by the executive in relation to private persons (including businesses), and provides systems of legal protection for these persons. The exercise of public powers changes the legal position of persons irrespective of their consent. Therefore these persons need special protection attuned to the one-sided use of public, administrative, powers. Administrative law provides the rules for the legitimate use of public power and offers procedures to deal with complaints. For instance, rules concerning the process of obtaining authorisation to bring novel foods to the market, or how to contest such an authorisation granted to a competitor, belong to the field of administrative law.

The European Union has its own administrative law. The examples in the description below will be taken from the EU.

The (administrative) decision is a central concept in administrative law. A large part of the work of the executive branch of government is to apply in individual cases the rules that are laid down in legislation.

In food businesses it is usually the responsibility of the regulatory affairs department to deal with public authorities acting under administrative law. The legal department of businesses concerns itself more often with contracts and other private law matters.

2.6.1 The decision

A special way of applying legislation is to give administrative authorities the power to apply the rules in individual cases to situations or persons whose identity is known. The rule can make certain actions unlawful but provide the opportunity to ask permission for those actions in individual cases.

Law text box 2.2. Article 288(4) TFEU on decisions.

A decision shall be binding in its entirety. A decision which specifies those to whom it is addressed shall be binding only on them.

The decision to give permission and the decision to refuse permission are examples of administrative decisions. Administrative law contains the rules that determine the types of administrative decisions, the rights and obligations they can prescribe, the way the authorities have to make these decisions, and the objections or appeals that can be lodged against the decision. Some legislation contains rules that can be applied to individual persons by administrative authorities. These authorities apply the legislation to the individual person when they take an administrative decision that specifies her/his rights or obligations. The legislation is not made to be applied only once, its rules have to be applied time and again, every time the conditions specified by the rules are fulfilled.

The administrative decision is made for one person whose name and address are known, and applies the general provisions to the individual case. The administrative decision can be made on request or on the initiative of the authority.

The decision is binding for the person to whom it is addressed. A decision can do many things. It can refuse a request by an individual person to apply the general legislation, it can impose an obligation and it can grant a right.

Diagram 2.5. Administrative decisions: contents, initiative and example.

Administrative decisions	Contents	Initiative	Example from book
License	Permits an identified individual to do what is otherwise forbidden	Interested person requests	Market authorisation of a food (Chapter 9)
Subsidy	A grant of money from a public authority for a specified activity	Interested person requests or initiative of authorities	
Enforcement action	The authorities restore the lawful situation by removing illegal objects or restoring a missing object	Initiative of the public authority; possibly at the request of a 3rd party	Suspension of an unsafe food; closure of a food business (Chapter 13)
Administrative sanctions	Take away a right as punishment	Initiative of the public authority	
Other administrative decision	Varies	Varies	

2.6.2 License

The license is made on request; it gives a person permission to do what is generally forbidden. The government wants to keep a close watch on some developments. The legislator makes a statutory regulation that declares specified activities illegal. But interested persons can request the executive power to grant an exception by a license. The license is an explicit decision that allows the activity to be carried out by the person to whom the decision is addressed. The granting of a licence is subject to the administrative law rules on decision making. The licence may be granted under conditions and for a specified period of time. The conditions give the authorities the opportunity to make prescriptions and terms that apply the general conditions set by the legislation to the specific situation that the authorities know from the request and their information gathering.

Chapter 9 of this book deals with pre-market approval schemes. For certain types of foods like food additives, food supplements, genetically modified foods and novel foods there is a ban on bringing them to the market. This ban can be lifted for certain foods after they have been proven to be safe. This can be done by a general rule (including an additive in the list of approved additives) or by licence (called 'authorisation') granted to the applicant (for example to bring a novel food or GMO to the market).

2.6.3 Subsidy

A subsidy is an example of an administrative decision that determines a right for a specific person. The subsidy is an entitlement to a sum of money provided by an administrative authority for a specific activity for the benefit of a specific group of people, or the general interest of society. The subsidy is not a payment for goods or services supplied to the authority. The granting of a subsidy may contain obligations.

2.6.4 Administrative enforcement

An enforcement action is a set of physical acts taken by the administrative authority to end an infringement of the law.

Such an infringement can be the presence of an object or situation that is not allowed. In food law this will mainly apply to unsafe foods.[21] The authorities will remove the object or change the situation. An infringement can be an omission to do what legislation prescribes. The public authorities will end the omission. The offender will have to pay the costs in either case.

[21] See generally Regulation 882/2004 on Official controls, discussed in Chapter 15.

The authorities have to put the decision to take enforcement action in writing. It is an administrative decision that has to be notified to the persons responsible for the infringement before the measures are taken. The decision contains the following information:[22]

- A specification of the action that will be taken.
- The legislation or statutory regulation that has been breached.
- The fact that the offender will have to pay the costs.
- The final period granted to the offenders to take the required measures themselves.

2.6.5 Injunction

An alternative for the enforcement action is to send the offender an administrative decision with the specification of his duty combined with a statement describing the sum of money that he will have to pay for each day he fails to do his duty (so called 'penalty payment'). Decisions of this type are sometimes called injunctions. Such a payment of money is not a sanction but a coercive sum of money. The offender can end the obligation to pay it by doing what s/he has to do. The sum of money will become larger for every period s/he fails to do that.

A sanction (or 'administrative fine') is a fixed sum of money that has to be paid for an offence already committed. The offender cannot undo the sanction by performing her/his duties.

2.6.6 General principles of good governance

In most legal systems, administrative authorities have to observe certain general principles of proper public administration or good governance in their contacts with persons and organisations. The principles are part of the legislation in some countries or have been created by administrative courts and accumulate in their case law. See Diagram 2.6. Examples of important principles and duties for administrative authorities.

2.6.7 Administrative procedure

The time limit for taking a decision about a request is determined by the legislation that the request wants to be applied. If that legislation does not provide a time limit a reasonable period must be respected. The authority must give the person who requests the decision the opportunity to state his views.

[22] See for example Article 54(3) of Regulation 882/2004.

Diagram 2.6. Examples of important principles and duties for administrative authorities.

Name	Content	Example
	Every administrative authority has the following obligations:	
Impartiality	Perform its duties without prejudice	
Integrity transparency	Ensure that a personal interest of the public authority's decision makers does not influence its decision making on the matter	
Good decision making	Gather the necessary information concerning the relevant facts and the interests to be weighed	
Detournement de pouvoir	The authority may not use the power to make a decision for a purpose other than that for which it was given	Article 5 TFEU
Proportionality	The adverse consequences of a decision for one or more interested parties may not be disproportionate to the purposes to be served by the order	TFEU Protocol (2) on the application of the principles of subsidiarity and proportionality
Accountability transparancy	Administrative decisions must be based on reasons that carry the decision and have to state those reasons	Article 296 TFEU; Article 41(2) 3rd indent of the Charter of Fundamental Rights of the European Union

2.6.8 Administrative review

In some countries, the person to whom the administrative decision is addressed has the right to submit a notice of objection to the administrative authority which made the decision. The same right can be used against a written refusal to make a decision and the failure to make a decision in a reasonable period of time.

Such an objection gives the authority the obligation to reconsider its decision. The authority will change its decision if it is convinced by the objections or a part of them. The authority can also change its decision based on arguments that are not part of the objections.

If the authority rejects the objections, it is sometimes possible to appeal to a higher administrative authority by submitting a notice of appeal to the appeals authority.[23] This authority has the power to make a decision that changes the decisions of lower authorities.

[23] For example, Article 36 of Regulation 1829/2003 on genetically modified foods gives the European Commission the power to review decisions and failures to act by the European Food Safety Authority.

2.6.9 Appeal to a judge

As a consequence of the rule of law it is always possible to appeal to an administrative court by submitting a notice of appeal to that court. The court is independent of the administrative authorities. It can declare the administrative decision null and void. The authority that made the decision will have to make a new one.

Law text box 2.3. Article 263(4) TFEU on the right to appeal an institution's decisions.

> Any natural or legal person may (...), institute proceedings against an act addressed to that person or which is of direct and individual concern to them, and against a regulatory act which is of direct concern to him or her (...).

A fundamental difference between an administrative decision and legislation is that an appeal (in common law often called 'a remedy') against a decision is guaranteed and an appeal against a particular act of legislation is usually refused. The separation of powers and the balance of power between the legislator and the judiciary makes it impossible for a judge to deal with general objections against an Act of the legislator. The judge has to decide on the internal consistency of the body of legislation, and has to strike down a piece of legislation that is in conflict with higher legislation. But this assistance of the judiciary to the legislator to preserve the intended meaning of two conflicting pieces of legislation is different from a court case that sets out to test all legislation.

The administrative decision must be tested against the legislation that was made to provide the rules for that kind of decision. Occasionally the test will reveal that the administrative authority made a mistake. In some of these cases it turns out that there are different opinions about what a rule means. The judge will select or construct the valid meaning of the rule from the history of the making of the rule, or from the place of the rule in the system of law. In some of these cases it will become clear that the rule is in contradiction to a higher rule. The judge will base her/his decision on the higher legislation which makes the disregarded rule meaningless.

The person for whom the decision is made has the rights of objection and appeal. The same rights are given to interested parties, persons whose interest is directly affected by the decision. The criterion of 'directly affected interest' separates the persons who are entitled to object from persons in the general public to whom this right is denied. The interests that qualify a person as directly affected are, for instance, his activities in the same line of business or property rights that will be affected. If the authority gives one person a licence and refuses to give the same licence to another person, this last person is an interested party if his circumstances are comparable.

2.7 Criminal law

Criminal law is the branch of law that defines crimes and fixes punishments for them. Criminal law also includes rules and procedures for preventing and investigating crimes and prosecuting suspected criminals, as well as provisions governing the composition of courts, the conduct of trials, the organisation of police forces and the administration of penal institutions. A crime is an offence against the collective interest, and is prosecuted by the state. Punishments for criminal acts include fines or a prison sentence, but also probation or community service which are seen as alternatives for a prison sentence.[24]

Countries in the civil law tradition have a Criminal Code. It is an inventory of crimes and punishments. It contains the definitions of crimes, for example fraud, and states the limits to the punishment that a judge can give. In several countries criminal rulings on poisoning and fraud provided a basis for early developments in food law.[25]

The Code does not contain all offences and crimes. The legislator makes laws on other issues and includes the definitions of new crimes for behaviour that is relevant for that issue and rejected by society.[26]

European Union legislation lays down rules on several aspects of food law. Some of these are enforced with the weapons of criminal justice. The Union does not have its own criminal law nor the authority to prosecute and try suspects. The Union relies on the Member States to provide these remedies. Article 17 of the General Food Law (Regulation 178/2002) on responsibilities lays the first responsibility to ensure that foods or feeds satisfy the requirements of food law on food and feed business operators. The Member States are ordered to enforce food law, and monitor and verify that the relevant requirements of food law are fulfilled. In addition, 'Member States shall also lay down the rules on measures and penalties applicable to infringements of food and feed law. The measures and penalties provided for shall be effective, proportionate and dissuasive'.

Traditionally criminal law has been used for the enforcement of legal obligations. Increasingly, however, enforcement powers are also granted under administrative

[24] In most countries capital punishment is considered contrary to human rights and has therefore been abolished. See for example Article 2(2) of the Charter of Fundamental Rights of the European Union: No one shall be condemned to the death penalty, or executed.

[25] See for example: G. Dannecker (ed.) Criminal Law on Food Processing and Distribution, and Administrative Sanctions in the European Union, Trier Academy of European Law series issue 10, Bundesanzeiger Verlag, Cologne, Germany, 1994.

[26] An example is the legislation on road traffic. To drive away after an accident without taking care of the possible victims, and without identifying yourself is a crime that is defined in the traffic legislation. The same legislator has made the Criminal Code and the traffic legislation, and there is no special procedure for Codes, so there is no difference in hierarchy between the two.

law as described above. Even private law may play a role in enforcement. For instance, the rule that unsafe food may not be brought to the market may be enforced under criminal law if in (national) legislation it is defined as a crime to sell unsafe food. The legislator may also choose to apply administrative law sanctions instead of criminal law sanctions. Regardless of the system of sanctioning that has been chosen, the victim of unsafe food may claim compensation for damages under private liability law (see Section 2.7.5 and Chapter 14, Section 14.7).

2.8 Private law

Private law[27] deals with the enjoyment of rights and obligations, relations and disputes between individuals or organisations without public power. Public authorities wielding public power, (lawyers use the terminology 'public authorities as such'), are not involved in these relations and disputes.[28] A public authority can be a party in a private law contract, for instance when a ministry of the national government hires a caterer to provide daily lunches for its civil servants. In such a case the state represented by the government is treated as any other private contract partner, and public powers cannot be used in this relationship. For private law, the role of public power is merely to provide the means and forum through which private disputes are resolved. If need be, the state provides an armed 'strong arm' (the police) to enforce the verdict of a judge in a private law suit.

The structure of private law can be described briefly by focussing on four complexes which each have their own character: the law of persons defining the status of a person, law of property defining the claims of persons on goods (including intellectual property on ideas), contract law defining legal agreements between persons, and non-contractual liability law (or tort law) for injuries and damages caused by other persons that have no other pre-existing relevant legal relation with the victim. In tort it is the act that caused the damage that created the legal relation.

2.8.1 Draft Common Frame of Reference

Private law is the national law of the Member States. It is not European Union Law. Certain European rules make requirements on the content of national private law, such as the directive on product liability. For its functioning in the Member States, European law to some extent depends on national private law. To ensure that European law is designed in a way that fits well with national private law, the European Commission has requested a team of experts to provide a summary of the common features of private law in the Member States. The experts provide a summary in the form of a code. It is known as the 'Draft Common Frame of

[27] Also referred to as 'civil law', but not to be confused with civil law as distinct from common law.
[28] The use of public power is ruled by constitutional or administrative law.

Reference' (DCFR).[29] This DCFR can be used as a model in education. Some believe that the European Commission hopes that in the long run it will develop into a real European Civil Code.

The DCFR covers topics relevant for the EU Internal market such as contract, property and liability. Other topics such as law of persons are not included. Here below, where applicable we select examples from the DCFR, rather than from national civil codes.

2.8.2 Law of persons

In law persons can have two distinctly different characters: natural persons and legal persons together represent all there is on personality. Human beings are called natural persons in contrast to the legal person which is an artificial construct made by law.

Natural persons

The status of persons is directly connected to their existence as human beings, even before they are born. The rights of the person are confirmed and protected by human rights law and find many of their aspects regulated by private law. The right to a name, inviolability, family life, marriage, upbringing, protection of a minor by limiting his capacities to enter into binding legal arrangements and recognition of the full legal capacities of the natural person who comes of age: these are all part of private law. Countries with a civil law system usually have a Civil Code that encompasses the law of persons.

Legal persons

Legal personality is a legal construction that enables organisations to act and be treated in law like a human being. An organisation with legal personality can own property, make contracts, be liable and so on. Examples of legal personality are an association, a foundation, or a limited company. The constructions that are possible and necessary to create a legal person are part of private law, usually a part of the Civil Code on persons.[30]

Diagram 2.7 depicts the essential building blocks of a legal person: the need for a recognisable design of the entity, the need for internal organisation, based on

---

[29] Study Group on a European Civil Code, Principles, Definitions and Model Rules of European Private Law. Draft Common Frame of Reference (DCFR), Sellier European Law Publishers, München, Germany, 2009.

[30] Council Regulation 2157/2001 on the Statute for a European Company has introduced the possibility to set up a legal person under EU law. So far this option has not gained much popularity.

Diagram 2.7. A legal person. The legal person is a fiction that enables organisation and the separation of rights and liabilities of the legal person from those of the (natural) persons who act on behalf of the legal person. The basic structure of the legal person is rather simple: replace the general meeting of the shareholders by the same meeting of the members of an association and the Diagram turns into a scheme of a society. Most Civil Code countries have similar basic sets of possible legal persons. Public law legal persons have the same structure, although they are founded by special public legislation. That legislation provides their charter for the organisation and the internal and external rules.

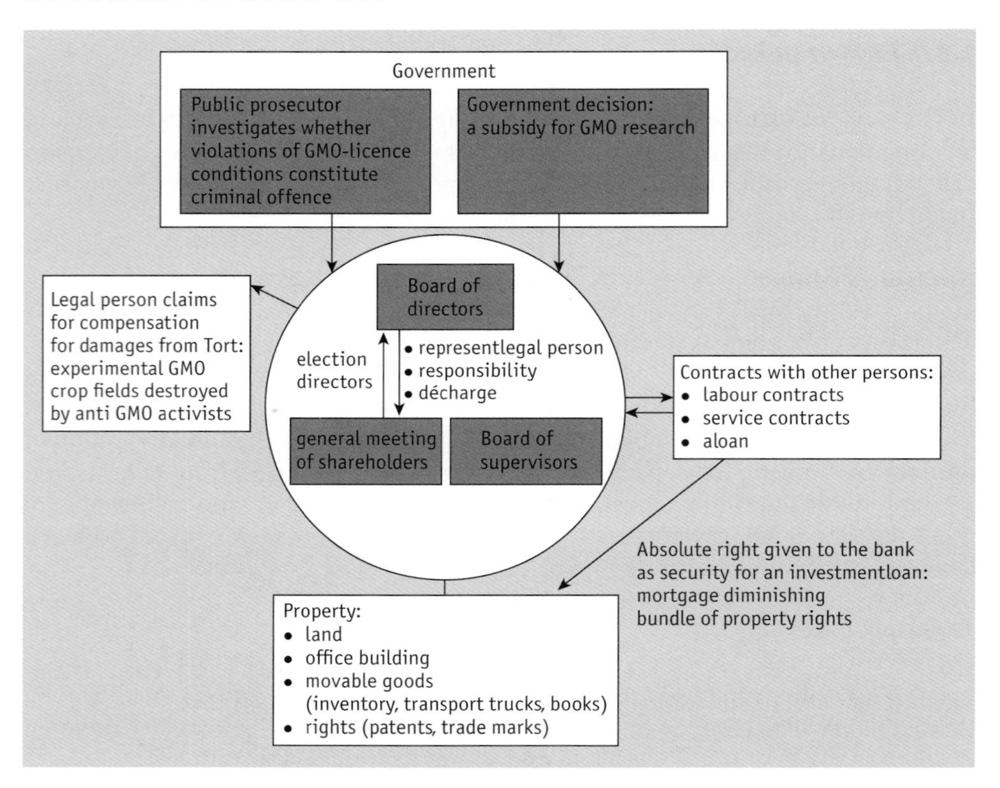

internal rules,[31] the need for representation and the need for rules on external relations and liabilities. Representation is a legal construct that defines that the acts and omissions of one person are deemed to be the acts and omissions of another person.[32] The law takes regard solely of the legal consequences for the person who is represented. The representative does not bind himself and so liabilities or contractual obligations do not entitle the creditors to demand compensation from the representative: their rights are rights in the relation to the (legal) person who was represented. Representation can be based on a contract. For the legal person representation is a necessity which has to be an element of the legal construct.

[31] The so-called Articles of Association, or by-laws.

[32] See for example Articles II-6:101-6:112 DCFR.

Although legal persons can be created or directed by other legal persons, ultimately there must be one or more natural persons to create the construction and act on behalf of it.

Public law has its own legal persons such as the local communities, intermediate bodies like counties or provinces, special bodies like universities (these can also be private law legal persons) or the states of a federation and the state itself. The European Union is a legal person.[33]

2.8.3 Property law

In civil law countries the Civil Code prescribes a closed system for property law. Contrary to contract law (discussed in Section 2.7.4), where the parties are free to use their creativity to design new types of contracts, the possible types of property rights are limited. They are circumscribed by the Code. Property of tangibles, material goods like land, this book or a delivery van, is an absolute right.[34] It is absolute because it gives the owner rights that s/he can maintain against all other persons. Everyone has the obligation not to disturb the owner in her or his enjoyment of her or his property. Property right is not absolute in the sense that it would provide limitless powers. There are and always have been laws that limit the rights of the property owner. If you own a piece of land, your property right does not automatically allow you to build a factory or plant an orchard on it.

The Civil Code provides rules for certain other absolute rights, which are split-offs of property law. You may own the land, but someone else may own the (private law) right to build on that land. You may create the absolute right of way for someone else to walk or drive over your land. Alternatively you could have made a contract with this person giving the same right to walk or drive over your land. Only it will turn out to be not quite the same right. The contract you made is a legal relation between the two persons involved. It is a personal obligation. If you sell your land and someone else becomes the owner, you are no longer able to allow your contract partner to pass over the land. If you create the absolute right of way, that right continues to exist even when you transfer the property of the land to someone else, or when you die. As an absolute right it has to be recognised by everyone, including the person who acquires the land after you. As a consequence, private law has attached heavy formal demands on absolute rights. If you want to transfer ownership, everyone has to have the opportunity to know about the transfer. Therefore the law requires a deed made by a notary public to decide which property is handed over. The transfer of property, as a result of a contractual obligation, can only be made by this deed and by the entry of the deed describing the transfer into a public register.

[33] See Article 47 of the Treaty on European Union.

[34] It is also called a 'real' right because it defines the relationship between a person and (not another person but) a thing (*'res'* in Latin).

Diagram 2.8. Rights included in property.

Property: a bundle of rights
Right to access the property
Right to use
Right to consume (e.g. destroy)
Right to retrieve from others (*revindicatio*)
Right to exclude use by others
Right to transfer

Law text box 2.4. DCFR on ownership.

VIII. – I:202: Ownership
'Ownership' is the most comprehensive right a person, the 'owner', can have over property, including the exclusive right, so far as consistent with applicable laws or rights granted by the owner, to use, enjoy, modify, destroy, dispose of and recover the property.

Special laws have created special property rights which are collectively known as intellectual property rights. Copyright, patent right, trademarks give the person who first entered them in the prescribed register the exclusive rights related to these creations. Like 'conventional' property rights, they give the owner absolute rights to be respected by everyone. Given its importance for the food sector, we have dedicated a separate chapter to intellectual property, therefore it will not be discussed in this chapter.

2.8.4 Contract law

A contract, in law, is an agreement that creates binding obligations between the parties involved. It is a mutual exchange of promises, which the law will enforce. In general, to make a valid contract, there must be two or more separate and definite parties, an offer, an acceptance, the intention to create legal relations, and a capacity to do so.[35] The protection of public morality, public policy or public security is part of the demands for a valid contract. In general, contracts may be either oral or written. Certain classes of contracts, however, must be concluded in writing in order to be enforceable. These may include contracts involving the sale and transfer of real estate.

The obligation to fulfil the promise made in the contract can be upheld in a court of law. In case of non-fulfilment the debtor (the one who made the promise) is

[35] In most jurisdictions young children are considered incapable of concluding contracts.

liable for the damages suffered by the creditor (the one to whom the promise was made).[36]

To a large extent food safety is ensured through so-called self-regulation or private standards. Contract law is a legal instrument for this self-regulation. Businesses may for instance make contracts with other businesses, or a foundation promoting food quality, to apply certain measures to prevent contamination of food during its production and to submit to third party audits to prove that they have adhered to their contracts.

2.8.5 Non-contractual liability law (tort law)

The fourth main complex of private law is tort law, the law of wrongful acts that have to be prevented, and if they occur have to be stopped and for which compensation of damages has to be paid.[37] In the civil law tradition this category has a name that varies with the language. The name indicates that it is an unlawful occurrence that must be restored by the use of private law, irrespective of the criminal law relevance of the act. Private law contains several types of liabilities, but the most prominent liability is non-contractual liability.

The name non-contractual liability for this separate category is the result of a negative selection. The common law tradition uses the word 'tort'. The crucial aspect of this category is the absence of a legal relation between the persons who get involved in a non-contractual liability situation. The event that brings them within the reach of this part of the law is itself the trigger of the sequence: event → unlawful → duty to restore lawful situation with compensation of damages.

Torts mainly encompass those obligations and duties of care which we all owe to others who may foreseeably be affected by our actions, balanced by a standard test of reasonableness that is applied so as not to extend torts to remote and generalised effects. To pass that test, a defendant must demonstrate that s/he has taken all reasonable care to avoid acts and omissions, which could reasonably be foreseen to cause injury or harm to others (due diligence). Torts are essentially civil wrongs that provide individuals with a course of action (or 'remedy') (that is, a wrong which the courts recognise and are willing to judge) for damages to atone for the breach of a legal duty of care. Damages for torts may include economic as well as physical damage, but will not include prison terms or fines, as such remedies are reserved for public authorities. A consumer who suffers ill health due to a food business neglecting its duty of care with respect to food safety could

[36] See Study Group on a European Civil Code, Principles, Definitions and Model Rules of European Private Law. Draft Common Frame of Reference (DCFR), Sellier European Law Publishers, München, Germany, 2009, book III, Chapter 3.

[37] The word tort comes via Old French from the Latin *tortus*, the past participle of *torquere* which means to twist. So tort is (something) twisted.

thus seek compensation both for damage to health, and for economic damage in the form of medical costs or income losses. A special type of tort is based on EU and national legislation on product liability.[38]

The same behaviour that caused the tort with its private law consequences may also be defined as a criminal act. Then the public prosecutor for crimes will start criminal proceedings to obtain a conviction and punishment.

Law text box 2.5. DCFR on liability.[39]

VI – I:101: Basic rule

(1) A person who suffers legally relevant damage has a right to reparation from a person who caused the damage intentionally or negligently or is otherwise accountable for the causation of the damage.

2.9 Substantive and procedural law

Many branches of law can be subdivided into substantive law and procedural law. Substantive law consists of the rules that define a particular right or duty. Property right is an example of substantive private law, theft is substantive criminal law. The civil law states separate substantive law from procedural law and present them in different law books. They have a Code Civil with a Code of Civil Procedure, and a Criminal Code with a Criminal Procedural Code.

2.9.1 Code of Civil Procedure

The Code of Civil Procedure provides the rules for court cases to solve conflicts about private law. The rules prescribe how to begin a case and how to proceed before the court of first instance and to appeal to higher courts. Civil procedures depend to a large extent on the action and reactions of the parties. One of them serves a writ of summons to the other to state her/his claims and to announce that s/he has put the case on the court's agenda. The other party has three options. S/he can end the conflict by doing what the writ of summons demands. S/he can contest the claims in court, or not react at all.

[38] See Chapter 16, Section 16.7.

[39] European Union law is mainly administrative law. The EU has no criminal and private (civil) law system. For criminal and civil issues the EU depends on the national legal systems in the Member States. For this reason EU legislation often relates to national civil law. To acquire a better understanding of the common aspects and differences in civil law in the Member States, a group of scholars has been asked to devise a common frame of reference. In 2008 these scholars published a draft common frame of reference (DCFR) setting out the common principles found in civil law in the Member States in the *form* of a code. This DCFR is not (EU) law but a description of national law in the Member States. For a hyperlink see the reference list.

The judge follows the steps taken by the parties:
- When the summoned party does not appear in court, the judge will decide the case on the basis of the writ, unless the demands are evidently unreasonable.
- When both parties present their case in court the judge will consider the claims and the offers of the parties to provide evidence. The judge will direct who will present evidence in court and will determine its value. The parties in the case can reach an agreement to settle the conflict out of court at any time.

The Code of Civil Procedure determines the rules for every step in court cases, but the parties decide first whether they want to take that step. They put each other in positions that require a reaction.

2.9.2 Execution of judgements in private law cases

The decision of the court becomes final when all possibilities for appeal are exhausted or parties decide to refrain from appeal. In most cases the parties will carry out the orders of the court. They know that the final decision of the court is an enforceable title that provides a writ of execution. This writ can be used by the party whose rights are recognised and affirmed by the court to force the other party to do what the judge ordered. The writ of execution is brought by a bailiff. Her/his position can vary according to the national system of law. A bailiff is either a public law officer or a private person who has made it his/her business to carry out the provisions of the Code of Civil Procedure to force obedience to court decisions.

The bailiff will serve the writ to the party that has to pay or surrender an object. When that party refuses to comply, the bailiff will take the step to execute the judgement. S/he will take possession of the assets and sell them to raise the money determined by the court. In some cases s/he can take the disputed object and hand it over to the winning party. Police forces will protect the bailiff if necessary.

2.9.3 Criminal Procedural Code

The substantive law on major and minor crimes can be found in the Criminal Code and other legislation in civil law countries. The way to determine whether a specific person has committed a particular crime is determined by procedural criminal law. The Criminal Procedural Code sets the rules for when and under what conditions a suspect can be apprehended, how long he may be interrogated, and when he has to be brought before a judge. The Code likewise contains the rights of an accused person to have the assistance of a lawyer (attorney), and access to a judge to guard the time limits set for the different periods of detention. It ensures that the court proceedings are properly run.

The judge can apply the substantive law (Is a crime committed? Is the defendant guilty? Which punishment is right?) only when the proceedings follow the prescripts of procedural criminal law. The judges who try criminal cases lead the proceedings actively.

2.10 International private law

The national private law of each country has its own characteristics that make it different from the other national private law systems. It develops in the national system of law, changes under its influence and exerts its influences on it. Society adapts to changing ideas, some lead to changes in legislation. Private law even if it was the same in different countries does not remain the same under diverging developments. Legal transactions that remain in one system are not involved with divergent law, and as far as their own system is changing, they adapt. But legal transactions that have effects in more than one legal system are bound to encounter an impossible situation.

2.10.1 Conflict rules

A contract that was negotiated in London between a Chinese and an American business, after the Chinese invitation to open negotiations during a business fair in Milan, was finally agreed on by an exchange of signed documents in Amsterdam. What contract law will be applied if any differences of opinion between the contract parties have to be settled? Who determines the choice of national private law, especially when the result depends on the chosen system? Such questions and the rules applied to answer them are the domain of international private law.

Such complications resulting from relations that go beyond the boundaries of national legal systems are very old and familiar and have led to a system of so-called conflict rules.[40] That central part of private international law has a set of solutions like the rule that the law on contract follows the place where the contract was concluded; i.e. when the last offer was accepted by the other party and this acceptance has been communicated to the party that made the last offer. The second part of the answer is that many contract parties who conclude contracts across national borders know this. They can accept it, but they cannot know in advance where the negotiations will end. So it is more likely that the choice of the national law system is a part of the contract itself. Private parties have this freedom because national governments generally accept the rule that the conflict rule will give way to a choice of the applicable law.

[40] In this expression the word 'conflict' does not refer to a conflict between the parties but between different legal systems. Which system applies and which has to give way? The conflict rule decides.

This is different when real estate is involved. When, for example, a French person buys a piece of land from a Polish person who lives in Paris. They have closed the deal in Spain and the land property lies in Poland. Here the conflict rule is simple: Polish law does not accept any choice of a legal system to transfer property rights on land in Poland. Only Polish law can be applied. Real estate is governed by the law of the land where it is situated (this principle is known as '*lex loci*').

See Law text box 2.6 Convention on the law applicable to contractual obligations for a treaty made to clarify the freedom of choice. See Law text box 2.7 Convention on the law applicable to contractual obligations, Title II Uniform rules, Article 4 Applicable law in the absence of choice, for another solution when the contract parties made no choice.

Law text box 2.6. Convention on the law applicable to contractual obligations.

Title II Uniform rules, Article 3 Freedom of choice

1. A contract shall be governed by the law chosen by the parties. The choice must be expressed or demonstrated with reasonable certainty by the terms of the contract or the circumstances of the case. By their choice the parties can select the law applicable to the whole or a part only of the contract.
2. The parties may at any time agree to subject the contract to a law other than that which previously governed it, whether as a result of an earlier choice under this Article or of other provisions of this Convention. Any variation by the parties of the law to be applied made after the conclusion of the contract shall not prejudice its formal validity under Article 9 or adversely affect the rights of third parties.
3. The fact that the parties have chosen a foreign law, whether or not accompanied by the choice of a foreign tribunal, shall not, where all the other elements relevant to the situation at the time of the choice are connected with one country only, prejudice the application of rules of the law at the country which cannot be derogated from by contract, hereinafter called 'mandatory rules'.
4. The existence and validity of the consent of the parties as to the choice of the applicable law shall be determined in accordance with the provisions of Articles 8, 9 and 11.

Law text box 2.7. Convention on the law applicable to contractual obligations.

Title II Uniform rules, Article 4 Applicable law in the absence of choice

1. To the extent that the law applicable to the contract has not been chosen in accordance with Article 3, the contract shall be governed by the law of the country with which it is most closely connected. Nevertheless, a separable part of the contract which has a closer connection with another country may by way of exception be governed by the law of that other country.
2. Subject to the provisions of paragraph 5 of this Article, it shall be presumed that the contract is most closely connected with the country where the party who is to effect the performance which is characteristic of the contract has, at the time of conclusion of the contract, his habitual residence, or, in the case of a body corporate or unincorporate, its central administration. However, if the contract is entered into in the course of that party's trade or profession, that country shall be the country in which the principal place of business is situated or, where under the terms of the contract the performance is to be effected through a place of business other than the principal place of business, the country in which that other place of business is situated.

2.10.2 Harmonisation of different national private law

In practice different systems of law operating simultaneously is inconvenient. It can make transactions quite complicated. A basic value of law is its function to settle conflicts by clarity of rules.

One of the ways to provide more transparent law is to make the law of different systems more alike. This can be done by negotiating treaties that determine the rules for a particular part of the law. There are many treaties on this subject, the Convention on the law applicable to contractual obligations' is one example, the 'United Nations Convention on Contracts for the International Sale of Goods (CISG)' is another.

The CISG provides legal rules on how to make a contract for the international sale of goods. It specifies the obligations of the buyer and seller, remedies for breach of contract and other aspects of the contract. The Convention entered into force on 1 January 1988. It is one of the results of the United Nations Commission on International Trade Law (UNCITRAL). This Commission was established by the UN General Assembly in 1966.[41]

The International Chambers of Commerce sponsored the Incoterms 2000, the INternational COmmercial TERMS like FOB (Free On Board) and CIF (Costs, Insurance, Freight) that determine who pays the costs.[42]

[41] UNGA Resolution 2205 (XXI) of 17 December 1966.
[42] ICC Official Rules for the Interpretation of Trade Terms, ICC Publication No. 560, 2000 Edition.

2.11 International public law

International public law is the subject of the next chapter.

2.12 European Union law

European law is not included in Diagram 2.4 representing the branches of law. It started as international law with special traits setting it apart on the edge between international law and the law of a new type of federation of national states. In the more than five decades of its existence, European law acquired a Union and several branches of law combining international, constitutional, administrative and a little private law; in short it is all over the place, a system of its own. This is the subject of Chapter 4.

2.13 Food law

The scheme representing the branches of law as applied in legal scholarship does not depict a part of law called 'food law'. Much of food law is national and European administrative law, the rules safeguarding the relationships between public authorities and private persons, when authorities apply public power to limit or enable certain actions by private persons. So food law might be presented as a subdivision of administrative law. However that classification would lose sight of other important traits of food law. Next to administrative law, food law can also be private law.[43] Private regulation by food producers to set quality standards or to give certain methods of producing food a special place in the market uses private law contracts or private law legal persons in addition to, or in lieu of, action by public authorities. Food law is concerned with intellectual property rights when patents or trademarks are used in the production or sale of feed or food. These rights are also part of private law. Food law is involved with state or constitutional law when a Food Authority is founded to perform certain functions essential for a proper food law system.[44] Food law provides instruments that make reality of human rights guaranteed by national constitutions and international treaties.[45] Food law is also assisted by criminal law when the government classifies certain actions detrimental to the common good (e.g. food safety) as punishable offences.[46] In short food law is a functional part of law that combines parts of several branches of law in a coherent body of law to serve its goals.

---

[43] See Chapter 23. See also: B.M.J. van der Meulen (ed.) Private Food Law. Governing food chains through contract law, self-regulation, private standards, audits and certification schemes, Wageningen Academic Publishers, Wageningen, the Netherlands, 2011.

[44] See Chapter 5.

[45] See Chapter 3. See also: B. Wernaart, The enforceability of the human right to adequate food. A comparative study, Wageningen Academic Publishers, Wageningen, the Netherlands, 2014.

[46] See Chapter 15. See also: G. Dannecker (ed.) Criminal Law on Food Processing and Distribution, and Administrative Sanctions in the European Union, Trier Academy of European Law series issue 10, Bundesanzeiger Verlag, Cologne, Germany, 1994.

To illustrate this in Diagram 2.4, food law would be a contorted line or some other bent figure collecting parts of several branches. The lighter shades in the Diagram indicate those branches that are most important for food law.

A functional area of law is defined not by the scholarly distinctions but by the function rules have in relation to a societal phenomenon regardless of the place of these rules within the legal system. All rules relating to environmental issues together form the functional domain of environmental law; all rules relating to food together form the functional domain of food law.

3. International food law

Hanna Schebesta, Bernd van der Meulen and Menno van der Velde[47]

3.1 Introduction

International law provides a high-level framework within which national or regional law has to fit. This is true for food law as well. Therefore, both national and EU food law have to take the respective international commitments into consideration respective international commitments. This chapter provides a brief general introduction to international public law in Section 3.2[48] and an overview of international food law in Sections 3.3 till 3.7. The chapter concludes with an observation regarding the relevance of international food law in general and for EU food law in particular in Section 3.8.

3.2 General introduction to international public law

3.2.1 The origins of international law: the nation state[49]

European nation state dominance left a profound mark on current international law. The Peace of Westphalia in 1648 marked the beginning of the reign of a system of sovereign states. Every state became sovereign; it would not recognise any higher authority that could bind it against its will with legal rules or measures. Every state was equal in this respect. The preponderant position of the nation state has been such as to determine the words and concepts that we use – and law that is valid all over the world is commonly referred to as 'inter-national law'.[50]

[47] This chapter elaborates on B.M.J. van der Meulen, The Global Arena of Food Law: Emerging Contours of a Meta-Framework, Erasmus Law Review 2010, pp. 217-240, and B.M.J. van der Meulen, International Food Law, Part I, in: I. Scholten-Verheijen *et al.*, Roadmap to EU Food Law, Eleven International Publishing, The Hague, the Netherlands, 2012. These texts in their turn have been based on Chapter 16 in: B.M.J. van der Meulen and M. van der Velde, European Food Law Handbook, Wageningen Academic Publishers, Wageningen, the Netherlands, 2008. The first section of this chapter is based on Chapter 3 in the European Food Law Handbook.

[48] This section can be understood as an extension of the general introduction to law in Chapter 2.

[49] P. Malanczuk, Akehurst's Modern Introduction to International Law, 7th ed., Routledge, London, UK, 1997.

[50] This classic conception of international law is of course challenged on various fronts. Realist views take into account the real geopolitical power position of states, and challenge the egalitarian nature of international law. It is also argued that 'international law' is not necessarily statist: legal theories such as global administrative law argue that the proliferation of organisations gives rise to a phenomenon of law which is not generated at inter-state level. In addition, the state-based conception of international law fails to accommodate the role of global private players as global rule makers.

3.2.2 Sources of public international law

An international system of sovereign states is based on the consensus of the participating states. However, even under this classic conception of international law, states could limit their absolute sovereignty and freedom to act by being bound by their own consent – as evidenced for example explicitly through the signature of a treaty or implicitly through the adherence to a practice over time.

The sources of international law are the following:
- *international treaties*, establishing rules recognised by the states;
- *international customary law*, as evidence of a general practice accepted as law;
- *general principles of law*, recognised by civilised nations; and to some extent the
- *opinio iuris*, that is, judicial decisions and the teachings of the most highly qualified publicists of the various nations, as subsidiary means for the determination of rules of law.[51]

3.2.3 Treaties

Of these, international treaties – being an *explicit* agreement by states to be bound to certain rules – are the most relevant for the purposes of this book. Basically, a treaty (agreement, charter, convention, covenant, protocol; many names are used) is a contract between states. Bilateral treaties are concluded between two States, while international agreements to which more than two states are parties are called multilateral treaties.

3.2.4 Treaty making

The life of a treaty is marked by several legally significant instances. The first is the signature of a treaty, a moment that is often highly ceremonial and indicates that the treaty text has become final and negotiations are deemed concluded. In addition, a treaty has to be ratified by the state. The ratification confirms the consent of the state to be bound and it takes place at domestic level through the necessary procedures, usually the involvement of the legislature. Finally, it is important to note the moment of entry into force of a treaty. This may be a fixed date in time, and in case of multilateral treaties it is usually subject to a requirement of a minimum number of ratifications. A treaty only enters into force once the treaty has been ratified by the necessary number of parties.

3.2.5 The relationship between international and national law

At the national level, the effect of international treaties is operationalised differently by country, depending on whether a national constitution sets out a dualistic or

[51] Article 38 Statute of the International Court of Justice.

monistic understanding of the relationship between international and national law. In a monistic state, such as the Netherlands or the USA, the legal system treats international and national law as a unitary legal order. An international treaty is then, once ratified, directly applicable in the state – for example a national law could be declared invalid if it were in conflict with international law. Dualist states such as Italy and the United Kingdom, on the other hand, require a national implementing act in order for international law to produce effects within the state.

3.2.6 Intergovernmental organisations

Next to states, intergovernmental organisations (IGOs) (often simply international organisations) have emerged as important players and law-makers (always dependent on state-based participation and consent) at the global level. They are founded by international treaties, such as the Charter of the United Nations, but to a certain extent they have a life of their own. The founding treaty vests organisations with a legal personality and typically sets up several institutions such as political decision-making institutions, an administrative apparatus, and sometimes dispute settlement mechanisms.

The difference between a treaty and an organisation is well illustrated by the General Agreement on Tariffs and Trade, which in 1995 was transformed from a Treaty to the World Trade Organization (WTO). Organisations can be global and hence open to membership by all states that meet the membership requirements or, for example, regional and hence limited to parties from a particular geographical location.

3.2.7 Players in the Global Arena

Before moving on to the substantive content of international food law, the following section presents a brief overview of the major players. At the global level, food law is embedded in the general international law structures dominated by the United Nations 'family' and the World Trade Organization. Some other organisations specifically address food and food-related issues, of which the most important ones are introduced below.[52]

3.2.8 The United Nations (UN)

The UN is the largest and most important international organisation. Virtually all states in the world are members (193). The main tasks of the UN are to ensure international peace, security and respect for human rights. Its headquarters are

[52] For a different selection, see: D.J. Shaw, Global Food and Agricultural Institutions, Routledge, London, UK, 2009, analysing the FAO, the WFP, the ARD (Agricultural and Rural Development Department of the World Bank), IFAD (International Fund for Agricultural Development) and the CGIAR (Consultative Group on International Agricultural Research).

in New York and Geneva. The UN employs some 40,000 staff. Diagram 3.1 depicts the organisation chart of the United Nations system. It is a simplification of a complicated structure with councils, commissions, committees, funds, programmes, specialised agencies, related organisations, subsidiary bodies and the six principal organs. See Diagram 3.2 for the principal organs of the United Nations' composition, functions, powers and place in the Charter.

The Security Council is responsible for dealing with threats of violence, the Economic and Social Council (ECOSOC) is responsible for social and economic security. These two organs representing two approaches to the common wealth are placed directly beneath the General Assembly, the weak main organ of the UN. See Diagram 3.2 for the principal organs of the United Nation's composition, functions, powers and place in the Charter.

The UN family possesses its own programmes and funds which are subsidiary organs of the UN, and specialised agencies, which are intergovernmental organisations performing different kinds of services for the world. Some members of this family are specialised agencies like the Food and Agriculture Organization (FAO) and the World Health Organization (WHO). They have been active in the field of food law since their establishment in 1945 and 1948. Their work for the international community is in some ways comparable to the work of national ministries for their national community. Together they founded the Codex Alimentarius Commission in 1963.[53] Membership of the Commission is open to Member States of FAO or WHO. One of the principal purposes of the Commission is the preparation of food standards and their publication in the Codex Alimentarius.

The specialised agencies have their own bases in the Charter and they report directly to the Economic and Social Council (ECOSOC). The Charter has two conditions for international organisations that are brought into a special relationship with the UN. They have to be well-established international organisations with a set of purposes that are specified in their own basic document. Each agency was admitted on the basis of a special agreement that stipulated the conditions of recognition.

Specialised UN agencies have the mandates to deal with specific themes. Next to the FAO and the WHO, these include the World Food Programme (WFP) and others like the United Nations Conference on Trade and Development (UNCTAD). Generally, the relations between the UN and the specialised agencies are much less intense than was intended. It is also important to note that although the World Trade Organization (WTO) cooperates with the UN, it is not a part of the UN system.

[53] See the Codex Alimentarius Commission. Website: www.codexalimentarius.net.

Diagram 3.1. The organisational chart of the United Nations system.

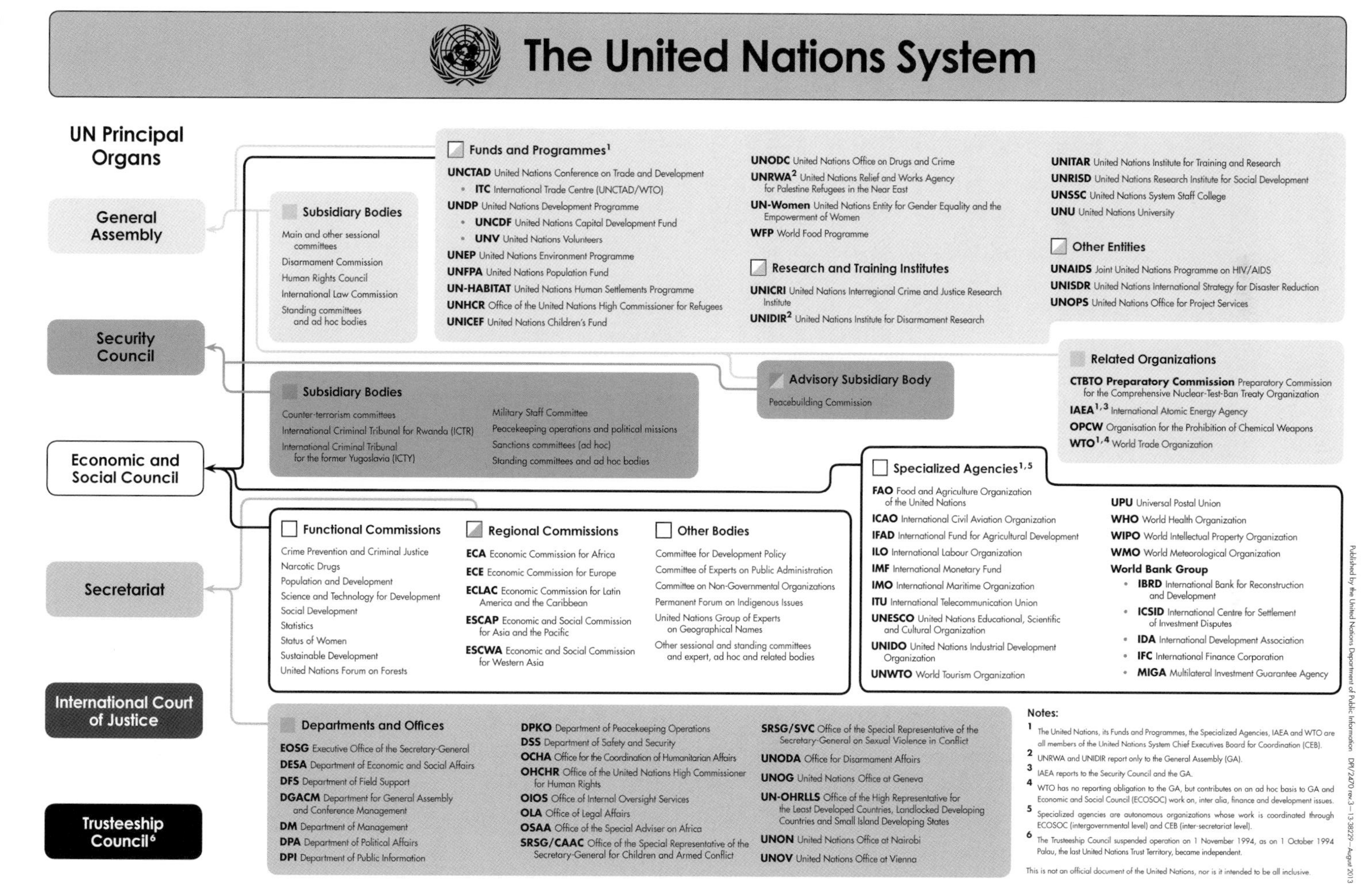

Diagram 3.2. The principal organs of the United Nations: composition, functions, powers and place in the Charter. Source: UN Charter.

Composition Chapter III	UN Charter	Functions and powers
General Assembly 192 Member States Session September-December Seat New York	Chapter IV 1. Article 10 2. 11(1) 3. 13 4. 14	Discussion and recommendations 1. Discuss any questions or any matters within the scope of the present Charter 2. Consider the general principles and cooperation to maintain international peace and security 3. Initiate studies and make recommendations to promote international co-operation 4. Recommend measures for the peaceful adjustment of any situation
Security Council 5 permanent members 10 chosen by UNGA from the regional groups	Chapter V	1. Primary responsibility for the maintenance of international peace and security 2. Acts on behalf of the Member States 3. Plan system to regulate armaments to promote international peace and security with the least diversion for armaments of the world's human and economic resources
Economic and Social Council 54 members, 18 elected each year for three years	Chapter X Article 62(1) Article 62(2) 6. Article 62(3) 7. Article 62(4) 8. Article 63(1)	With respect to international economic, social, cultural, educational, health, and related matters: Make or initiate studies and reports Make recommendations to a. UNGA b. UN Members c. Specialised agencies concerned Promote respect for, and observance of, human rights and fundamental freedoms for all. Prepare draft conventions for submission to UNGA Call international conferences Make agreements to bring Specialised agencies in the UN system Coordination of the Specialised agencies through consultation
Specialised agencies	Chapter IX Article 57 Chapter X Article 63	19 Specialised agencies Food and Agriculture Organization of the United Nations (FAO)" World Health Organization. The World Trade Organisation is not a Specialised agency

▶▶

Diagram 3.2. Continued.

Composition Chapter III	UN Charter	Functions and powers
International Court of Justice	Chapter XIV	The principal judicial organ Comply with ICJ decision Recourse to the Security Council Advisory opinion ICJ Statute determines the competence of the Court Only States can be parties in Court case Parties decide to put a dispute before the Court
Secretariat	Chapter XV	Secretary-General is the chief administrative officer Appointed by General Assembly after Security Council recommendation SG in all meetings of UNGA, Security Council, and ECOSOC

3.2.9 Food and Agriculture Organization (FAO)

The FAO is now a UN specialised agency that was set up on 16 October 1945,[54] a date commemorated every year as 'World Food Day'. The FAO's objective is to eradicate hunger and to make high quality food accessible to all. It focuses on both developed and developing countries. The FAO supports the elaboration of agreements and policies by providing a neutral platform for negotiation and information. It aims to improve nutrition, increase agricultural production and contribute to the world economy.

The FAO is governed by a Conference of the Member States that meets every second year to evaluate the work done and approve the budget. 49 Member States are chosen from the Conference to act as a temporary Council. The FAO consists of eight departments that focus on specific topics such as Agriculture and Consumer Protection, Economic and Social Development and Technical Cooperation. Its headquarters are in Rome with a total staff of about 3,600 in a considerable number of regional, sub-regional and national offices around the world.

3.2.10 World Health Organization (WHO)

The UN established the WHO Organization in 1948[55] to monitor global health trends, coordinate health care activities and promote the health of the world's population. It has 193 Member States. Its secretariat employs 8,000 people, working at the organisation's headquarters in Geneva and in regional and country offices. Its most

[54] See generally: www.fao.org.
[55] See generally: www.who.int.

important institution is the 'World Health Assembly', which meets once a year in Geneva to determine the policy and the programme budget of the organisation. The Executive Board, which consists of 34 members, implements WHO policy.

The WHO plays a central role in the case of global crises threatening public health, such as large-scale food safety incidents like the melamine crisis. The WHO derives powers vis-à-vis the Member States from the International Health Regulation 2005 (IHR).[56] The WHO has set up a global information network for the rapid exchange of information in food safety crises, namely the International Food Safety Authorities Network (INFOSAN).

3.2.11 Codex Alimentarius Commission (CAC)

In 1961 the FAO and the WHO established the Codex Alimentarius Commission. Over the years the CAC has established specialised committees. These committees are hosted by Member States all over the world. Some 175 countries, representing about 98% of the world's population, participate in the work of the Codex Alimentarius. The CAC formulates internationally recognised standards, codes of practice, guidelines and other recommendations relating to foods, food production and food safety.

For over 40 years, the EU was not a member of the Codex Alimentarius Commission. It exercised its influence through its Member States that were in the Codex. On 17 November 2003[57] the logical next step was taken in the process of increasing recognition of the Codex. The Council applied on behalf of the European Union for membership of the Codex Alimentarius Commission. The EU became a full member of the CAC.

3.2.12 World Food Programme (WFP)

The WFP is the food aid arm of the United Nations. It is a UN organisation that provides food to refugees, in long-term development projects in the Third World and in situations where people are without adequate food due to natural or man-made disasters.

The WFP's objective is to save lives and to protect livelihoods in emergency situations, to be prepared for such situations and to re-establish food security when emergencies have passed. Furthermore, the WFP works to reduce hunger and malnutrition anywhere in the world and to reinforce the capacity of countries to reduce hunger.

[56] See: whqlibdoc.who.int/publications/2008/9789241580410_eng.pdf.

[57] Council Decision 2003/822. In this way point 83 in the Action Plan on Food Safety (Annex to the White Paper on Food Safety) was implemented.

The WFP is governed by its Executive Board, consisting of 36 Member States of the UN or the FAO. The Board meets four times a year to formulate the WFP's short- and long-term policies on food supply. The organisation's headquarters are in Rome.

3.2.13 World Trade Organization (WTO)

As a successor to the General Agreement on Tariffs and Trade (GATT) of 1947, the World Trade Organization started its operations on 1 January 1995.[58] The WTO is a platform for international negotiations on the liberalisation of world trade. These negotiations take place in rounds named after the location of the kick-off meeting. The WTO itself resulted from the so-called Uruguay round. Since 2001, the WTO has been involved in the Doha development round. The negotiations have so far only resulted in the Bali Package in 2013, in which all WTO members reached agreement on four selected areas that were part of the Doha agenda – including food security.

These treaties concluded between the WTO members are legally binding. The Dispute Settlement Understanding provides an adjudication procedure to resolve conflicts.[59] Diagram 3.3 presents the structure of WTO law. The WTO is not a large international organisation with institutions as is, for example, the EU. It only has a permanent secretariat in Geneva. Decisions at top-level are taken by the

Diagram 3.3. Structure of the WTO.

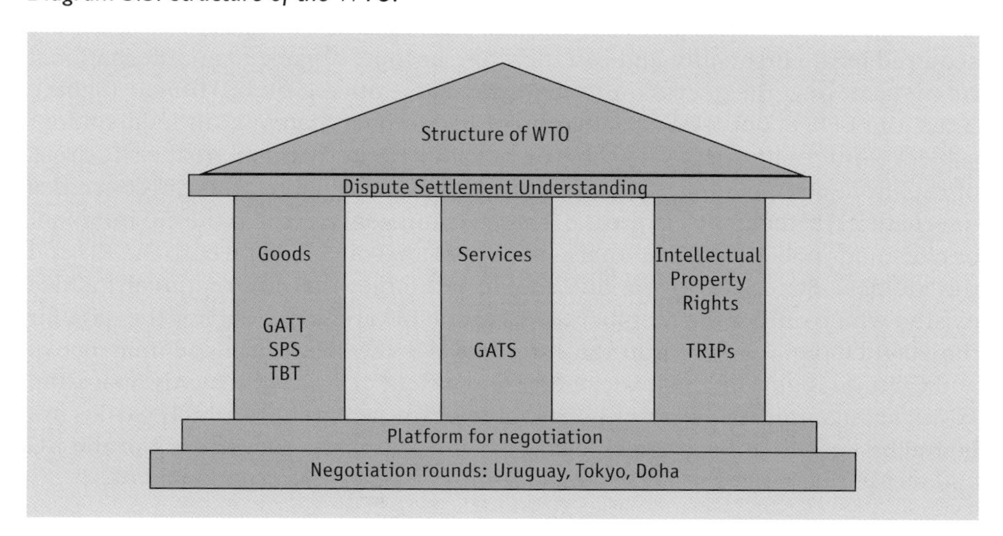

[58] See generally: www.wto.org.

[59] See the Understanding on Rules and Procedures Governing the Settlement of Disputes, available at: http://tinyurl.com/ow5hgtf.

Ministerial Conference that meets at least once every two years. Other decisions fall under the competence of the General Council. The General Council consists of the ambassadors and heads of delegation in Geneva. It meets several times a year in the periods other than the year of the Ministerial Conference and is entitled, where necessary, to take decisions on behalf of the Ministerial Conference. The General Council may also act as a Dispute Settlement Body (DSB).[60]

3.2.14 Supranational organisations

International organisations that have real powers to limit their Member States' sovereignty are called supranational organisations. The limitation of state power has been more intensive on a regional scale. Supranational international organisations often allow other representatives from interests inside the Member State to participate in decision making, in addition to the representatives of the national government. Decisions can be taken by (qualified) majority vote. Many decisions are binding and may have a direct effect inside the Member States without the intervention of the national government. The organisation has institutions and a corps of civil servants who are empowered to carry out parts of its policies inside the Member States. The national administrations have to co-operate with the supranational organisation and carry out its policies. Supranational organisations have independent institutions to supervise the execution of the organisations' policies in the Member States and to decide conflicts.

The EU can be characterised as a supranational organisation, emphasising the fact that the Member States of the EU have to some extent transferred their sovereign powers to a supranational level. Through the Treaty of Lisbon, the EU acquired legal personality and can therefore become a party to an international treaty itself (e.g. the accession to the European Convention of Human Rights). Thus, the EU (as opposed to the group of its Member States) is an independent subject of international law, capable of concluding international treaties. In areas of exclusive competence, such as international trade policy, the EU concludes international agreements. In areas of shared competence, such as the environment or consumer policy, international agreements are concluded by both the EU and the Member States. However, one should not forget that international treaties exist to which only some Member States of the EU are party. Such is the case for the 2009 Optional Protocol to the International Covenant on Economic, Social and Cultural Rights on the justiciability of the right to food, which only Spain and Portugal have ratified so far. It is crucial to examine carefully which parties are bound by specific international legal instruments, as the relationship of the EU and its Member States is a complex one with respect to international law.

[60] And as Trade Policy Review Mechanism (TPRM), not further discussed in this book.

3.3 International food law

Food comes within the ambit of international law from a variety of angles. A number of international treaties have collateral effects on food security, although their main immediate concern is a different policy area such as the environment or intellectual property protection. Examples are the International Union for the Protection of New Varieties of Plants (on intellectual property rights protection and agricultural initiatives on plant breeders' rights)[61] or the Convention on Biological Diversity (environmental focus).

The following section focuses on international law and food through the lens of the Human Right to food. The latter constructs a global framework which formulates the main objectives in delivering food security and provides important stimulants for policy. The food and trade law section focuses on the WTO framework, and the legal framework dealing with food regulation as a trade issue. The international standards section explains how standards for food are agreed upon, thereby setting the substantive content of international food regulations.

3.4 The Human Rights dimension of international food law

3.4.1 Human Rights in international law

The third great change for classical international law, after the creation of international organisations and the endowment of supranational powers, was the reappearance of the individual human being. Old international law had exiled human beings to the inside of the nation state. In the twentieth century, human beings became the subjects of international law again through international treaties recognising human rights or individual criminal responsibility.

3.4.2 United Nations

The original UN's 'Bill of Rights' was the 1948 Universal Declaration of Human Rights. In 1966, it was considerably strengthened by enshrining the rights mentioned in this declaration in two binding treaties: the International Covenant on Civil and Political Rights (ICCPR) and the International Covenant on Economic, Social and Cultural Rights (ICESCR); both entered into force in 1976. These treaties are important for food law because in addition to the classic freedoms from government oppression (freedom of religion, the press, association), they recognise entitlements to the essentials of life.[62]

[61] On this topic see Chapter 22.

[62] President Franklin D. Roosevelt formulated this as freedom from want in his Four Freedoms' Address to the USA Congress on January 6, 1941. Source: www.ourdocuments.gov.

These instruments are complemented by treaties that guarantee respect for the rights of specific groups (such as the UN Convention on the Elimination of All Forms of Discrimination against Women and the UN Convention on the Rights of the Child), elaborate certain rights (such as the Convention on the Prevention and Punishment of the Crime of Genocide) or generate regional structures for human rights (European Convention for the Protection of Human Rights and Fundamental Freedoms (ECHR) and the European Social Charter (ESC).

The recognition of food as a human right by the UN and its Member States is the most fundamental dimension of international food law. Article 11 of the International Covenant on Economic, Social and Cultural Rights (ICESCR) explicitly refers to the right to adequate food as part of the right to an adequate standard of living.

Law text box 3.1. Article 11 of the ICESCR.

Article 11[1]

(1) The States Parties to the present Covenant recognize the right of everyone to an adequate standard of living for himself and his family, *including adequate food*, clothing and housing, and to the continuous improvement of living conditions. The States Parties will take appropriate steps to ensure the realization of this right, recognizing to this effect the essential importance of international co-operation based on free consent.

(2) The States Parties to the present Covenant, recognizing the fundamental *right of everyone to be free from hunger*, shall take, individually and through international co-operation, the measures, including specific programmes, which are needed:

(a) To improve methods of production, conservation and distribution of food by making full use of technical and scientific knowledge, by disseminating knowledge of the principles of nutrition and by developing or reforming agrarian systems in such a way as to achieve the most efficient development and utilization of natural resources;

(b) Taking into account the problems of both food-importing and food-exporting countries, to ensure an equitable distribution of world food supplies in relation to need.

[1] Emphasis added.

The Committee on Economic, Social and Cultural Rights has elaborated the meaning of this article in a General Comment. According to General Comment No. 12,[63] food is considered adequate if it is accessible 'in a quantity and quality sufficient to satisfy the dietary needs of individuals, free from adverse substances, and acceptable within a given culture'. In other words, adequacy encompasses nutritional quality, availability, acceptability and safety.

[63] Available at: http://tinyurl.com/kfno2ou.

With regard to the realisation of the human right to food, or food security, Member States have three types of obligations: an obligation to respect – this is an obligation not to interfere with people's means to feed themselves; an obligation to protect – from interference by third parties; and an obligation to fulfil – provide food in situations like natural disasters where people are unable to take care of themselves.

These state obligations can be seen as the human rights foundation of food and agricultural law. Food law aims to ensure the safety ('free from adverse substances') of available food as well as, to a certain extent, its cultural acceptability (through labelling requirements with regard to ethical sensitive subjects like genetic modification and irradiation). The availability of food is a subject of agricultural policy and development cooperation.

It has been years since the adoption of the 'Voluntary Guidelines to Support the Progressive Realization of the Right to Adequate Food in the Context of National Food Security' under the auspices of the FAO.[64] However, FAO Member States remain unwilling to accept obligations under the right to food beyond the obligations of the state towards its own population. If, however, the world is serious about the Millennium Development Goals (in particular about halving world hunger by 2015) such reluctance cannot continue.[65]

Of the institutions introduced above, the World Food Programme, in particular, helps to ensure the availability of food in emergency situations (obligation to provide). The FAO focuses on ensuring sustainable production in the long run. However, the topic most frequently addressed in international food law is food safety, in combination with the state obligation to protect against food safety risks through regulatory systems.

Despite all efforts at the UN level, many countries refuse to recognise social, economic and cultural human rights like the right to food as justiciable rights,

---

[64] Available at: http://tinyurl.com/pynyxrt.

[65] For further reading on the right to food, see B.M.J van der Meulen, The Right to Adequate Food: Food Law between the Market and Human Rights, Elsevier, The Hague, the Netherlands, 2004; B.M.J. van der Meulen, 'The Freedom to Feed Oneself: Food in the Struggle for Paradigms in Human Rights Law', in: O. Hospes and I. Hadiprayitno, Governing Food Security, Wageningen Academic Publishers, Wageningen, the Netherlands, 2010, pp. 81-104; W.B. Eide and U. Kracht (eds.), Food and Human Rights in Development, Vol. I: Legal and Institutional Dimensions and Selected Topics, Intersentia, Antwerp, Belgium, 2005; W.B. Eide and U. Kracht (eds.), Food and Human Rights in Development, Vol. II Evolving Issues and Emerging Applications, Intersentia, Antwerp, Belgium, 2007; G. Kent, Freedom from Want: The Human Right to Adequate Food, Georgetown University Press, Washington DC, USA, 2005.

that is to say, rights that can be invoked in a court of law.[66] Notable exceptions are India,[67] South Africa[68] and Switzerland.[69] Generally speaking, human rights is the area of international law that addresses individuals most directly. For all practical purposes, the consequence of the reluctance of states to recognise the justiciability of the right to food is that, even with regard to human rights, the international framework functions only as a framework for state regulation, not as a framework for business practices. Some progress has been made through the 2009 Optional Protocol to the International Covenant on Economic, Social and Cultural Rights, which entered into force in May 2013. The protocol makes the right to food justiciable at the international level, although only 13 states have ratified the protocol to date (2014).

3.4.3 Council of Europe

The Council of Europe is a regional international organisation with a mainly intergovernmental character. It is both older and larger than the EU. It was founded in 1949 and currently has 45 Member States. It adopted the European Convention for the Protection of Human Rights and Fundamental Freedoms, drawn up in Rome in 1950. On a regional scale, this instrument presented a powerful means of securing these rights. The treaty contains precise formulations of human rights and it established the international European Court of Human Rights. This court adjudicates complaints from individuals about failures of national states to secure their rights under the Convention. The notion that an individual could start an international court case against a sovereign state was so sensitive in the first decades that the individual had access only to the independent Commission of Human Rights. This Commission investigated the case and tried to find an amiable solution using the methods of silent diplomacy. However, if the Commission

[66] See generally: M. Vidar, State Recognition of the Right to Food at the National Level, UNU-WIDER Research Paper No. 2006/61, available at: http://tinyurl.com/qxma5pz. For an analysis of the situation in the Netherlands, see: B.M.J. van der Meulen and O. Hospes (eds.), Fed up with the right to food? The Netherlands' policies and practices regarding the human right to adequate food, Wageningen Academic Publishers, Wageningen, the Netherlands, 2009. For the Netherlands and Belgium see: B. Wernaart, The enforceability of the human right to adequate food: a comparative study, Wageningen Academic Publishers, Wageningen, the Netherlands, 2013.

[67] See Human Rights Law Network, Right to Food, 4th ed., Human Rights Network, New Delhi, India, 2009; P. Ahluwalia, The Implementation of the Right to Food at the National Level: A Critical Examination of the Indian Campaign on the Right to Food as an Effective Operationalization of Article 11 of ICESCR, Center for Human Rights and Global Justice Working Paper, Economic, Social And Cultural Rights Series, No. 8 (2004).

[68] See Constitutional Court of South Africa, 4 October 2000, *Government et al.* v. *Grootboom et al.*, case CCT 11/00, available at: http://tinyurl.com/n9cum82. See also Article 27 of the South African Constitution. On the right to food in South Africa, see: http://tinyurl.com/pzfsgkg and S. Kozah, Realising the Right to Food in South Africa: Not by Policy Alone – A Need for Framework Legislation, South African Journal on Human Rights 2002, pp. 664-683.

[69] See Urteil der II. öffentlichrechtlichen Abteilung vom 27. October 1995, *i.S. V. gegen Einwohnergemeinde X. und Regierungsrat des Kantons Bern (staatsrechtliche Beschwerde)*, BGE 121 I 367. For an abstract, see: http://tinyurl.com/p8znj49.

found indications of a serious breach of the Treaty, it could decide to start a court case. This could only be a case between the Commission and the accused state. The individual who complained was not a party in this conflict. In 1998, this was changed and the individual is now the claiming party in her/his own court case.

Through the EU Charter of Fundamental Rights, the EU has vested itself with a fundamental rights basis. Although proclaimed in 2000, it became binding only through the Lisbon Treaty in 2009. In addition, the Lisbon Treaty includes the undertaking that the EU will accede to the ECHR. There is, however, no mention of the right to food in this new human rights 'constitution' of the EU; neither in the Charter nor in the ECHR.

3.5 The trade dimension of international food law

3.5.1 The World Trade Organization

The WTO is a political platform for world trade negotiations, but also a system of law. This system of law comprises a so-called 'single undertaking'. This means that once accord has been reached on certain agreements, states have to accept the whole package or nothing at all. They cannot pick and choose. There are three major domains: trade in goods, governed by the General Agreement on Tariffs and Trade (GATT), trade in services, governed by the General Agreement on Trade in Services (GATS) and intellectual property rights, governed by the Agreement on Trade Related Aspects of Intellectual Property Rights (TRIPs).[70] The GATT, together with the Agreement on the application of sanitary and phytosanitary measures (SPS Agreement) and the Agreement on technical barriers to trade (TBT Agreement), are the most important WTO agreements for the area of food law.[71]

3.5.2 General exceptions under the GATT

The original GATT Treaty, which entered into force in 1947, is the legal predecessor to the WTO. In 1994, the WTO was created by means of an umbrella treaty and the GATT 1947 was included as an annex in the WTO Agreement.[72] GATT aims to liberalise international trade by setting equal treatment of all trading partners as the norm.[73] It recognises, however, the necessity to make exceptions. The most important exceptions can be found in Article XX (general exceptions) and Article XXI (security exceptions).

[70] On the TRIPs agreement, see Chapter 22.
[71] Another relevant agreement for food law in a broad sense is the Agreement on Agriculture, which contains three pillars regulating domestic support, market access and export subsidies. It is the WTO counterpart to the EU Common Agricultural Policy. Food production and subsidisation, however, has been excluded from the scope of this book.
[72] WTO Agreement, Annex 1A: Multilateral Agreements on Trade in Goods.
[73] Most favoured nation clause (Article I GATT).

Food law aims to protect consumer health. As a consequence, the most important exception to international free trade from a food law point of view is the protection of health. This exception can be found in Article XX(b) of the GATT. See Law text box 3.2. The exception given for the protection of human, animal or plant life or health has been further elaborated in the SPS Agreement.

Law text box 3.2. Article XX(b) GATT on (phyto)sanitary measures.

Article XX: General Exceptions

Subject to the requirement that such measures are not applied in a manner which would constitute a means of arbitrary or unjustifiable discrimination between countries where the same conditions prevail, or a disguised restriction on international trade, nothing in this Agreement shall be construed to prevent the adoption or enforcement by any contracting party of measures:

(a) (...);

(b) necessary to protect human, animal or plant life or health;

(...)

3.5.3 The Agreement on the application of sanitary and phytosanitary measures

The SPS Agreement is very important from a food safety point of view. It addresses the fears that sanitary and phytosanitary measures (i.e. measures to protect the health of people, animals or plants) might be used as covert methods of protectionism, to make imports from other WTO Member States more difficult. The SPS Agreement on the one hand recognises the right of the parties to take foods safety and animal or plant health measures. On the other hand it provides conditions with which such measures must comply. Most importantly, they must be scientifically justified and they may not be discriminating, nor constitute disguised barriers to international trade (Law text box 3.3).

Law text box 3.3. Article 2(2) SPS Agreement on science.

Agreement on the application of sanitary and phytosanitary measures
Article 2
Basic Rights and Obligations

1. Members have the right to take sanitary and phytosanitary measures necessary for the protection of human, animal or plant life or health, provided that such measures are not inconsistent with the provisions of this Agreement.
2. Members shall ensure that any sanitary or phytosanitary measure is applied only to the extent necessary to protect human, animal or plant life or health, is based on scientific principles and is not maintained without sufficient scientific evidence, except as provided for in paragraph 7 of Article 5.

The scope of protection of the SPS Agreement extends to measures which are intended to protect:

- the territory of a country from pests and diseases;
- human or animal health from risks arising from additives, contaminants, toxins or disease-causing organisms in food, drink and feed;
- human health from animal- or plant-borne diseases and pests;
- animals and plants by limiting damage caused by pest.

Typical SPS measures are rules on the presence of additives, contaminants or toxins in foods, but also certification, processing methods and labelling if the respective rules concern food safety. Measures that deal with human health but not food safety or animal- and plant-borne diseases do not fall within the scope of the SPS Agreement. An example of this is cigarette labelling, which is outside of the SPS Agreement, but a product characteristics measure under the TBT Agreement.

Law text box 3.4. Sanitary and phytosanitary measures.

Paragraph 1 of Annex A to the Agreement on the application of sanitary and phytosanitary measures:
[s]anitary or phytosanitary measures include all relevant laws, decrees, regulations, requirements and procedures including, inter alia, end product criteria; processes and production methods; testing, inspection, certification and approval procedures; quarantine treatments including relevant requirements associated with the transport of animals or plants, or with the materials necessary for their survival during transport; provisions on relevant statistical methods, sampling procedures and methods of risk assessment; and packaging and labelling requirements directly related to food safety.

In particular, the SPS Agreement covers not only technical regulations, but measures that encompass both the setting of the regulatory system through general rules (regulation) and the application of the system through individual decisions.

Governments' SPS measures – in addition to having the purpose of protecting human, animal and plant life or health – must be based on science and a scientific assessment of risks. In cases where relevant scientific evidence is insufficient, a member may provisionally adopt sanitary or phytosanitary measures on the basis of available pertinent information, including that from the relevant international organisations as well as from sanitary or phytosanitary measures applied by other members. In such circumstances, members shall seek to obtain the additional information necessary for a more objective assessment of risk and review the sanitary or phytosanitary measure accordingly within a reasonable period of time.[74]

[74] Article 5(7) SPS Agreement. This provision probably stood as a model for the wording of the precautionary principle in Article 7 of the General Food Law.

If, however, the measures are in conformity with international standards, no scientific proof of their necessity is required. These measures are by definition considered to be necessary (Law text box 3.5).

Law text box 3.5. Article 3(2) SPS Agreement on international standards.

Article 3
Harmonisation

1. To harmonise sanitary and phytosanitary measures on as wide a basis as possible, Members shall base their sanitary or phytosanitary measures on international standards, guidelines or recommendations, where they exist, except as otherwise provided for in this Agreement (...).
2. Sanitary or phytosanitary measures which conform to international standards, guidelines or recommendations shall be deemed to be necessary to protect human, animal or plant life or health, and presumed to be consistent with the relevant provisions of this Agreement and of GATT 1994.

(...)

The SPS Agreement explicitly recognises international standards set by the so-called 'three sisters of the SPS Agreement': the Codex Alimentarius Commission (CAC) discussed hereafter; for animal health the International Office of Epizootics (OIE)[75]; and for plant health the Secretariat of the International Plant Protection Convention (IPPC).[76] The standards on food and on food safety are mainly to be found in the so-called Codex Alimentarius.[77]

If no international standards apply or if measures are chosen that are stricter than the applicable international standards, the burden of proof is on the authority taking the measure. Scientific research will have to provide the evidence required.

The European Union was confronted with this requirement in a dispute about meat from American cattle that had been treated with hormones. Based on health concerns, the Union refused to admit such meat to its market through EU directives prohibiting the importation and sale of meat and meat products treated with certain growth hormones. The USA and Canada filed a complaint under the WTO Dispute Settlement, and were found to be in the right.[78] The Codex Alimentarius allows the use of a limited number of hormones under certain restrictions in cattle. The EU had not provided sufficient scientific evidence to substantiate the health

[75] The abbreviation follows the French spelling.
[76] See SPS Agreement Annex A(3) Definitions. The Agreement on Technical Barriers To Trade, (TBT Agreement) has similar articles.
[77] In this book we will not discuss the other two sisters.
[78] The WTO's appellate body ruled on 16 January 1998 (WT/DS48/AB/R). For a hyperlink see reference list. On this case see: D. Bevilacqua, The Codex Alimentarius Commission and its Influence on European and National Food Policy, European Food and Feed Law Review 2006: 3-16.

risks that it claimed the hormones posed to humans. The EU continued to apply the measures and refused to comply with the WTO ruling. As a consequence, the USA and Canada were authorised to take so-called counter-veiling measures (i.e. economic sanctions) in the form of higher customs duties on specific EU products. In 2004, the EU initiated new proceedings on the matter, arguing that new scientific evidence supported the ban on hormone beef. The EU, however, did not succeed in convincing the Appelate Body (AB). Consequently, the counter-veiling measures against the EU remained operative. The beef hormones 'war' was finally resolved by a political deal in 2009: the EU committed to raise the quality beef import quota, in return for which the USA would gradually phase out the sanctions.

3.5.4 The Agreement on Technical Barriers to Trade

The original TBT Agreement dates from the Tokyo Round of trade negotiations. This 1979 'Standards Code' as it was also called, dealt with all technical barriers to trade and included measures which now fall under the SPS Agreement. It was a plurilateral agreement as signatories were limited to selected WTO members. Only through the Uruguay round and the creation of the WTO was the TBT Agreement revised and integrated into the 'single undertaking', i.e. applicable to all WTO members.

Scope of application

The TBT Agreement covers all technical regulations, voluntary standards and the procedures to ensure that these are met, except when these are sanitary or phytosanitary measures with the purpose of health protection falling under the scope of the SPS Agreement. The TBT is therefore subsidiary to the SPS Agreement. It is important to note that the scope of the SPS Agreement, however, is determined based on the purpose of a measure (health protection), while the TBT Agreement application is limited depending on the type of measure undertaken.

To illustrate, a country's regulations on the treatment of imported fruit to prevent pest spreading are measures related to health from plant or animal disease and therefore fall under the SPS Agreement. Other regulations concerning product characteristics such as the quality, grading and labelling of imported fruit are mere product requirements and therefore within the scope of the TBT Agreement.

Justifications

The assessment of whether TBT regulations are necessary differs in another important respect from the SPS Agreement. They can be justified if they are needed to fulfil a number of purposes, not merely the protection of health. TBT measures can have the purpose, for example, of preventing deceptive practices.

Many labelling, nutrition, quality and packaging requirements for foodstuffs therefore fall within the scope of the TBT Agreement when they are not strictly concerned with human health. Consequently, the justification for such measures is not limited to potential health risks that need scientific substantiation. Justifications can be based on technological reasons or geographical factors.[79]

Diagram 3.4. SPS or TBT? Which agreement does a measure come under? (source: WTO).

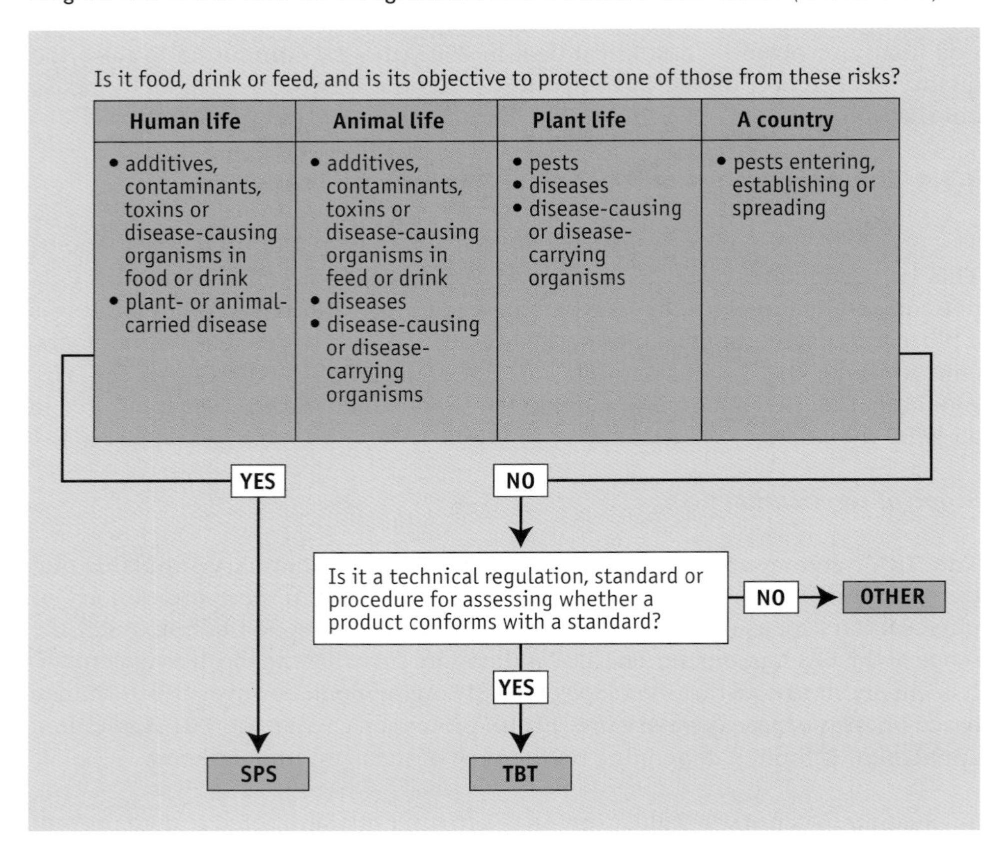

[79] WTO, 'Understanding the WTO Agreement on Sanitary and Phytosanitary Measures', available at http://tinyurl.com/oa3m762.

Diagram 3.5. Examples and summary of SPS and TBT measures (source: WTO).

Examples		
Fertilizer	Regulation on permitted fertilizer residue in food and animal feed	SPS
	Specifications to ensure fertilizer works effectively	TBT
	Specifications to protect farmers from possible harm from handling fertilizer	TBT
Food labelling	Regulation on permitted food safety: health warnings, use, dosage	SPS
	Regulation on size, construction/structure, safe handling	TBT
Fruit	Regulation on treatment of imported fruit to prevent pests spreading	SPS
	Regulation on quality, grading and labelling of imported fruit	TBT
Bottled water: specifications for the bottles	Materials that can be used because safe for human health	SPS
	Requirements: no residues of disinfectant, so water not contaminated	SPS
	Permitted sizes to ensure standard volumes	TBT
	Permitted shapes to allow stacking and displaying	TBT
Cigarette packets	Government health warning: 'Smoking can seriously damage your health': the label's objective is health but it is not about food, so it is not SPS	TBT

To summarise	
SPS measures typically deal with:	TBT measures typically deal with:
• additives in food or drink	• labelling of food, drink and drugs
• contaminants in food or drink	• grading and quality requirements for food
• poisonous substances in food or drink	• packaging requirements for food
• residues of veterinary drugs or pesticides in food or drink	• packaging and labelling for dangerous chemicals and toxic substances
• certification: food safety, animal or plant health	• regulations for electrical appliances
• processing methods with implications for food safety	• regulations for cordless phones, radio equipment, etc.
• labelling requirements directly related to food safety	• textiles and garments labelling
• plant/animal quarantine	• testing vehicles and accessories
• declaring areas free from pests or disease	• regulations for ships and ship equipment
• preventing disease or pests spreading to a country	• safety regulations for toys
• other sanitary requirements for imports (e.g. imported pallets used to transport animals)	

Types of measures

The TBT Agreement applies to WTO Member States' 'technical regulations' and 'standards'. Technical regulations are defined in the TBT[80] as 'a document which lays down product characteristics or their related processes and production methods, including the applicable administrative provisions' and they are mandatory. 'Standards' as understood in the TBT are voluntary documents, 'approved by a recognized body, that provides, for common and repeated use, rules, guidelines or characteristics for products or related processes and production methods'.[81]

For example in the US-Tuna II case,[82] Mexico challenged the conditions of use of a US 'dolphin-safe' label. Mexico alleged that these measures were inconsistent with the TBT Agreement, because the conditions depended on the area where tuna was harvested in combination with the permissible fishing methods. Mexico complained that the result of the label conditions was to exclude most Mexican, but not other nations', tuna.

First of all, the conditions laying down the use of a 'dolphin-safe' label were of an environmental and conservationist motivation and not health-related. Therefore such labels fell within the material scope of the TBT Agreement.

Secondly, it was questioned whether the conditions for the use of the dolphin-safe label were to be considered as a 'technical regulation' within the meaning of the TBT Agreement. The Appellate Body concluded in particular that the conditions for the use of the dolphin-safe label were to be considered as a 'technical regulation' within the meaning of the TBT Agreement. The measures were 'composed of legislative and regulatory acts of the US federal authorities and include[d] administrative provisions' and goods (i.e. the tuna) had to comply with the statutory requirements or could not be marketed as 'dolphin safe', but could only be marketed as 'tuna'.

3.5.5 Outlook

One of the new issues discussed in the SPS committee is how to deal with private standards for food. The concern was raised in 2005 by Saint Vincent and the Grenadines concerning a GlobalGAP[83] standard on bananas. No consensus has been reached in particular on the applicability of Article 13 SPS to private actors, and – even more fundamentally – how to define a private standard.[84] It is therefore highly questionable as to how far WTO provisions can be applied to

[80] In Annex 1 to the TBT.

[81] Outside the context of TBT, this definition of standard does not apply in this book.

[82] United States-Measures concerning the importation, marketing and sale of tuna and tuna products. Report of the Appellate Body of 16 May 2012, WT/DS381/AB/R (*US-Tuna II*).

[83] At that time called 'EurepGAP'.

[84] See also Chapter 23 on private food law.

private standardisation and certification bodies.[85] Similarly, the TBT committee struggles with certification, accreditation and developing countries.[86]

These current issues deal with the rise of private powers to become global actors and their inclusion in an arena that was dominated by state actors, and later on international organisations. The international law framework is still struggling with the position of individuals and the challenge of how to deal with increasingly powerful non-state private actors and processes.

3.6 Codex Alimentarius

The Codex Alimentarius Commission (CAC) is one of the most important bodies in which actual substantive food issues are agreed upon at international level. It establishes standards, codes of practices, guidelines and other recommendations relating to foods, food production and food safety through an elaborate procedure of international negotiations. In principle, these standards are voluntary. However, their wide reception and recognition through the WTO grant them such important authority as to make them factually almost mandatory (see Diagram 3.6).

Beside the food standards, Codex Alimentarius includes advisory provisions called codes of practice or guidelines. These codes of practice and guidelines mainly address food businesses. All standards taken together are called '*Codex Alimentarius*'.

Diagram 3.6. Procedure for adopting Codex standards (Codex Alimentarius website).

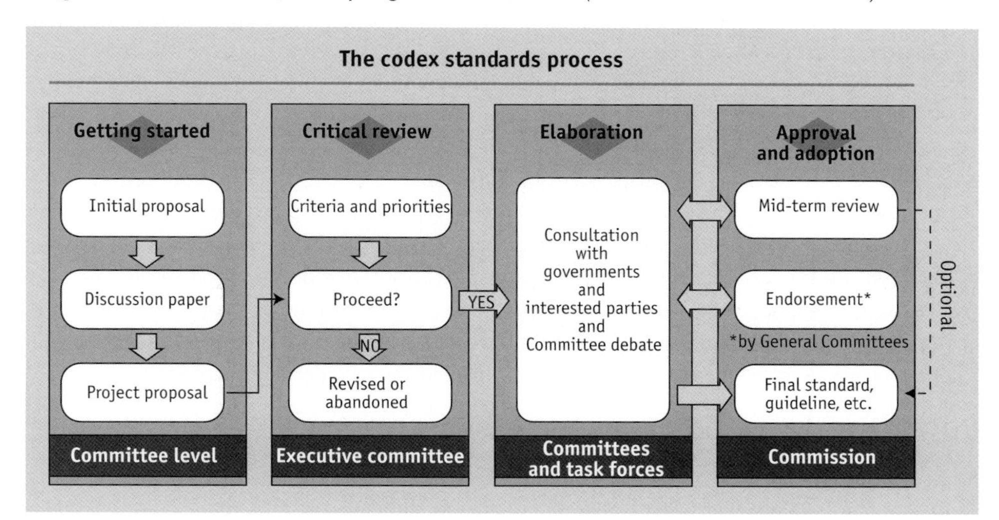

[85] See p 24 WTO Agreement Series Sanitary and Phytosanitary. Available at: http://tinyurl.com/k42z6f9.

[86] For details, see: B.M.J. van der Meulen (ed.), Private Food Law. Governing food chains through contract law, self-regulation, private standards, audits and certification schemes, Wageningen: Academic Publishers, Wageningen, the Netherlands, 2011.

In Latin this means 'food code'. It can be seen as a virtual book filled with food standards. The food standards represent models for national legislation on food.

The Codex comprises more than 200 standards, close to 50 food hygiene and technological codes of practice, some 60 guidelines, over 1000 food additives and contaminants evaluations and over 3,200 maximum residue limits for pesticides and veterinary drugs. Finally, the Codex Alimentarius includes requirements of a horizontal nature on labelling and presentation and on methods of analysis and sampling.[87] As will become apparent in the second part of this book, the examples given here below show a certain familiarity from the point of view of EU food law.

3.6.1 Procedural Manual

The 'constitution' of the Codex Alimentarius is the Procedural Manual. The Procedural Manual not only gives the procedures and format for setting Codex Standards and Guidelines, but also some general principles and definitions (Law text box 3.6). The principles relate among other things to the scientific substantiation of the work of Codex Alimentarius and the use of risk analysis for food safety (Law text box 3.7).

Law text box 3.6. Some definitions in the Codex Alimentarius Procedural Manual.

Food means any substance, whether processed, semi-processed or raw, which is intended for human consumption, and includes drink, chewing gum and any substance which has been used in the manufacture, preparation or treatment of 'food' but does not include cosmetics or tobacco or substances used only as drugs.
Food hygiene comprises conditions and measures necessary for the production, processing, storage and distribution of food designed to ensure a safe, sound, wholesome product fit for human consumption.
Food additive means any substance not normally consumed as a food by itself and not normally used as a typical ingredient of the food, whether or not it has nutritive value, the intentional addition of which to food for a technological (including organoleptic) purpose in the manufacture, processing, preparation, treatment, packing, packaging, transport or holding of such food results, or may be reasonably expected to result, (directly or indirectly) in it or its by-products becoming a component of or otherwise affecting the characteristics of such foods. The term does not include 'contaminants' or substances added to food for maintaining or improving nutritional qualities.

[87] S. Navarro and R. Wood, Codex Deciphered, Leatherhead, Surry UK, 2001, p. 1. On Codex Alimentarius, see also: FAO, Understanding The Codex Alimentarius, FAO, Rome, Italy, 2005; E. Kimbrell, What is Codex Alimentarius, 2000, AgBioForum 3(4): 197-202; FAO/WHO, Report of the Evaluation of the Codex Alimentarius and other FAO and WHO Food Standards Work. For hyperlinks see reference list. The most important book on the Codex Alimentarius is probably the dissertation of: M. D. Masson-Matthee, The Codex Alimentarius Commission and Its Standards. An examination of the legal aspects of the Codex Alimentarius Commission, Asser Press, The Hague, the Netherlands, 2007.

Law text box 3.7. Some principles in the Codex Alimentarius Procedural Manual.

Statements of Principle concerning the role of science in the Codex decision-making process and the extent to which other factors are taken into account

1. The food standards, guidelines and other recommendations of Codex Alimentarius shall be based on the principle of sound scientific analysis and evidence, involving a thorough review of all relevant information, in order that the standards assure the quality and safety of the food supply.
2. When elaborating and deciding upon food standards Codex Alimentarius will have regard, where appropriate, to other legitimate factors relevant for the health protection of consumers and for the promotion of fair practices in food trade.
3. In this regard it is noted that food labelling plays an important role in furthering both of these objectives.
4. When the situation arises that members of Codex agree on the necessary level of protection of public health but hold differing views about other considerations, members may abstain from acceptance of the relevant standard without necessarily preventing the decision by Codex.

3.6.2 Standards

The work of the CAC has resulted in a vast collection of internationally agreed food standards that are presented in a uniform format. Most of these standards are of a vertical nature. They address all principal foods, whether processed, semi-processed or raw. Standards of a horizontal nature are often called 'general standards', like the General Standard for the Labelling of Prepackaged Foods.[88]

According to this general standard, the following information shall appear on the labelling of prepackaged foods:
- the name of the food; this name shall indicate the true nature of the food;
- list of ingredients (in particular if one of a list of 8 allergens is present);
- net contents;
- name and address of the business;
- country of origin where omission could mislead the consumer;
- lot identification;
- date marking and storage instructions;
- instructions for use.

3.6.3 Codes of practice

As indicated above, in addition to the formally accepted standards the Codex includes recommended provisions called codes of practice or guidelines. There

[88] CODEX STAN 1-1985 (Rev. 1-1991).

is, for example, a 'Code of Ethics for International Trade in Food',[89] and a set of hygiene codes like the 'Recommended International Code of Practice General Principles of Food Hygiene' and the 'Hazard Analysis and Critical Control Point (HACCP) System and Guidelines for its Application' (Law text box 3.8).

Law text box 3.8. The principles of HACCP according to Codex Alimentarius.

Principle 1	Conduct a hazard analysis.
Principle 2	Determine the Critical Control Points (CCPs).
Principle 3	Establish critical limit(s).
Principle 4	Establish a system to monitor control of the CCP.
Principle 5	Establish the corrective action to be taken when monitoring indicates that a particular CCP is not under control.
Principle 6	Establish procedures for verification to confirm that the HACCP system is working effectively.
Principle 7	Establish documentation concerning all procedures and records appropriate to these principles and their application.

3.6.4 Legal force

The Codex standards do not represent legally binding norms. Member States undertake to transform the Codex standards into national legislation. However, no sanctions apply if they do not honour this undertaking.

One might ask what is the use of such non-binding standards. The answer comprises different elements. Generally nation states are reluctant to enter into internationally binding agreements because these limit their sovereignty. For this reason it turns out to be easier to agree on non-binding 'soft law' standards than on binding 'hard law'. By agreeing to non-binding standards, the participating states develop a common 'language' or nomenclature. All states and other subjects of international law will mean the same thing; when they meet to negotiate about 'food'; they mean 'food' as defined in the Codex. The same holds true for 'milk' and 'honey' and all the standards that have been agreed upon. The notion of HACCP has been developed – and is understood – within the framework of Codex Alimentarius.[90] In this way the Codex Alimentarius provides a common frame of reference, but there is more.

The mere fact that national specialists on food law enter into discussion on these standards will influence them in their work at home. A civil servant drafting a piece of legislation will look for examples. As regards food s/he will find examples

[89] CAC/RCP 20-1979 (Rev. 1-1985).

[90] Recommended International Code Of Practice General Principles Of Food Hygiene Cac/Rcp 1-1969, Rev. 3-1997, Amd. (1999).

in abundance in the Codex. In these subtle ways the Codex Alimentarius is likely to have a major impact on the development of food law in many countries even without a strict legal obligation to implement.

It turns out more than once that soft law has a tendency to solidify. Once agreements are reached parties tend to put more weight on them than was initially intended. In the following sections it will be shown that this is true for Codex standards as well. As a result of several developments they are well on their way to acquiring at least a quasi-binding force.

WTO/SPS

As we have seen above, the inclusion of the Codex Alimentarius in the SPS Agreement greatly enhances its significance. WTO members who follow Codex standards are liberated from the burden to prove the necessity of the sanitary and phytosanitary measures they take. If they cannot base their measures on Codex, they have to prove that their measures are science-based.

For the food sector the practical result is that they have access to the majority of the world's markets if their products are up to Codex standard. It is much easier to apply Codex than to study and apply a whole range of differing national standards. The catch is that industry depends on their national governments to take action within WTO if they have to face trade barriers that do not comply with Codex. The EU provides the option for its businesses to call on the European Commission for support if they are confronted by trade barriers in third countries. See Section 3.7.

3.6.5 European Union

Also in the context of European food law the Codex is increasing its legal impact. In its case law the Court of Justice of the European Union uses Codex standards as an interpretation aid for open standards in European law. In the so-called Emmenthal cheese case, for instance, the Court was called upon to judge whether the definition of Emmenthal cheese in French legislation constituted a barrier to trade as interdicted in Article 30 of the EC Treaty.[91] The case concerned a French cheese trader who was prosecuted for attempting to sell a product as Emmenthal cheese that did not have a brown rind as required in the French standard for Emmenthal cheese. The Court decided that the French law was indeed in breach of Article 30 of the EC Treaty. In its reasoning the Court considered that the definition of Emmenthal in the Codex does not in all situations require a rind (Law text box 3.9). In other words: in cases like this, the Codex standard helps to define the limits that European law sets to national legislators in the EU Member States.

[91] Now Article 34 TFEU. On the topic see Chapters 6 and 7.

Law text box 3.9. Codex Alimentarius used by the Court of Justice of the European Communities to define Emmenthal cheese, judgment of 5 December 2000, Case C-448/98.

32. In the case at issue in the main proceedings it should be noted that according to the Codex Alimentarius referred to in paragraph 10 of this judgment, which provides indications allowing the characteristics of the product concerned to be defined, a cheese manufactured without rind may be given the name Emmenthal since it is made from ingredients and in accordance with a method of manufacture identical to those used for Emmenthal with rind save for a difference in treatment at the maturing stage. Moreover it is undisputed that such an Emmenthal cheese variant is lawfully manufactured and marketed in Member States other than the French Republic.
33. Therefore even if the difference in the maturing method between Emmenthal with rind and Emmenthal without rind were capable of constituting a factor likely to mislead consumers it would be sufficient whilst maintaining the designation Emmenthal for that designation to be accompanied by appropriate information concerning that difference.
34. In those circumstances the absence of rind cannot be regarded as a characteristic justifying refusal of the use of the Emmenthal designation for goods from other Member States where they are lawfully manufactured and marketed under that designation.
35. The answer to the question referred for a preliminary ruling must therefore be that Article 30 of the Treaty precludes a Member State from applying to products imported from another Member State where they are lawfully produced and marketed a national rule prohibiting the marketing of a cheese without rind under the designation Emmenthal in that Member State.

3.6.6 Codex in the General Food Law

The significance of the Codex Alimentarius is recognised in the General Food Law (Regulation 178/2002[92]). In Article 13 the General Food Law lays emphasis on the importance of the development of international standards. More important however is Article 5, paragraph 3 of the General Food Law. Although this paragraph leaves a wide margin of appreciation for the national and EU legislators, it introduces an obligation to take international standards like Codex into account (Law text box 3.10).

Law text box 3.10. International standards for food law, Article 5(3) General Food Law.

Article 5(3) Regulation 178/2002
Where international standards exist or their completion is imminent, they shall be taken into consideration in the development or adaptation of food law, except where such standards or relevant parts would be an ineffective or inappropriate means for the fulfilment of the legitimate objectives of food law or where there is a scientific justification, or where they would result in a different level of protection from the one determined as appropriate in the Community.

[92] See Chapter 9.

This comes close to an obligation under European law for both the Union and the Member States to include standards like the Codex in national and European Union food legislation. If this Article is indeed interpreted and applied in this way, it will mean a boost for the legal position of Codex standards in the EU. The General Food Law itself gives the proper example. The definition of food, which is the GFL's foundation, is based on the Codex Alimentarius (Law text box 3.11).[93]

Law text box 3.11. Compare definition of 'food' in Codex Alimentarius Commission Procedural Manual with definition of 'food' in Article 2 General Food Law.

Codex Alimentarius – Procedural Manual, definitions for the Purposes of the Codex Alimentarius, page 49:

For the Purposes of the Codex Alimentarius:

Food means any substance, whether processed, semi-processed or raw, which is intended for human consumption, and includes drink, chewing gum and any substance which has been used in the manufacture, preparation or treatment of 'food' but does not include cosmetics or tobacco or substances used only as drugs.

Article 2 Regulation 178/2002
Definition of 'food'

For the purposes of this Regulation, 'food' (or 'foodstuff') means any substance or product, whether processed, partially processed or unprocessed, intended to be, or reasonably expected to be ingested by humans.

'Food' includes drink, chewing gum and any substance, including water, intentionally incorporated into the food during its manufacture, preparation or treatment. It includes water after the point of compliance as defined in Article 6 of Directive 98/83/EC and without prejudice to the requirements of Directives 80/778/EEC and 98/83/EC.

'Food' shall not include:

(a) feed;
(b) live animals unless they are prepared for placing on the market for human consumption;
(c) plants prior to harvesting;
(d) medicinal products within the meaning of Council Directives 65/65/EEC and 92/73/EEC;
(e) cosmetics within the meaning of Council Directive 76/768/EEC;
(f) tobacco and tobacco products within the meaning of Council Directive 89/622/EEC;
(g) narcotic or psychotropic substances within the meaning of the United Nations Single Convention on Narcotic Drugs, 1961, and the United Nations Convention on Psychotropic Substances, 1971;
(h) residues and contaminants.

[93] On the concept of food in the General Food Law, see Chapter 9 Section 9.3.1.

This example is followed in later food legislation as well. The General hygiene regulation (852/2004),[94] for example, states in its 15th consideration: *'The HACCP requirements should take account of the principles contained in the Codex Alimentarius'.*

Also the principle laid down in Article 6 of the General Food Law that food law shall be based on risk analysis can be understood in this context. If requirements of European food safety law are not in conformity with the Codex Alimentarius, sooner or later they will be contested under the SPS Agreement as barriers to international trade. They will only stand in the WTO forum if they are science-based, that is, based on risk analysis.

3.7 Businesses and international trade disputes

3.7.1 The Trade Barriers Regulation

The WTO dispute settlement procedure is only accessible for the members of the WTO. Only states are members. The procedure is therefore not directly available for businesses. The EU, however, provides some relief.

On 1 January 1995 (the starting date of the WTO), Regulation 3286/94 the so-called Trade Barriers Regulation (TBR) came into force.[95] This regulation gives businesses,[96] their associations and EU Member States the right to lodge a complaint with the European Commission against barriers to trade adopted or maintained by third countries. The complaint must show the existence of the measure, that it is contrary to international obligations and that it has adverse effects. If it is warranted in the interests of the Union, the Commission will investigate the case and – if the Commission agrees that an infringement of international (WTO) law has taken place – try to come to a solution either through negotiations or, if necessary, a dispute settlement procedure.

3.7.2 Damages

Economic sanctions in response to infringements of WTO law usually take the form of punitive import duties on products from the country that was found at fault in the dispute settlement procedure. The products concerned are not necessarily those that profited from the trade barrier at issue. The businesses producing and exporting the products that are subject to these import duties are likely to suffer damages . One could argue that these damages are caused by the actions of their

[94] Discussed in Chapter 13.

[95] Council Regulation (EC) No 3286/94 of 22 December 1994 laying down Community procedures in the field of the common commercial policy in order to ensure the exercise of the Community's rights under international trade rules, in particular those established under the auspices of the World Trade Organization. More information about the TBR is available at http://tinyurl.com/nmmuerw.

[96] Called 'community enterprise' if they complain about barriers to export and 'community industry' if they complain about barriers to import.

authorities that constituted the illegal barrier to trade. In theory, the EU can be held liable for both unlawful and lawful conduct of its institutions. However, WTO rules are not rules by reference to which the legality of EU institutions' conduct may be reviewed. Unlawfulness of the institutions' action can be established by the Court of Justice of the European Union, and hence give rise to liability, only where they have infringed a provision of EU law. However, economic operators can obtain compensation if they bear a disproportionate part of the burden resulting from a restriction of access to export markets under non-contractual EU liability. They have to establish that they suffered actual damage, a causal link between the institutions' conduct and the damage and the unusual and special nature of the damage. In a case concerning US retaliation measures for the incompatibility of the EU regime governing the import of bananas with the WTO agreements, the Court of Justice of the EU has ruled that the EU was not liable for damage caused to an economic operator. The possibility of being hit by such import duties was considered to be part of the normal risk of engaging in international trade.[97]

3.8 Concluding remark

Unlike EU food law, international food law mostly exercises its influence on businesses (and consumers) in an indirect way. It sets requirements on states including the EU and provides procedures for states. Natural and legal persons depend on their states to comply and to take action. International food law is therefore rather a meta-framework: a framework of frameworks.

The International Covenant on Economic, Social and Cultural Rights establishes objectives to respect, protect and ensure access to adequate food, understood in terms of nutrition, availability, acceptability and safety. The WTO adds to this the objective of free trade. Together, these objectives place an emphasis on protection against unsafe food.

The methodology to be applied is risk analysis. Measures aimed at protecting the health of consumers must be based on science either directly, through the application of the risk analysis methodology, or indirectly, by conforming to international standards such as the Codex Alimentarius (which in turn are based on risk analysis). These standards are not binding upon citizens and businesses but provide models for national and regional food law. For this book the most important point is that EU food law should take international food law into account.

In this way, the Codex Alimentarius provides a model for the content of national and EU food legislation, encompassing vertical standards for a wide range of products, lists of synthetic additives that may be used in food for a technological function and maximum limits for pesticide residues and contamination with

[97] See Judgement of 14 December 2005, Case T-383/00, Beamglow Ltd vs. European Parliament, Council of the European Union and Commission of the European Communities [2005] ECR II-05459.

microorganisms or chemical substances. It also provides codes of practice for food hygiene and general standards on food labelling.

Finally, the WHO supports its members with a system of rapid information exchange when dealing with large-scale food safety incidents.

The risk analysis methodology has been adopted in the WTO's policies to liberalise trade and in its procedures for dispute settlement. Applying the Codex Alimentarius is recognised as a harmonised way for Member States (including the EU) to fulfil their commitment to basic sanitary measures on risk analysis. The common understanding of what food safety means also underlies the emerging structure of crisis management made available by the WHO.

All in all, these elements result in a global system of food safety governance. Even though this system is created by different, more or less independent players, it shows a certain coherence in that the elements mutually reinforce rather than contradict each other.

4. The foundations of the European Union

Hanna Schebesta and Menno van der Velde

4.1 Introduction

The first substantive step towards the creation of the European Union, as we know it today, was taken with the signature of the Treaty Establishing the European Coal and Steel Community (ECSC) in 1951. This Treaty entered into force on 23 July 1952 and basically placed the Belgian, Dutch, French, German, Italian and Luxembourgian coal and steel industries (which were crucial to the production of arms – it all started shortly after the Second World War) under the control of an independent supranational authority. In 1957, the same six European states went on to sign the two Rome Treaties. One established the European Atomic Energy Community (Euratom) which remained largely unchanged until today. Perhaps more importantly, a second treaty established the European Economic Community (EEC) which, in particular, would create a common market amongst these six states where goods, services, persons and capital could move freely.[98] Over the years this community of states has witnessed very considerable changes.[99] Thus, today we no longer speak about a European (Economic) Community, but about a European Union which deals with tasks that go well beyond purely economic ones. Moreover, the European Union now counts 28 Member States that are all joined in a community of law.[100]

European Union legislation plays a key role in food law, both when EU law deals expressly with the regulation of food as such, and when it deals with other issues, like the internal market, which have an important influence on food law. EU law has certain characteristics that make it a unique specimen of international law. EU law is the supreme law, irrespective of the place the individual Member State constitution offers to international law within the Member State's national legal system. The supremacy of EU law means that if Member State law appears to be contrary to EU law the Member State law must either be interpreted so as to

[98] See also Chapter 7. The EEC Treaty came into force on 1 January 1958.
[99] On 1965 the so-called Merger Treaty integrated the institutions set up under the different treaties. Today, the European Commission, the Council and the European Parliament continue to govern both the European Union and Euratom.
[100] Official EU documents can be found on the internet: http://europa.eu. Choose from the 24 official languages for instance the EN-Gateway to the European Union, then choose the tab Documents. The official source for EU documents is the Official Journal of the European Union, published every working day. It consists of two series, L (all binding measures) and C (information, preparatory work, notices and recommendations), and a supplement S (tenders on EU scale for Member State public projects to enable all relevant businesses in the EU to take part in the bidding to get the contract. Member States are not allowed to give preferential treatment to enterprises in their own country). For all practical purposes, the best entry to EU legislation and case law is the Eur-Lex database which is available at eur-lex.europa.eu. For a more detailed explanation of this database, see the annex (Chapter 25) to this book.

conform to EU law, or Member State law must be set aside. European Union law in other words lays down the limits to what is possible and what is impossible in law, both on the level of the Union and on the level of the 28 EU Member States. Consequently, to understand food law it is necessary to know the foundations of European Union law.

4.2 The basic structure of the EU and its law

European integration led to the establishment of several international organisations, which are collectively known as the European Union. This conglomerate of organisations and political arrangements is shaped by deliberate design but also by the day-to-day operations of this complicated machinery. The day-to-day work not only deals with the tasks on the agenda, but by doing so in a particular way also changes the design of the machinery through practice. This combination of design and experience creates the EU as it actually is. The result is known as the '*acquis communautaire*' which is the interlocking complex of the original treaties, the changes to the treaties, EU secondary legislation, the customs developed both in practice and in case law of the Court of Justice of the European Union. All states that join the European Union must accept the '*acquis communautaire*' as it stands on the day they join. The European Union has substantial supranational elements, but certain areas – in particular the Common Foreign and Security Policy – continue to be largely intergovernmental.

The concept of the European Union rightly conveys that the EU is not a normal international organisation. Even though its characteristics are not in themselves unique because certain other international organisations are similar in nature, the intensity and the combination of these characteristics in the EU are unique and set the EU apart from other international organisations. To mention but a few of the characteristics: the EU conveys its own citizenship on all the citizens of the Member States, which is additional to their national citizenship.[101] The EU can create legislation, which is binding for its Member States, all its citizens and all organisations within it spheres of authority. This EU legislation works directly within the legal systems of the Member States and when it has been adopted there is no longer a need for consent by the national governments. EU law is supreme in the sense that it trumps national law that is contrary to EU law. Moreover, this is determined by independent judges who have to apply the law independently from the opinions and wishes of the Member State governments. The rule of EU law is carried out both by the European Union's own judiciaries and by all the Member State judiciaries. Today voting in the European Union is primarily based on majority voting which means that Member States that have voted against a given adopted measure will still be bound by the measure (no veto right), the European Union has its own sources of income and determines its own budget.

[101] Citizenship of the Union and the rights that follow are regulated by Articles 20-25 TFEU.

From time to time conflicts arise between the European Union's own institutions (e.g. between the European Parliament and the Council of Ministers) and these conflicts can be settled by the Court of Justice of the European Union. The same goes for disputes between the Member States and between the European Union and the Member States. An independent institution, the European Commission, supervises the faithful execution of EU obligations and EU law by the Member States. If there are differences of opinion on these issues, the Commission can advise a Member State to change its behaviour. If the Member State does not agree, it is up to the Commission to decide to bring the state before the Court of Justice. The Court decides the issue, and if the judgment goes against the Member State it will usually comply with the ruling. However, it has been the case that Member States have had to appear a second time before the Court of Justice because they have failed to duly carry out the first verdict of the Court. Article 260(2) TFEU makes it possible to impose a fine on Member States that fail to comply with a judgment by the Court of Justice.

The decisive first step in the creation of this complex Union was taken by a French Minister of Foreign Affairs with a German name; Robert Schuman. On the 9th of May 1950[102] he read a declaration to the press which heralded a new approach of France's policies that sought a solution to the age-old problem of how to live peacefully with Germany. In dealing with EU law it is essential always to realise this highly political aspect of EU law, even when buried under the oncoming avalanche of the technical details of food law. European Union law inherently deepens the integration process and contributes to the overall goals of preventing war and providing peace, security and prosperity. Law text box 4.1 presents the text of Schuman's proposal to create a common future under a High Authority.

[102] For this reason, the 9th of May is celebrated as European Union Day. The EU now also has an anthem of its own: die Ode an die Freude (Beethoven's 9th).

Law text box 4.1. The Schuman Declaration of 9 May 1950.[103]

'World peace cannot be safeguarded without the making of creative efforts proportionate to the dangers which threaten it. The contribution which an organised and living Europe can bring to civilisation is indispensable to the maintenance of peaceful relations. In taking upon herself for more than 20 years the role of champion of a united Europe, France has always had as her essential aim the service of peace. A united Europe was not achieved and we had war.

Europe will not be made all at once, or according to a single plan. It will be built through concrete achievements which first create a de facto solidarity. The coming together of the nations of Europe requires the elimination of the age-old opposition of France and Germany. Any action taken must in the first place concern these two countries.

With this aim in view, the French Government proposes that action be taken immediately on one limited but decisive point. It proposes that Franco-German production of coal and steel as a whole be placed under a common High Authority, within the framework of an organisation open to the participation of the other countries of Europe. The pooling of coal and steel production should immediately provide for the setting up of common foundations for economic development as a first step in the federation of Europe, and will change the destinies of those regions which have long been devoted to the manufacture of munitions of war, of which they have been the most constant victims.

The solidarity in production thus established will make it plain that any war between France and Germany becomes not merely unthinkable, but materially impossible. The setting up of this powerful productive unit, open to all countries willing to take part and bound ultimately to provide all the member countries with the basic elements of industrial production on the same terms, will lay a true foundation for their economic unification. This production will be offered to the world as a whole without distinction or exception, with the aim of contributing to raising living standards and to promoting peaceful achievements. [...]

In this way, there will be realised simply and speedily that fusion of interest which is indispensable to the establishment of a common economic system; it may be the leaven from which may grow a wider and deeper community between countries long opposed to one another by sanguinary divisions. By pooling basic production and by instituting a new High Authority, whose decisions will bind France, Germany and other member countries, this proposal will lead to the realisation of the first concrete foundation of a European federation indispensable to the preservation of peace.

To promote the realisation of the objectives defined, the French Government is ready to open negotiations on the following bases.

[103] Source: Historical Archives of the European Union: http://europa.eu, choose tab Documents à Historical Archives of the European Union à Research by subject à Keywords.

Law text box 4.1. Continued.

The task with which this common High Authority will be charged will be that of securing in the shortest possible time the modernisation of production and the improvement of its quality; the supply of coal and steel on identical terms to the French and German markets, as well as to the markets of other member countries; the development in common of exports to other countries; the equalisation and improvement of the living conditions of workers in these industries.

To achieve these objectives, starting from the very different conditions in which the production of member countries is at present situated, it is proposed that certain transitional measures should be instituted, such as the application of a production and investment plan, the establishment of compensating machinery for equating prices, and the creation of a restructuring fund to facilitate the rationalisation of production. The movement of coal and steel between member countries will immediately be freed from all customs duty, and will not be affected by differential transport rates. Conditions will gradually be created which will spontaneously provide for the more rational distribution of production at the highest level of productivity. In contrast to international cartels, which tend to impose restrictive practices on distribution and the exploitation of national markets, and to maintain high profits, the organisation will ensure the fusion of markets and the expansion of production.

The essential principles and undertakings defined above will be the subject of a treaty signed between the States and submitted for the ratification of their parliaments. The negotiations required to settle details of applications will be undertaken with the help of an arbitrator appointed by common agreement. He will be entrusted with the task of seeing that the agreements reached conform with the principles laid down, and, in the event of a deadlock, he will decide what solution is to be adopted.

The common High Authority entrusted with the management of the scheme will be composed of independent persons appointed by the governments, giving equal representation.

A chairman will be chosen by common agreement between the governments. The Authority's decisions will be enforceable in France, Germany and other member countries. Appropriate measures will be provided for means of appeal against the decisions of the Authority.

A representative of the United Nations will be accredited to the Authority, and will be instructed to make a public report to the United Nations twice yearly, giving an account of the working of the new organisation, particularly as concerns the safeguarding of its objectives.

The institution of the High Authority will in no way prejudge the methods of ownership of enterprises. In the exercise of its functions, the common High Authority will take into account the powers conferred upon the International Ruhr Authority and the obligations of all kinds imposed upon Germany, so long as these remain in force'.

The Schuman Declaration gave the impetus for the creation of the first organisations for integration in Europe, the European Coal and Steel Community in 1952, Euratom and the European Economic Community in 1958. The original six Member States transferred some of their national sovereignty to these Communities, thus 'limiting their sovereignty and creating a Community based on law and binding them and their citizens, creating rights and obligations (...)' as the Court of Justice stated in 1962 in its ruling in the Van Gend en Loos case[104] and in the Costa versus ENEL case in 1964.[105] The Court stated its fundamental position that 'the Community constitutes a new legal order, for the benefit of which the states have limited their sovereign rights, albeit within limited fields, and the subjects of which comprise not only the Member States but also their nationals'.[106] These judgments by the Court of Justice have become corner stones of EU constitutional law. Individuals next to states as direct participants in this new legal order clearly set EU law apart from general international law, certainly at that time. However, the Member States were determined to specify exactly which parts of their sovereignty were transferred to the new organisations, and to what extent.

4.3 The Lisbon Treaty reform

The Lisbon Treaty entered into force on 1 December 2009 and introduced the Treaty on European Union (abbreviated TEU) and the Treaty on the Functioning of the European Union (abbreviated TFEU). Today these two Treaties form the legal foundation of the European Union. The TEU and the TFEU have the same legal value and shall be referred to as 'the Treaties'.[107] The changes that the Lisbon Treaty brought to those Treaties that were in force prior to 1 December 2009 are more on the organisation of the integration and less on the substance. This holds particularly for food law.

In 1992 the original European Union was 'founded on the European Communities, supplemented by the policies and forms of cooperation established by this Treaty'.[108] With the Lisbon Treaty the European Union assumes the identity, place and legal personality that were previously held by the European Community.[109] The policies are redistributed among the TEU and the TFEU. With the introduction of the Lisbon Treaty, the numbering of the Treaty articles has changed. Articles

[104] Case 26/62 NV Algemene Transport en Expeditie Onderneming Van Gend en Loos *v.* Nederlandse Administratie der belastingen. Rulings of the JCEU are accessible on the website of the Court of Justice at 'curia.europa.eu' and in the European Union's legislation database EURLEX.

[105] Case 6/64 Costa *v.* ENEL [1964] ECR 585.

[106] Case 28/67 Firma Molkerei-Zentrale Westfalen Lippe GmbH *v.* Hauptzollamt Paderborn [1968] ECR 143 at 152.

[107] Article 1(2) TFEU, and Article 1, paragraph 3 TEU.

[108] Article 1 of the 1992 EU Treaty.

[109] The EU 'shall replace and succeed' the European Community, see Article 1 TEU. In other words, the EEC and the EC no longer exist. This book makes reference to them only where history is addressed and in quotes.

referring to 'TEC' refer to the old Treaties – it is important to bear this in mind and make the conversion when looking at legal texts predating 2009.[110]

More than ever, the changes brought about by the Lisbon Treaty centre on the division of powers between the Member States and the European Union, between national interests expressed through intergovernmental bargaining and supranational decision-making.

The TEU contains the main provisions on intergovernmental integration. The external and security policies now form part of these outer reaches of the integration process. At the same time the TFEU treaty contains its own external action provisions in Part five of TFEU. See Diagram 4.1 about The Two Treaties.

The transfer of the main articles on tasks and policies from the former EC Treaty to the new TEU pushes that Treaty to the centre of the integration process. However, while the TEU in a legal sense sets up the institutions of the European Union, it does little more than anchor their existence. It is the TFEU that provides the essential details of the institutions' functioning and activities. As the name of this Treaty suggests, all the rules on the composition, tasks and powers of the EU institutions, and the instruments and procedures they need to exercise those powers, are the substance of the Treaty on the Functioning of the European Union.

The TFEU part of the new constellation remains the most supranational part, although its intergovernmental underwiring is self-evident. The Member States are still preoccupied with the extent of the powers of the Union. This has led to a Treaty text that in much greater detail than before spells out the division of powers in five lengthy articles. The main principles which govern the division of power between Member States and the EU level are the 'principle of conferral', 'subsidiarity' and 'proportionality'.

4.3.1 The principle of conferral and the duty of loyal cooperation

The European Union has powers only if these are specified in the Treaties and transferred to them. In a national legal system the Constitution specifies the sovereignty assumed by the state and assigns specific parts of this national sovereignty to specific organs of the state, for example the legislature, the executive and the judiciary. These organs use their powers in a general way, subject to the conditions set by the Constitution. Usually emergency powers are also foreseen. The founders of the European Union wrote into the original Treaties that the Union has power only if the Treaties confer this particular power on the Union. The EU therefore adheres to a system of limited and enumerated powers. If the Treaties do not confer a particular power on the European Union, this power

[110] The consolidated versions of the TEU and TFEU are available at 'europa.eu'. It contains a 'Table of Equivalences' which provides an overview of the old and new article numbering.

*Diagram 4.1. Two EU Treaties. Policies directly relevant for food law are indicated by an asterisk *.*

Treaty on European Union

Preamble

Title I Common provisions

Title II Democratic Principles

Title III Institutions

Title IV Enhanced Cooperation

Title V External Action Common Foreign and Security Policy from:

Second pillar:
common foreign
and security policy

Foreign policy
- Cooperation common positions and measures
- Peacekeeping
- Human rights
- Democracy
- Aid to non-member countries

Security policy
- Drawing on the WEU: questions concerning the security of the EU
- Desarmament
- Financial aspects of defence
- Long-term: Europe's security framework

Title VI Final Provisions

Treaty on the Functioning of the European Union

Preamble

Part one Principles

Title I Union Competence

First pillar:
the European
Communities

Part two Non-discrimination and citizenship of the Union

- EU citizenship

Part three Union policies and internal actions

Title I Internal market

Title II Free movement of goods

EC
- Customs union* and single market*
- Agricultural policy*
- Structural policy
- Trade policy*

Title IV Free movement of persons, services and capital

Title V Area of Freedom, Security and Justice from:

Third pillar:
Police and Judicial
Co-operation in
Criminal Matters

- Establish progressively an area of freedom, security and justice
  - no internal borders
  - free movement of persons
  - asylum policy
  - external borders
  - immigration policy

remains in the hands of the Member States. This basic principle is technically called the conferral of power.[111]

The Treaty of Lisbon has formally strengthened the position of Member States, by re-affirming that the powers not explicitly conferred to the EU are to remain with the Member States, in particular respecting their national sovereignty concerns. At the same time, Member States remain under a duty of 'loyal cooperation' to the community of states. See Law text box 4.2.

Article 4 TEU raises the issue of the division of powers together with another subject, the Member States' duty to do everything possible to further the interests of the Union, and refrain from doing anything that might harm the interests of the Union. The Court of Justice has increasingly referred to this double allegiance to determine the obligations of the Member States. It is still not clear whether the Lisbon Treaty's allocation of these principles in the powers section of the TEU will influence the Court's opinion.

Law text box 4.2. Remaining competences, equality of Member States and the principle of sincere cooperation.

Treaty on European Union
Article 4

1. In accordance with Article 5, competences not conferred upon the Union in the Treaties remain with the Member States.
2. The Union shall respect the equality of Member States before the Treaties as well as their national identities, inherent in their fundamental structures, political and constitutional, inclusive of regional and local self-government. It shall respect their essential State functions, including ensuring the territorial integrity of the State, maintaining law and order and safeguarding national security. In particular, national security remains the sole responsibility of each Member State.
3. Pursuant to the principle of sincere cooperation, the Union and the Member States shall, in full mutual respect, assist each other in carrying out tasks which flow from the Treaties.
   The Member States shall take any appropriate measure, general or particular, to ensure fulfilment of the obligations arising out of the Treaties or resulting from the acts of the institutions of the Union.
   The Member States shall facilitate the achievement of the Union's tasks and refrain from any measure which could jeopardise the attainment of the Union's objectives.

[111] See Article 5(1) and (2) TEU.

4.3.2 Subsidiarity

The concept of subsidiarity had also been part of EC treaty law to act as an assurance against the usurpation of too much power by the EU vis-à-vis the Member States. The principle prescribes an attitude for the EU institutions and the leaders of those institutions on how to use power. It operates only in those situations where the EU has a power and a possibility to share that power with others. If the EU has exclusive power in a given field, it cannot step down from its responsibility and surrender the power to another entity (the Member States). However, the EU can delegate some of its powers as long as its responsibility is not diminished. See Article 2 TFEU in Law text box 4.3 on the categories and areas of Union competence.

Law text box 4.3. Limits and use of Union competences: the principles of conferral, subsidiarity and proportionality.

Treaty on European Union
Article 5

1. The limits of Union competences are governed by the principle of conferral. The use of Union competences is governed by the principles of subsidiarity and proportionality.
2. Under the principle of conferral, the Union shall act only within the limits of the competences conferred upon it by the Member States in the Treaties to attain the objectives set out therein. Competences not conferred upon the Union in the Treaties remain with the Member States.
3. Under the principle of subsidiarity, in areas which do not fall within its exclusive competence, the Union shall act only if and in so far as the objectives of the proposed action cannot be sufficiently achieved by the Member States, either at central level or at regional and local level, but can rather, by reason of the scale or effects of the proposed action, be better achieved at Union level.
   The institutions of the Union shall apply the principle of subsidiarity as laid down in the Protocol on the application of the principles of subsidiarity and proportionality. National Parliaments ensure compliance with the principle of subsidiarity in accordance with the procedure set out in that Protocol.
4. Under the principle of proportionality, the content and form of Union action shall not exceed what is necessary to achieve the objectives of the Treaties.
   The institutions of the Union shall apply the principle of proportionality as laid down in the Protocol on the application of the principles of subsidiarity and proportionality.

Subsidiarity does not question the existence of EU powers, but acts as a restraint. It requires an authority to consider whether the goals for which a competence was conferred on the EU are not equally well served by the exercise of powers by institutions that are closer to the communities they serve. The EU could conclude that it would serve additional ends of society if it shared its power with national, regional or local authorities.[112]

The official interpretation of the principle of subsidiarity (as well as the principle of proportionality) is laid down in a special Protocol.[113] The implementation in practice demands an explicit statement by the EU institution that exercises its power that it has considered refraining from using this power. When the test against the subsidiarity criterion does not lead to the conclusion to benefit decentralised authorities by not using the EU power, the next test is proportionality – see Section 4.3.3 below. Law text box 4.4 on the practice of subsidiarity and proportionality shows the public announcement about the weighing of these aspects and the consequences it had.

Law text box 4.4. The practise of subsidiarity and proportionality.

Regulation (EC) No 1185/2009 of the European Parliament and of the Council of 25 November 2009 concerning statistics on pesticides.
The European Parliament and the Council of the European Union,
Having regard to the Treaty establishing the European Community, and in particular Article 285(1) thereof,
...
Whereas:
(16) Since the objective of this Regulation, namely the establishment of a common framework for the systematic production of Community statistics on the placing on the market and use of pesticides, cannot be sufficiently achieved by the Member States and can therefore be better achieved at Community level, the Community may adopt measures, in accordance with the principle of subsidiarity as set out in Article 5 of the Treaty. In accordance with the principle of proportionality, as set out in that Article, this Regulation does not go beyond what is necessary in order to achieve that objective.

4.3.3 Proportionality

Arguably, in practice the European Union's principle of proportionality is the most important of all the Union's law principles. Today we find the principle in Article 5 TEU (see Law text box 4.3) and the official interpretation of the principle of proportionality is laid down in the special Protocol that also covers the principle of subsidiarity. In contrast to the principle of subsidiarity, the EU's principle of proportionality applies in all situations of application of EU law. Thus it also applies

[112] Originally, the principle of subsidiarity was developed in Catholic social theories about the relations between society and state. These theories favoured the exercise of public power by authorities that are situated as closely as possible to the society they govern. Better still in their midst.
[113] Protocol on the application of the principles of subsidiarity and proportionality, OJ C 310, 16 December 2004, pp. 207-209.

when the European Union legislates in a field where it has exclusive competence, and it equally applies where a Member State acts in a field that is covered by EU law. In other words, the principle of proportionality permeates all areas of EU law.

The European Union's principle of proportionality is made up of three cumulative requirements:

- Firstly, the measure must be capable of achieving the stated objective (the suitability test).
- Secondly, the Member State measure must constitute the least disruptive measure that is capable of achieving the stated lawful objective (the necessity test).
- Thirdly, even if the two first requirements are met, it is still necessary to consider whether the measure is proportionate vis-à-vis the lawful objective (the proportionality *stricto sensu* test).

The proportionality principle has often played a main part in cases involving food law, not only with regard to the European institutions but also with regard to the Member States. For example, if a Member State has banned a food product because it finds that the product constitutes a danger to the health of its population, this ban is only lawful if it complies with the proportionality principle. Protecting the population's health is clearly a lawful objective, and a ban against an unhealthy food product is an efficient way of attaining this protection. This means that a ban fulfils the first of the three conditions. However, it may be that a ban is a much too draconian measure and that instead the Member State could merely have introduced a requirement that the food products must be clearly labelled about the health issue so as to warn the consumer. In such a case the necessity requirement has not been met, and it will not even be necessary to consider whether or not the third condition has been met; the ban will be held unlawful under EU law.

4.4 The powers of the EU

From a legal point of view, how is it determined when the EU may take action? At the most abstract level, the EU always acts in order to achieve one of its objectives, which are listed in Article 3 TEU (Law text box 4.5).

The fine print of the division of powers can be found in Title I 'Categories and Areas of Union Competence', Articles 2 to 6 of the TFEU (reproduced in Law text box 4.6). They are likely to play a continuing part in the discussions about what the Union can do, may do, should do, ought to do, or not.

The Lisbon Treaty has resulted in an increased number of clauses about powers. This approach stands in opposition to the original main line of the treaties that refrained from sweeping statements and instead measured precisely the combination of institutions, voting requirements, instruments and procedures

Law text box 4.5. Objectives of the European Union, Article 3 TEU.

Article 3 TEU

1. The Union's aim is to promote peace, its values and the well-being of its peoples.
2. The Union shall offer its citizens an area of freedom, security and justice without internal frontiers, in which the free movement of persons is ensured in conjunction with appropriate measures with respect to external border controls, asylum, immigration and the prevention and combating of crime.
3. The Union shall establish an internal market. It shall work for the sustainable development of Europe based on balanced economic growth and price stability, a highly competitive social market economy, aiming at full employment and social progress, and a high level of protection and improvement of the quality of the environment. It shall promote scientific and technological advance.
   It shall combat social exclusion and discrimination, and shall promote social justice and protection, equality between women and men, solidarity between generations and protection of the rights of the child.
   It shall promote economic, social and territorial cohesion, and solidarity among Member States.
   It shall respect its rich cultural and linguistic diversity, and shall ensure that Europe's cultural heritage is safeguarded and enhanced.
4. The Union shall establish an economic and monetary union whose currency is the euro.
5. In its relations with the wider world, the Union shall uphold and promote its values and interests and contribute to the protection of its citizens. It shall contribute to peace, security, the sustainable development of the Earth, solidarity and mutual respect among peoples, free and fair trade, eradication of poverty and the protection of human rights, in particular the rights of the child, as well as to the strict observance and the development of international law, including respect for the principles of the United Nations Charter.

available for every separate policy. Now the TEU and TFEU provide both: a general statement and a precise allocation.

Under the 'principle of conferral', the EU may act only where it has competence to do so. The different policy areas are matched with respective competence categories. The categories of competences are:

- Exclusive competencies of the EU (Article 3 TFEU);
- Shared competencies with Member States (Article 4 TFEU);
- Supplementary competencies to powers of Member States (Article 6 TFEU).

Law text box 4.6. The categories and areas of Union competence.

The Treaty on the Functioning of the European Union
Title I. Categories and Areas of Union Competence
Article 2

1. When the Treaties confer on the Union exclusive competence in a specific area, only the Union may legislate and adopt legally binding acts, the Member States being able to do so themselves only if so empowered by the Union or for the implementation of Union acts.
2. When the Treaties confer on the Union a competence shared with the Member States in a specific area, the Union and the Member States may legislate and adopt legally binding acts in that area.
   The Member States shall exercise their competence to the extent that the Union has not exercised its competence. The Member States shall again exercise their competence to the extent that the Union has decided to cease exercising its competence.
3. The Member States shall coordinate their economic and employment policies within arrangements as determined by this Treaty, which the Union shall have competence to provide.
4. The Union shall have competence, in accordance with the provisions of the Treaty on European Union, to define and implement a common foreign and security policy, including the progressive framing of a common defence policy.
5. In certain areas and under the conditions laid down in the Treaties, the Union shall have competence to carry out actions to support, coordinate or supplement the actions of the Member States, without thereby superseding their competence in these areas.
   Legally binding acts of the Union adopted on the basis of the provisions of the Treaties relating to these areas shall not entail harmonisation of Member States' laws or regulations.
6. The scope of and arrangements for exercising the Union's competences shall be determined by the provisions of the Treaties relating to each area.

Article 3

1. The Union shall have exclusive competence in the following areas:
   (a) customs union;
   (b) the establishing of the competition rules necessary for the functioning of the internal market;
   (c) monetary policy for the Member States whose currency is the euro;
   (d) the conservation of marine biological resources under the common fisheries policy;
   (e) common commercial policy.
2. The Union shall also have exclusive competence for the conclusion of an international agreement when its conclusion is provided for in a legislative act of the Union or is necessary to enable the Union to exercise its internal competence, or in so far as its conclusion may affect common rules or alter their scope.

Law text box 4.6. Continued.

Article 4

1. The Union shall share competence with the Member States where the Treaties confer on it a competence which does not relate to the areas referred to in Articles 3 and 6.
2. Shared competence between the Union and the Member States applies in the following principal areas:
   (a) internal market;
   (b) social policy, for the aspects defined in this Treaty;
   (c) economic, social and territorial cohesion;
   (d) agriculture and fisheries, excluding the conservation of marine biological resources;
   (e) environment;
   (f) consumer protection;
   (g) transport;
   (h) trans-European networks;
   (i) energy;
   (j) area of freedom, security and justice;
   (k) common safety concerns in public health matters, for the aspects defined in this Treaty.
3. In the areas of research, technological development and space, the Union shall have competence to carry out activities, in particular to define and implement programmes; however, the exercise of that competence shall not result in Member States being prevented from exercising theirs.
4. In the areas of development cooperation and humanitarian aid, the Union shall have competence to carry out activities and conduct a common policy; however, the exercise of that competence shall not result in Member States being prevented from exercising theirs.

Article 5

1. The Member States shall coordinate their economic policies within the Union. To this end, the council shall adopt measures, in particular broad guidelines for these policies. Specific provisions shall apply to those Member States whose currency is the euro.
2. The Union shall take measures to ensure coordination of the employment policies of the Member States, in particular by defining guidelines for these policies.
3. The Union may take initiatives to ensure coordination of Member States' social policies.

Article 6

The Union shall have competence to carry out actions to support, coordinate or supplement the actions of the Member States. The areas of such action shall, at European level, be:

(a) protection and improvement of human health;
(b) industry;
(c) culture;
(d) tourism;
(e) education, vocational training, youth and sport;
(f) civil protection;
(g) administrative cooperation.

This conferral of power provides but the skeleton and remains an abstract toolkit. The precise allocation of powers is detailed in Part III of the TFEU, which unfolds the Union policies in 24 so-called titles. Each title provides a set of specifications attuned to the characteristics of its policy field. It is among this elaboration of each policy that the necessary powers are attributed. The TFEU provides a separate transfer of power for each separate policy. Each transfer takes from the abstract toolkit which is deemed to fit the particular policy and assigns the institutions that have to carry out these tasks, in accordance with the procedures that are also given.

4.4.1 Example of the conferral of power

Each policy area therefore enjoys a particular article that regulates the conferral of power in that specific field. All legal instruments must be based on such a power conferral article, which is therefore also called a 'legal basis'. Diagram 4.2 shows the conferral of power for agriculture policy. The power clause indicates the attribution to the Council and the European Parliament to co-decide in the ordinary legislative procedure.

Diagram 4.2. Conferral of power: agriculture.

Common agricultural policy making: Articles 38 to 44 TFEU	
Article 4 (TFEU)	TITLE III AGRICULTURE AND FISHERIES
1. The Union shall share competence with the Member States where the Treaties confer on it a competence which does not relate to the areas referred to in Articles 3 and 6. 2. Shared competence between the Union and the Member States applies in the following principal areas: (...) (d) agriculture and fisheries, excluding the conservation of marine biological resources;	Article 38 1. The Union shall define and implement a common agriculture and fisheries policy (...) Article 39 1. The objectives of the common agricultural policy shall be (...) Article 43 (...) 2. The European Parliament and the Council, acting in accordance with the ordinary legislative procedure and after consulting the Economic and Social Committee, shall establish the common organisation of agricultural markets provided for in Article 40(1) and the other provisions necessary for the pursuit of the objectives of the common agricultural policy and the common fisheries policy.

The attribution of powers for food policy and food law is more complicated. This attribution has several bases in the TFEU of which agriculture is only one, and today not the most important one. Diagram 4.3 indicates the different parts of the TFEU that attribute the powers needed to make food policy and food law.

Diagram 4.3. EU powers on food law.

- Free movement of goods (Part 3, Title II TFEU)
- Agriculture (Part 3, Title III TFEU)
- Approximation of laws (Part 3, Title VII)
- Public health (Part 3, Title XIV)
- Consumer protection (Part 3, Title XV)
- Common commercial policy (Part 5, Title II)
- Atomic energy policy (Euratom Treaty)

The European Atomic Energy Community, established by the less prominent Euratom Treaty, is not without relevance for food law. Safety limits regarding radioactive contamination of foods find (at least in part) their basis here.[114]

The institutions of the European Union have to prove that the necessary power resides with the Union each time they want to act. Therefore each authoritative act of the European Union starts by mentioning the legal basis, i.e. the specific Treaty article(s) that give(s) the powers that are used: 'Having regard to Article (...)'.

For example, Regulation (EC) No 178/2002, usually called the General Food Law (GFL) states: 'Having regard to the Treaty establishing the European Community, and in particular Articles 37, 95, 133 and Article 152(4)(b) thereof'. This article refers to the old Treaty numbering (under the Treaty establishing the European Community, TEC). The articles correspond to Articles 43 (agriculture), 114 (approximation of laws), 207 (common commercial policy) and 168 (public health) of the TFEU. If called upon, the Court of Justice will verify the existence of the powers used by the EU to adopt a legal measure. If the Court finds that the legal measure has no basis in the Treaties, it will render the measure illegal which means that the Union's actions based on the measure will also be illegal, because it is an exercise of power that exceeds the powers that have been conferred on the Union.[115]

[114] See Chapter 11, Section 11.9.

[115] Article 263 TFEU lists four grounds for annulment of binding legal instruments: lack of competence is the first one.

4.4.2 Additional powers

Although the TFEU contains an extensive list of policies, it is not always clear under what heading a particular measure must be subsumed. In other words: which Treaty Article or Articles have to be 'regarded'. Debates on the choice of the correct Treaty Article flare up every now and then. They are important when the choice makes a difference to the power of the institutions or to the type of instruments that can be used.

An example: during the preparation of the Regulation on Genetically Modified Food,[116] the question was debated as to whether the regulation should be based on what today is Article 114 TFEU (harmonisation) or on what today is Article 352 TFEU (provision that is used when specific powers to attain a Union objective is missing). The predecessor to Article 114 TFEU – which was ultimately agreed upon – prescribes what today is called the ordinary legislative procedure that basically gives the European Parliament as much influence as the Council whilst allowing for majority voting in the Council (so that no Member State has a veto right). In contrast, Article 352 TFEU prescribes a procedure that requires unanimity in the Council and thus gives the individual Member State more influence through a veto right.[117]

Given the strict rules on the conferral of powers, a problem could surface due to the fact that on the one hand the EU has to achieve the goals specified in the Treaties while on the other hand it must have an explicit attribution of power for every step it takes. It might turn out that the EU lacks the power that is needed to carry out the task. The drafters of the original Treaty did not assume that they would be able to foresee all the necessities that might arise in the complicated process of integrating national societies. Nor did they pretend that all possible developments of the different national societies and their economies could be locked into the Treaties. They therefore included an 'escape clause' in the Treaties to meet this type of situation. Law text box 4.7 presents the 'escape clause'.

Article 352 TFEU has served the European Union with a way out that has been used more than one hundred times. It has been used to integrate trade in processed agricultural products into the common agricultural policy and later to give what was then the European Economic Community a common regional policy, a social policy and an environmental policy. At that time the applicable Treaty did not contain these policies. Subsequently, these policies have been incorporated into the Treaty system with the Single European Act which amended the EEC Treaty

[116] Regulation 1829/2003 of the European Parliament and of the Council of 22 September 2003 on genetically modified food and feed (with subsequent amendments). See Chapter 10 for discussion of the GM food regulation.
[117] See Chapter 5, in particular **Section 5.2**.

Law text box 4.7. Safety valve for attribution of powers, Article 352 TFEU.

1. If action by the Union should prove necessary, within the framework of the policies defined in the Treaties, to attain one of the objectives set out in the Treaties, and the Treaties have not provided the necessary powers, the Council, acting unanimously on a proposal from the Commission and after obtaining the consent of the European Parliament, shall adopt the appropriate measures. Where the measures in question are adopted by the Council in accordance with a special legislative procedure, it shall also act unanimously on a proposal from the Commission and after obtaining the consent of the European Parliament.
2. Using the procedure for monitoring the subsidiarity principle referred to in Article 5(3) of the Treaty on European Union, the Commission shall draw national Parliaments' attention to proposals based on this Article.
3. Measures based on this Article shall not entail harmonisation of Member States' laws or regulations in cases where the Treaties exclude such harmonisation.
4. This Article cannot serve as a basis for attaining objectives pertaining to the common foreign and security policy and any acts adopted pursuant to this Article shall respect the limits set out in Article 40, second paragraph, of the Treaty on European Union.

in 1986.[118] Article 352 TFEU can still be used for its purpose, but it is now much less likely that this will happen because the Treaties today contain a significantly greater array of policies and instruments.

4.4.3 Implied powers

To complete the picture of the EU's powers, mention must also be made of what is referred to as the 'implied powers'. These powers are not explicitly mentioned in the Treaties, but follow by necessity from the powers that can be found therein. For example, if the TFEU gives the EU the powers to regulate fisheries in general, it follows from this express internal power that the Union also has implied external power to deal with international aspects of fisheries outside the EU. The Court of Justice of the European Union has decided several times that the power to act within the EU includes the power to act outside the EU if the policy so requires.[119]

4.5 EU policy instruments

The EU has several instruments to carry out its tasks and policies.[120] These instruments are available for the institutions in different settings: for the European Parliament acting jointly with the Council on the basis of an initiative of the Commission; for the Council acting alone on the basis of an initiative of the

[118] 1986 Single European Act. OJ L 169, 29 June 1987.
[119] Joined Cases 3, 4 and 6/76, Kramer. Conservation of certain species of fish. EU party to a Treaty to limit fishing in the North East Atlantic.
[120] Article 288 TFEU defines some important instruments.

Commission; and for the Commission acting on the basis of the powers given by the Treaties, and even more on the basis of the powers delegated to it in secondary legislation (such as a regulation).

Advice, recommendations and opinions are not binding policy statements. They may give guidance, rally support and lead to what is normally called 'soft law'. Whilst binding law is often referred to as 'hard law', soft law is non-binding law, but is nonetheless taken into account when a question of interpretation is decided – for example it may provide persuasive arguments with respect to the interpretation of 'hard law'.

The legally binding EU instruments are the regulation, the directive and the decision. They have quite different forms, and lead to different results. The regulation and the directive are general legislative acts, the decision is an act made for one particular situation.

4.5.1 Primary law and secondary law

Treaties, regulations, directives and decisions embody EU law. The founding Treaties are the primary law as they created the European Union. Primary law can only be changed by other primary law, such as new treaties or basic principles of international law (*ius cogens*). According to Article 48 TEU amendments to the Treaties must be carried out by the Member States in cooperation with some of the EU institutions.

Legislation based on the two Treaties (TEU and TFEU) carries out what was foreseen in the primary law, so in relation to the Treaties all other instruments are secondary law. Of these, regulations are the instruments to create legislation that resemble most the legislative instruments used by the Member States in their national systems. Directives are especially created for the peculiar EU combination of legislative activities of the EU in tandem with the Member States. Decisions stand apart from the legislative acts; they are for individual cases and can be based on the Treaties or on secondary legislation.

4.5.2 The regulation

Regulations are defined in the TFEU in this way: 'A regulation shall have general application. It shall be binding in its entirety and directly applicable in all Member States'.[121]

The elements 'general application', 'binding' and 'directly applicable', construct the character of the regulation: it is a legally binding rule, it is legislation binding

[121] Article 288 TFEU.

all natural and legal persons, that fall under the jurisdiction of the EU, including the Member States, the EU itself and hence its institutions and bodies, if they are in the situation defined in the rule. The jurisdiction of the EU extends over the combined territories of the Member States and beyond when the EU takes legal action against persons or companies that influence the internal market from abroad.

The regulation cuts the matter for which it is made as legislation right out of the national legislation of the Member States. This matter is now legislated solely on the higher level of EU law. The regulation does not need national law to operate inside the Member States. The regulation secures that its law is the same everywhere on the territory of the Union. Only when the regulation itself allows it, can different laws be applied for certain regions or groups. Only when the regulation allows it, can national law make choices presented in the regulation or make more detailed rules.[122] Most EU regulations are enforced by the authorities of the Member States.

These kinds of legal acts have to be made according to strict procedures,[123] of which publication in the Official Journal of the European Union is one.[124] Under the rule of law legally binding instruments have to be knowable before they can be binding. The Member States are not allowed to present these EU regulations in their national law books or official journals as national measures. This is strictly prohibited to avoid confusion about the EU nature of this law. A regulation is binding because it is EU law, not because it is national law of the Member States. Therefore the Member States may not create the false impression that a regulation is national law, or that it can be changed by a new national law made on a later date.

The regulation is binding as long as it is made in accordance with the Treaties, which prescribe the conditions that have to be met if and when the EU institutions exercise their powers. Many conditions have to be met; they have been dealt with under the heading of the conferral of power. The TFEU prescribes four grounds for annulment, which the Court of Justice must consider when a regulation is challenged before it. One of these grounds is 'infringement of the Treaties or of any rule of law relating to their application'.[125]

[122] Regulation 1169/2011 of the European Parliament and of the Council of 25 October 2011 on the provision of food information to consumers, amending Regulations (EC) No 1924/2006 and (EC) No 1925/2006 of the European Parliament and of the Council, and repealing Commission Directive 87/250/EEC, Council Directive 90/496/EEC, Commission Directive 1999/10/EC, Directive 2000/13/EC of the European Parliament and of the Council, Commission Directives 2002/67/EC and 2008/5/EC and Commission Regulation (EC) No 608/2004 (discussed in Chapter 14) devotes an entire chapter (VI articles 38-45) to national measures.

[123] Articles 293-299 TFEU.

[124] Article 297 TFEU.

[125] Article 263 TFEU.

The basic regulation

A distinction is made in practice between a basic regulation and other regulations. The basic regulation is based immediately on the relevant TFEU article(s). It contains the basic provisions for a particular EU policy. It sets the broad outlines and empowers the Council or the Commission to make more detailed regulations on certain issues. The TFEU and the basic regulation together determine the options for the more detailed legislation.

Regulation (EC) No 178/2002 of the European Parliament and of the Council of 28 January 2002 laying down the general principles and requirements of food law, establishing the European Food Safety Authority and laying down procedures in matters of food safety, for short the General Food Law, is in many ways the basic regulation for food policy and food law.

The framework regulation

The framework regulation is the instrument made by the legislator to determine the most important elements and formulate the conditions that have to guide the future legislator on how to fill in the framework. The framework regulation can also formulate the conditions that the more detailed regulations have to meet. The Council working together with the European Parliament will in many cases be the legislator of a framework regulation. The legislator who made the framework regulation can also make the legislation to fill in the framework. It can also be another legislator to whom this power is delegated, usually the Commission. In this book, for example, the discussions of food additives in Chapter 10 and of contaminants in Chapter 11 begin by introducing the framework directive and framework regulation, respectively, and then move on to discuss the legislation set in place to fill in the framework.

4.5.3 The directive

The directive as a legal measure was expressly designed as an additional instrument to facilitate the relationship between the law of the EU and the national laws of the Member States. The TFEU defines a directive as follows: 'A directive shall be binding, as to the result to be achieved, upon each Member State to which it is addressed, but shall leave to the national authorities the choice of form and methods'.[126]

The directive can be addressed to one or more Member States, but it is usually directed to all. It prescribes the objectives that have to be met, and the deadline by which this has to be achieved. It leaves to the Member States the choice of

[126] Article 288 TFEU.

means to fulfil the obligation. The Member States are the best experts on their own systems of national laws. They are best suited to find the ways to fit their national legal systems to the EU objectives and the rules prescribed by the directive.

The directive is used if the EU finds it necessary to harmonise the national laws of the Member States: the laws in the Member States do not have to be identical in all respects (in that case the regulation would be the required instrument), but must be sufficiently comparable to fit in with EU principles and policies. The original European Economic Community (EEC) Treaty[127] foresaw the need especially for the common market to have large-scale harmonisation of national laws dealing with all aspects of goods that the Treaty prescribes to be in free movement. On the other hand, nobody could foresee how many national laws had to be harmonised, nor to what extent. To solve this problem, the EEC Treaty offered in Article 100 (now Article 115 TFEU) the directive as a bridge between EU law and national law, and specified that unanimity was required for each harmonisation directive. In that way all Member States could participate with confidence in the uncharted harmonisation effort, knowing they would be their own master at all times.

In practice the difficulties arising from different national laws with an effect on the common market proved to be even tougher. To remedy a jammed decision-making process, the Treaty that is normally called the Single European Act in 1986 introduced the internal market and amended a number of articles of the then applicable EEC Treaty. One of these changes was the introduction of Article 100A (now Article 114 TFEU) allowing other instruments alongside directives, and prescribing decision making by majority vote. A quarter of a century of trying with unanimity had shown the impossibility of harmonising when every Member State has to agree with every detail. This was no surprise for the men who laid the foundations of the EEC Treaty. They wrote in the report commissioned by the governments of the six ECSC Member States (the Spaak Report)[128] that the new Community (now the Union) had to have majority decision making if it were to work. A notorious example of the effects of not voting but aiming at unanimity is the Directive for fruit jams and marmalades that took eighteen years of negotiation; and after that landmark the negotiators moved to tackle the other products on the table.

A directive is binding EU law for the Member States to which it is addressed. Usually it leads these states to change their national laws. These national laws (with changes originating in the European Union) are binding on those who live or work in this national law system.

[127] The EEC Treaty is the founding document made in 1957 for the original European Economic Community, the precursor of the European Union.

[128] Spaak Report: Rapport des Chefs de Délégations aux Ministres des Affaires Etrangères. Secretariat of the Intergovernmental Conference, Brussels, 21 April 1956.

In this case there are two sets of rules: the changed national law on the level of the Member State and the directive on the EU level. The national rules must stay within the bounds specified in the directive. If the Member State has fulfilled its obligation the national law will carry out the rules of the directive. However, opinions may differ. Then those who have to apply the national law (judges, government agencies, regional or local governments) are obliged to apply the national law as it is supposed to be, that is in the way the directive prescribes it. They have to examine the national law for compatibility with the EU directive. This is an important legal obligation that is one of the consequences of Article 4(3) TEU, which requires the Member States to loyally carry out all tasks that flow from the Treaties – including loyally implementing and fully complying with directives.

There are several requirements for the correct implementation of a directive:
- Firstly, the Union legislator has to enact directives that are clearly formulated[129] and leave little room for misinterpretation.
- Secondly, the national legislator must transform the directive in the right way and before the deadline laid down by the EU.
- Thirdly, if doubts arise as to the compatibility of a national law with a directive, the national judge has to interpret the relevant national law assuming that the national legislator did transform the directive to the best of his ability into national law. That is to say, the national judge must use the interpretation that brings national law closest to the content of the directive. This can eliminate a broad spectrum of possible doubt. But where forced interpretation would be stretching the national law into something it clearly is not, other methods are needed.

A clear-cut case is when the Member State fails to make the required national law before the implementation deadline specified in the directive, a situation that occurs frequently. In these circumstances, with states that fail to comply with the directive, the question becomes urgent as to how the prescriptions of the directive can become binding law. EU law has to provide a solution for this problem. Part of this solution is provided by the Court of Justice. It is called direct effect. Direct effect is discussed in Section 4.5.5.

4.5.4 The decision

As a rule, regulations and directives create rules that will be applied again and again to an unknown number of persons, each time the situation occurs that is described in these rules. The rules are binding in general and they spell out the consequences for those persons who fulfil the conditions laid down in the rules.

[129] Political compromise, unfortunately, often leads to a compromised text that can be understood in different ways. This is sometimes forgivingly referred to as 'constructive ambiguity'.

A decision is essentially different, as it normally formulates the law for one specific situation and only for that particular situation. It is directed to a known addressee (person, business, Member State) and defines a particular aspect of the legal situation for this particular addressee.[130] A decision constitutes not a rule, but the application of a rule to one particular event.[131]

The TFEU does not define the decision, but prescribes its consequences in Article 288 TFEU. The Court of Justice defined the decision as 'a measure emanating from the competent authority, intended to produce legal effects and constituting the culmination of procedure within that authority, whereby the latter gives its final ruling in a form from which its nature can be identified'.[132] This definition stresses the difference between a decision creating rights, and earlier communications during the preparation of the decision, such as a letter or a recommendation that does not create rights because it is not binding. Decisions binding natural and legal persons have to be based on rules, on legislation.[133] This is one of the consequences of the rule of law, the principle that government itself is bound by the law it has made. Arbitrary decisions are not allowed, especially when decisions in comparable cases turn out to be different without any argumentation that could explain the difference. Hence a prior law must spell out the conditions for each decision. This enables the executive to take the right decision, and gives the courts the yardstick with which to check the validity of the decision. The rule of law requires the possibility for every decision to be checked on its legality by a court of law.

Most decisions to carry out EU policies are taken by national authorities of the Member States. Only a part of these EU policy decisions are made at the EU level, and then in most cases by the European Commission. Appeals against a decision taken by an EU institution or an EU body are brought directly before the Court of Justice of the European Union or the lower General Court of the European Union. The following is an example of a Commission decision that at the same time sheds light on the extent of the Union's jurisdiction even outside its territory. The Commission has repeatedly declared cartels involving companies outside the EU illegal and presented these 'foreign' companies with administrative sanctions of millions of euros. They have to pay first, and only then can they start a case

[130] It is beyond the scope of this book to delve into the subtleties of decisions that address specific situations which cannot readily be connected to particular (legal) persons.

[131] An unfortunate mixup of different meanings of the word 'decision' occurs because it is used in the strict sense described so far, but is also used more loosely for all kinds of determinations that do not have the legal consequences of the decision. They do not create rights or obligations for one particular individual. Government ministries and agencies take many decisions as steps in the process of preparation or administration that have no legal effects for individuals outside the administration because these are internal affairs for this particular branch of government.

[132] Case 54/65 Compagnie des Forges de Châtillon v High Authority [1966] ECR 185.

[133] This separates decisions with effects outside the authority from internal decisions, which do not have immediate effects for the outside world.

before the General Court of the European Union. An example: the administrative sanctions applied by the Commission on the Swiss company Hoffmann-La Roche & Co. AG for abusing a dominant position in several markets for vitamins. Hoffmann-La Roche initiated a court case against the decision of the Commission, but the Court found for the Commission.[134]

Appeals against decisions on EU matters taken by national authorities must be submitted to the national courts. These must apply EU law. EU law contains a mechanism called preliminary ruling to assist the national judges in their task of applying EU law (see Chapter 5.6). The national law systems of the Member States must provide adequate means for a just procedure and for enforcement of the judgment.

4.5.5 Direct effect

The Treaties contain many obligations for the Member States and the institutions. In the first phase of the development of the European Economic Community, the Member States thought that they were the sole subjects of the rights and obligations spelled out for Member States in the Treaty. The Van Gend en Loos case, mentioned earlier as one of the cornerstones of EU constitutional law, originated as a court case of the Dutch transport company Van Gend en Loos against the Dutch customs authority. Van Gend en Loos refused to pay an increase in a customs duty levied for goods imported from Germany. In these early years of the EEC the old customs duties were abolished in three steps, with each step lasting four years. The EEC Treaty contained a stand-still clause specifying that the Member States would not increase existing duties while they were busy abolishing them.

Due to a commitment to the Benelux, based on a separate customs duties treaty, the Dutch government introduced a regrouping of existing duties. This included the increase in certain duties for goods that were brought into a different grouping. Van Gend en Loos contested the increase it had to pay, basing its claim on the stand-still clause in the EEC Treaty. The Dutch government maintained that this obligation was a matter for the Member States and the institutions only. This was a remarkable opinion for the government of a country that insists on the supremacy and direct effect of international law in general. The Dutch government maintained that natural and legal persons had nothing to do with this article, and could certainly not derive rights from this article written for the Member States. The Court of Justice ruled that the stand-still clause was binding on the Member States and at the same time created rights for their nationals. In this case the rule was easily applied: no increase meant that every national law that did increase duties was to be set aside by the EEC Treaty clause. Every person who was affected by this EEC law could base her/his claims directly on this Treaty;

---

[134] Commission Decision 76/642/EEC of 9 June (IV/29.020 – Vitamins). OJ L 223, 16 August 1976, pp. 27-38. Case 85/76 *Hoffmann-La Roche & Co. AG v. Commission*.

national authorities were not needed to create rights derived from this part of the Treaty. In other words, this article of the Treaty had direct effect. Ultimately, the case law of the Court of Justice identified many articles of the EEC Treaty that had direct effect.

An article of a binding text can have direct effect if it meets the following criteria:
1. The obligation imposed on Member States is clear and precise;
2. The obligation is unconditional and,
3. If implementing measures are provided for, the institutions or the Member States are not allowed any margin of discretion.

Basically, this means that the relevant part of the article's text is clear enough on its own to be implemented (see for example Law text box 4.8).

Law text box 4.8. Some Articles of the TFEU with direct effect.

Article 30. Customs duties and charges having equivalent effect prohibited between Member States.
Article 34. Quantitative restrictions on imports and all measures having equivalent effect prohibited between Member States.
Article 45(1). Freedom of movement for workers secured within the Union.
Article 49. Within the framework of the provisions set out below, restrictions on the freedom of establishment of nationals of a Member State in the territory of another Member State shall be prohibited.
And so on, all through the policies of the EU. One last example:
Article 157(1). Each Member State shall ensure that the principle of equal pay for male and female workers for equal work or work of equal value is applied.

[1] Note: the preference for direct effect of TFEU articles, giving private persons directly effective rights in the Treaty, was so strong that even the principle of Article 157 was considered to have direct effect. It could be argued that especially a principle cannot have direct effect because it needs additional legislation before it can be operative. But the Court of Justice preferred the right of persons, even when that had huge consequences for the economies of the Member States: retro-active payment of due equal payment? (See: Case 43/75 *Defrenne v. SABENA*.)

In this way supervision of the Treaties has become a concern for millions of natural and legal persons: each time one of their rights is involved, they can initiate a court case using the parts of the Treaties with direct effect. The constitutions of the Member States can make no difference for direct effect of the Treaties. The distinction between monism and dualism is irrelevant.[135] The Court of Justice ruled that the European Union and the internal market cannot be affected by national constitutions. The rules laid down by the national constitutions for the relations between international law and national law are valid only for the powers

[135] See Chapter 3, Section 3.2.5.

that remain with the Member State. The sovereignty that the Member States have transferred to the EU can no longer be diminished by national constitutions or any other national law.

4.5.6 Direct effect is not limited to Treaty articles

A regulation has direct effect (is immediately applicable) by definition. This instrument is made to rule all, without any interference of any other legislative authority. The regulation shall be applied to all situations that fall under its rules, and this application is immediate. The EU regulation has the same effects on national law as a fully self-executive treaty.[136]

Treaties got direct effect through case law, regulations by definition and directives by surprise. Over the decades, the experience with the directive as an instrument of EU law turned up two problems. Sometimes the national law made to implement the directive in national law contains phrases that (appear to) contradict the directive itself. Frequently Member States are not capable of introducing the implementing national laws in time. This leads to a situation where some Member States have made the required law, while others have not. The result is a substantial difference in legal conditions across the EU.

Again, EU law has to provide a solution for these problems. The Court of Justice has developed case law to enable EU law and the institutions to work as well as possible, given the difficulties of integration.[137] In this vein the Court ruled that if a national law contradicts the directive, the provisions of the directive will have direct effect if they meet the criteria. In this way the national judge or a national authority must overrule national law. In the case of a directive that was not implemented in national law within the agreed time-frame, the directive has itself as much direct effect as possible. The extent of this direct effect depends on how the directive is formulated. The criteria for direct effect have to be met. Public authorities of the Member State that is at fault have the legal obligation to apply the directive itself, as long as the required national law is not made. This is the how the directive, an instrument especially designed not to have direct effect, got direct effect anyway. See Law text box 4.9.

Direct effect of a directive creates a situation whereby the parts of the directive with direct effect replace the faulty national legislation, and substitute the higher directly effective rules of the directive itself. The national judge then has to apply at the same time a national law and a higher EU directive with a different meaning, leaving him no choice but to disregard the national law and to apply the directly effective directive. Leaving it to the national legislator to put right the mistakes. However, directives were invented and joined to the Union's arsenal

---

[136] See Chapter 3, Section 3.2.5 on the relation between international and national law.

[137] Case 148/78 *Pubblico Ministero v. Ratti*.

Law text box 4.9. Direct effect of directives, the Ratti case. Pubblico Ministero v. Ratti case 148/78.

Directive 73/173 required Member States to introduce into their national laws rules on the packaging and labelling of solvents. Italy failed to implement the directive and kept in force its national legislation. Ratti produced the solvents in accordance with Directive 73/173. In 1978 he was prosecuted in Milan for breaking the Italian law. In a preliminary procedure[1] from the Italian court to the European Court of Justice the latter answered the questions of the Italian judge as follows:

> 'This question raises the general problem of the legal nature of the provisions of a directive adopted under Article 189 of the Treaty.[2]
>
> In this regard the settled case law of the Court, last reaffirmed by the judgment of 1 February 1977 in Case 51/76 Nederlandse Ondernemingen 1977 1 ECR 126, lays down that, whilst under Article 189 regulations are directly applicable and, consequently, by their nature capable of producing direct effects, that does not mean that other categories of acts covered by that article can never produce similar effects.
>
> It would be incompatible with the binding effect which Article 189 ascribes to directives to exclude on principle the possibility of the obligations imposed by them being relied on by persons concerned.
>
> Particularly in cases in which the Community authorities have, by means of directives, placed Member States under a duty to adopt a certain course of action, the effectiveness of such an act would be weakened if persons were prevented from relying on it in legal proceedings and national courts prevented from taking it into consideration as an element of Community law.
>
> Consequently a Member State which has not adopted the implementing measures required by the directive in the prescribed periods, may not rely, as against individuals, on its own failure to perform the obligations which the directive entails.
>
> It follows that a national court requested by a person who has complied with the provisions of a directive not to apply a national provision incompatible with the directive not incorporated into the national legal order of a defaulting Member State, must uphold that request if the obligation in question is unconditional and sufficiently precise.
>
> Therefore the answer to the first question must be that after the expiration of the period fixed for the implementation of a directive a Member State may not apply its internal law – even if it is provided with penal sanctions – which has not yet been adopted in compliance with the directive, to a person who has complied with the requirements of the directive'.

[1] See Chapter 5, Section 5.6.1.
[2] Article 189 TEEC is now Article 288 TFEU.

in order not to have direct effect. The Union already has a legislative instrument with direct effect, the regulation. Yet the European Court of Justice decided that the fact that the Treaties deliberately provided two legislative instruments, one designed to have direct effect, and the other one designed not to have this effect, did not preclude the necessity to give the second instrument direct effect in cases where the Member States failed to transform the directive into correct and fitting national law. The negligence or failure of the Member State is not to be allowed to have a disruptive effect on EU law by removing from their jurisdiction the consequences of a legally made directive. Moreover, as the Court of Justice had decided earlier that the citizens of the Member States were direct participants in the Union's legal order, not to apply (a part of) a directive would rob them of their rights under EU law. So the directive is allowed to have direct effect as much as possible in the situation where the Member State failed to implement the directive in the period prescribed, and as much as necessary to compensate for defective national legislation made to implement the directive. The Court of Justice continues to create as much room as possible for the efficient and correct working of EU law. It also continues to guard over the rights that natural and legal persons have under EU law, and to deal with failing Member States as sternly as possible.

The direct effect of directives is not the same as the direct effect of regulations. The existence of the intermediate function of the Member State makes the law more complicated. The fact that the applicable law is different from the national law may come as a surprise, at any rate it will come late, after a period in which the national law was applied. It is difficult to require that private persons have so much insight and knowledge of EU law that they know better than their government what the correct application of EU law means in this particular case. The Court of Justice finds it impossible to confront private persons, companies large and small, with obligations stemming from a directive that their national government failed to implement properly. So the direct effect of directives can create rights, entitlements for natural and legal persons against public authorities, but it cannot create obligations for private parties. The Court of Justice seeks, on the other hand, to find obligations for the Member State as much as possible. The failing state is not allowed to reap profits from failure. Directives that create obligations are applied to public law authorities, as well as legal persons where a public authority has a deciding power. To take an example from case law, if a directive gives rights for workers, private employers cannot be forced to pay for these rights nor to allow them in cases where they have no direct monetary value. At the same time, public authorities who are employers do have to grant the directly effective directive rights to their employees. Even private law enterprises where a public authority has a deciding say (for instance, 51% of the shares) do have the obligation to grant the rights. The curious outcome of this case law is that a worker, who also cannot be required to have more EU legal expertise than her/his government, does obtain the rights from the directive if s/he happens to work for an organisation with public authority dominance, while s/he will not have the same rights if s/he works for a private sector employer.

The basic norm of the Court of Justice's case law on the failure of the Member States to implement directives, no obligations, only rights for private persons, does not have a solution for the situation where the right of one private person is the obligation of another. The Court of Justice has tried to remedy this situation with the decision in the Francovich case,[138] holding that Member States may have to pay compensation for damages that are the result of the fact that the Member State did not adopt legislation to implement a directive.

The aim of stimulating the Member States as much as possible to do their job properly and to apply EU law as it is meant to be, is a laudable objective, and is reflected in several other actions of the European Court of Justice to further the rule of law in the European Union. The Member States cannot invoke or rely on a directive which they did not implement. In a criminal law case a Dutch prosecutor sought to punish the manager of a restaurant where tap water was sold as high quality mineral water. This type of fraudulent handling of food with the use of the designation mineral water was not a criminal offence in Dutch law. However, it was an offence according to an EU directive meant to regulate the use of the designation mineral water. And in the usual way of operation of the combination of EU and national law, the national legal system has to provide for the prosecution of suspects and their punishment if found guilty. The public prosecutor in the Netherlands had no difficulty in providing evidence that the restaurant had indeed used the designation mineral water. So the prosecution based its case on the directive. However, the Dutch government had failed to implement that directive. In a preliminary procedure the Court of Justice decided that the directive cannot be used by the state who failed to implement it. In addition to the firm case law of the Court of Justice not to reward states for their misbehaviour, there is also the criminal law requirement that an action can only be a crime if there is a law that has formulated exactly what behaviour is regarded as a crime. In addition to precise formulation, this law has to precede the behaviour of the accused. In other words one can only be accused of committing a crime, if it was defined as a crime before one acted. In the case of a directive that was valid and working in the EU law system at the time of the alleged actions, but still had to be transformed into national law, there is enough uncertainty to exclude any working of this directive to construct a crime of any kind.[139]

[138] Joined Cases C-6/90 and C-9/90 *Francovich and Others v. Italian State.*

[139] Case 80/86 *Officier van Justitie (Public Prosecutor) v. Kolpinghuis Nijmegen.*

The principled difference between a regulation and a directive, which is the directive's reason for being, becomes more gradual in practice. Some regulations leave several choices to the Member States, making additional national legislation necessary, while some directives cover so much detail as to leave no real choice to the Member States. Yet, even then the difference remains that the regulation produces law on the EU level, while even a detailed blueprint directive must be implemented into 'national law'.

In food law the Product liability directive 85/373 discussed in Chapter 16 provides an excellent example of a directive so detailed that national legislators have little else to do than to copy and paste the provisions as set out in the directive. Regulation 178/2002 on the other hand contains examples of regulation provisions addressing the Member States in a way similar to a directive.[140]

[140] See in particular Article 4 of Regulation 178/2002.

5. The institutions of the European Union

Morten Broberg and Menno van der Velde

The European Union has seven institutions: the European Parliament, the European Council, the Council, the European Commission, the Court of Justice of the European Union, the European Central Bank and the Court of Auditors.[141] In addition the EU makes use of specialised agencies such as the European Food Safety Authority.

5.1 The European Council

With the Lisbon Treaty the European Council was made an institution of the European Union. It consists of the presidents of those Member States that have a politically active and responsible president, the prime ministers from the Member States where the head of state is merely ceremonial, together with the President of the European Council and the President of the European Commission. They are assisted by The High Representative of the Union for Foreign Affairs and Security Policy. In this composition the European Council meets twice every six months, to decide on the most important and politically controversial subjects. The idea is that the European Council shall provide the Union with the necessary impetus for its development and that it shall define the Union's general political directions and priorities. In contrast, the European Council shall not exercise legislative functions. The European Council must present a report about each of its meetings to the European Parliament.[142]

As part of the Lisbon Treaty, the European Council was given its first president, a position that in its present appearance is not equal to a presidency of the European Union, but is seen by some as its beginning.[143]

5.2 The Council of the European Union

While the Commission is officially named European Commission with permission to use the simple Commission, and the Court's formal name is the Court of Justice of the European Union, the Council is officially merely named the 'Council'.

Together with the European Council, the Council is the most intergovernmental institution of the EU.[144] It consists of a representative of each of the Member States, who can bind her/his government to the decisions made in Council. The representative has to be from a ministerial level. If general issues are on the

---

[141] Article 13 TEU.

[142] Article 15(6)(d) TEU.

[143] Article 15(5) and (6) TEU.

[144] Article 16 TEU.

agenda, it is usually the minister of Foreign Affairs. If more specialised subjects are to be debated, the Council will consist of the ministers with that specialisation: agriculture, Ecofin (economy and financial matters), social policies and so on. The presidency over the Council is held in turn by the Member States for a period of six months. The order in which the Member States hold this office is determined until July 2020 by Council Decision 2007/5/EC, Euratom.

The Council meets in the following ten different configurations:
- General Affairs
- Foreign Affairs
- Economic and Financial Affairs ('Ecofin')
- Justice and Home Affairs (JHA)
- Employment, Social Policy, Health and Consumer Affairs
- Competitiveness (Internal Market, Industry, Research and Space)
- Transport, Telecommunications and Energy
- Education, Youth, Culture and Sport
- Agriculture and Fisheries
- Environment

Food law can be on the agenda of the two last Councils. See Law text box 5.1 for the tasks of the Council.

Law text box 5.1. Tasks of the Council, Article 16(1) TEU.

Article 16(1) TEU

1. The Council shall, jointly with the European Parliament, exercise legislative and budgetary functions. It shall carry out policy-making and coordinating functions as laid down in the Treaties.

The Council has an important role in creating legislation. However, it needs an initiative proposal from the Commission before it can act and today it also needs the consent of European Parliament in a very large number of cases.

The Council empowers the Commission to negotiate treaties with other countries or international organisations. Where the European Union's Treaties provide for intergovernmental decision making the Council will be the one to decide. This is particularly relevant with regards to the European Union's Common Foreign and Security Policy (CFSP) for external relations.

Work in the Council is prepared by COREPER,[145] a committee of national representatives (senior civil servants) at the level of ambassadors who are permanently present in Brussels, or by the Special Committee for Agriculture (SCA).

5.2.1 Voting in the Council of the European Union

Voting requirements in the Council vary with the Treaty articles that attribute the powers for each specific policy. Usually the Council can decide with a qualified majority.[146] For this, each Member State is given a number of votes, with the more populous Member States having more votes than the less populous ones.

Voting requirements have been, and still are, an important issue for a number of Member States. The question 'voting or consensus (not voting)' has played a major role in the development of the European Union, and the voting system used by the Council is very important because it is copied in all committees that administer the policies of the Union.

Most policies require the formation of a committee composed of civil servants from the Member States as their representatives, and a representative of the Commission as chairman. For food law this is the Standing Committee on the Food Chain and Animal Health, established in Article 58 GFL.[147]

These are the so-called comitology committees that make a large number of comitology regulations every year, especially about agriculture, food safety and consumer protection. In these committees national civil servants participate in decision making. Where the committee is to issue an opinion that has more than advisory power, it does so with the same number of votes as their respective ministers in the Council. In comitology committees the representative of the Commission has no vote, but is armed with some detailed procedural rules that can force the Member States to take a decision. These precise comitology techniques are analysed in Section 5.5. But the voting rules are explained here.

Not least the increase in the number of Member States from originally six to 28 (2013) has been an unavoidable stimulant for decision making by a majority. The original six Member States brought the EU to a stand-still from 1966 until 1986, because they disregarded the Treaty articles that prescribe decision making by majority vote. The French boycotted the EU institutions for half a year in 1965. They refused to accept the essential obligation written into the original EEC Treaty that the third and last transition period of four years, which was to begin in 1966, would introduce majority voting. That would be unacceptable for the

[145] Many of the acronyms and abbreviations commonly used in the EU are based on the French names (in this case: Comité de Représentants Permanent).
[146] Article 238 TFEU.
[147] See Section 5.10.

French Gaullist government. It could not tolerate the vital interests of France being decided by a majority of other nation states. The other Member States could only see one way to end the crisis. They reverted to the classic diplomatic solution for a conflict between parties that refuse to come to terms: the agreement to disagree.

The Member States agreed on the Luxembourg Accord.[148] They agreed that the French would return to the Council and the EU system could resume its normal business. The French remained convinced that no Treaty could bring about a decision on vital French interests by a majority that did not include France itself. The other Member States restated their position that the EEC Treaty enabled majority decisions. They declared themselves willing to seek a solution that would be acceptable to all Member States, if one or more of them would declare that their vital interests were at issue. This solution would have to be found within a reasonable period of time. This effort had to respect the interest of the states and of the EU. The French delegation reiterated that the search for a solution acceptable for all would have to be continued until it was found. All of the then six Member States agreed to disagree on this last point. However, that did not impede the resumption of EU activities according to the normal procedures (!). Rather, after this statement was made and majority voting prescribed by the applicable EEC Treaty started on 1 January 1966, all Member States refrained from asking for a vote to determine the outcome of a discussion in the Council. Discussions went on indefinitely and the EU sank into a quagmire of words. At the end of this period other Member States began to invoke the vital national interest principle as well. Intergovernmentalism is a contagious affliction. In 1985 Germany blocked a decision on the price for wheat because it did not agree with the outcome of the discussions. So the German minister invoked his country's vital national interests that were endangered by this proposed decision. The long periods of negotiations without decisions had a great effect on the functioning of the common market and most profoundly on food law.

The British, who joined the EU in 1973, well after the Luxembourg Accord, were glad to seize the opportunity to use this hidden controversy in the *acquis communautaire*[149] to further their cause to limit the powers of the EU as much as possible. This led to a new accord even after the creation of the European Union in 1993; the 1994 Ioannina Compromise.[150] It declared that the Member States had to take notice of the fact that the Council had to decide to continue discussions when a minority of 23-26 votes (the votes for each Member State are weighted) declared that they were opposed to the decision that the majority was about to make in accord with the majority decision rule of the EC Treaty. The Council would continue discussions to find a solution that was acceptable to at least 68 votes in a reasonable period of time while respecting (!) the EC Treaty

[148] Text of the Luxembourg Accord published in the EEC Bulletin no. 3, 1966, page 5.
[149] For a definition of the *acquis communautaire*, see Chapter 4, Section 4.2.
[150] Text of the Ioannina Compromise published in the EU Bulletin no. 3 1994, pages 63f.

articles. Where the Treaty prescribed that 62 votes were sufficient, the Council had raised the majority of 71.3% required by the Treaty to 78.2% required by the Council. Nothing gave (or gives) the Council the power to change the EC Treaty single-handedly. To get a good perspective on this required majority: several states change their constitution with 66.7% of the votes and take other decisions by 50.1% when that is the available majority.

With the Lisbon Treaty, Article 16 TEU has introduced a new definition of the rule of qualified majority which applies from 1 November 2014 onwards. However, between that date and 31 March 2017, it will be possible for each Member State to require the previous weighting rules to be applied, and it is also possible to make the 'Ioannina compromise' applicable. See Law text box 5.2.

5.3 The European Commission

5.3.1 Day-to-day administration

The Commission (or European Commission)[151] consists of 28 Commissioners, one from each Member State. The Commissioners must be independent. First, the President is appointed by the European Parliament upon a proposal by the European Council (i.e. in practice a proposal by the Member States). The other commissioners are selected by the Council (i.e. in practice by the Member States) in consultation with the President-elect. The full list of the commissioners will thereupon be put before the European Parliament for a vote of consent. If this consent is obtained, the European Council may appoint the Commission. The term of office is five years. It is synchronised with the five-year term of the European Parliament, taking regard of the power the Parliament has to dismiss the Commission collectively.

The Commission is responsible for the day-by-day administration of the Union. See Law text box 5.3 for its tasks. Some of the administrative powers are given by the Treaties directly; most are given by the Council.[152] Comitology decision making is applied to the powers that the Council has delegated to the Commission.

Each Commissioner has one or more specific policy fields as her/his primary responsibility. Decisions are as a rule taken by the Commission as a college. The Commission staff for each policy field is organised in a Directorate General (often simply referred to as a 'DG'). Food safety is the responsibility of DG Sanco (Health and Consumers). In the clear majority of cases, the Commission has the exclusive right of initiative for EU legislation. This means that the Council and the European Parliament can only adopt a law if the Commission has first presented a proposal.

[151] Articles 17 TEU and 244-250 TFEU.
[152] See Articles 16 TEU and 290-291 TFEU, Law text box 5.1.

Law text box 5.2. Voting (by qualified majority) in the Council, Art. 16(2)-(5) TEU and 238 TFEU.

Article 16(2)-(5) Treaty on European Union

1. ....
2. The Council shall consist of a representative of each Member State at ministerial level, who may commit the government of the Member State in question and cast its vote.
3. The Council shall act by a qualified majority except where the Treaties provide otherwise.
4. As from 1 November 2014, a qualified majority shall be defined as at least 55% of the members of the Council, comprising at least fifteen of them and representing Member States comprising at least 65% of the population of the Union.
   A blocking minority must include at least four Council members, failing which the qualified majority shall be deemed attained.
   The other arrangements governing the qualified majority are laid down in Article 238(2) of the Treaty on the Functioning of the European Union.
5. The transitional provisions relating to the definition of the qualified majority which shall be applicable until 31 October 2014 and those which shall be applicable from 1 November 2014 to 31 March 2017 are laid down in the Protocol on transitional provisions.

Article 238 Treaty on the Functioning of the European Union

1. Where it is required to act by a simple majority, the Council shall act by a majority of its component members.
2. By way of derogation from Article 16(4) of the Treaty on European Union, as from 1 November 2014 and subject to the provisions laid down in the Protocol on transitional provisions, where the Council does not act on a proposal from the Commission or from the High Representative of the Union for Foreign Affairs and Security Policy, the qualified majority shall be defined as at least 72% of the members of the Council, representing Member States comprising at least 65% of the population of the Union.
3. As from 1 November 2014 and subject to the provisions laid down in the Protocol on transitional provisions, in cases where, under the Treaties, not all the members of the Council participate in voting, a qualified majority shall be defined as follows:
   (a) A qualified majority shall be defined as at least 55% of the members of the Council representing the participating Member States, comprising at least 65% of the population of these States.
   A blocking minority must include at least the minimum number of Council members representing more than 35% of the population of the participating Member States, plus one member, failing which the qualified majority shall be deemed attained;
   (b) By way of derogation from point (a), where the Council does not act on a proposal from the Commission or from the High Representative of the Union for Foreign Affairs and Security Policy, the qualified majority shall be defined as at least 72% of the members of the Council representing the participating Member States, comprising at least 65% of the population of these States.
4. Abstentions by Members present in person or represented shall not prevent the adoption by the Council of acts which require unanimity.

Law text box 5.3. Tasks of the Commission, Article 17 TEU.

Article 17 TEU

1. The Commission shall promote the general interest of the Union and take appropriate initiatives to that end. It shall ensure the application of the Treaties, and of measures adopted by the institutions pursuant to them. It shall oversee the application of Union law under the control of the Court of Justice of the European Union. It shall execute the budget and manage programmes. It shall exercise coordinating, executive and management functions, as laid down in the Treaties. With the exception of the common foreign and security policy, and other cases provided for in the Treaties, it shall ensure the Union's external representation. It shall initiate the Union's annual and multiannual programming with a view to achieving interinstitutional agreements.
2. Union legislative acts may only be adopted on the basis of a Commission proposal, except where the Treaties provide otherwise. Other acts shall be adopted on the basis of a Commission proposal where the Treaties so provide.
3. The Commission's term of office shall be five years.
   The members of the Commission shall be chosen on the ground of their general competence and European commitment from persons whose independence is beyond doubt.
   In carrying out its responsibilities, the Commission shall be completely independent. Without prejudice to Article 18(2), the members of the Commission shall neither seek nor take instructions from any Government or other institution, body, office or entity. They shall refrain from any action incompatible with their duties or the performance of their tasks.
4. The Commission appointed between the date of entry into force of the Treaty of Lisbon and 31 October 2014, shall consist of one national of each Member State, including its President and the High Representative of the Union for Foreign Affairs and Security Policy who shall be one of its Vice-Presidents.
5. As from 1 November 2014, the Commission shall consist of a number of members, including its President and the High Representative of the Union for Foreign Affairs and Security Policy, corresponding to two thirds of the number of Member States, unless the European Council, acting unanimously, decides to alter this number.
   The members of the Commission shall be chosen from among the nationals of the Member States on the basis of a system of strictly equal rotation between the Member States, reflecting the demographic and geographical range of all the Member States. This system shall be established unanimously by the European Council in accordance with Article 244 of the Treaty on the Functioning of the European Union.
6. The President of the Commission shall:
   (a) lay down guidelines within which the Commission is to work;
   (b) decide on the internal organisation of the Commission, ensuring that it acts consistently, efficiently and as a collegiate body;
   (c) appoint Vice-Presidents, other than the High Representative of the Union for Foreign Affairs and Security Policy, from among the members of the Commission.

   A member of the Commission shall resign if the President so requests. The High Representative of the Union for Foreign Affairs and Security Policy shall resign, in accordance with the procedure set out in Article 18(1), if the President so requests.

Law text box 5.3. Continued.

7. Taking into account the elections to the European Parliament and after having held the appropriate consultations, the European Council, acting by a qualified majority, shall propose to the European Parliament a candidate for President of the Commission. This candidate shall be elected by the European Parliament by a majority of its component members. If he does not obtain the required majority, the European Council, acting by a qualified majority, shall within one month propose a new candidate who shall be elected by the European Parliament following the same procedure.
   The Council, by common accord with the President-elect, shall adopt the list of the other persons whom it proposes for appointment as members of the Commission. They shall be selected, on the basis of the suggestions made by Member States, in accordance with the criteria set out in paragraph 3, second subparagraph, and paragraph 5, second subparagraph.
   The President, the High Representative of the Union for Foreign Affairs and Security Policy and the other members of the Commission shall be subject as a body to a vote of consent by the European Parliament. On the basis of this consent the Commission shall be appointed by the European Council, acting by a qualified majority.
8. The Commission, as a body, shall be responsible to the European Parliament. In accordance with Article 234 of the Treaty on the Functioning of the European Union, the European Parliament may vote on a motion of censure of the Commission. If such a motion is carried, the members of the Commission shall resign as a body and the High Representative of the Union for Foreign Affairs and Security Policy shall resign from the duties that he carries out in the Commission.

5.3.2 Guardian of the Treaties

The Commission is often referred to as the 'guardian of the Treaties'. It monitors the Member States' implementation of EU law. It initiates infringement proceedings in the event of any suspected violation of EU law by a Member State.[153] The infringement procedure has two distinct phases. In the first phase the main aspect of the procedure is the clarification of the situation and the attempt at reaching an agreement between the parties. This phase is confidential. The Commission confronts the Member State with its perception of a breach of EU law. The Member State has ample opportunity to present its view on the issue. Information is shared. Perhaps the Commission is persuaded by the Member State that there is no breach of EU law. Or the Commission remains convinced that the Member State is acting wrongly. In the latter case the Commission can advise the Member State on how to adapt its behaviour so as to end the breach. The state will reconsider and perhaps accept the Commission's point of view and correct the wrongdoing. Or the state may reject the advice. If the Member State does not adapt its behaviour it will be for the Commission to decide whether or

[153] Article 258 TFEU.

not to initiate phase two. The Commission is not obliged to proceed further with the infringement procedure. It can decide to accept the position of the Member State. If the Commission is convinced that the wrong should be righted, it will start a public open court case by summoning the Member State to appear before the Court of Justice of the European Union to defend itself against the accusation that it has infringed specific articles of the Treaties. Lawyers of both parties will present their clients' views and evidence, the judges will ask their questions and finally the Court of Justice will render its judgment. The Court's decision is binding, the Member State's position will be accepted or rejected, and state and Commission must act accordingly.

The independent position of the European Commission, its power to supervise the Member States, the power to administrate the Union's policies and issue binding decisions are a reflection of the fact that the Commission is a supranational element in the EU; i.e. without being a state itself, the Commission has taken over power that normally belongs to the states and thus the power and tasks of the Commission go beyond the normal international cooperation between states. However, the Commission's position is balanced, inter alia, by the Council's close surveillance of the Commission's activities. In this regard, the comitology mechanism is more than an arrangement to co-ordinate EU policies with those of the Member States; in practice it is designed to serve the interests of the Member States by giving the final say to these states themselves through their representatives in the committees and in the Council.[154]

5.4 The European Parliament

The European Parliament,[155] which is located in Brussels, Strasbourg and Luxembourg, is always on the move. It represents the peoples of the Member States. Every state has a fixed number of members of Parliament (so-called MEPs). They are chosen on the bases of 28 national systems of direct general elections, which are held every five years.[156]

Since the inception of the European Union, the European Parliament has acquired more and more legislative powers. Following the entry into force of the Lisbon Treaty co-decision between the Parliament and the Council have become the 'ordinary legislative procedure' where both institutions are on an equal footing. The Parliament plays an important role in the budgetary procedure and it has supervisory powers over the Commission. These powers are exercised mainly through the fact that the Commission must answer parliamentary questions, must defend its proposals before Parliament and must accept the powers of the European Ombudsman (who is appointed by the European Parliament). The

[154] Regarding comitology, see Section 5.5 below.
[155] Article 14 TEU.
[156] Article 14(3) TEU.

Commission presents its annual report on the activities of the Union. Parliament and Commission discuss it in a general policy debate. Parliament can, by a two-thirds majority of its members, pass a motion of censure and thereby compel the Commission to resign as a body.[157] So far the Parliament has used this power once.

The Parliament is empowered to set up temporary Committees of Inquiry[158] to look specifically at alleged cases of infringement of EU law or mismanagement. A committee of this kind was used, for example, to look into the Commission's responsibility for the delay in responding to the BSE crisis. In this context a motion of censure was introduced. Although it was not passed, it exercised considerable influence as will be elaborated on later.[159]

The work of the Parliament is prepared in committees. The committees that are most involved in matters related to food law are 'ENVI: Environment, public health and food safety' and 'AGRI: Agriculture and rural development'. The European Parliament has become a supranational element as an institution directly elected to represent the peoples of the Union, and equipped with real legislative powers.

5.5 Council and Commission – Delegating competence to the Commission, and controlling it

5.5.1 Introduction

Most EU policies are executed by the administrations of the Member States, working in these cases under EU law. For this work the obligations contained in Article 4(3) TEU are important: the Member States have to take all appropriate measures, whether general or particular, to ensure fulfilment of the obligations which are based on the Treaties. The same rule applies to the work the states have to do as a result of actions taken by the EU institutions. The Member States have to facilitate the achievement of the EU's tasks. They have to abstain from any measure which could jeopardise the attainment of the objectives of the Treaties.

Between the formulation of policy objectives and the regulations that serve these objectives at the EU level and the administration of these policies in the Member States there are many implementing measures.

Over the years the European Union developed a so-called comitology system to be the primary one to take care of this implementation at the EU level. Essentially this meant that the Member States would put together comitology committees made up of civil servants from all Member States and headed by a civil servant from the Commission. A number of different comitology procedures were established

[157] Article 234 TFEU.
[158] Article 226 TFEU.
[159] See Chapter 7, Section 7.4.

– where some would give the comitology committees (i.e. the Member States) more power than others.

The comitology committee system first of all meant that the Member States were able to control the implementation and administration of EU legislation. Whilst the Member States (and the Council) supported this system, both the Commission and the European Parliament were very critical towards it. With the Lisbon Treaty the comitology committee system has been amended. The first matter to observe in this regard is that directives and regulations that are adopted after the entry into force of the Lisbon Treaty (i.e. after 1 December 2009) must fall into one of the following categories:

- Legislative acts.[160]
- Non-legislative acts of general application – or delegated acts.[161]
- Implementing acts.[162]

Only the legislation that has been qualified as an 'implementing act' allows for comitology procedures.

5.5.2 Delegated Acts

Under Article 290 TFEU (see Law text box 5.4), the Commission can be given the power to adopt 'delegated acts' – also referred to as 'non-legislative acts of general application' – intended to 'supplement or amend certain non-essential elements' of a basic legal act. The procedure by which the Commission prepares and draws up delegated acts is not specified in Article 290 TFEU. Instead it must be specified in the individual basic legal act and may thus differ from case to case.

Both the European Parliament and the Council are given extensive scrutiny powers over delegated acts. The Commission is required to notify the European Parliament and the Council simultaneously as soon as it adopts a delegated act.

In the basic legal act, the European Parliament or the Council may exercise the right of opposition to a particular delegated act on any grounds and within a period determined in the basic act. In order to reject the proposed delegated act an absolute majority is needed in the European Parliament and a qualified majority in the Council. The delegated act will only enter into force after the expiry of the period for objection. If the Parliament or the Council object, the delegated act cannot enter into force. Instead the Commission must either adopt a new proposal or choose to do nothing at all.

[160] Article 289 TFEU.
[161] Article 290 TFEU.
[162] Article 291 TFEU.

Law text box 5.4. Article 290 TFEU.

Article 290 TFEU

1. A legislative act may delegate to the Commission the power to adopt non-legislative acts of general application to supplement or amend certain non-essential elements of the legislative act. The objectives, content, scope and duration of the delegation of power shall be explicitly defined in the legislative acts. The essential elements of an area shall be reserved for the legislative act and accordingly shall not be the subject of a delegation of power.
2. Legislative acts shall explicitly lay down the conditions to which the delegation is subject; these conditions may be as follows:
   (a) the European Parliament or the Council may decide to revoke the delegation;
   (b) the delegated act may enter into force only if no objection has been expressed by the European Parliament or the Council within a period set by the legislative act.

   For the purposes of (a) and (b), the European Parliament shall act by a majority of its component members, and the Council by a qualified majority.
3. The adjective 'delegated' shall be inserted in the title of delegated acts.

The European Parliament and the Council also have the power to revoke the delegation of powers given to the Commission at any time. If so, the delegated powers may partly or fully be taken away from the Commission.

5.5.3 Implementing Acts (Comitology)

Comitology committees

Comitology committees have already been introduced as committees where 28 civil servants, one from each Member State and in reality representing that state, cooperate under the chairmanship of a civil servant from the Commission. The committee members have the votes and the chairman has the procedures that the committee has to apply in its meetings. The comitology committees have two levels of procedures: the procedures that determine how the meetings of the committee are organised, how much time they have to cooperate, debate or negotiate before they have to take a decision, how decisions are made, how the votes are counted and how many votes are needed to take a decision. Most of these rules are recorded and the chairman applies them.

The second level of procedures is the set of rules that determines the committee's powers to decide. One type of procedure concerns the provision of advice to the Commission. Another type allows the Commission to take a decision but if the committee does not agree the Commission cannot adopt the measure. If, however, the committee has rejected the Commission's proposal for an implementing act, but such an act is nonetheless necessary, the Commission must enter a new procedure in order to try to have the act adopted.

Comitology measures

The committees can draft different measures like regulations and decisions. When they draft regulations, they can make more detailed rules or they have the power to replace rules from the basic regulation by rules that reflect scientific or technical developments. When the measures are decisions, the comitology committees apply EU rules to make EU decisions that must be applied by the Member States. Comitology has developed into a flexible and unique way in which to make policies and law. Unique in the literal sense, since there are no other states or organisations with procedures and measures like this.

The list of examples of possible measures is endless. A comitology measure can contain the details that determine which substances are added to, or removed from, the list of substances that may not be added to food. The same goes for the list of substances that are allowed, but only to specified maximum levels. It may be necessary to change this list not only for the substances but also for the levels. Rules about the registration of food businesses are changed to specify the units that have to be registered or the methods of registration and the access to this information. Some food businesses have to be approved before they begin their activities. The evaluation of the approval system can lead to changes that have to be decided at the Union level. The government organisations that must inspect premises where food is prepared need detailed and up-to-date EU instructions to ensure effective inspections, the use of the best available techniques, and the adoption of new methods or new standards.

EU food law is used to secure a high level of food safety for food and feed going through all the stages of primary production, preparation, processing, manufacturing, packaging, storing, transportation, distribution, handling and offering for sale or supply to the consumer. Food hygiene rules have to be applied at all these diverse stages. The movement of food and feed ingredients through all these stages must be traceable in case something has gone wrong somewhere; in this event, it is imperative to find out who else might be at risk.

Who registers what, and which records have to be kept? Which changes to logistics and registration techniques have consequences for food law that are too minor to cause an essential change in legislation, but significant enough to make changes necessary?

If you expected the Commission – the institution created by the Treaties as the Union's day-to-day authority – to be the institution to take care of these details, you would be failing to take into account the strong opinions of the Council on this issue. The Council has held the key positions from the beginning of the EEC Treaty and has made basic legislation with built-in mechanisms to exert its influence during the execution of EU policies.

Over the years a very considerable – and complicated – comitology committee system was created. However, with the Lisbon Treaty this system underwent an extensive overhaul. The consequence is that today the comitology committee system differs appreciably from the one that existed pre-Lisbon. The legal basis for the new regime can be found in Article 291 TFEU, see Law text box 5.5.

Law text box 5.5. Article 291 TFEU and Article 1 of the Comitology Regulation

Article 291 TFEU

1. Member States shall adopt all measures of national law necessary to implement legally binding Union acts.
2. Where uniform conditions for implementing legally binding Union acts are needed, those acts shall confer implementing powers on the Commission, or, in duly justified specific cases and in the cases provided for in Articles 24 and 26 of the Treaty on European Union, on the Council.
3. For the purposes of paragraph 2, the European Parliament and the Council, acting by means of regulations in accordance with the ordinary legislative procedure, shall lay down in advance the rules and general principles concerning mechanisms for control by Member States of the Commission's exercise of implementing powers.
4. The word 'implementing' shall be inserted in the title of implementing acts.

Article 1
Subject-matter

This Regulation lays down the rules and general principles governing the mechanisms which apply where a legally binding Union act (hereinafter a 'basic act') identifies the need for uniform conditions of implementation and requires that the adoption of implementing acts by the Commission be subject to the control of Member States.

The Council and the Parliament may adopt a so-called 'basic act', i.e. a legally binding Union act which identifies the need for uniform conditions of implementation. This basic act vests the Commission with the power to adopt implementing acts. Article 291(3) TFEU requires that the Council and the Parliament shall adopt a regulation which lay down the rules and general principles concerning mechanisms that enable the Member States to control the Commission's exercise of these implementing powers. Therefore, on 16 February 2011 the European Parliament and the Council adopted Regulation 182/2011 which establishes mechanisms for Member States' control of the Commission's exercise of implementing powers.[163] This regulation essentially embodies the new comitology system; Article 1 of Regulation 182/2011 is reproduced in Law text box 5.5.

[163] Regulation 182/2011 of 16 February 2011 laying down the rules and general principles concerning mechanisms for control by Member States of the Commission's exercise of implementing powers.

The Comitology Regulation from 2011 has replaced the 1999-Comitology-decision that previously laid down rules and procedures for comitology committees. This new regime not only applies to new legislation, but also replaces the former comitology regime with regards to legislation that was in force prior to the entry into force of the new regime. Rules and procedures for the very large number of comitology committees that had been established prior to the entry into force of the Comitology Regulation are therefore now to be found in this regulation.

The comitology rules are very important because they are used on a large scale in the day-to-day administration of EU policies. Basic regulations that contain the main rules for a policy prescribed by the Treaties also contain an article that creates a committee for the day-to-day management of that policy, or to make rules in certain situations and under certain conditions. Within the field of food law the General Food Law (Regulation 178/2002) is of particular relevance. The comitology committee system established by the General Food Law is presented below in Section 5.5.7.

5.5.4 Characteristics of the comitology procedure

There are four elements in the comitology procedure that explain its special character:

1. The Member States want to limit the power of the European Union. They want to have control also over the very detailed legislation and they do not want to leave this power exclusively to the Commission. Indeed, this is now explicitly reflected in the name of the 2011 Comitology Regulation: Regulation 182/2011 of 16 February 2011 laying down the rules and general principles concerning mechanisms for control by Member States of the Commission's exercise of implementing powers.
2. The procedures have tight time schedules which will often force the Member States to take a decision.
3. The voting requirements are tied with the type (and necessity) of the different measures to be adopted. If it is important to adopt a given measure it will generally be more difficult for the Member State representatives in the comitology committee to reject it.
4. The fourth element is the empirical fact that a majority that opposes the Commission proposal is not necessarily the majority that supports another proposal. The fact that there will be a decision anyway will influence the negotiation position and tactics of the Member States.

The Comitology Regulation in Article 2 provides for two distinct procedures, namely (1) the advisory procedure, and (2) the examination procedure – see Law text box 5.6. These two procedures are further presented in Sections 5.5.5 and 5.5.6.

Law text box 5.6. Article 2 of the Comitology Regulation.

Article 2
Selection of procedures

1. A basic act may provide for the application of the advisory procedure or the examination procedure, taking into account the nature or the impact of the implementing act required.
2. The examination procedure applies, in particular, for the adoption of:
   (a) implementing acts of general scope;
   (b) other implementing acts relating to:
      (i) programmes with substantial implications;
      (ii) the common agricultural and common fisheries policies;
      (iii) the environment, security and safety, or protection of the health or safety, of humans, animals or plants;
      (iv) the common commercial policy;
      (v) taxation.
3. The advisory procedure applies, as a general rule, for the adoption of implementing acts not falling within the ambit of paragraph 2. However, the advisory procedure may apply for the adoption of the implementing acts referred to in paragraph 2 in duly justified cases.

5.5.5 The advisory procedure

The advisory procedure applies where the nature or the impact of the implementing act makes the advisory procedure more adequate than the examination procedure, cf. Article 2(3) of the Comitology Regulation reproduced in Law text box 5.6. This essentially means that, as a general rule, the advisory procedure applies in those situations where none of the criteria laid down in Article 2(2) of the Comitology Regulation, which would normally point towards an examination procedure, are present. However, these criteria are merely guidelines and there may therefore be situations where the advisory procedure applies even though it fulfils one or more of the examination procedure criteria.

Where the advisory procedure is used, the committee shall deliver an opinion. This may be by consensus, or – if no consensus can be reached – by simple majority vote.

When the committee has delivered its opinion the Commission shall take 'the utmost account of the conclusions drawn from the discussions within the committee and of the opinion delivered'. In other words, the Commission is not bound by an opinion that has been reached on the basis of the advisory procedure. See further Article 4 of the Comitology Regulation reproduced In Law text box 5.7.

Law text box 5.7. Article 4 of the Comitology Regulation.

Article 4 of the Comitology Regulation
Advisory procedure

1. Where the advisory procedure applies, the committee shall deliver its opinion, if necessary by taking a vote. If the committee takes a vote, the opinion shall be delivered by a simple majority of its component members.
2. The Commission shall decide on the draft implementing act to be adopted, taking the utmost account of the conclusions drawn from the discussions within the committee and of the opinion delivered.

5.5.6 The examination procedure

The examination procedure applies where the nature or the impact of the implementing act makes this procedure more adequate than the advisory procedure. In practice this means that where one or more of the criteria set out in Article 2(2) of the Comitology Regulation is present, as a general rule, the examination procedure shall be used. These criteria are:
(a) implementing acts of general scope;
(b) other implementing acts relating to:
- (i) programmes with substantial implications;
- (ii) the common agricultural and common fisheries policies;
- (iii) the environment, security and safety, or protection of the health or safety, of humans, animals or plants;
- (iv) the common commercial policy;
- (v) taxation.

It may be noted that in the field of food law it is likely that implementing acts will fall within one of the following three criteria: (1) implementing acts of general scope, (2) implementing acts relating to the common agricultural and common fisheries policies, and (3) implementing acts relating to the protection of the health or safety, of humans, animals or plants. This means that within the field of food law the examination procedure is likely to be widely used with regards to implementing acts.

Where the examination procedure applies, the comitology committee delivers its opinion by majority in accordance with the Treaty provisions that otherwise apply to the voting in the Council (Article 16(4) and (5) TEU and Article 238(3) TFEU). If the committee delivers a positive opinion, the Commission shall adopt the draft implementing act. In contrast, if the committee delivers a negative opinion, the Commission shall not adopt the draft implementing act. If the implementing act is deemed to be necessary, the Commission may either submit an amended version of the draft implementing act to the same committee, or it may instead

Law text box 5.8. Article 5 of the Comitology Regulation.

Article 5 of the Comitology Regulation
Examination procedure

1. Where the examination procedure applies, the committee shall deliver its opinion by the majority laid down in Article 16(4) and (5) of the Treaty on European Union and, where applicable, Article 238(3) TFEU, for acts to be adopted on a proposal from the Commission. The votes of the representatives of the Member States within the committee shall be weighted in the manner set out in those Articles.
2. Where the committee delivers a positive opinion, the Commission shall adopt the draft implementing act.
3. Without prejudice to Article 7, if the committee delivers a negative opinion, the Commission shall not adopt the draft implementing act. Where an implementing act is deemed to be necessary, the chair may either submit an amended version of the draft implementing act to the same committee within 2 months of delivery of the negative opinion, or submit the draft implementing act within 1 month of such delivery to the appeal committee for further deliberation.
4. Where no opinion is delivered, the Commission may adopt the draft implementing act, except in the cases provided for in the second subparagraph. Where the Commission does not adopt the draft implementing act, the chair may submit to the committee an amended version thereof.
   Without prejudice to Article 7, the Commission shall not adopt the draft implementing act where:
   (a) that act concerns taxation, financial services, the protection of the health or safety of humans, animals or plants, or definitive multilateral safeguard measures;
   (b) the basic act provides that the draft implementing act may not be adopted where no opinion is delivered; or
   (c) a simple majority of the component members of the committee opposes it.
   In any of the cases referred to in the second subparagraph, where an implementing act is deemed to be necessary, the chair may either submit an amended version of that act to the same committee within 2 months of the vote, or submit the draft implementing act within 1 month of the vote to the appeal committee for further deliberation.
5. By way of derogation from paragraph 4, the following procedure shall apply for the adoption of draft definitive anti-dumping or countervailing measures, where no opinion is delivered by the committee and a simple majority of its component members opposes the draft implementing act.
   The Commission shall conduct consultations with the Member States. 14 days at the earliest and 1 month at the latest after the committee meeting, the Commission shall inform the committee members of the results of those consultations and submit a draft implementing act to the appeal committee. By way of derogation from Article 3(7), the appeal committee shall meet 14 days at the earliest and 1 month at the latest after the submission of the draft implementing act. The appeal committee shall deliver its opinion in accordance with Article 6. The time limits laid down in this paragraph shall be without prejudice to the need to respect the deadlines laid down in the relevant basic acts.

choose to submit the draft implementing act to a special appeal committee (see Law text box 5.8).

If the comitology committee does not deliver an opinion, as a rule the Commission may (but is not obliged to) adopt the draft implementing act.

Moreover, an Appeal Committee is created which shall have the power to change, adopt or reject the Commission's proposal. If this Appeal Committee cannot reach a qualified majority vote, the Commission 'may' (but is not obliged to) adopt the measure.

5.5.7 Comitology procedure applicable to the General Food Law

The General Food Law (Regulation 178/2002), in Articles 58 and 59, has established the Standing Committee on the Food Chain and Animal Health (Law text box 5.9).

Law text box 5.9. The Standing Committee on the Food Chain and Animal Health, Articles 58 and 59 General Food Law.

Regulation 178/2002
Chapter v Procedures and final provisions
Section 1 Committee and mediation procedures
Article 58 Committee

1. The Commission shall be assisted by a Standing Committee on the Food Chain and Animal Health, hereinafter referred to as the 'Committee', composed of representatives of the Member States and chaired by the representative of the Commission. The Committee shall be organised in sections to deal with all relevant matters.
2. Where reference is made to this paragraph, the procedure laid down in Article 5 of Decision 1999/468/EC shall apply, in compliance with Articles 7 and 8 thereof.
3. The period provided for in Article 5(6) of Decision 1999/468/EC shall be three months.

Article 59 Functions assigned to the Committee
The Committee shall carry out the functions assigned to it by this Regulation and by other relevant [Union] provisions, in the cases and conditions provided for in those provisions. It may also examine any issue falling under those provisions, either at the initiative of the Chairman or at the written request of one of its members.

Article 58 refers, in particular, to Article 5 of the former comitology decision; Council Decision 1999/468/EC of 28 June 1999 laying down the procedures for the exercise of implementing powers conferred on the Commission. According to Article 13 of the Comitology Regulation from 2011 this reference now means that the examination procedure referred to in Article 5 of the Comitology Regulation shall apply with regards to the Standing Committee on the Food Chain and Animal

Health. Moreover, according to the Comitology Regulation Article 5(4)(2)(b) the General Food Law (i.e. the basic act) shall be deemed to provide that, in the absence of an opinion from the Standing Committee on the Food Chain and Animal Health, the Commission may not adopt the draft implementing act (see Law text box 5.9).

In other words, the Standing Committee on the Food Chain and Animal Health established under the GFL must apply the examination procedure.

5.5.8 Special procedures

Article 7 of the 2011 Comitology Regulation (see Law text box 5.10) provides a special procedure for 'exceptional cases', namely where the draft implementing act needs to be adopted without delay in order to avoid creating a significant disruption of the markets in the area of agriculture or a risk for the financial interests of the Union. If the Commission adopts the implementing act on this basis, it must immediately submit the adopted implementing act to the appeal committee – and if the appeal committee delivers a negative opinion, the Commission shall repeal that act immediately.

Moreover, Article 8 of the Comitology Regulation (see Law text box 5.10) lays down that a basic act may provide that, on duly justified imperative grounds of urgency, the Commission shall adopt an implementing act which shall apply immediately, without its prior submission to a committee, and shall remain in force for a period that may last up to six months unless the basic act provides otherwise. After the adoption the Commission must submit the urgently adopted act to the relevant committee in order to obtain its opinion. If the examination procedure applies and if the comitology committee delivers a negative opinion, the Commission must repeal the implementing act.

5.6 The Court of Justice of the European Union

The Court of Justice is the highest judicial authority in matters of EU law.[164] In general terms, its task is to '... ensure that in the interpretation and application of the Treaties the law is observed'.[165] The Court of Justice has one judge per Member State; the judges have to be independent from the Member States and the institutions.[166] They are appointed for a period of six years by common accord of the governments of the Member States. The period in office is renewable. The

[164] Articles 19 TEU and 251-281 TFEU.
[165] Article 19(1) TEU.
[166] Article 19(2) TEU.

Law text box 5.10. Articles 7 and 8 of the Comitology Regulation.

Article 7
Adoption of implementing acts in exceptional cases

By way of derogation from Article 5(3) and the second subparagraph of Article 5(4), the Commission may adopt a draft implementing act where it needs to be adopted without delay in order to avoid creating a significant disruption of the markets in the area of agriculture or a risk for the financial interests of the Union within the meaning of Article 325 TFEU.

In such a case, the Commission shall immediately submit the adopted implementing act to the appeal committee. Where the appeal committee delivers a negative opinion on the adopted implementing act, the Commission shall repeal that act immediately. Where the appeal committee delivers a positive opinion or no opinion is delivered, the implementing act shall remain in force.

Article 8
Immediately applicable implementing acts

1. By way of derogation from Articles 4 and 5, a basic act may provide that, on duly justified imperative grounds of urgency, this Article is to apply.
2. The Commission shall adopt an implementing act which shall apply immediately, without its prior submission to a committee, and shall remain in force for a period not exceeding 6 months unless the basic act provides otherwise.
3. At the latest 14 days after its adoption, the chair shall submit the act referred to in paragraph 2 to the relevant committee in order to obtain its opinion.
4. Where the examination procedure applies, in the event of the committee delivering a negative opinion, the Commission shall immediately repeal the implementing act adopted in accordance with paragraph 2.
5. Where the Commission adopts provisional anti-dumping or countervailing measures, the procedure provided for in this Article shall apply. The Commission shall adopt such measures after consulting or, in cases of extreme urgency, after informing the Member States. In the latter case, consultations shall take place 10 days at the latest after notification to the Member States of the measures adopted by the Commission.

Court has jurisdiction in several quite different branches of law. The four most common types of cases are:

1. requests for a preliminary ruling;
2. proceedings for failure to fulfil an obligation (also called the infringement procedure);
3. proceedings for annulment;
4. proceedings for failure to act.

5.6.1 The preliminary ruling procedure

EU law contains a mechanism called the preliminary procedure[167] which enables the Court of Justice to assist the national courts in their tasks of applying EU law. The same mechanism is designed to prevent national courts, which can be separated by almost a continent, from developing differing interpretations of EU law. That would break the unity of EU law and lead to different conditions on the internal market and in the execution of Union policies. Every national court from the Member States may ask the opinion of the Court of Justice on how to interpret EU law in the case to be tried by the national court. When the Court of Justice receives such a preliminary question, it does not take over the case from the national court, nor will it pronounce statements on the interpretation of national law that is to be applied in the proceedings. The Court of Justice answers the questions asked by the national court on the interpretation of EU law in order to assist the national judge as a colleague. When the Court of Justice has rendered a preliminary ruling, this ruling will provide guidance to all national courts in all Member States that are faced with a similar question of interpretation of EU law. The opinions expressed by the Court of Justice are binding. The importance of this preliminary ruling procedure is reflected in the fact that several Member States regularly use the possibility given by EU law to intervene in the proceedings before the Court of Justice to present their opinions.

Whereas all national courts in the EU will normally have a right to ask the Court of Justice's opinion, national courts of last instance (often supreme courts) are under an obligation to ask the advice of the Court of Justice if it is necessary to interpret EU law to decide the case before this national court. In this way EU law is likely to be interpreted in the same way everywhere. If the Member State courts fail to correctly interpret EU law, in principle the Commission can use its supervising power that can lead to a procedure against the Member State for failure to fulfil an EU obligation (as set out in the next section). The persistent wrong interpretation of EU law by an independent national court adds up to a failure of this court's Member State. The question as to whether the Member State is committing a wrong because one of its national courts is acting against EU law will be decided by the independent Court of Justice. However, the Commission has been reluctant in the past to open proceedings against a Member State when the alleged failure was caused by a court. A well-known case was the French Conseil d'État that had very strong, but wrong, ideas about how to apply EU law. To questions posed by members of the European Parliament on this matter, the Commission answered that it would carefully consider the French position, taking note of the independence of the judiciary that needs special attention. In the course of a few years the Commission did nothing in public and the French Conseil d'État changed its way of applying EU law. In the Cassis de Dijon case, the

[167] Article 267 TFEU.

Court of Justice formulated the principle of mutual acceptance of the food safety decisions of other EU Member States. The Court did this in answering questions asked by a German court using the preliminary ruling procedure.[168]

5.6.2 Proceedings for failure to fulfil an obligation (also called the infringement procedure)

So-called infringement proceedings[169] are described in the section on the work of the Commission as supervisor, i.e. 'guardian of the treaties' (Section 5.3.2). In Case 178/84, *Commission v. Germany (the German Beer Case)*, the Commission asked the Court to render illegal the German action of hindering foreign beers from accessing the German market. The German government maintained that in order for a beverage to be called 'beer' it had to be made according to a centuries old recipe, called the *Reinheitsgebot* which basically meant that beer could only be made from malted barley, hops, yeast and water. If the beer was made from other ingredients – as is normal outside Germany – it could not be sold as beer in Germany. The Commission charged that the German requirements amounted to a measure with an effect equivalent to a quantitative restriction of imports of beer into Germany.[170]

The Treaties, normally, prohibit measures that hinder the free movement of goods. The German government countered by relying on the exception made in the Treaties; a Member State is allowed to hinder the free movement of goods if this is necessary to protect the population against a health risk. The Court of Justice did not accept the German argument, and thus found for the Commission. By refusing access to the German market to foreign beers, which had not been made according to a German recipe, Germany had failed to fulfil the obligation not to hinder the free movement of beers.

Experience has shown that several Member States have been very slow in complying with a judgment of the Court of Justice. To remedy this situation a procedure has been introduced whereby the Commission can bring the failing state before the Court of Justice and demand execution of the judgment as well as sanctioning the Member State's failure by requiring that a 'fine' is imposed. It is for the Court of Justice to decide whether such a 'fine' shall be imposed on the Member State. If the Court of Justice renders a judgment against the Member State, the Commission can use the judgment to force the Member State to pay.[171]

---

[168] See Chapter 7, Sections 7.3.2 and 7.3.3 for an analysis of this case.
[169] Article 258 TFEU.
[170] See Chapter 7, Section 7.3.4 for an analysis of this concept.
[171] Article 260 TFEU

5.6.3 Proceedings for annulment

If one or more of the Member States, the Council, Commission or Parliament believes that a particular EU law is illegal they may ask the Court to annul it.[172] The Court reviews the legality of binding acts made by the institutions. There are four grounds to annul the act concerned:

- lack of competence;
- infringement of an essential procedural requirement;
- infringement of the Treaties or of any rule of law relating to its application;
- misuse of powers.

The first ground, the lack of power has been explained in the previous chapter under 'The Powers of the EU: attribution'.[173] An example of the second ground is a regulation on agriculture made by the Council, without the involvement of the European Parliament even though this is required. The third ground is illustrated in the previous chapter under 'The regulation'.[174] An example of the fourth ground is to use a regulation as an instrument to decide the legal position of one person only. The proceedings to obtain a review must be instituted within two months of the publication of the measure. However, everyone can at all times invoke the illegality of a legal measure of general application, on the same grounds for annulment, if that measure is used to determine the outcome of a case.[175]

5.6.4 Proceedings for failure to act

The Treaties require the European Parliament, the Council and the Commission to make certain decisions under certain circumstances. If they fail to do so, the Member States and the other Union institutions can lodge a complaint with the Court of Justice to have this violation officially recorded. However, such an action may be brought only after the institution has been given the opportunity to remedy the situation. When the failure to act is held to be unlawful, it is for the institution concerned to put an end to the failure by appropriate measures.

5.6.5 The General Court of the European Union

To cope with the tremendous workload of the Court of Justice, a lower General Court (originally called the Court of First Instance)[176] was established in 1989. This lower court can have jurisdiction in the same subjects as the Court of Justice with the exception of those subjects that have been reserved for the Court of Justice in the Court's Statute. Appeal from the General Court to the Court of Justice against

[172] Article 263 TFEU

[173] See Section 4.4.

[174] See Section 4.5.2 at the end of the first paragraph.

[175] Article 277 TFEU

[176] Article 256 TFEU

a decision given by the former can be based on points of law only. This means that the facts of the case are determined only once: by the lower judiciary.

5.6.6 Proceedings initiated by natural or legal persons

In certain situations the proceedings for annulment and for failure to act can also be initiated by private parties (natural or legal persons). According to Article 263, paragraph 4, TFEU they can go to EU courts only 'against an act addressed to that person or which is of direct and individual concern to them, and against a regulatory act which is of direct concern to them and does not entail implementing measures'. The object of this formula is to prevent private persons from being able to generally initiate proceedings against general legal rules, legislation contained in regulations or directives as such. That power is reserved for the Member States and the institutions. The cases initiated by natural or legal persons come before the General Court.

Both the Court of Justice and the General Court are independent and have real powers to decide conflicts in a legal order where Member States and persons have rights. Moreover, through the preliminary procedure the Court of Justice assists all national judges with the correct interpretation of EU law they have to apply. It is thus clear that both courts are powerful supranational elements in the EU.

5.7 The European Court of Auditors

The European Court of Auditors[177] was added to the EU institutions by the 1992 Treaty on European Union. It consists of one national from each Member State. The members of the Court of Auditors are appointed for a term of six years, which is renewable. The Council makes a list of candidates in accordance with the proposals made by each Member State. The Council after consulting with the European Parliament adopts the list of members drawn up in accordance with the proposals made by each Member State. The task of the Court of Auditors is to examine the income and the expenditure of the Union. It examines whether financial management has been sound and it will report on any irregularity found. The Court of Auditors reports annually, and this report is published in the Official Journal, together with the reactions of the institutions. The Court of Auditors can publish special reports at any time. It has the power to carry out on-the-spot investigations in the offices of the institutions, and in the Member States.

[177] Articles 285-287 TFEU

5.8 The European Central Bank

The European Central Bank (ECB),[178] together with the national central banks of the Member States whose currency is the euro, is responsible for the monetary policy of the Union. The primary objective of the ECB (and the national central banks in the Euro-zone) is to maintain price stability. In addition, the ECB shall support the European Union's general economic policies. Only the ECB may authorise the issue of the euro. It shall be independent in the exercise of its powers and in the management of its finances.

When an EU act is to be adopted within an area of the ECB's responsibilities, the ECB must be consulted. It shall address an annual report on the activities and on the monetary policy of both the previous and current year to the European Parliament, the Council, the Commission and the European Council.

5.9 The European Food Safety Authority

5.9.1 EFSA's mission

The European Food Safety Authority (EFSA) is established in Article 22 of Regulation 178/2002 (the General Food Law, or: GFL) as an independent agency responsible for risk assessment and risk communication. The GFL gives EFSA (called 'the Authority') the mission presented in Law text box 5.11.

Law text box 5.11. Article 22(2) GFL on the mission of EFSA.

Article 22(2) GFL
The Authority shall provide scientific advice and scientific and technical support for the [Union's] legislation and policies in all fields which have a direct or indirect impact on food and feed safety. It shall provide independent information on all matters within these fields and communicate on risks.

The quality and accessibility of scientific advice is considered to be of paramount importance to ensure effective, timely and appropriate decision-making. EFSA must cover all parts of the food chain. Scientific matters are considered to be part of a continuum from primary production through the food production chain to the consumer. The operation of EFSA as an independent entity is intended to ensure that there is a functional separation of the scientific assessment of risk from risk management decisions. Scientific risk assessment should not be swayed by policy or other external considerations. The independent position is designed to guarantee impartiality and objectivity.

[178] Articles 282-284 TFEU

Risk management on the other hand, in its widest possible sense, remains within the domain of the Commission, the Parliament and the Council. Risk management decisions take into account all relevant aspects, not only science but also many other matters – for example economic, societal, traditional, ethical and environmental factors, as well as the feasibility of controls. Risk management is a function of accountable, political decision makers. Close collaboration between the independent EFSA and those charged with the responsibility of risk management can inform them with scientific risk assessment.

EFSA's Advisory Forum provides central co-ordination to the efforts and resources of the national food authorities and agencies in Europe. EFSA is also the centre for such networking.

5.9.2 The organisation of EFSA

An agency is a public law legal person, made to carry out a well-defined part of government policy, usually of a technical nature. The agency works outside the regular civil service management lines, although the statutes, charter or other documents made for the founding of the legal person can contain rules on the relations between the responsible public authority and the agency that can be quite close. It is often maintained by the Commission that allowing agencies do a part of the job for the European Union is a way of decentralising the power of the EU institutions involved, the Council and the Commission. However, whether there is in fact delegation of power depends on the charter given to each particular agency. Beyond that, there are good reasons why decentralisation cannot use agencies as a vehicle. The Treaties determine the responsibilities of each institution, and this cannot be changed by creating an agency outside the 'normal' means. It is more pertinent to describe the spreading of agencies across the combined territories of the Member States as de-concentration. Instead of centralisation of all supporting civil service units, the executive spreads its services throughout the European Union.

The EFSA has four bodies: the Management Board, the Executive Director (and her staff), the Advisory Forum with the Member States, and the Scientific Committee with the Scientific Panels.[179] The permanent staff supports the work of these bodies and contributes its own work to realise the tasks. See Diagram 5.1 for EFSA's organisational chart.

5.9.3 The Management Board

The Management Board is composed of one representative of the Commission and fourteen members who are appointed by the Council in consultation with

[179] Article 24 GFL.

Diagram 5.1. Organisational chart EFSA (www.efsa.eu).

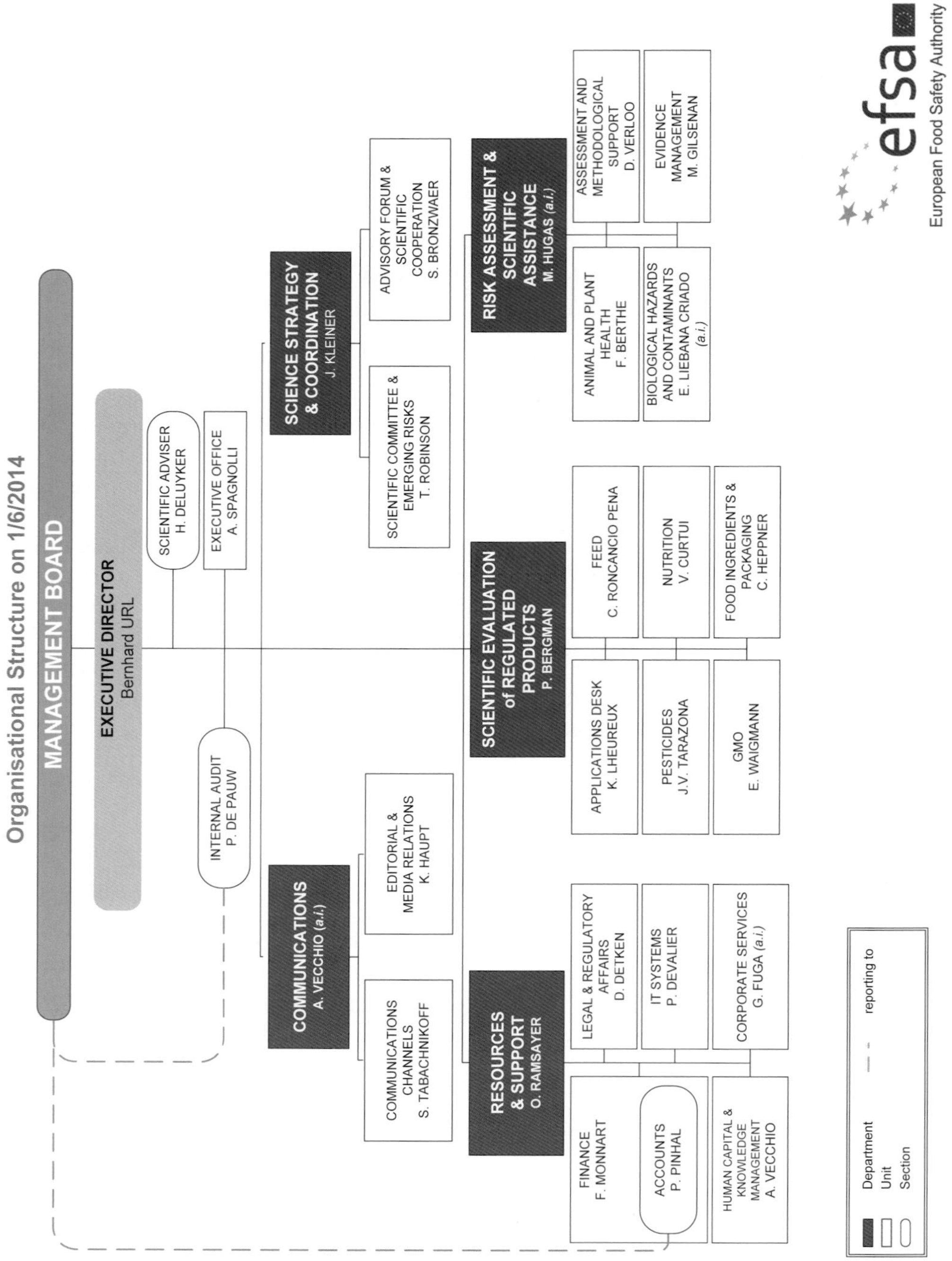

Organisational Structure on 1/6/2014

MANAGEMENT BOARD

EXECUTIVE DIRECTOR
Bernhard URL

SCIENTIFIC ADVISER
H. DELUYKER

EXECUTIVE OFFICE
A. SPAGNOLLI

INTERNAL AUDIT
P. DE PAUW

COMMUNICATIONS
A. VECCHIO (a.i.)

COMMUNICATIONS CHANNELS
S. TABACHNIKOFF

EDITORIAL & MEDIA RELATIONS
K. HAUPT

SCIENCE STRATEGY & COORDINATION
J. KLEINER

SCIENTIFIC COMMITTEE & EMERGING RISKS
T. ROBINSON

ADVISORY FORUM & SCIENTIFIC COOPERATION
S. BRONZWAER

RESOURCES & SUPPORT
O. RAMSAYER

FINANCE
F. MONNART

ACCOUNTS
P. PINHAL

HUMAN CAPITAL & KNOWLEDGE MANAGEMENT
A. VECCHIO

LEGAL & REGULATORY AFFAIRS
D. DETKEN

IT SYSTEMS
P. DEVALIER

CORPORATE SERVICES
G. FUGA (a.i.)

SCIENTIFIC EVALUATION of REGULATED PRODUCTS
P. BERGMAN

APPLICATIONS DESK
K. LHEUREUX

PESTICIDES
J.V. TARAZONA

GMO
E. WAIGMANN

FEED
C. RONCANCIO PENA

NUTRITION
V. CURTUI

FOOD INGREDIENTS & PACKAGING
C. HEPPNER

RISK ASSESSMENT & SCIENTIFIC ASSISTANCE
M. HUGAS (a.i.)

ANIMAL AND PLANT HEALTH
F. BERTHE

BIOLOGICAL HAZARDS AND CONTAMINANTS
E. LIEBANA CRIADO (a.i.)

ASSESSMENT AND METHODOLOGICAL SUPPORT
D. VERLOO

EVIDENCE MANAGEMENT
M. GILSENAN

Department
Unit
Section
reporting to

efsa
European Food Safety Authority

the European Parliament from a list drawn up by the Commission. The fourteen members are selected on the basis of their competence and expertise. The principle of the broadest possible geographic distribution within the Union comes second to their excellent personal qualities. Four of the members must have a background in organisations representing consumers and other interests in the food chain.[180] The members have a term of office of four years that can be renewed once.

Half of the members of the Management Board were appointed or reappointed with a mandate for the period 2014 to 2018. The new term of office of the remaining half runs from 1 July 2012 to 30 June 2016 to strike a balance between new appointments every four years and the experience and knowhow of the members already in office.

The Management Board adopts EFSA's internal rules, see Diagram 5.7 for some examples.

Decisions are made by a simple majority. The Board elects the Chair from its members for a period of two years, which may be renewed. Meetings will be held at the invitation of the Chair or at the request of at least one-third of the members. The Executive Director must be present at the meetings of the Management Board and takes care of the Secretariat. The Chair of the Scientific Committee will be invited to attend the meetings. Neither of these two persons have any voting rights. See Diagram 5.1 for the Management Board's position in EFSA's organisational chart.

The Management Board tasks are to:

- ensure that EFSA carries out its mission and performs the tasks assigned to it;
- adopt EFSA's annual work programme;
- adopt a revisable multi-annual programme.

Both must be consistent with the Union's legislative and policy priorities in the area of food safety. An annual general report on EFSA's activities has to be adopted before 30 March of the next year and published by 15 June.[181]

5.9.4 Regulation of EFSA

The creation of the European Food Safety Authority on 1 January 2002 and its expansion to a staff of approximately 450 civil servants, over 200 experts in panels and working groups, 28 Member States, as well as Norway and Iceland are connected in EFSA's Advisory Forum (with Switzerland, candidate countries and the European Commission having observer status) and similar developments over the whole range of tasks have their bases in legislation and are expressed

[180] Article 25 GFL.

[181] Article 25(8), second paragraph, read together with Article 26(3)(b), second paragraph, GFL.

in legal documents that create more rules or decisions about plans, programmes and the construction and maintenance of the organisation. To give an impression of the many documents that are part of the project, Diagrams 5.2 to 5.10 and Law text boxes 5.12 to 5.15 give the tasks and a selection of these documents. There are several types of documents: those directly related to the tasks of providing scientific opinions and risk communication, other indirectly related to the tasks as service documents shaping the facilities or conditions to enable the work on the tasks. Usually the content is indicated by the title. It is always accessible at EFSA's website (www.efsa.eu).

Diagram 5.2. EFSA main tasks.

Article in GFL	Tasks described in Section 5.9.12	EFSA main tasks
29	1.	Scientific opinions
30	2.	Diverging scientific opinions
31	3.	Scientific and technical assistance
32	4.	Scientific studies
33	5.	Data collection
33(5), 33(5)(a) 33(5)(b)	5.	Inventory of Data collecting systems at Union level before 1 March 2003
33(6)	5.	EFSA forwards results data collection to European Parliament, Commission and Member States
34	5.	Identification of emerging risks
35	6.	Rapid alert system
36	7.	Networking organisations
36	7.	Financing of work group projects
22(2), 40	8.	EFSA communications in general and risk communication
	8.	The EFSA Register of Questions
41	8.	Access to documents
42	8.	Consumers, producers and other interested parties

Diagram 5.3. Six categories of EFSA's core management documents.

Policy and strategy papers	Diagram 5.4
Operating rules	Diagram 5.6
Selection of experts	
Transparency and access to documents	Diagram 5.8
Administrative conduct and anti-corruption	Diagram 5.7
Finance	

Diagram 5.4. Examples of EFSA Core Management Documents: policy and strategy papers.

Strategy for Cooperation and Networking with Member States	Articles 22(7), 23(g) and 36 GFL
Risk Communications strategy and plans	Articles 22(2), 23(j), 40 and 42 GFL

Diagram 5.5. Examples of policy planning documents.

Management Plan of the European Food Safety Authority for 2013, MB 13.12.12 Document providing the work plan of the Authority during 2013
Annual Activity Report 2012 Article 25(8) (second paragraph) GFL: before 30 March following year
Annual work programme (AWP) Article 25(8) GFL: before 31 January each year Multi-annual work programme Article 25(8) GFL
Finance Article 43 GFL Financial Regulation of the European Food Safety Authority Implementing Rules for the Financial Regulation

Diagram 5.6. Examples of EFSA Core Management Documents: operating rules.

Decision of the Management Board (MB) of the European Food Safety Authority		
Subject	**MB Decision**	**Articles GFL**
Rules of Procedure of the Management Board of the European Food Safety Authority, adopted in Parma, 20 October 2011.	20.10.2011	25(5)
Decision of the Management Board of the European Food Safety Authority concerning the establishment and operations of the Scientific Committee, Scientific Panels and of their Working Groups, adopted in Parma, 15 March 2012.	15.03.2012	28(9), 29(7)
Decision concerning the operation of the Advisory Forum. Done at Pafos, 27 March 2008.	27.03.2008	25(3), 27(5)
Decision on implementing rules concerning the tasks, duties and powers of the Data Protection Officer, Implementing rules. Done at Parma, 24 January 2006.	24.1.2006	25(3), Chapter III

5.9.5 Integrity, independence and transparency

The integrity and independence of EFSA's staff and the members of its organisational bodies are of paramount importance to the Authority. The General Food Law gives clear instructions. The articles on independence and transparency are not only at the centre of EFSA's corporate identity but are also core values of science generally.

Transparency is about the consistent and traceable conduct of the organisation and its functionaries and personnel. The openness of the organisation hinges partly on the crystal-clear commitments of the people involved, whose personal interests do not trouble the clear course of the organisation. Openness about potential relations between the job in the organisation, or one particular item on an agenda, and (the semblance of) personal interests connect the declarations of interests to transparency. See Law text box 5.12 for transparency, one of the central values for public law entities.

Law text box 5.12. EFSA transparency, Article 38 GFL.

Article 38 GFL

1. The Authority shall ensure that it carries out its activities with a high level of transparency. It shall in particular make public without delay:
   (a) agendas and minutes of the Scientific Committee and the Scientific Panels;
   (b) the opinions of the Scientific Committee and the Scientific Panels immediately after adoption, minority opinions always being included;
   (c) without prejudice to Articles 39 and 41, the information on which its opinions are based;
   (d) the annual declarations of interest made by members of the Management Board, the Executive Director, members of the Advisory Forum and members of the Scientific Committee and Scientific Panels, as well as the declarations of interest made in relation to items on the agendas of meetings;
   (e) the results of its scientific studies;
   (f) the annual report of its activities;
   (g) requests from the European Parliament, the Commission or a Member State for scientific opinions which have been refused or modified and the justifications for the refusal or modification.
2. The Management Board shall hold its meetings in public unless, acting on a proposal from the Executive Director, it decides otherwise for specific administrative points of its agenda, and may authorise consumer representatives or other interested parties to observe the proceedings of some of the Authority's activities.
3. The Authority shall lay down in its internal rules the practical arrangements for implementing the transparency rules referred to in paragraphs 1 and 2.

Article 38 GFL shows that the transparency principles are active on two different levels. All instructions are addressed to EFSA as an organisation and its activities, except Article 38(1)(d) that addresses individual persons. The content is different too: transparency, as used in Article 38(1)(d), does not require a contribution to the corporate transparency in the work attitude and activities; it requires integrity, the absence of personal interests and the prevention of every likelihood and appearance of personal graft and gain. EFSA has laid down an internal policy on independence and on scientific decision-making processes to apply the GFL obligations (Diagram 5.7).

Diagram 5.7. Policy on Independence and Scientific Decision-Making Processes of the European Food Safety Authority.

Guidance documents	MB Decision number
Policy on Independence and Scientific Decision-Making Processes of the European Food Safety Authority. Adopted in Warsaw on 15 December 2011.	15.12.2011

The Declarations of Interests (DOI) of every person related to EFSA who is required to make these declarations can be accessed at EFSA's web site on a first name basis. The Specific Declaration of Interests (SDOI) is required in addition to the general annual declaration (ADOI) for each separate occasion with regard to an agenda item where personal interests might (seem to) be present.

The personal transparency requirements are linked to requirements of independence and good administrative behaviour. See Law text box 5.13 on independence and Law text box 5.14 on the EFSA Code of good administrative behaviour. See Diagram 5.8 for aspects of transparency in relation to the organisation.

Law text box 5.13. Article 37 GFL on independence.

Article 37 GFL

1. The members of the Management Board, the members of the Advisory Forum and the Executive Director shall undertake to act independently in the public interest.
   For this purpose, they shall make a declaration of commitment and a declaration of interests indicating either the absence of any interests which might be considered prejudicial to their independence or any direct or indirect interests which might be considered prejudicial to their independence. Those declarations shall be made annually in writing.
2. The members of the Scientific Committee and the Scientific Panels shall undertake to act independently of any external influence. For this purpose, they shall make a declaration of commitment and a declaration of interests indicating either the absence of any interests which might be considered prejudicial to their independence or any direct or indirect interests which might be considered prejudicial to their independence. Those declarations shall be made annually in writing.
3. The members of the Management Board, the Executive Director, the members of the Advisory Forum, the members of the Scientific Committee and the Scientific Panels, as well as external experts participating in their working groups shall declare at each meeting any interests which might be considered prejudicial to their independence in relation to the items on the agenda.

Diagram 5.8. Examples of EFSA Core Management Documents: Management Board decisions on transparency and access to documents.

Subject	MB Decision number	Articles GFL
Agreed openness, transparency and confidentiality – general principles.	16.09.2003 – 13	25(3), 37, 38, 38(3), 39 and 39(4)
Implementing measures of transparency and confidentiality requirements.	10.03.2005 – 10	GFL
Decision concerning access to documents.	16.09.2003 – adopted	25(3), 41, 41(2)

5.9.6 EFSA Code of good administrative behaviour

EFSA's Management Board adopted the EFSA Code of good administrative behaviour in 2003.[182] It is a catalogue of the general principles of good governance described in Chapter 2 Introduction to law, Section 2.6.6. Law text box 5.15 gives a concise impression of its content with an invitation to read the EFSA Code as a model for good administration and courtesy. As one of its bases the Code invokes Article 41 of the European Union's Charter of Fundamental Rights. It precedes highlights of the Code in Law text box 5.14.

Law text box 5.14. The Right to good administration, Article 41 Charter of Fundamental Rights of the European Union.

Article 41 Right to good administration

1. Every person has the right to have his or her affairs handled impartially, fairly and within a reasonable time by the institutions and bodies of the Union.
2. This right includes:
   (a) the right of every person to be heard, before any individual measure which would affect him or her adversely is taken;
   (b) the right of every person to have access to his or her file, while respecting the legitimate interests of confidentiality and of professional and business secrecy;
   (c) the obligation of the administration to give reasons for its decisions.
3. Every person has the right to have the Union make good any damage caused by its institutions or by its servants in the performance of their duties, in accordance with the general principles common to the laws of the Member States.
4. Every person may write to the institutions of the Union in one of the languages of the Treaties and must have an answer in the same language.

[182] MB Decision 16.09.2003-11-Adopted. Management Board of the European Food Safety Authority, EFSA Code of good administrative behaviour, Done at Brussels, 16 September 2003.

Law text box 5.15 lists the considerations preceding the articles and the titles of the articles indicating their subject from the EFSA Code of good administrative behaviour.

Law text box 5.15. EFSA Code of good administrative behaviour.[183]

The management board
Having regard to the Treaty of the European Union, and in particular Articles 21 and 195 thereof,
Having regard to Article 41 of the Charter of Fundamental Rights, (...)

WHEREAS the Amsterdam Treaty has explicitly introduced the concept of openness into the Treaty on European Union by stating that it marks a new stage in the process of creating an ever closer union in which decisions are taken as openly as possible and as closely as possible to the citizen,

WHEREAS the Charter of fundamental rights proclaimed at the Nice Summit in December 2000 includes as fundamental rights of citizenship the right to good administration and the right to complain to the European Ombudsman against maladministration,

WHEREAS, in order to bring the administration closer to the citizens and to guarantee a better quality of administration, a Code should be adopted which contains the basic principles of good administrative behaviour for agents and other servants of the Authority when dealing with the public,

Considering it therefore desirable to establish a Code governing the principles of good administrative behaviour which the agents and other servants of the Authority should respect in their relations with the public, and to make this Code publicly available,

Has decided as follows:

Article	Title
1.	General provision
2	Personal scope of application
3.	Material scope of application
4.	Lawfulness
5.	Absence of discrimination
6.	Proportionality
7.	Absence of abuse of power
8.	Impartiality and independence
9.	Objectivity
10.	Legitimate expectations and consistency
11.	Fairness

▶▶

[183] The document is accessible at http://tinyurl.com/nu7prpy.

Law text box 5.15. Continued.

12.	Courtesy
13.	Reply to letters in the language of the citizen
14.	Acknowledgement of receipt and indication of the competent agent or other servant
15.	Obligation to transfer to the competent service of the Authority
16.	Right to be heard and to make statements
17.	Reasonable time-limit for taking decisions
18.	Duty to state the grounds of decisions
19.	Indication of the possibilities of appeal
20.	Notification of the decision or recommendation
21	Data protection
22.	Requests for information
23.	Requests for public access to documents
24.	Keeping of adequate records
25.	Public access to the Code
26.	Right to complain to the European Ombudsman
27.	Revision
28.	Entry into force

5.9.7 The Executive Director

The Executive Director is EFSA's legal representative[184] and is responsible for:
a. the day-to-day administration of EFSA;
b. drawing up a proposal for EFSA's work programmes in consultation with the Commission;
c. implementing the work programmes and the decisions adopted by the Management Board;
d. ensuring the provision of appropriate scientific, technical and administrative support for the Scientific Committee and the Scientific Panels;
e. developing and maintaining contact with the European Parliament, and ensuring a regular dialogue with its relevant committees.

The Executive Director is responsible for the drafts of the programmes, reports and accounts that the Management Board has to decide on. The Executive Director takes care that the adopted programmes are sent to the European Parliament, the Council, the Commission and the Member States. The general report is sent to the European Parliament, the Council, the Commission, the Court of Auditors, the European Economic and Social Committee and the Committee of the Regions. The work programmes and the general report are published. See Diagram 5.1 for the Executive Director's position in EFSA's organisational chart.

[184] Article 26 GFL.

5.9.8 The Advisory Forum

The Advisory Forum is the link between EFSA and the Member States.[185] It is composed of representatives from competent bodies in the Member States which have tasks similar to those of EFSA. There will be one representative designated by each Member State. Members of the Advisory Forum may not be members of the Management Board. The Executive Director convenes and chairs the Advisory Forum. A third of its members can also initiate a meeting. The Forum has to meet at least four times per year. Representatives of the Commission's departments can participate in the work of the Advisory Forum. The Executive Director can invite representatives of the European Parliament as well as from other relevant bodies to take part.

The Advisory Forum gives the Executive Director advice on all aspects of the director's tasks, and in particular on drawing up a proposal for EFSA's work programme. The Executive Director may also ask the Advisory Forum for advice on the prioritisation of requests for scientific opinions. The Advisory Forum is a mechanism for an exchange of information on potential risks and the pooling of knowledge. It ensures close cooperation between EFSA and the competent bodies in the Member States. See Diagram 5.1 for the Advisory Forum's position in EFSA's organisational chart and the position of administrative support for the Advisory Forum under 'Scientific Committee and Advisory Forum'.

5.9.9 Scientific opinions

Several scientific committees were attached to the European Commission in the period preceding the GFL. Today they have been reorganised and have become part of EFSA. One Scientific Committee and ten permanent Scientific Panels are responsible for providing EFSA's scientific opinions, each within their own spheres of competence. The scientific panels have their legal base in Article 28(4) GFL. EFSA began with eight scientific panels each indicated by name in the GFL. The Commission has the authority to change the number and names of the panels. This authority can be used at the request of EFSA's Managing Board when changes are necessary 'in the light of technical and scientific development'. The Commission can exercise this authority only by using the procedure of Article 58 GFL (i.e. the comitology procedure) with the representatives of the Member States in the Standing Committee on the Food Chain and Animal Health. Today (2014) EFSA has 10 scientific panels and a scientific committee.

[185] Article 27 GFL.

5.9.10 The Scientific Panels

The comitology procedure was used in 2006 when the Panel on plant health was split from the Panel on plant health, plant protection products and their residues.[186] Experience showed that the Panel on additives, flavourings, processing aids and materials in contact with food (AFC) had to deal with almost half of the requests for a scientific opinion. It was anticipated that this workload would become even heavier. It was therefore decided to divide the Panel on additives, flavourings, processing aids and materials in contact with food (AFC) in two: the Panel on food additives and nutrient sources added to food (ANS) and the Panel on food contact materials, enzymes, flavourings, and processing aids (CEF).[187] See Diagram 5.9 for an overview of EFSA's scientific panels.

The procedures for the operation and cooperation of the Scientific Committee and the Scientific Panels have to be laid down in EFSA's internal rules. The Management Board used this delegation of power to adopt the Decision of the Management Board of the European Food Safety Authority concerning the establishment and operations of the Scientific Committee, Scientific Panels and of their Working Groups.[188] One of the Operating rules is given in Law text box 5.13. See Diagram 5.1 for the position of administrative support for the Scientific Panels in EFSA's organisational chart under Risk Assessment.

Diagram 5.9. EFSA's scientific panels, their acronyms, base in the GFL and mandates.

Panel on food additives and nutrient sources added to food (ANS), based on Article 28(4)(a) GFL.

Mandate: the safety in use of:

- food additives;
- nutrient sources added to food and to food supplements; and
- associated subjects concerning the safety of other deliberately added substances including substances added for other than a technological purpose, e.g. with functional properties but excluding flavourings and enzymes.

[186] Commission Regulation (EC) No 575/2006 of 7 April 2006 amending Regulation (EC) No 178/2002 of the European Parliament and of the Council as regards the number and names of the permanent Scientific Panels of the European Food Safety Authority.

[187] MB 11.09.07 – 4.2 Proposal to divide the tasks of the AFC panel and Commission Regulation (EC) No 202/2008 of 4 March 2008 amending Regulation (EC) No 178/2002 of the European Parliament and of the Council as regards the number and names of the Scientific Panels of the European Food Safety Authority (Text with EEA relevance).

[188] EFSA Management Board Decision concerning the establishment and operations of the Scientific Committee, Scientific Panels and of their Working Groups, 15 March 2012.

Diagram 5.9. Continued.

Panel on food contact materials, enzymes, flavourings, and processing aids (CEF), based on Article 28(4)(j) GFL.

Mandate: the safety in use of:

- food contact materials;
- flavourings and processing aids, including enzymes;
- associated subjects concerning the safety of other substances indirectly added to food; and
- questions related to the safety of new processes (e.g. irradiation).

Panel on animal health and welfare (AHAW), based on Article 28(4)(h) GFL.

Mandate: all aspects of animal health and animal welfare, primarily relating to food producing animals including fish.

Panel on biological hazards (BIOHAZ), based on Article 28(4)(f) GFL.

Mandate: biological hazards relating to food safety and food-borne disease, including food-borne zoonoses and transmissible spongiform encephalopathies, microbiology, food hygiene and associated waste management.

Panel on contaminants in the food chain (CONTAM), based on Article 28(4)(g) GFL.

Mandate: contaminants in food and feed, associated areas and undesirable substances such as natural toxicants, mycotoxins and residues of non authorised substances not covered by another panel.

Panel on additives and products or substances used in animal feed (FEEDAP), based on Article 28(4)(b) GFL.

Mandate: safety for the animal, the user/worker, the consumer of products of animal origin, the environment and to the efficacy of biological and chemical products / substances intended for deliberate addition/use in animal feed.

Panel on Genetically Modified Organisms (GMO), based on Article 28(4)(d) GFL.

Mandate: genetically modified organisms, such as micro-organisms, plants and animals, relating to deliberate release into the environment and genetically modified food and feed, including products deriving from GMOs.

Panel on dietetic products, nutrition and allergies (NDA), based on Article 28(4)(e) GFL.

Mandate: dietetic products, human nutrition and food allergy, and other associated subjects such as novel foods.

Panel on plant protection products and their residues (PPR), based on Article 28(4)(c) GFL.

Mandate: safety of plant protection products for the user/worker, the consumer of treated products and the environment.

Panel on Plant health (PLH), based on Article 28(4)(i) GFL.

Mandate: plant health especially with regard to organisms that pose a risk to plant health. This panel was created to enlist a wide range of expertise in the various fields relevant to plant health, such as entomology, mycology, virology, bacteriology, botany, agronomy, plant quarantine and epidemiology of plant diseases.

5.9.11 The Scientific Committee

The Scientific Committee is the institutionalised meeting of the Chairs of the Scientific Panels and six independent scientific experts who do not belong to any of the Scientific Panels. The six independent experts are appointed by the Management Board. The Scientific Panels are composed of independent scientific experts appointed by the Management Board. Ten panels have been established with a specified mandate so far (2013).[189] Decisions are made with a simple majority vote, minority opinions are recorded with the name of the author and the argumentation. The representatives of the Commission's departments have a right to be present in the meetings of the Scientific Committee, the Scientific Panels and their working groups. They assist in clarifying issues or providing information, but are not allowed to influence discussions.

Both the Scientific Committee and the Scientific Panels can set up Working Groups. Their expertise will be used to prepare draft scientific opinions. External experts, who possess particular and relevant scientific knowledge, can be invited to participate in a Working Group.[190] The Scientific Committee and permanent Scientific Panels can, where necessary, organise public hearings. Commission Regulation 1304/2003 of 23 July 2003 prescribes the procedure that EFSA has to apply to requests for scientific opinions.[191]

The Scientific Committee is responsible for the general co-ordination necessary to ensure the consistency of the scientific opinion procedure, in particular with regard to the adoption of working procedures and harmonisation of working methods. It provides opinions on multi-sector issues falling within the competence of more than one Scientific Panel, and on issues which do not fall within the competence of any of the Scientific Panels. Where necessary, and particularly in the case of subjects which do not fall within the competence of any of the Scientific Panels, the Scientific Committee will set up Working Groups. In such cases, it shall draw on the expertise of those Working Groups when establishing scientific opinions. The GFL lays great stress on the openness and transparency of EFSA's activities, as far as the confidential aspects of its work permit. See Diagram 5.1 for the position of administrative support for the Scientific Committee in EFSA's organisational chart under the Scientific Committee and Advisory Forum.

[189] See Diagram 4.9.

[190] Articles 5 and 6 MB Decision, 11.09.07.

[191] Commission Regulation (EC) No 1304/2003 of 23 July 2003 on the procedure applied by the European Food Safety Authority to requests for scientific opinions referred to it.

5.9.12 Tasks of EFSA

The following eight tasks will be briefly described:
- presenting scientific opinions;
- finding a solution for diverging scientific opinions;
- providing scientific and technical assistance to the European Commission;
- commissioning scientific studies;
- building a data collection;
- establishing monitoring procedures;
- promoting networking and (risk) communication;
- processing the information delivered by the rapid alert system.

Task 1. Presenting scientific opinions

One of EFSA's main tasks is to prepare and present scientific opinions. This work is done by the Scientific Committee or the Scientific Panels. Requests for a scientific opinion are assigned by the Executive Director to one of the Scientific Panels according to its mandate, or to the Scientific Committee for questions on multi-sector issues falling within the competence of more than one Panel, and on issues which do not fall within the competence of any of the Panels. Article 29 GFL specifies that opinions can be requested by the European Parliament, the Member States and the Commission. EFSA can give opinions on its own initiative.

Regulation 1304/2003[192] on the procedure that EFSA has to apply to requests for scientific opinions in Article 3(1) explicitly orders EFSA to refuse to give an opinion to anyone else. The Commission can ask EFSA to give opinions in addition to those based on the GFL in all specific cases where EU legislation grants this right.

EFSA has to operate a register of requested opinions and own initiative opinions. The register can be accessed by everyone and gives information on the progress of EFSA work. Each scientific opinion has to contain the question, the background to the request, the information considered, the scientific reasoning and the opinion of the Scientific Committee or Panel. The full scientific opinion is published on EFSA's website.

Task 2. Diverging scientific opinions

In the case of diverging scientific opinions, the Advisory Forum has to contribute to find a solution when disagreement occurs between EFSA and an organisation working on the same topics in a Member State. When both parties identify a substantive divergence of opinion over scientific issues, Article 30(4) GFL orders them to tackle this problem. Their co-operation has to result in resolving the

[192] Article 2 of Regulation 1304/2003.

differences or a joint document clarifying the contentious scientific issues and identifying the relevant uncertainties in the data.

Task 3. Provide scientific and technical assistance

EFSA has to provide scientific and technical assistance to the European Commission if requested. This assistance will be scientific or technical work involving the application of well-established scientific or technical principles. The scientific evaluation by EFSA is not needed. The assistance can be given without compromising the independent position of EFSA. The Commission will enlist this assistance especially for the establishment or evaluation of technical criteria and the development of technical guidelines.

Task 4. Commission scientific studies

EFSA must commission scientific studies necessary for the performance of its mission, using the best independent scientific resources available. EFSA has to avoid duplication with research programmes of the Member States or the EU, and shall foster co-operation through appropriate co-ordination. The Advisory Forum must assist EFSA in this respect, giving substance to its role to ensure pooling of knowledge and close co-operation between EFSA and the competent bodies in the Member States.

Task 5. Data collection

EFSA shall build up a collection of data in particular relating to:

a. food consumption and the exposure of individuals to risks related to the consumption of food;
b. incidence and prevalence of biological risks;
c. contaminants in food and feed;
d. residues.

EFSA will collect data in close co-operation with all organisations operating in the field of data collection, including those from applicant countries, third countries or international bodies.

Task 6. Monitoring procedures

EFSA will establish monitoring procedures for the systematic collection of information to identify emerging risks in the fields within its mission. Where EFSA has information leading it to suspect a serious risk, the Member States, other Union agencies and the Commission have to co-operate to find and deliver any relevant information to EFSA. To enable it to perform its task of monitoring the health and nutritional risks of foods as effectively as possible, EFSA shall receive all messages forwarded via the rapid alert system.[193] EFSA analyses this information and gives it to the Commission and the Member States for risk analysis.

Task 7. Networking and communication

EFSA has to promote networking between European organisations that work in the same fields as EFSA. The aim of such networking is, in particular, to facilitate a scientific co-operation framework by the coordination of activities, the exchange of information, the development and implementation of joint projects, and the exchange of expertise and best practices in the fields within EFSA's mission.[194]

Task 8. Risk communication

EFSA has to contribute to risk communication as an independent source of information with a solid reputation based on scientific excellence. EFSA's risk communication activities must respect the Commission's risk communication about risk management decisions. Risk communication was one of the central issues in the evaluation of EFSA's first external evaluation.[195]

EFSA's Register of Questions is an example of the free flow of information on food relevant issues supported by the independent Authority. The register stood at 13271 questions on 21 May 2013. See Diagram 5.10 on the EFSA Register of Questions to get an impression of the information it provides.[196]

Diagram 5.10. The EFSA Register of Questions.

Question number	Subject	Panel	Status	Details
EFSA-Q-2008-428	EU-wide collective scientific expertise on the possible causative factors of CCD affecting bee colonies.[1]	AMU	Finished	

[1] CCD stands for Colony Collapse Disorder: the adult working bees leave the colony en masse for an unknown destination. The hive is left with a live queen, honey, immature brood, no dead bees and too few working bees. The cause or causes of this no-future scenario are not known. Vanishing bee populations are part of the human experience with bee keeping but reported cases have risen remarkably in the EU and the US since 2006. See the website of the Agricultural Research Service of the US Department of Agriculture for current research and estimates of $15 billion lost added value a year and the estimate that one third of the diet in the US depends directly and indirectly on bees. And for them these are only the pollinators from the old world. Google CCD bee or visit http://www.ars.usda.gov/News/docs.htm?docid=15572.

[193] On the rapid alert system for food and feed (RASFF) see Chapter 15, Section 15.4.
[194] Article 36 GFL.
[195] Article 61 GFL. The report: European Food Safety Authority, Evaluation of EFSA, Final Report, Contract FIN0105, Bureau van Dijk Ingénieurs Conseils with Arcadia International EEIG, Brussels, 5 December 2005. EFSA's reaction: Management Board of the European Food Safety Authority, Proposal for Management Board Conclusions of the External Evaluation of EFSA and Recommendations Arising from The Report, MB 20.06.06-4.
[196] Available at: http://registerofquestions.efsa.europa.eu/roqFrontend/questionsListLoader?panel=ALL.

5.9.13 The organisation of the EFSA staff

The EFSA staff supports the work of the Scientific Panels, the Work Groups, the Advisory Forum and takes care of the other EFSA tasks mentioned above. In 2011 EFSA initiated a reorganisation. In 2013 the staff organisation consisted of five directorates: Resources & Support Directorate, Communications Directorate and three scientific directorates, namely the Risk Assessment and Scientific Assistance Directorate, the Scientific Evaluation of Regulated Products Directorate, and the Science Strategy and Coordination Directorate.

5.9.14 Legal protection

Article 47 GFL attributes jurisdiction on the Court of Justice of the European Union to decide on both contractual and non-contractual liability of EFSA.

5.10 The Standing Committee on the Food Chain and Animal Health

The Standing Committee on the Food Chain and Animal Health (SCFCAH by some referred to as SCoFCAH, SCoFoCAH or SCoFoChAH) was established to ensure a more effective and comprehensive approach to the food chain aspects of food production.[197] It replaced the Standing Veterinary Committee, the Standing Committee for Foodstuffs, the Standing Committee for Feedingstuffs and the Standing Committee on Plant Health. The unified committee assists the Commission and the EFSA. See Section 5.5 on comitology.

The Standing Committee on the Food Chain and Animal Health is organised in the following sections:

- General Food Law;
- biological safety of the food chain;
- toxicological safety of the food chain;
- controls and import conditions;
- animal nutrition;
- phytopharmaceuticals;
- animal health and animal welfare.

[197] Article 58 GFL

5.11 The Advisory group on the food chain and animal and plant health

In 2004 the European Commission created an advisory group on the food chain and animal and plant health.[198] This group is attached directly to the Commission to give it advice in the following fields of the Commission's programme of work:

- food and feed safety;
- food and feed labelling and presentation;
- human nutrition, in relation to food legislation;
- animal health and welfare;
- matters relating to crop protection, plant protection products and its residues, and conditions for the marketing of seed and propagation material, including biodiversity, and including matters pertaining to industrial property.[199]

The group brings together 45 stakeholder organisations of representative European organisations selected by the Commission. Each organisation has to coordinate consultation and information activities within its own organisation to present views that are as representative as possible. The selection criteria for membership of the advisory group are:

- the organisation's objective to protect interests in the fields of the Commission's work programme where the advisory group will be asked to advise;[200]
- the general nature of the protected interests;
- a representation covering all or most Member States; and
- permanent existence at Union level; with
- direct access to members' expertise to permit swift and coordinated reactions.[201]

The list of the 45 organisations selected by the Commission is published in the Official Journal of the European Union.[202] The advisory group is chaired by the Commission; it has two regular plenary meetings each year at the premises of the Commission and whenever the Commission considers a meeting necessary. The Commission takes care of the publicity for the work of the group.[203]

---

[198] Commission Decision of 6 August 2004 concerning the creation of an advisory group on the food chain and animal and plant health (2004/613/EC), pp. 17-19. Draft: Communication from the Commission, concerning the creation of an advisory group on the food chain and animal and plant health and the establishment of a consultation procedure on the food chain and animal and plant health through representative European bodies.

[199] Article 2(1) Commission Decision 2004/613/EC.

[200] Commission of the European Communities, Communication from the Commission to the European Parliament, the Council, the European Economic and Social Committee and the Committee of the Regions, Commission Legislative and Work Programme 2008, COM(2007) 640 final, Brussels, 23 October 2007.

[201] Article 3(1) Commission Decision 2004/613/EC.

[202] Commission Decision of 14 April 2011 on the members of the advisory group on the food chain and animal and plant health established by Decision 2004/613/EC. OJ L 101, p. 126.

[203] Article 4 Commission Decision 2004/613/EC.

5.12 Stakeholders' participation

DG SANCO wants to reinforce stakeholders' participation in the preparation and evaluation of policies.[204] One of the activities is the Healthy Democracy process. Its aim is 'to improve stakeholder involvement and participation. In the long term, the aim is to establish a solid network of stakeholders and research bodies to improve its substantive performance.'[205]

A Peer Review Group was established in early 2006 to assist DG SANCO's efforts to improve stakeholder involvement. The Group selected a top ten of recommendations, see Law text box 5.16.

Law text box 5.16. Stakeholder participation in SANCO's activities: Recommendations from the Peer Review Group.

I:	Establish a 'Stakeholder Dialogue Group' to get advice on processes rather than on content
II:	Improve Transparency through better 'Forward Planning'
III:	More and Better Feedback
IV:	Engage the Un-engaged & Going Local
V:	Driving Up Data Quality
VI:	Definition of Representativeness
VII:	Be Aware of Stakeholder Asymmetries
VIII:	More Flexible and Longer Consultation Timeframe
IX:	Improvement of Inter-DG Coordination
X:	More Transparent Comitology

The fourth recommendation 'IV: Engage the Un-engaged & Going Local' addresses two major weaknesses of all international organisations, the remote or fragmented civil society. See Law text box 5.17.

This recommendation led to an interesting proposal: that the Commission can use its permanent delegations in the Member States as a platform for contacts with the civil societies of the Member States to construct a European civil society.

---

[204] European Commission, Health & Consumer Protection Directorate-General, Stakeholder Involvement Event, The Report, Brussels, 23 May 2007. European Commission, Health & Consumer Protection Directorate-General, 03 - Science and stakeholder relations, Joint meeting of the Stakeholder Dialogue Group and the Advisory Group on the Food Chain and Animal and Plant Health, summary record, Brussels, SANCO 03, 30 November 2007.

[205] European Commission, Health & Consumer Protection Directorate-General, Healthy democracy. Conclusions and Actions following the DG SANCO 2006 Peer Review Group on Stakeholder Involvement.

Law text box 5.17. Stakeholder participation in SANCO's activities: Engage the Un-engaged & Going Local.

(Recommendation IV)
The Peer Review Group noted that the challenges of achieving representativeness and engaging hard-to-reach groups are exacerbated at the European level where there are few tangible connections between citizens and the Brussels institutions.
In order to ensure a better engagement, DG SANCO should:
1. consider the use of Commission's delegations in Member States as platform for the debate;
2. work together with existing stakeholders (in particular NGOs) to identify the 'unengaged'.

The following observation is true for all international organisations: the challenges of all political entities to achieve representativeness and engage hard-to-reach groups 'are exacerbated at the European level where there are few tangible connections between citizens and the Brussels institutions'.

This is directly relevant for the stakeholder asymmetries of 'recommendation VII': those stakeholder groups that are connected with the EU have a disproportionate influence. The EU should be well aware of this phenomenon: agricultural stakeholders occupied half its budget and 60% of its legislation for decades.

6. The embedding of food law into substantive EU law

Morten Broberg and Menno van der Velde

6.1 Introduction

If there were no EU law, there would still be food law. If there were no food law, however, there certainly could not be EU law as we know it today. This is because food law is a necessity on its own and because food law is embedded in EU law that covers the essentials of the European integration process. Without food law, the EU internal market would not be the same. The common market, which was set as a goal in 1958, and the internal market, which was set as a goal in 1986, were designed to be drivers of the integration process, as much as, or even more than, motors of economic development. The intimate relationship between the EU's internal market and food law caused them both to suffer during the low tides of European integration in the period from 1966 to 1986. During this period the Court of Justice of the European Communities (now the Court of Justice of the European Union) played a major part in the integration process, and a surprisingly large number of the Court of Justice's important rulings were rendered in the field of food law. Food law has therefore been a major factor in keeping integration alive in the ebb times and booming in the renaissance that began in 1986. See Chapter 7, Section 7.3 Advancement through case law in this book.

In addition to the formal structure of EU law, which determines what types of measures are possible, and what their effects are on the EU itself, on the EU institutions, on the Member States and on private persons, as well as on the relations between these various actors, substantive EU laws determine to a considerable extent the conditions under which food law has to operate.

6.2 The Customs Union

Essential building blocks of the Treaty of the European Economic Community, which came into force on 1 January 1958, were the customs union and the common market. The customs union required the abolition of customs duties and customs borders between the Member States. This was linked with the creation of one common external customs frontier that created the same conditions all along the external border of the EEC for entry of goods from the world outside the EEC. In principle the GATT agreement[206] required that if a signatory state to the agreement offered lower customs duties to one other state, all the GATT signatories

[206] General Agreement on Tariffs and Trade, i.e. an agreement aimed at diminishing customs duties and other barriers to international trade. See Chapter 3, Section 3.5.2 in this book.

should be able to obtain the same lower customs duty.[207] The GATT agreement, however, provides an explicit exception with regards to customs unions so that GATT signatories that participate in a customs union are not required to offer the same low (customs union) duties to other GATT signatories. During and after its creation the European Union participated in several international negotiation rounds in GATT as well as in its successor the World Trade Organisation (WTO).

Inside the customs union the borders separating the Member States had to be abolished over a transition period of twelve years, starting at the beginning of the EEC on 1 January 1958. This period was divided into three rounds each of four years. Decision making was to be by unanimous vote during the first two periods and by majority vote in the last period. So from 1966 onwards majority decision making was prescribed by the Treaty, and the customs union would be completed by 1970.

6.3 Ban on all customs levies and all measures with an equivalent effect

The prohibition of customs duties on imports and exports included charges having equivalent effect. To comply with this obligation Member States had to abandon the legal instruments that they had developed in the course of the preceding decades. Thus, originally several national legal instruments were used at the national borders on goods entering the state and sometimes also on goods exported from the state. For example, persons transporting incoming goods were subjected to payment to obtain permission to cross the border with the goods. This generated income for the national government, but simultaneously it interfered in the market economies of both the exporting and importing states. Allowing desired goods to enter without a customs levy did not increase their price on the internal market; subjecting the products of competing foreign producers to high customs levies protected the national producers. All kinds of levies were created with different names, different purposes, different ways of collection and with the common denominator that they all required a payment for entry of goods originating from abroad. With regards to intra-EU trade, the EEC Treaty ordered the removal of all customs levies and all measures with an effect equivalent to that of a customs levy.

6.4 Ban on quantitative restrictions

A second group of measures on customs borders had been created: quantitative restrictions. In a given year a certain amount of goods were allowed to be imported. When the imports reached the quota-limit, it was no longer possible to import more of the product in question until a new year (and a new quota) had begun. The quantitative restrictions have their own group of measures with equivalent

[207] The so-called 'Most Favoured Nation Clause'.

effect. Although they did not take the form of a straight limitation of a quantity, practically speaking they had the same results. An example is the measure used by the Dutch government in the 1950s. Bakers had to bake bread with a certain percentage of Dutch-grown wheat. This was another way of limiting the amount of imported wheat. The EEC Treaty required the removal of quantitative restrictions and all measures with equivalent effect.

It is with these measures that food law can become involved. Many rules of food law are made to serve non-economic interests, like human health. Under certain conditions the Treaty on the Functioning of the European Union (TFEU) allows the Member States to introduce measures that pursue these non-economic interests even if they create hindrances to free trade.[208] But even with these exceptions the food law measures are only lawful under EU law if they pass a severe test. The first question to be asked is whether the matter has already been the subject of EU harmonisation. If the European Union has exhaustively regulated the matter the Member States will normally be barred from retaining or introducing differing legislative measures on the same matter. If the European Union has not introduced harmonisation measures with regards to the matter, the next question is whether the Member State measures are capable of negatively affecting trade between the Member States. In other words, the question is whether the disputed measure is capable of hindering access to the Member State's market. If so the third question that must be considered is whether the Member State measure pursues a lawful objective. The TFEU provides a non-exhaustive list of lawful objectives and this includes public health. Food safety falls under public health and therefore is a lawful objective. Finally, it must be asked whether the trade hindrance caused by the measure is proportionate to the objective pursued (see Chapter 4 for a presentation of the proportionality principle).

The above may be illustrated by a situation where a Member State introduces a prohibition against the selling of bread that contains more than 2% salt. Assume that the question has not been the subject of EU harmonisation. We first observe that the prohibition hinders the sale of bread that contains more than 2% salt and thus is a barrier to entering the market in question. The question therefore is whether the prohibition pursues a lawful objective. Now, assume that the limitation has been introduced in order to keep the population's salt intake down and thereby reduce the risk that they will fall victim to a number of diseases. This clearly is a lawful objective. But does the prohibition also overcome the proportionality test? There probably is no doubt that the prohibition may help reduce the population's salt intake, so it is 'suitable'. But is it also 'necessary' to introduce an outright prohibition? After all, instead of introducing a prohibition the authorities could merely have introduced a labelling obligation so that sellers of bread were required to label the salt contents. In this way the population would be able to take the

[208] Article 36 of the TFEU.

salt contents into due account when purchasing bread; without having to ban the sale of bread containing more than 2% salt.[209]

If a Member State fails to comply with the above rules it may be brought before the Court of Justice of the European Union. See Diagram 6.1 on the different effects of directives and regulations on national law.

To get a good perspective on the problems connected to the question as to whether a particular legislative act is a measure that constitutes a barrier to the market of the Member State in question, or is covered by one of the exceptions to free movement of goods and therefore must be accepted, it is illuminating to look at the way situations may get out of hand when such rules do not function properly: e.g. a trade war.

6.5 Trade wars

The relations between the nation states and their economies can be characterised by a certain roughness if there are no intermediate or reconciling structures like the EU or WTO. Take the relations between the United States and the European Union in recent years as an example. Even after decades of negotiating customs reduction treaties, and concluding them successfully, every now and then a trade war disrupts these relations. A disagreement about support given by the EU to the steel industry in the Union leads the US to install barriers for the import of cheese and flowers. First of all it is not quite clear why American aficionados of European cheeses should be punished by their government who increases the prices of these cheeses by putting a punitive levy on them when they are imported into the USA. The reason for the US government installing these extra charges is that years of complaining about European governments' subsidies to the European steel industries have not changed EU policies. And even if there is a good reason why American cheese eaters must suffer for the losses incurred by American steel companies, it is still unclear why European cheese makers have to suffer for the advantages given by their governments to steel industries.

These imbalances are taken for granted, because the US government felt the need to do something, anything, in the steel conflict with the EU. It is an indication of the basic characteristics of international relations acting outside of intermediate structures. There are many reasons why the steel dispute may be more the outcome of an escalation process than of real differences in the treatment of the steel market. To name but a few: the American government might be misinformed, it might be overstating the effects of EU policies, it might have neglected the fact that the European steel policy itself is a reaction to an earlier American policy to support their steel industry that is still in place, and so on. The basic issue is that there is

[209] A situation very similar to the one in this example may be found in Case C-17/93 *van der Veldt*.

a lack of strong mechanism functions that would stimulate restraint, conciliation, or even impartial examination of the facts. A conflict like this occurred in 1998 between the US and the EU about steel with cheese as a bargaining chip, and in 2002 when textiles and orange juice were drawn into the conflict.

In the past trade conflicts regularly divided European nation states, but this changed with the creation of what we today call the European Union. The transfer of powers from the European nation states to the European Union, the creation of common policies and the establishment of the common market make the recurrence of this type of conflict between the different Member States nearly impossible, simply because they no longer have the instruments that are necessary to be able to have a dispute like this. Even more important is the fact that when disagreement about the common policy arises between the Member States, as will inevitably happen from time to time, there are mechanisms that de-escalate the conflict instead of firing the blast furnace of dissent. The Member States and the institutions are interlocked by the economic consequences of integration, and they are interlocked in the complicated web of competences, instruments, procedures and decisions of EU policy formation and execution. They are forced to negotiate. And if this turns out not to be enough to disband the conflict, they have recourse to the Court of Justice of the European Union: an independent institution that will establish the facts of the case from a position outside of the conflict and with the prestige and deep roots of judges in the political and legal systems and rule of law in Europe. These impartial judges will not only assist the conflicting, quarrelling states to see the facts as they are, they will also pronounce the outcome of the confrontation of these facts with the rules that were agreed on before. In short, a judge will decide an economic (political?) conflict on the basis of law.

Such is the past and the background for the law on the common market, the customs union and the common commercial policy for economic relations between the EU and countries outside of the Union. It is necessary to understand the problems that are connected with trade in food. Indeed, a very considerable proportion of all cases decided by the Court of Justice of the European Union with regard to questions relating to the common market have concerned food products. In many instances Member States have invoked the protection of public health in situations that in reality concerned illegitimate protection of national food producers against competition from imported food products. It is therefore only natural that health issues that are raised for legitimate reasons have to face the suspicion created by a long history of appeals on health issues that actually were a front to hide a trade interest. This suspicion has even entered Treaty on the Functioning of the European Union where it grants in Article 36 exceptions to the rule of free movement of goods on the common market 'on grounds of public morality, public policy or public security; the protection of health and life of humans, animals or plants ...' and then expresses the need to shackle the exceptions for these meritorious causes with two chains. Firstly, that 'such prohibitions or restrictions shall not, however, constitute a means of arbitrary discrimination', nor secondly 'a disguised restriction on trade between Member States'. The first limitation of the exceptions to the rule of free movement of goods is an expression of the need to ban any discrimination on grounds of nationality, Article 18 TFEU, which is a necessity if the EU is to build and maintain a common market out of former national markets. The second limitation of the exceptions is a reflection on the experience with customs law. Apparently experience had led to caution if not suspicion. On the level of the worldwide international community, the same experiences have led to two WTO agreements: the Agreement on Technical Barriers to Trade (the TBT Agreement) and the Agreement on the Application of Sanitary and Phytosanitary Measures (the SPS Agreement). See Chapter 3 of this book.

6.6 The common market

Together with the customs union, the common market is another essential building block of integration. It prescribes free movement of goods across the whole territory of the Union, once the goods have been lawfully put on the market in any of the Member States. This can be goods from countries outside the Union which have been placed on the market in another Member State, or goods which were produced and marketed in one of the Member States. Free movement of goods is one of the European Union's so-called 'four freedoms' regarding free movement (the other three are free movement of persons, services and capital). These freedoms are crucial for the EU's creation of one unitary market, like the domestic market of a single state. The common market is not just about free movement, but also involves an active government policy to keep the market functioning as a true unitary market. Thus, several policies support the common market: competition policy, private enterprises are not allowed to divide amongst themselves the market that the Member States have unified; and public procurement policy, the Member State governments and other public authorities may not give preferential treatment to enterprises of their own nationality when they are buying goods or investing in major projects. Member state authorities also have to follow EU rules on the support given to companies.

The central problem for the Union is that while the customs borders were dismantled, and later the borders checking the movement of persons were also widely abolished,[210] this does not change the fact that the Member States' different national law systems still leave their imprint on the territory of the Union. Travelling from one Member State to another is moving from one system of law to another. These different legal systems have different rules for many aspects of the same goods, and this is not least true with regards to food products.

[210] In particular in those Member States (and third countries!) that adhere to the Schengen agreement.

6.7 Technical barriers to trade

With the creation of the common market, another barrier to free movement of goods had to be faced, namely the technical barriers to trade. A good illustration of a technical barrier to trade is the earlier Belgian law that prescribed that butter being sold by retailers to consumers had to be packed as a cube. The reason for this requirement was to protect the consumer from confusing butter and margarine. Outside Belgium butter (and margarine) was normally not sold in cubes, but instead in rectangular blocks. Whatever the ideas behind the Belgian legislation, the effect on producers outside Belgium was clear. If they wanted to export their butter to Belgium, they would have to install the machinery for packing their butter in a cube form. If they happened to be in a country where the traditional way to present butter is in rectangular blocks, this meant additional expenditure in machinery and space, and the logistics of a separate stream of products to Belgium. The Belgian national 'cubic' legislation on butter was therefore a measure of equivalent effect to a quantitative restriction. The aim of preventing confusion between butter and margarine was not proportionate to the objective of protecting the consumer from confusing margarine with butter since this objective could be achieved with less drastic measures: for instance, a label indicating the content.

There were two ways in which the Belgian law was enforced. At the time when the case was brought up, dismantling the internal customs borders had not yet removed customs civil servants from the national borders. At that time the Member States still had controls on transport of goods across their borders, to test whether the imported goods met the various technical requirements produced by national legislation, introduced to serve a wide variety of interests. Belgian civil servants patrolling the borders to other Member States could intercept the rectangular butter transport before it became illegal because it brought the wrong butter form into Belgium. In this period it was a well-known complaint from transporters that while the customs barrier had been removed, they still had to comply with all the different demands of the national law systems they crossed on their routes in integrating Europe. The transporter, who had to bring with him 72 different documents that were required by the successive national authorities on his route, acquired almost legendary status. The effort to obtain all these documents from all these national administrations and the time lost with all the long delays at the internal borders between the Member States gave rise to calculations on how much economic capacity could be freed by a real common market without technical barriers to trade.

The second way to enforce the Belgian 'cubic' butter law was to have inspectors visit the retail shops regularly to check whether various food laws were being obeyed. The cubic packaging requirement would be infringed if the retailer offered imported rectangular blocks of butter for sale. The inspector would then confiscate these illegal packages of butter and would institute prosecution for breaching the 'cubic' butter law. Depending on the rules of Belgian criminal enforcement law, this could be a fine to be paid directly, or the responsible retailer might be obliged to appear before a court that would try him/her as a suspect of the crime of offering non-cubic butter for sale in a Belgian shop. If the court based the case solely on Belgian law there could be no other outcome than to punish the offender with the sanctions provided by Belgian law. The consequence would be that no Belgian retailer would dare to import butter in rectangular blocks. The Belgian territory of the common market would hereby be effectively closed for these products. This effect of the Belgian requirement on butter packaging was thus equivalent to a quantitative import restriction. And such state measures with equivalent effect to import restrictions were (and still are) forbidden under EU law which is supreme to Belgian law. So the free movement of goods which originally was regulated in the EEC Treaty (and subsequently in the EC Treaty and today the Treaty on the Functioning of the European union) prescribed that the Belgian court give no effect to the Belgian Cubic Butter Law, and acquit the defendant retailer. To make sure that this was the correct thing to do, the Belgian court, who had to decide this case, asked the advice of the Court of Justice of the European Union.[211]

The Court of Justice ruled that the law demanding cubic butter was a technical barrier to trade, an unnecessary obstacle to free movement of goods. The consequence was that Belgian law could not be applied to imports from other Member States. So in the case of cubic butter, the Belgians would get the luxury of having both cubic and rectangular butter. In contrast, Belgian producers probably started to wonder why *they* were limited to the production of cubes.

6.8 From common to internal market

In the 1985 White Paper Completing the Internal Market[212] the European Commission took note of the consequences of the failure of the harmonisation efforts for the common market. Although many directives had been adopted, there were still major obstacles on the road to creating one single market. The Commission's Internal Market White Paper gave an inventory of more than 300 legislative measures that were necessary to have a real common market with no internal borders. The program of the internal market was to achieve that objective by the last day of 1992. The Single European Act Treaty introduced the internal market in 1987 in the EEC Treaty and created a special harmonisation provision to make it possible to adopt the 300 measures before the target date. The new Article 100a of the EEC Treaty (now Article 114 TFEU) introduced the possibility of using other instruments for the effort to make common rules, and it allowed for majority decision-making for these measures. See Law text box 6.1.

[211] Case 261/81 *Walter Rau v. De Smedt* [1982] ECR 3961.

[212] COM(85) 310 Final, presented to the public by EC Commissioner Lord Cockfield on 15 June 1985.

Law text box 6.1. The two harmonisation articles of theTreaty on the Functioning of the European Union: Article 114 for the internal market and Article 115 for the common market.

Treaty on the Functioning of the European Union
TITLE VII Common rules on competition, taxation and approximation of laws
Chapter 3 Approximation of laws
Article 114

1. Save where otherwise provided in the Treaties, the following provisions shall apply for the achievement of the objectives set out in Article 26. The European Parliament and the Council shall, acting in accordance with the ordinary legislative procedure and after consulting the Economic and Social Committee, adopt the measures for the approximation of the provisions laid down by law, regulation or administrative action in Member States which have as their object the establishment and functioning of the internal market.
2. Paragraph 1 shall not apply to fiscal provisions, to those relating to the free movement of persons nor to those relating to the rights and interests of employed persons.
3. The Commission, in its proposals envisaged in paragraph 1 concerning health, safety, environmental protection and consumer protection, will take as a base a high level of protection, taking account in particular of any new development based on scientific facts. Within their respective powers, the European Parliament and the Council will also seek to achieve this objective.
4. If, after the adoption of a harmonisation measure by the European Parliament and the Council, by the Council or by the Commission, a Member State deems it necessary to maintain national provisions on grounds of major needs referred to in Article 36, or relating to the protection of the environment or the working environment, it shall notify the Commission of these provisions as well as the grounds for maintaining them.
5. Moreover, without prejudice to paragraph 4, if, after the adoption of a harmonisation measure by the European Parliament and the Council, by the Council or by the Commission, a Member State deems it necessary to introduce national provisions based on new scientific evidence relating to the protection of the environment or the working environment on grounds of a problem specific to that Member State arising after the adoption of the harmonisation measure, it shall notify the Commission of the envisaged provisions as well as the grounds for introducing them.
6. The Commission shall, within six months of the notifications as referred to in paragraphs 4 and 5, approve or reject the national provisions involved after having verified whether or not they are a means of arbitrary discrimination or a disguised restriction on trade between Member States and whether or not they shall constitute an obstacle to the functioning of the internal market.
   In the absence of a decision by the Commission within this period the national provisions referred to in paragraphs 4 and 5 shall be deemed to have been approved.
   When justified by the complexity of the matter and in the absence of danger for human health, the Commission may notify the Member State concerned that the period referred to in this paragraph may be extended for a further period of up to six months.

Law text box 6.1. Continued.

7. When, pursuant to paragraph 6, a Member State is authorised to maintain or introduce national provisions derogating from a harmonisation measure, the Commission shall immediately examine whether to propose an adaptation to that measure.
8. When a Member State raises a specific problem on public health in a field which has been the subject of prior harmonisation measures, it shall bring it to the attention of the Commission which shall immediately examine whether to propose appropriate measures to the Council.
9. By way of derogation from the procedure laid down in Articles 258 and 259, the Commission and any Member State may bring the matter directly before the Court of Justice of the European Union if it considers that another Member State is making improper use of the powers provided for in this Article.
10. The harmonisation measures referred to above shall, in appropriate cases, include a safeguard clause authorising the Member States to take, for one or more of the non-economic reasons referred to in Article 36, provisional measures subject to a Union control procedure.

Article 115

Without prejudice to Article 114, the Council shall, acting unanimously in accordance with a special legislative procedure and after consulting the European Parliament and the Economic and Social Committee, issue directives for the approximation of such laws, regulations or administrative provisions of the Member States as directly affect the establishment or functioning of the internal market.

The Internal Market White Paper introduced a new approach to harmonisation. The case law of the Court of Justice prescribed mutual recognition of the laws of the Member States if a good had been lawfully produced and put on the market in a Member State. This case law is discussed in Chapter 7, Section 7.3. Exceptions to the principle of mutual recognition that are compatible with the internal market can be made in Union law only. The Commission therefore proposed to concentrate the harmonisation effort on those national measures that created barriers to trade and were allowed by EU law. The legislative effort in the internal market programme had to exchange the many different legitimate Member State trade barriers (typically safeguards of health and safety) for one single EU law aimed at safeguarding the same interests as those originally safeguarded by the different Member State legislations. In other words, the European Union would increase the efforts to harmonise the differing legislative measures in the field. The approach to regulating these concerns was changed and was (and still is) referred to as 'the new approach'. Instead of making detailed rules for all aspects of the matter, the detailed technical provisions would no longer be a part of the EU legislation. Instead, the directives or regulations would to a considerable extent refer to standards made by European standardisation organisations. These are international non-profit organisations that are established to develop unified European product standards and to rationalise the market. CEN (Comité Européen de Normalisation), CENELEC

(Comité Européen de Normalisation Électrotechnique) and ETSI (The European Telecommunications Standards Institute) are standardisation organisations for products in which all 28 EU Member States are amongst the members. These organisations keep track of technical progress to keep the European standards up to date. The EU legislation follows the same developments by referring to these standards, without the need to go through the process of law-making. The reference in the directive or regulation to the technical standards does not make these standards part of the binding law. They remain voluntary, but industry is encouraged to apply the standards because if the producer of the good can prove that the good conforms to the standards which the EU harmonisation measure refers to, the Member State authorities must assume that goods made according to these standards meet the requirements of EU legislation. In other words, if the producer can show that the products comply with the voluntary standard, the products will be allowed to move freely in the internal market. In contrast, for goods that are not made according to the standards it is for the manufacturer to prove that the product meets the requirements of the directive.[213]

6.9 Strengthening of substantive EU law

The BSE food crisis echoed in the Inter-Governmental Conference on the 1997 Treaty of Amsterdam. This treaty amended the 1992 European Union Treaty and the 1992 European Community Treaty. A number of articles were changed to ensure better policies on, among other issues, food. Already, what is now Article 114 of the TFEU (i.e. the special harmonisation provision introduced by the Single European Act as part of the creation of the internal market) in its third paragraph included a commitment that the Commission will base its harmonisation proposals on a high level of protection of public health, safety, environmental protection and consumer protection. To this the Amsterdam Treaty added: 'taking account in particular of any new development based on scientific facts'. This reference to new developments based on scientific facts is of obvious importance to food law and the work of the European Food Safety Agency (EFSA). See Law text box 6.1 above for the text of the two harmonisation treaty articles that have been and still are of major importance for food law.

The Amsterdam Treaty also changed what is now Article 168(1) of the TFEU into: 'A high level of human health protection shall be ensured in the definition and implementation of all Union policies and activities. Union action, which shall complement national policies ...'. In the original article, EU policies and activities merely had to contribute to a high level of human health protection. Furthermore, the same article in paragraph 4(b) gives the Council and the Parliament acting

[213] The question to what extent this also applies to private food safety standards is discussed in: N. Coutrelis, EU 'new approach' also for food law? in: B.M.J. van der Meulen (ed.), Private Food Law. Governing food chains through contract law, self-regulation, private standards, audits and certification schemes, Wageningen: Academic Publishers, Wageningen, the Netherlands, 2011, pp. 381-390.

together in accordance with the ordinary legislative procedure the power to adopt 'measures in the veterinary and phytosanitary fields which have as their direct objective the protection of public health' in order to ' meet common safety concerns'.

Finally, the Amsterdam Treaty added to the consumer protection provision in what is now Article 168 of the TFEU the clause '...the Union shall contribute to protecting the health, safety and economic interests of consumers ...'. This provision has, inter alia, formed part of the legal basis for the adoption of 'Regulation 178/2002 laying down the general principles and requirements of food law, establishing the European Food Safety Authority and laying down procedures in matters of food safety'.

7. Food law: development, crisis and transition

Bernd van der Meulen

7.1 History of European food law

European Food Law has developed in several stages. As with any subject area, current situations and reactions can best be understood if one understands how the situation has developed. Therefore, it is useful to take historical developments into account in order to gain a better understanding of what remains from past structures, of mistakes made, lessons learned and probable future developments.

From the beginnings of the European Union in 1958 until the eruption of the BSE crisis in the mid-1990s, European food law was principally directed at the creation of an internal market for food products in the EU.[214]

This market-oriented phase can be subdivided into two stages. During the first, emphasis was on harmonisation through vertical directives. This stage ended with the 'Cassis de Dijon' case law. During the second stage emphasis shifted to harmonisation through horizontal directives.[215]

The BSE crisis, and other food scares that followed soon after, brought to light many serious shortcomings in the existing body of European food law. It became evident that fundamental reforms would be needed. In January 2000, the European Commission announced its vision for the future development of European food law in a 'White Paper on Food Safety'.[216]

The 'White Paper on Food Safety' emphasised the Commission's intent to change its focus in the area of food law from the development of an internal market to assuring high levels of food safety. In the years since its publication, a great deal of important legislation has been passed, and further proposals are in development or under consideration. As this book looks toward the future, we will focus principally on food law as it has developed since the White Paper, and on proposals for continuing reform in coming years. See Diagram 7.1 on the development of European food law as approached in this book.

---

[214] During different stages of its development, different names have been used for the market in the EU. In this chapter we use its latest name: 'internal market'. The earlier name 'common market' is only used in quotes.

[215] The distinction between horizontal and vertical directives will be discussed hereafter.

[216] COM(1999) 719 def. Commission White Papers traditionally contain numerous proposals for Community action in specific areas, and are developed in order to launch consultation processes at the European level. If White Papers are favourably received by the Council, they often form the basis of later 'Action Programs' to implement their recommendations.

Diagram 7.1. Development of European food law.

Phase	Turning point	Orientation	Main instruments
First	Cassis de Dijon (1979)	Market	Vertical directives
Second	BSE crisis (1997)	Market	Horizontal directives
Third		Safety first; market second	Horizontal regulations

7.2 Creating an internal market for food in Europe

When the six original members of what is today the European Union signed the Treaty of Rome in 1957, they created a community with an economic character. This was reflected not only in its original name – the European Economic Community – but also in the original goal to create a common market.

At the heart of the instruments to achieve this goal are the famous four freedoms of the European Union: the free movement of labour, the free movement of services, the free movement of capital and the free movement of goods. The free movement of goods (Article 26 TFEU) has been vital to the development of food law.

Although all Member States agreed generally about the desirability of a free market unimpeded by national borders, in practice the tendency remained for each state to protect its own markets and enterprises wherever and however possible, and to trust the judgements of its own institutions before those of the Union or other members. Consequently, the Member States preferred to trust their own laws and standards before those of the Union or other Member States.

This limitation in objective and the struggle for its fulfilment left a clear mark on the initial development of European food law. Irrespective of how attractive the idea of a free market seemed, there still existed countless national provisions setting different standards for products and thus impeding the free movement of goods. As we have already seen in the example of the Belgian standard for margarine (Chapter 6, Section 6.7), different standards mean additional costs and create de facto obstacles to the free movement of goods for producers who want to export their products.

During the first years of implementing the ambitious idea of trade without frontiers, European Union legislation aimed primarily at facilitating the internal market through the harmonisation of national standards. Agreement about the quality and identity of food products was considered necessary. To reach such agreement directives were issued on the composition of certain specific food products. This is called vertical (recipe, compositional or technical standards) legislation. This creation of similar standards in all the Member States is called positive harmonisation.

Example of vertical legislation

A well-known example of recipe-legislation is the discussion about how much cocoa a food product should contain to be called chocolate, and whether vegetable fats other than cocoa butter may be used. The Directive relating to Cocoa and Chocolate products Intended for Human Consumption[1] set the cocoa percentage at 35 for the entire European Union. Since 3 August 2003[2] up to a maximum of 5% of the total weight may be added in other vegetable fats to certain chocolate products.

[1] Directive 73/241, meanwhile replaced by Directive 2000/36.
[2] The date on which the term of implementation of Directive 2000/36 expired. Since then this rule from the Directive has direct effect if it is not yet transformed into national law.

Early attempts to establish an internal market for food products in the European Union by prescribing harmonised product compositions along such lines faced two substantial obstacles. Firstly, at that time all legislation required unanimity in the Council, which gave each Member State a virtual right of veto over new legislation.[217] Secondly, there was the sheer scale of the task. Browse through a supermarket in any EU Member State and consider the variety of products on the shelves. There are, as the Union institutions soon realised, simply too many food products to deal with. Creating compositional standards for each product would have been a mission impossible, and the Commission wisely chose to seek alternatives. Nevertheless quite a few products remain subject to European rules on compositional standards.[218] These compositional standards form the legacy of the first phase of EU food law. They are being updated or replaced when necessary but no new products are being added. Compositional standards are outside the scope of this book. For some examples see Diagram 7.2.

[217] See Chapter 5.
[218] E.g. sugar, honey, fruit juices, milk, spreadable fats, jams, jellies, marmalade, chestnut puree, coffee, chocolate, natural mineral waters, minced meat, eggs, fish. Wine legislation is a body of law in itself.

Diagram 7.2. Vertical legislation: examples of compositional standards.

Directive 2001/114	partly or wholly dehydrated preserved milk for human consumption
Directive 2001/113	fruit jams, jellies and marmalades and sweetened chestnut purée intended for human consumption
Directive 2001/112	fruit juices and certain similar products intended for human consumption
Directive 2001/111	certain sugars intended for human consumption
Directive 2001/110	honey
Directive 2000/36	cocoa and chocolate products intended for human consumption
Directive 1999/4	coffee extracts and chicory extracts
Regulation 122/94 / Regulation 251/2014	definition, description, presentation, labelling and the protection of geographical indications of aromatised wine products
Directive 93/45	the manufacture of nectars without the addition of sugars or honey

7.3 Advancement through case law

It was the Court of Justice of the European Union that 'restarted the engine' of European co-operation in the area of food law[219], in this case through new, broad, interpretations of the prohibition of quantitative restrictions on imports and all measures having equivalent effect.[220] This prohibition is a key provision for the free movement of goods on the common market.[221] See Law text box 7.1.

Law text box 7.1. Article 34 TFEU.

Article 34 TFEU
Quantitative restrictions on imports and all measures having equivalent effect shall be prohibited between Member States.

This prohibition should be read in connection with Article 36 TFEU which lists possible exceptions to the free movement of goods. These exceptions are: public morality, public policy or public security; the protection of health and life of humans, animals or plants; the protection of national treasures possessing artistic, historic or archaeological value; or the protection of industrial or commercial property. Of particular relevance to food safety law will be, of course, exceptions justified on grounds of the protection of health and life of humans, animals or plants.

[219] And more in general.
[220] On the relevance of Article 25 EC Treaty banning customs duties and charges having equivalent effect, see Morten P. Broberg, Transforming the European Community's Regulation of Food Safety, SIEPS 2008:5, freely available on the Internet.
[221] At that time Article 30 of the EC Treaty; now Article 34 TFEU.

7.3.1 Dassonville

In the 1974 Dassonville case,[222] the European Court of Justice gave a broad definition of the concept of 'measures having equivalent effect'. The case concerned parallel imports of Scotch whisky into Belgium. Belgian law required that a certificate of origin from the British authorities accompany such products. Mr. Dassonville bought Scotch whisky in France for re-importation into Belgium. It was cheaper there, as Scottish exporters had been trying to gain a share in the French market with reduced prices. As Mr. Dassonville could not obtain British certificates of origin in France, he created his own and was subsequently charged with fraud by the Belgian authorities. The Belgian criminal court needed to determine whether convicting Mr. Dassonville for fraud was compatible with European Law. To settle this issue, the Belgian court submitted questions of interpretation to the European Court of Justice in Luxemburg. The Court of Justice of the European Union found that the Belgian law could not be applied as it constituted a restriction on trade and was thus prohibited under what is now Article 34 TFEU. The Court held 'that all trading rules enacted by Member States which are capable of hindering, directly or indirectly, actually or potentially, intra-Community trade are to be considered as measures having an effect equivalent to quantitative restrictions' and are thus prohibited in the absence of a specific allowable justification. Such justifications could only be the protection of human, animal or plant health. From this ruling it follows that mere disparities between national laws can be held against an importing country as measures having equivalent effect to quantitative restrictions. In later rulings the Court detailed how such situations should be handled in practice.[223]

7.3.2 Cassis de Dijon

The Cassis de Dijon[224] ruling was seminal in this respect. A German chain of supermarkets – price fighter Rewe – sought to import Cassis de Dijon, a fruit liqueur, from France. The German authorities, however, refused to authorise the import because the alcohol content was lower than allowed by German national product standards, which stipulated that such liqueurs should contain at least 25% alcohol. Cassis de Dijon contained just 20% alcohol.

The German authorities acknowledged that this was a restriction on trade, but sought to justify it on the basis that beverages with too little alcohol pose several risks. The German authorities argued that alcoholic beverages with low alcohol content could induce people to develop tolerances for alcohol more quickly than beverages with higher alcohol content, and that consumers trusting the (German) law might feel cheated if they purchased such products with the expectation of

[222] EC Court of Justice 11 July 1974, Case 8/74 (Dassonville).
[223] Some other relevant cases not discussed here are; 407/58, 53/80, 132/80, 124/81, 249/81, 261/81, 177-178/82, 202/82, 222/82, 21/84, 90/86, 274/87, 277-318-319/91, 292/92 123/00.
[224] CJEU 20 February 1979, Case 120/78 (Cassis de Dijon).

higher alcohol content. Finally, Germany submitted that in the absence of such a law, beverages with low alcohol content would benefit from an unfair competitive advantage because taxes on alcohol are high, and beverages with lower alcoholic content would be saleable at significantly lower prices than products produced in Germany according to German law.

The Court held that the type of arguments presented by the German authorities would be relevant, even where they did not come under the specific exceptions contained in the TFEU, provided that those arguments met an urgent need. This is known as the rule of reason. The Court's broad interpretation of the prohibition against trade restrictions should be applied within reason.

The Court found that Germany's public health argument did not meet this standard of reasonableness. The Court specifically cited the availability of a wide range of alcoholic beverages on the German market with alcohol content of less than 25%. As to the risk of consumers feeling cheated by lower than expected alcohol content, the Court suggested that such a risk could be eliminated with less effect on the internal market by displaying the alcohol content on the beverages label.

For cases such as this one, in which there are no specific justifications for restrictions on the trade between Member States, the Court introduced a general rule: products that have been lawfully produced and marketed in one of the Member States, may not be kept out of other Member States on the grounds that they do not comply with the national rules. This is called the principle of mutual recognition (Law text box 7.2).[225]

Law text box 7.2. CJEU in Cassis de Dijon on mutual recognition.

> There is (...) no valid reason why, provided that they have been lawfully produced and marketed in one of the Member States, alcoholic beverages should not be introduced into any other Member State; the sale of such products may not be subject to a legal prohibition on the marketing of beverages with an alcohol content lower than the limit set by the national rules.
>
> (...) The concept of 'measures having an effect equivalent to quantitative restrictions on imports' contained in [Article 34 TFEU] is to be understood to mean that the fixing of a minimum alcohol content for alcoholic beverages intended for human consumption by the legislation of a Member State also falls within the prohibition laid down in that provision where the importation of alcoholic beverages lawfully produced and marketed in another Member State is concerned.

[225] Some other relevant cases in this line not discussed in this book are: 216/84, 17/93, 267-268/91, 391/92, 425-427/97, 405/98, 443/98, 37/99, 416/00, 497/03.

7.3.3 Mutual recognition

In essence, the Court's ruling was that, within the context of the internal market, what is good enough for consumers in one Member State is good enough for consumers across the Union. This ruling broke the deadlock in the development of a free market resulting from the existence of different national standards. The principle of mutual recognition signalled a giant leap forwards in European law, and 'Cassis de Dijon' has become a symbol of this stage in the development of the European understanding of free movement of goods.

With its ruling the Court in Luxemburg laid the legal foundation for a well-functioning internal market. Food products that comply with the statutory requirements of the Member State where they are brought on the market must, in principle, be admitted to the markets of all other Member States. This form of harmonisation, where products are legal regardless of which national legal system they comply with, is known as negative harmonisation, as opposed to positive harmonisation which is ensuring that national legislation is similar.

Several commentators expressed concern that the Cassis de Dijon decision would lead to product standards based on the lowest common denominator. It is clear that manufacturers established in Member States with the most lenient technical requirements or legal procedures do gain a competitive advantage.

The limitations and drawbacks of the principle of mutual recognition highlighted the need for further harmonisation of product requirements at the European level. For Member States with more stringent national standards, European-level legislation became the best hope for raising neighbours' standards. The Cassis de Dijon ruling marked a significant change in the perception of the benefits of harmonisation. Before Cassis, harmonisation was seen merely as a condition for the functioning of the internal market. Afterwards, emphasis shifted to the need to alleviate the consequences of the internal market. In legal terms, too, the wave of harmonisation that followed Cassis differed from earlier efforts. Emphasis shifted from product-specific, vertical (recipe) legislation, to horizontal legislation, meaning general rules addressing common aspects for all foodstuffs, or at least for as many foodstuffs as possible.

Mutual recognition remains the rule to this day. The consequence bears repeating; foodstuffs that have legally come to the market in any Member State, may in principle be sold without restrictions across the whole territory of the European Union.[226]

[226] The European Commission still grapples with the consequences of this case law. Immediately after the ruling, the Commission gave its interpretation: Communication from the Commission concerning the consequences of the judgment given by the Court of Justice on 20 February 1979 in Case 120/78 ('Cassis de Dijon') (Official Journal C 256, 3 October 1980, pp. 2-3). More recently a Regulation on mutual recognition entered into effect: Regulation 764/2008.

Example of horizontal legislation.

A prime example of this development is the Labelling Directive,[1] which included[2] a prohibition on misleading statements on foodstuff packaging[3] and a requirement to display specific indications, including a list of ingredients. Another example is the Hygiene Directive,[4] which required manufacturers of foodstuffs to identify the critical points in the production process and to draw up procedures to ensure safety.[5]

[1] Directive 2000/13.Now repealed by Regulation 1169/2011.
[2] Meaning: requires Member States to implement such prohibition.
[3] Article 2(a).
[4] Directive 93/43. Late 2002 the European Commission announced that the Hygiene directive 93/43 and 17 other Directives will be joined and harmonised with other directives containing rules on hygiene for different products (such as fresh meat, milk and fish) and be simplified. A change of instrument was also foreseen. The Community exchanged the directives for one regulation. On 30 April 2004 Regulation 853/2004 was published, which entered into force on 1 January 2006. This regulation is discussed in Chapter 13.
[5] The so-called HACCP system, 'Hazard analysis and critical control points'.

7.3.4 Reinheitsgebot

In line with 'Cassis de Dijon' is a later ruling, also concerning Germany. Germany had – and still has – special legislation on beer. A famous provision is the so-called 'Reinheitsgebot'. This German law on the 'purity' of beer is a historic law on taxes levied on beer. It dates back to 1516 when it was enacted by Duke William of Bavaria. Going back some 500 years, it is probably the oldest piece of food legislation in force today. In short it says that beer may only be produced using four ingredients: barley, hops, yeast and water. In the early 1980s, this provision was applied in such a way that imported products not complying with it could not be brought to the German market under the name of 'Bier'.

The European Commission took the view that those provisions were contrary to the ban on measures having an effect equivalent to quantitative restrictions on imports and brought infringement proceedings against the Federal Republic of Germany on two grounds, namely the prohibition on marketing under the designation '*Bier*' (beer) beers lawfully manufactured by different methods in other Member States and the prohibition on importing beers containing additives. With the judgment of 12 March 1987 in Case 178/84 Commission vs. Germany,[227] the Court held that the prohibition on marketing beers imported from other Member States which did not comply with the provisions in question was incompatible with Article 34 TFEU.[228]

[227] Commonly known as 'Reinheitsgebot' or 'German beer purity law'.
[228] At that time Article 30 of the EEC Treaty.

This case is one of many examples in which the Court applied its 'Cassis de Dijon' principles. It stands out from the other examples, however, mainly due to the follow up it enjoyed in the case discussed here below.

7.3.5 Brasserie du Pêcheur

After the Reinheitsgebot ruling Brasserie du Pêcheur, a French company based at Schiltigheim (Alsace),[229] brought an action against the Federal Republic of Germany for reparation of the loss suffered by it as a result of that import restriction between 1981 and 1987, seeking damages to the sum of DM 1,800,000,[230] representing a fraction of the loss actually incurred. Before the German national court, Brasserie du Pêcheur claims that it was forced to discontinue exports of beer to Germany in late 1981 because the competent German authorities considered that the beer it produced did not comply with the Reinheitsgebot. The German Court sitting on the case (the Bundesgerichtshof) asked the EU Court of Justice for a preliminary ruling on the question as to whether indeed Germany could be held liable for this type of damages on the grounds that its national legislation infringed the principle of mutual recognition as enshrined in Article 34 TFEU.

The Court of Justice answered (as quoted in Law text box 7.3)[231] that Member States are liable when their legislation infringes EU law if three conditions are met:

- the EU law infringed confers rights on the individuals (that is to say the claimant's citizens as well as businesses);
- the breach of EU law is sufficiently serious, and
- there is a direct causal link between the breach and the damage sustained.

Law text box 7.3. CJEU 5 March 1996 Brasserie du Pêcheur.

CJEU on Member State liability

'where a breach of Community law by a Member State is attributable to the national legislature acting in a field in which it has a wide discretion to make legislative choices, individuals suffering loss or injury thereby are entitled to reparation where the rule of Community law breached is intended to confer rights upon them, the breach is sufficiently serious and there is a direct causal link between the breach and the damage sustained by the individuals'.

Subsequently the Bundesgerichtshof made itself a laughing-stock in circles of EU law professionals by judging that in this case Brasserie du Pêcheur had not proven the required causal relation. While it did not help Brasserie du Pêcheur

[229] One of the well-known brands Brasserie du Pêcheur holds is Kronenbourg beer.

[230] About one million euros.

[231] CJEU 5 March 1996, Brasserie du Pêcheur SA v Bundesrepublik Deutschland and The Queen v Secretary of State for Transport, ex parte: Factortame Ltd and others, Joined cases C-46/93 and C-48/93.

in this case, the ruling of the CJEU is very important as it shows that EU Member States can be held liable for infringements on EU (food) law, even if they occur in their legislation.

The legacy of the second phase of development of EU food law is the principle of mutual recognition and the horizontal approach to food legislation.

Diagram 7.3. Case law on free movement of goods.

Case	Type of procedure	National procedure	Content
C-8/74 Dassonville	Preliminary ruling	Criminal	All trading rules potentially hindering intra-Community trade are measured having equivalent effect
C-120/78 Cassis de Dijon	Preliminary ruling	Administrative (tax) law	Mutual recognition
C-178/84 Reinheitsgebot	Infringement procedure	-	Mutual recognition
C-46/93 and C-48/93 Brasserie du Pêcheur	Preliminary ruling	Civil (tort) law	Member state is liable for damages if national legislation infringes on EC law

7.4 Breakdown

The heyday of market-oriented food law based on mutual recognition ended in tears. The food and agricultural sectors in the European Union emerged deeply traumatised from the 1990s. A series of crises resulted in a breakdown of consumer confidence in public authorities, industry and science. The – current – third phase of EU food law can only be truly fathomed if the trauma to which it responds is understood.

Although the bovine spongiform encephalopathy (BSE) crisis was not the first and, in terms of death toll, not the worst[232] food safety crisis in the EU it caused an earthquake in the legal and regulatory landscape of Europe (see Diagram 7.4

[232] A. Borda I *et al.*, Toxic Oil Syndrome Mortality: The First Thirteen Years, International Journal of Epidemiology 1998, p. 1057;
E. Gelpí *et al.*, The Spanish Toxic Oil Syndrome Twenty Years After Its Onset: A Multidisciplinary Review of Scientific Knowledge, Environmental and Health Perspectives 2002, p. 457 (finding that the toxic oil syndrome (TOS) epidemic that occurred in Spain in the spring of 1981 caused approximately 20,000 cases of a new illness. Researchers identified 1,663 deaths between 1 May 1981 and 31 December 1994 among 19,754 TOS cohort members. Mortality was highest during 1981). The poisoning was caused by fraud consisting of mixing vehicle oil with consumption oil.

Diagram 7.4. Background on BSE.[1]

Bovine spongiform encephalopathy (BSE) is a transmissible, slowly progressive degenerative disease of the central nervous system of adult cattle. This disease has a prolonged incubation period in cattle following oral exposure (2 to 8 years) and is always fatal. BSE is a type of transmissible spongiform encephalopathy (TSE), a fatal progressively degenerative disease of the central nervous system. TSEs occur in several animal species, like BSE in cattle, scrapie in sheep and goats, and some other. Consumption of contaminated feed transmits BSE. Infectious products include rendered animal proteins and compound animal feed containing meat-and-bone meal. Only a small amount of infectious material, perhaps as little as a gram of brain tissue, can transmit BSE. BSE was first identified as neurological disease in cattle in the United Kingdom in 1986. BSE has been reported in a large number of countries, but the majority of confirmed cases have occurred in the UK.

Variant Creutzfeldt-Jakob disease (vCJD) is a rare and invariably fatal disease in humans first described in the UK in March 1996. Most victims of the disease are from the UK or have lived there for significant periods of time. The UK Department of Health indicated that 156 deaths from confirmed or probable vCJD had occurred between 1995 and 1 September 2006. vCJD, like BSE, is a TSE, characterised by a spongy degeneration of the brain and its ability to be transmitted. Clinical signs of the disease include psychiatric symptoms (e.g. depression), followed by neurological signs that include unsteadiness, difficulty with walking and involuntary movements, ending with immobility and muteness. Victims of vCJD succumb to the disease at an average age of 29, after an illness with a median duration of 14 months.

Epidemiologists in the UK first reported the link between BSE and vCJD in 1996. After some initial uncertainty, the World Health Organization recognised the connection: 'Considerable epidemiological, neuropathological, and experimental data are consistent with the hypothesis that the agent that causes BSE in cattle also causes vCJD in humans. The most plausible route of human exposure is through the consumption of food contaminated with the BSE agent, although this has not been conclusively proven'. Tissue from the central nervous system of infected cattle is the most likely food contaminant.

[1] Taken from M.R. Grossman, Animal Identification and Traceability under the US National Animal Identification System. Journal of Food Law & Policy, 2/2/2006, pp. 231-315. See this publication for sources.

for background information on BSE). Subsequent food safety scares,[233] outbreaks of animal diseases[234] and scandals over fraudulent practices, added to a sense

[233] One example is the Belgian dioxin crisis. It was caused by industry oil that had found its way into animal feed and subsequently into the food chain. C. Whitney, Brussels Journal, Food Scandal Adds to Belgium's Image of Disarray, N.Y. Times, 9 June 1999. Another example is the introduction of merroxyprogesterone acetate (MPA) into pig feed in 2002. J. Graff, One Sweet Mess, Time, 21 July 2002. Sugar discharges from this production of MPA, a hormone used in contraceptive and hormone replacement pills, were used in pig feed and by that route MPA entered the food chain. In 2004 a dioxin crisis broke out in the Netherlands.

[234] Like Food and Mouth Disease, SARS and Avian Influenza.

of urgency to take protective measures. These fraudulent practices included the discharge of waste in animal feed[235] and the underworld involvement in the supply and employment of growth hormones[236] resulting in the murder of the veterinarian who brought the use of these illegal substances to the attention of the authorities and the public.[237]

Public awareness of the BSE epidemic, and the time it had taken British and European authorities to address it, presented a major challenge to European co-operation in the area of food safety. When the extent of the crisis became public, the European Union issued a blanket ban on British beef exports. In response, Britain adopted a policy of non-co-operation with the European institutions, and sought to deny the extent and seriousness of the BSE problem.[238]

The European Parliament played a crucial role in defusing this crisis. Although often accused of being a debating society of little consequence, during the BSE crisis the Parliament proved itself capable of political decisiveness and effective democratic oversight. A temporary Enquiry Committee was instituted to investigate the actions of the national and European agencies involved in the crisis.[239] The Enquiry Committee presented its report in early 1997 (See Diagram 7.5).[240] The report

[235] Probably the cause of the first dioxin crisis. See Whitney supra footnote 234.

[236] Union and national legislatures in the EU have been battling the use of artificial hormones – DES (diethylstilbestrol) in particular – for years. When it turned out to be impossible to separate their use from body-proper hormones and to get them under control, all hormones were finally banned. The legislation on the use and application of hormones started with Directive 81/602 (prohibiting certain matters with hormonal effects and of stuffs with thyrostatic effects). Directive 81/602 has been supplemented by Directive 85/358/EEC and replaced by Directive 88/146/EEC (prohibition of applications of certain stuffs with hormonal effect in the cattle breeding sector). A subsequent one, Directive 88/299, aims at the trade in animals and meat treated with stuffs with hormonal effect referred to in Directive 88/146.

[237] See K. Butler, Why the Mafia is into Your Beef: The EU Ban on Growth Hormones for Cows has Created a Lucrative Black Market, The Independent (London), 19 March. On Monday evening 20 February 1995 close to his home in Wechelderzande (Belgium) K. van Noppen, a Belgian veterinarian and inspector, was shot after exposing cattle breeders who illegally used hormones, their suppliers and civil servants who turned a blind eye to these practices. He had received several threats to discourage him from going after the 'hormone-mafia'. One of the inspectors accused by Van Noppen of corruption committed suicide in March 1996. In 2002 four people were found guilty by a jury, two of them cattle breeders. K. Butler, Four Men Found Guilty of Contract Hit on Vet, Independent (London), June 5, 2002, p. 11. The instigator of the assassination received a life sentence, the others 25 years each. Belgian Hormone Killers Jailed, BBC News, June 5, 2002.

[238] TV footage of the then Secretary of State, John Gummer feeding his young daughter a hamburger, to convince the public that nothing was wrong with British beef, became symbolic (16 may 1990 BBC). Text, picture and video available at: http://news.bbc.co.uk/onthisday/hi/dates/stories/may/16/newsid_2913000/2913807.stm.

[239] OJ C 261, 1996, p. 132.

[240] Report of the Temporary Committee of Inquiry into BSE, set up by the Parliament in July 1996, on the alleged contraventions or maladministration in the implementation of Community law in relation to BSE, without prejudice to the jurisdiction of the Community and the national courts of 7 February 1997, A4-0020/97/A, PE 220.544/fin/A. The report is often referred to by the name of the chairman of the Enquiry Committee, namely the Medina Ortega report.

strongly criticised the British government as well as the European Commission. The Commission was accused of wrongly putting industry interests ahead of public health and consumer safety; science had been biased and transparency had been lacking.

Diagram 7.5. Highlights from the Medina Ortega enquiry report on BSE.

Report on alleged contraventions or maladministration in the implementation of Community law in relation to BSE, 7 February 1997 A4-0020/97/A.

The preponderance of UK scientists and officials meant that the Scientific Veterinary Committee tended to reflect current thinking at the British Ministry of Agriculture, Fisheries and Food.

It made a partial and biased reading of the advice and warnings of the scientists. The views of certain scientists who could be considered as more critical were not taken into account. Some members of the Southwood Committee have said publicly that the minutes of its meetings were drawn up by a UK Ministry of Agriculture official and contain omissions and discrepancies.

In the BSE affair there has been, at the very least, a lack of transparency. This should appear clearer if one recalls that BSE has been the subject of an ongoing analysis by the BSE Subgroup of the Scientific Veterinary Committee. This subgroup has, almost throughout, been chaired by a UK national (first Mr Plowright and then Mr Bradley), and has included a substantial number of British scientists. Mr Bradley, who from 1969 to 1991 was head of the UK's Central Veterinary Laboratory and was subsequently an adviser to the British Ministry of Agriculture, has acted as rapporteur on BSE at the full meetings of the Scientific Veterinary Committee; it emerges, furthermore, from some of the minutes of the committee meetings that a number of members have suggested that Mr Bradley may have withheld information.

The Commission has failed to guarantee the proper functioning of veterinary controls within the internal market, thus breaching Directive 89/662, under which the Commission is empowered and obliged to carry out on-the-spot inspections to ensure the effectiveness of veterinary controls. Under Article 16 of this directive, the Commission has, in general terms, to monitor and oversee the national authorities' control programmes. The failure of the Commission to undertake initiatives in this field vis-à-vis the British Government could be interpreted as implying an unacceptable lack of vigilance ('culpa in vigilando') on its part and a deliberate resolve to leave management of this issue in the hands of the British, all the more so since, at that time, they were the only ones affected by the disease.

The Commission's actions may be characterised as follows:
It has given priority to the management of the market, as opposed to the possible human health risks existing in the light of the numerous scientific uncertainties concerning the possible effects of BSE on humans, and has neglected the principle of preventive action. There is a considerable body of material confirming this attitude.

Diagram 7.5. Continued.

The testimonies of present and former Commissioners and senior Commission officials have made it obvious that BSE has consistently had major political implications; this may be explained in view of the huge economic interests at stake in the meat, feeding stuff and animal residue processing industries, but, in terms of principles, it cannot justify the management of the BSE crisis.

By virtue of the opaqueness, complexity and anti-democratic nature of its workings, the existing system of comitology seems to be totally exempt from any supervision, thereby enabling national and/or industrial interests to infiltrate the Community decision-making process. This phenomenon is particularly serious where public health protection is at stake.

The Commission has to date failed to submit any convincing proposals by which it could regain credibility and begin genuinely to exercise its responsibilities in the field of health and consumer protection.

On the other hand, transparency in connection with BSE policy is closely bound up with the conditions of functioning, transparency and facilities for expression characterising the work of the scientists attached to the advisory scientific committees, at both Commission and Member State level.

Equally, in the interests of the transparency of anti-BSE measures, provision should be made to change the criteria for recruiting scientific staff to the European institutions.

Also at issue here is the degree of transparency required for the state of scientific knowledge on a subject to be determined, as well as the political use made of that knowledge by national governments or Commissioners, in relation to the perceived level of risk at a given moment.

On these points, the committee proposes:
The rules governing the work of the scientific committees which advise the Commission should be reformed in such a way as to professionalise their organisation. The appointment of their members should in future be strictly based on objective professional qualifications. There should be improved guarantees concerning their independence, the recording of dissenting views and the funding available to them. The administrative procedures for reimbursing participants' expenses should be revised, to make it easier for independent scientists to attend. When animal health matters are discussed, doctors of human medicine should also be present, so as to improve the assessment of possible public health risks.

The Enquiry Committee did not confine itself to an analysis and critical comments. The report went on to make concrete recommendations for improvements to the structure of European food law. Paradoxically, this reproachful report followed by a motion of censure proposed to the European Parliament provided the

Commission with the impetus it had hitherto lacked, indeed with a window of opportunity, to take the initiative for restructuring European food law in a way that considerably strengthened its own powers. The Commission's then President, Jacques Santer, undertook far-reaching commitments to implement the Committee's recommendations.

Progress was made along institutional lines as well as policy lines. The Directorate General (DG) XXIV 'Consumer Policy' created two years earlier, was reinforced and renamed 'Consumer and Health Protection Policy' and included the scientific advisory committees from the Directorate Generals for Industry and Agriculture.[241] A Scientific Steering Committee was created to bring wider scientific experience and overview to consumer health questions. The internal market 'product warning system' was also transferred from DG III (Agriculture) to DG XXIV. As of 1997, the centre of gravity in food legislation moved from DG Agriculture to DG XXIV, now called 'SANCO'.

As early as May 1997, the Commission[242] published a Green Paper on the general principles of food law in the EU.[243] It set out the structure of a legal system capable of getting a firm grip on food production. Consumer protection was made the main priority. The Commission committed to strengthening its food safety control function. This led directly to the establishment of the Food and Veterinary Office (FVO) in Dublin in 1997.[244] The FVO was charged with carrying out the Commission's control responsibilities in the food safety sector, to include monitoring animal health and welfare. Furthermore, the Commission announced the establishment of an independent[245] food safety authority.[246] At the European summit in Luxemburg at the end of the same year, the European Council adopted a statement on food safety.[247]

The Commission kept the pressure on beyond 1997, eventually gaining the support of the Court of Justice for the measures that had been taken against Great Britain

[241] G. Knudsen and M. Matikainen-Kallström, Joint Parliamentary Committee Report on food safety in the EEA, 1999; European Parliament Fact Sheets 4.10.1. Consumer Policy: principles and instruments, Chapter 3 Reform in the wake of the BSE crisis, Green Paper on general principles of food law.

[242] Interestingly, DG Industry was the instigator.

[243] Commission Green Paper on the General Principles of Food Law in the European Union, COM (1997) 176.

[244] See generally European Commission's Health and Consumer Protection DG, FVO at Home and Away, 7 Consumer Voice September 2002 and Health & Consumer Voice November 2007, Safer Food for Europe – over a decade of achievement for the FVO.

[245] It seems that right now more faith is placed in autonomous technocratic agencies than in politically responsible administrative bodies.

[246] Communication of the European Commission, Consumer Health and Food Safety COM(97) 183 fin. of 30 April 1997. See also: P. James, F. Kemper and G. Pascal, A European Food and Public Health Authority. The future of scientific advice in the EU, 1999.

[247] Conclusions of Luxembourg Summit 1997, 13 December 1997, nr. 57.

at the climax of the crisis.[248] Meanwhile, public attention had turned to a new food safety scare: the Belgian dioxin crisis. The Commission proved it had learned valuable lessons from its experience with BSE, and moved quickly and efficiently to protect consumers from the dioxin crisis. Nonetheless, this second crisis brought to light further shortcomings in European food law.

Despite the resignation of Santer's Commission (which was succeeded by the Commission led by Romano Prodi), food safety remained a priority issue. On 12, January 2000 the Commission published its famous White Paper on Food Safety.[249]

7.5 The White Paper: a new vision on food law

The Commission's vision on the future shape of EU food law, its blueprint so to speak, was laid down in the White Paper on Food Safety. Before the BSE crisis, European food safety law was subordinated to the development of the internal market. The shortcomings in the handling of the crisis clearly revealed a need for a new, integrated approach to food safety. The Commission aimed to restore and maintain consumer confidence. While the White Paper nowhere mentions it in so many words, this includes restoring the trustworthiness of the institutions concerned. To achieve this the Commission proposed an ambitious legislative programme. Eighty-four legislation and policy initiatives[250] were scheduled for the near future, and implementation commenced with unprecedented drive.

The White Paper focused on a review of food legislation in order to make it more coherent, comprehensive and up to date, and to strengthen enforcement. Furthermore, the Commission backed the establishment of a new European Food Safety[251] Authority, to serve as the scientific point of reference for the whole Union, and thereby contribute to a high level of consumer health protection.

7.5.1 Planning a European Food Safety Authority

The establishment of an independent European Food (Safety) Authority was considered by the Commission to be the most appropriate response to the need to guarantee a high level of food safety. According to the White Paper, this Authority should be entrusted with a number of key tasks embracing independent scientific advice on all aspects relating to food safety, operation of rapid alert systems, communication and dialogue with consumers on food safety and health issues as well as networking with national agencies and scientific bodies. The European Food

[248] See Case C-157/96, The Queen v. Ministry of Agric., Fisheries and Food; Case C-180/96, U.K. Gr. Brit and N. Ir. v. Commission, 1996; Case C-209/96 U.K. Gr. Brit and N. Ir. v. Commission.
[249] COM(1999) 719 def. Unlike a Green Paper that is intended mostly as a basis for public discussion, a White Paper contains concrete policy intentions.
[250] Listed in the Annex to the White Paper. This Annex bears the title Action Plan on Food Safety.
[251] In the White Paper the Commission speaks of a European Food Authority. The word 'safety' was inserted later in the General Food Law. See hereafter.

(Safety) Authority had to provide the Commission with the necessary scientific evaluation of risks associated with a product or technology, but it would remain the Commission's responsibility to decide on the appropriate response to that scientific evaluation.[252]

7.5.2 Planning new food safety legislation

Besides setting up the independent Authority, the White Paper called for a wide range of other measures to improve and bring coherence to the corpus of legislation covering all aspects of food products from 'farm to fork'.

The Commission identified measures necessary to improve food safety standards. Considering the developments described above, it is clear that, in a number of areas, existing European legislation had to be brought up to date. A new legal framework was proposed to cover the whole of the food chain, including animal feed production, the establishment of a high level of consumer health protection and clearly attributed primary responsibility for safe food to industry, producers and suppliers, including appropriate official controls at both national and European level. The ability to trace products through the whole food chain was considered a key issue.

The use of scientific advice should underpin food safety policy, whilst the precautionary principle[253] shall be used where appropriate. The ability to take rapid, effective, safeguarding measures in response to health emergencies throughout the food chain was recognised as an important element.

Proposals for the animal feed sector were to ensure that only suitable materials would be used in its manufacture, and that the use of additives would be more effectively controlled. Certain food quality issues, including food additives and flavourings and health claims, should be addressed, and controls over novel foods improved.[254]

7.5.3 Planning improvement of food safety controls

The experience of the Food and Veterinary Office, the Commission's own inspection service, which visits Member States on a regular basis, has shown that there are wide variations in the manner in which Union legislation is implemented and enforced. According to the White Paper, this means that consumers cannot be sure of receiving the same level of protection across the Union, which makes

[252] As we have seen in Chapter 5, the European Food Safety Authority was called into being in Regulation 178/2002.
[253] This principle indicates caution in the face of scientific uncertainty. See Section 9.7.
[254] Two years after the publication of the White Paper new pieces of legislation started to appear in the Official Journal. The flow continues to this day.

it difficult to evaluate the effectiveness of national measures. The White Paper proposed to create a European framework for the development and operation of national control systems, in co-operation with the Member States. This would take account of existing best practices, and the experience of the FVO, and be based on agreed criteria for the performance of these systems, leading to clear guidelines on their operation.

In support of Union level controls, more rapid, easier-to-use, enforcement procedures in addition to existing infringement actions were to be developed. Controls on imports at the external borders of the Union should be extended to cover all feed and foodstuffs, and action taken to improve co-ordination between inspection posts.[255]

In the White Paper national authorities are held responsible for ensuring that food safety standards are respected by food and feed business operators. They need to establish control systems to ensure that Union rules are being respected and, where necessary, enforced. In the opinion of the Commission these systems should be developed at a Union level, so that a harmonised approach is followed. To ensure that national control systems are effective, the Commission, through the FVO, carries out a programme of audits and inspections to evaluate the performance of national authorities against their ability to deliver and operate effective control systems, and are supported by visits to individual premises to verify that acceptable standards are actually being met. The Commission envisioned a Union framework consisting of three core elements:

1. Operational criteria set up at Union level, which national authorities would be expected to meet. These criteria would form the key reference points against which the competent authorities would be audited by the FVO, thereby allowing it to develop a consistent, complete, approach to the audit of national systems.
2. The development of Union control guidelines. These would promote coherent national strategies, and identify risk-based priorities and the most effective control procedures. A Union strategy would take a comprehensive, integrated, approach to the operation of controls. These guidelines would also provide advice on the development of systems to record the performance and results of control actions, as well as setting Union indicators of performance.
3. The enhancement of administrative cooperation in the development and operation of control systems. There would be a reinforced Union dimension to the exchange of best practice between national authorities. This would also include promoting mutual assistance between the Member States by integrating and completing the existing legal framework.

Furthermore, this would cover issues such as training, information exchange and longer term strategic thinking at Community level.[256]

---

[255] In Chapter 15 the regime on enforcement is discussed.

[256] White Paper, p. 30.

7.5.4 Planning improvement of consumer information

If consumers are to be satisfied that the actions proposed in the White Paper will lead to a genuine improvement in food safety standards, they must be kept well informed. The Commission, together with the new European Food Safety Authority, will promote a dialogue with consumers to encourage their involvement in the new food safety policy. At the same time, consumers need to be kept better informed of emerging food safety concerns, and of risks to certain groups from particular foods.

Consumers have the right to expect information on food quality and constituents that is helpful and clearly presented, so that informed choices can be made. Proposals on the labelling of foods, building on existing rules, have been brought forward in line with the White Paper proposals.[257] The importance of a balanced diet, and its impact on health, should also be presented to consumers.

7.5.5 International dimension

The European Union is the world's largest importer and exporter of food products. The actions proposed in the White Paper need to be effectively presented and explained to the EU's trading partners. An active role for the Union in international bodies is an important element in explaining European developments in food safety.

7.6 Intermezzo: documents

In this chapter we encountered a horde of different types of documents, with different legal relevance. For clarification Diagram 7.6 gives an overview.

[257] The principle of informed choice is discussed in Chapter 14.

Diagram 7.6. Types of documents within the Community.

Policy documents preceding legislation	**Legal acts mentioned in Article 288 TFEU**				**Policy documents following legislation**
Green paper *Discussion oriented*			Proposal *Usually from the Commission to the Parliament and the Council*	Proposal *Usually from the Commission to the Parliament and the Council*	Guideline *Interpretation of specific acts*
White Paper *Decision oriented*	Recommendation/opinion *Not binding*	Decision *Individually binding*	Directive *Binding on Member States*	Regulation *Generally binding on all addressees*	

Legal acts are defined in Article 288 TFEU:

- A *regulation* shall have general application. It shall be binding in its entirety and directly applicable in all Member States.
- A *directive* shall be binding, as to the result to be achieved, upon each Member State to which it is addressed, but shall leave to the national authorities the choice of form and methods.
- A *decision* shall be binding in its entirety. A decision which specifies those to whom it is addressed shall be binding only on them.
- *Recommendations* and *opinions* shall have no binding force.
- Policy documents come under various names. No formal definitions apply. The Glossary on the EU website describes Green and White Papers.
- *Green Papers* are documents published by the European Commission to stimulate discussion on given topics at European level. They invite the relevant parties (bodies or individuals) to participate in a consultation process and debate on the basis of the proposals they put forward. Green Papers may give rise to legislative developments that are then outlined in White Papers.
- Commission *White Papers* are documents containing proposals for Community action in a specific area. In some cases they follow a Green Paper published to launch a consultation process at European level. When a White Paper is favourably received by the Council, it can lead to an action programme for the Union in the area concerned.
- After a piece of legislation has entered into force, sometimes the Commission publishes *Guidelines* to explain how this legislation is understood by the Commission and to help other stakeholders to deal with it.
- *Communication* is a name used for a wide variety of policy documents both preceding and following legislation.

- The Green Paper and the White Paper discussed above are not law. They are not binding on anyone, but they do contain the intention to create a lot of legislation.

7.7 Follow up

The Annex to the aforementioned White Paper is the Action Plan on Food Safety, a list of 84 legislative steps that the Commission deemed necessary to create a regulatory framework capable of ensuring a high level of protection of consumers and public health. The turn of the millennium saw the beginning of the planned overhaul of European food law. The first new regulation took effect in 2002 and within a decade most of the 84 steps have been taken.[258] The new regulatory framework is based on regulations rather than directives.

Only two years after the White Paper was published, the cornerstone of new European food law was laid: Regulation 178/2002 of the European Parliament and of the Council of 28 January 2002 laying down the general principles and requirements of food law, establishing the European Food Safety Authority and laying down procedures in matters of food safety. This regulation is often referred to in English as the 'General Food Law'. The Germans speak of it as a 'Basisverordnung' – perhaps a more precise phrase given that the regulation is in fact the basis upon which European and national food laws are now being re-constructed.[259] The notion of a basic regulation has been discussed in Chapter 4. The main objective of the General Food Law is to secure a high level of protection of public health and consumer interests with regard to food products. It does so by stating general principles, establishing the European Food Safety Authority and giving procedures to deal with emergencies.

The General Food Law will be discussed in more detail in the Chapter 9. After the General Food Law, whole packages of new legislation followed. See Diagram 7.7 for an overview.

---

[258] See: K. Knipschild, Lebensmittelsicherheit als Aufgabe des Veterinär- und Lebensmittelrechts (diss.) Nomos Verlagsgesellschaft Baden-Baden, Germany, 2003; U. Nöhle, Risikokommunication und Risikomanagement in der erweiterten EU, ZLR 3/2005, pp. 297-305 and G. Berends and I. Carreno, Safeguards in food law – ensuring food scares are scarce, European Law Review 2005, pp. 386-405.

[259] Modern European food law displays several characteristics in which it is different from its predecessor: more emphasis on horizontal regulations (than on vertical legislation), more emphasis on regulations that formulate the goals that have to be achieved, so-called objective regulations, than on means regulations, increased use of regulations (rather than directives) and thus increasing centralisation.

Diagram 7.7. Highlights in the overhaul of EU food law.

2002	Regulation 178/2002 (GFL)	Chapter 90
2003	GMO package	Chapter 11
2004/2005	Regulations 852-854/2004 Hygiene package	Chapters 13 and 15
	Regulation 882/2004 Official controls	Chapter 15
	Regulation 1935/2004 Food contact materials	Section 11.8
2006/2007	Regulation 1924/2006 nutrition & health claims	Chapter 14
2007	White Paper A Strategy for Europe on Nutrition, Overweight and Obesity related health issues	Chapter 20
2008	Regulation 1331-4 on additives, flavourings and enzymes	Chapter 10
2011	Regulation 1169/2011 on Food information to consumers	Chapter 14
Ongoing	Nutrition and obesity policy	Chapter 20
	Modernisation legislation on novel foods	Chapter 10
	Modernisation of legislation on official controls and enforcement	Chapter 15

7.8 Future

It is next to impossible to predict how long we will remain in the third phase of EU food law and what will come afterwards. The window of opportunity for large-scale legislative projects on food that opened after the animal health and food safety scares of the 1990s seems to have closed. Some finalising proposals are underway. If no major crisis sparks new action, it seems unlikely that more legislation of a fundamental nature will be undertaken in the near future. The EU legislator may instead feel prompted to make an attempt at simplification and reduction of burdens for the food sector.

A pressing issue likely to remain on the agenda for years to come is overweight and obesity. So far the EU legislator has not found suitable instruments to deal with this problem. Measures are currently limited to providing consumers with information directly and on food product labels.

The melamine crisis in 2008 and the horse meat scandal in 2013 have shown that the current system of food safety law is designed to deal will food safety accidents. It does not seem very well prepared to deal with criminal intent. It may well be that this will result in new repressive instruments and approaches targeting people rather than problems.[260] At the moment of writing, however, this is mere speculation. What does seem certain is that food fraud is receiving increased attention.[261]

---

[260] See for an indication in this direction: Gilles Boin, Horsemeat Crisis about to Tighten French Food Law, EFFL 2013, pp. 247-249.

[261] See generally on this topic: Maree Gallagher and Ian Thomas, Food Fraud: The deliberate Adulteration and Misdescription of Foodstuffs, EFFL 2010, pp. 347-353.

II. Systematic analysis of food law

8. Systematic analysis of food law

Bernd van der Meulen

8.1 Introduction

From the developments discussed in the previous chapter, the food sector emerged as one of the most heavily regulated economic sectors in the European Union. At one point the European Commission's DG Enterprise and Industry (DG ENTR) took the position that the food sector had become the third most heavily regulated economic sector in the European Union.[262] It is not immediately apparent how one would measure the extent to which a sector of industry has been regulated. However, a good first step could be to simply count the number of official publications that relate to the sector at issue. Until its revision in the spring of 2014, the official database of EU legislation – Eurlex[263] – featured a thesaurus (Eurovoc) that grouped the included publications by domain.

The most recent date available in the Eurlex archive is the 1st of April 2014. On that date with 15,278 hits (out of 70,901 for the category Agri-Foodstuffs[264]) the subcategory foodstuffs[265] was the biggest, far ahead of sectors such as chemistry[266] with 8,849 (out of 40,429 for industry[267]). Some of these 15,278 documents consist of just a few pages, others of several dozens or even more than a hundred.[268]

Whatever the exact figures are, the conclusion is inevitable that European food law encompasses a larger quantity of sources than can possibly be grasped by simply reading them all. In other words, European food law cannot be 'known' in its entirety.[269]

However, as with legal studies in general, 'knowing' as in memorising all component parts is not the objective of this book. To understand food law, we need to analyse it. That is to break it down into components small enough and sufficiently homogenous to be characterised and scrutinised as elements. These elements in their relation

[262] After automobiles and chemicals. European Commission, Enterprise Directorate General. 'Invitation to Tender No. ENTR/05/075 'Competitiveness Analysis of the European Food Industry', 2005.

[263] On Eurlex see Annex 1.

[264] Eurovoc category 60.

[265] Eurovoc subcategory 6026.

[266] Eurovoc subcategory 6811.

[267] Eurovoc category 68.

[268] The consolidated version of Regulation 396/2005 on maximum residue levels of pesticides in or on food and feed of plant and animal origin, for example is no less than 1,849 pages long.

[269] Our efforts can be compared with that of a chess player. Even though the number of positions and moves on a chessboard must be finite given the fact that there is a fixed number of squares and pieces, this number is somewhere out in the billions and billions. The chess player cannot make it her or his business to know all possible moves and positions, but only to understand patterns and connections.

to and interaction with each other – in their structure so to speak – represent the system of European food law. The framework of analysis applied in this part of the book, has been set out in Diagram 8.1. The Diagram shows several layers of groupings of elements.[270]

8.2 A structure of food law as framework of analysis

The system of European food law as set out in this chapter and used as the basis for the structure of this book does not – at least not entirely – relate to a blueprint that has consciously been applied in creating the legislation, but is analytically superimposed on a situation that has grown organically,[271] to help make sense of it. It consists of the common features and typical characteristics identified through scholarly analysis in a host of legislation and other sources of law.

We organised the provisions on the basis of 'who, what and how' questions. That is to say, who do the provisions address, which problem do they take on, and how do they deal with the problem in terms of rights and obligations they assign to the addressees? We have taken the pattern emerging from this analysis to design a framework that enables researchers, students and practitioners to understand EU food law. Understanding here means to comprehend EU food law in its totality at an aggregate level. This framework can be used for research and teaching but also to acquire an understanding of EU food law with a view for application in legal practice.[272] The framework is present in graphic form in Diagram 8.1 and explained here below.

In designing the graphic representation we have included a few didactic triggers such as:

- ABC for the stakeholders: authorities, businesses and consumers;
- Words with a 'p': powers, principles, product, process, presentation and premises;
- Powers with an 'e': executive, enforcement, emergency and incident management.

---

[270] This framework has gradually developed. See: B.M.J. van der Meulen and M. van der Velde, Food Safety Law in the European Union, Wageningen Academic Publishers, Wageningen, the Netherlands, 2004, p. 146; B.M.J. van der Meulen and M. van der Velde, European Food Law Handbook, Wageningen Academic Publishers, Wageningen, the Netherlands, 2008, p. 251. We discussed the concept as such in B.M.J. van der Meulen, The system of food law in the European Union, 2009, Deakin Law Review 14(2), 305-339 and in B.M.J. van der Meulen, The Structure of European Food Law, Laws 2013, 2, pp. 69-98. We have taken it beyond a framework for analysis in: I. Scholten-Verheijen, T. Appelhof, R. van den Heuvel and B.M.J. van der Meulen, Roadmap to EU Food Law, Eleven international publishing/ Sdu, The Hague, the Netherlands, 2011.

[271] As set out in Chapter 7.

[272] In our experience, food regulatory affairs managers in particular feel more confident for better understanding the context of the rules they work with on a daily basis.

Diagram 8.1. Framework for the analysis of EU food law.

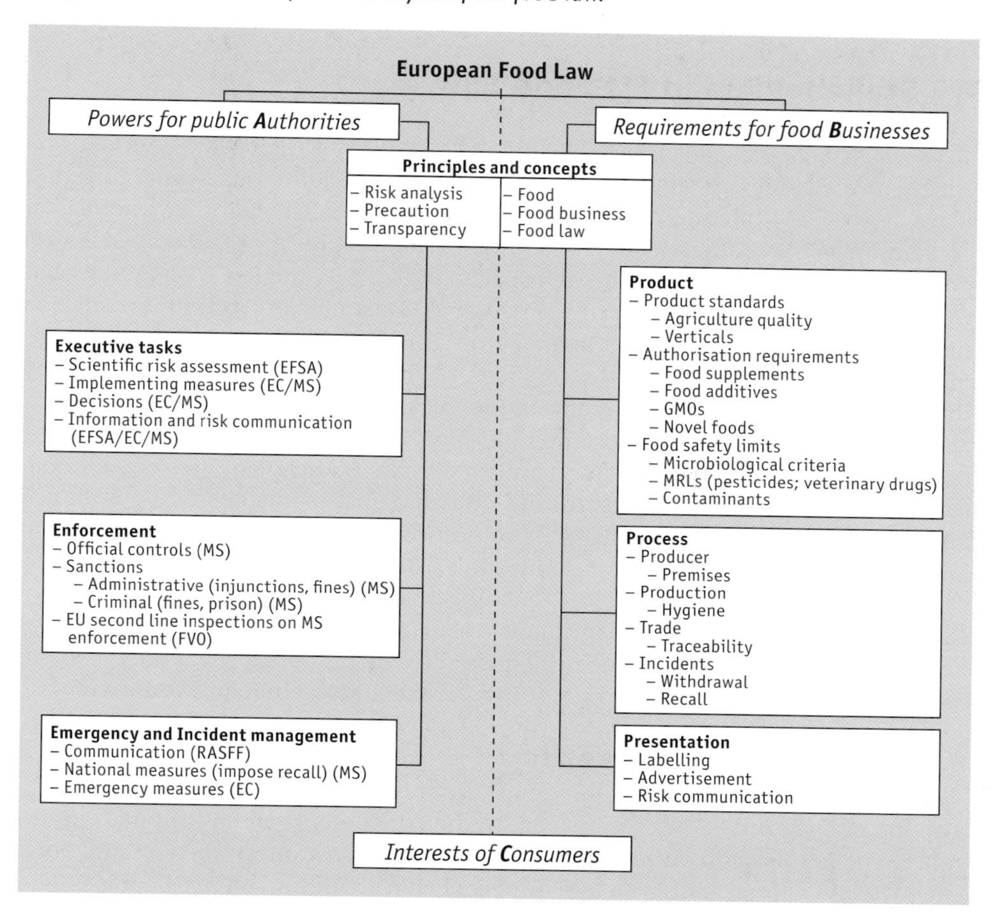

As explained below, in some situations other words might have captured the meaning slightly more accurately, but we believed that the chosen words will be easiest for students to remember.

In the Diagram, three layers can be distinguished that reflect the 'who, what and how' questions. The first level ('who') is visible in the three branches of the Diagram grouping the picture into provisions that address public authorities, food businesses and consumers. The second layer ('what') is represented by the blocks within the branches showing that authorities are addressed in terms of execution, enforcement and incident management, and businesses in terms of product, process and presentation. The third layer ('how') is shown within the blocks. For example, the block on the right-hand side on 'product' shows that the legislator deals with products by setting standards for them, by requiring authorisation and by setting limits on levels of contamination. Beyond this third layer we find the specific provisions of EU food law.

8.3 Stakeholders in EU Food Law

To which stakeholders does food law apply? The 'ABC' of EU food law is its focus on *A*uthorities, *B*usinesses, and *C*onsumers. The three are addressed in very different ways, however. While the protection of the life and health and other interests of consumers is the main objective of food law (see Chapter 9), EU food legislation does not provide consumers with any specific rights or remedies. Consumers that want to take legal action must rely on general consumer protection law such as product liability legislation (see Chapter 16). The consumers are shown at the basis of Diagram 8.1. The key to food safety is in the hands of the businesses handling the food. The most important requirements regarding food have the businesses as addressees. These are shown on the right-hand side of the Diagram. Obligations of public authorities – both at Union and at Member State level – are secondary to the obligations of businesses. Authorities have to ensure businesses' compliance and they have to deal with situations of non-compliance. The powers of public authorities are shown at the left-hand side of Diagram 8.1.

One could argue that the EU legislator is also addressing itself and the national legislators active in food law. The principles and concepts depicted at the top apply as much to the legislature as to the (other) stakeholders, indeed even more so.

8.4 Principles and concepts

The top of Diagram 8.1 shows principles and concepts. Food law applies certain general principles and definitions. Most of these are set out in the General Food Law. The most important principles and concepts are discussed in Chapter 9. The Diagram is not exhaustive but only shows some selected examples. One concept is of special significance. This is the concept of food safety.

8.5 Obligations of businesses

The quality of the food – in the widest sense of the word – largely depends on the businesses dealing with it. Therefore, it should not come as a surprise that most of the requirements of food law have food business operators as their addressee. These requirements are shown at the right-hand side of Diagram 8.1. We have subdivided them into three main categories: requirements on the product, requirements on the process and requirements on the communication (or presentation). The three categories address, in other words, what the product has to be, what the business has to do and what the business has to say.

8.5.1 Product requirements

The requirements food law holds regarding the food as such, that is to say the product, can be further subdivided into product standards, authorisation requirements and food safety limits. Product standards are the rules that dictate how a product should be. These are also known as verticals. They have been mentioned in Chapter 7 Section 7.2. We will also touch upon them in Chapter 14.However, they have not been included to any great degree in this book.

Some products may not be used in foods unless they have been explicitly authorised. This approach is discussed in Chapter 10.

The next approach covered in this book, is where the legislator sets limits to the presence of certain substances or micro-organisms in foods. Here the requirements are often expressed in numbers. Chapter 11 discusses limits to chemical and physical contaminants, Chapter 12 limits to the presence of micro-organisms.

8.5.2 Process requirements

The EU food legislator has not only regulated the qualities a food should possess as such, but also what has to be done to achieve this. The process requirements are subdivided into prevention (hygiene including the business' premises), preparedness and response. These topics are treated in Chapter 13.

8.5.3 Communication requirements

The third major group of requirements on food businesses regards the communication towards consumers. This group includes labelling and advertisement discussed in Chapter 14 and risk communication addressed briefly in Chapter 13.

8.6 Powers of public authorities

The left-hand side shows powers of public authorities. These authorities can be European and national. The institutions of the EU have been introduced in Chapter 5. Many of the powers are implicitly touched upon in the discussion of the obligations of businesses. For example, in the context of authorisation requirements discussed in Chapter 10, risk assessment by EFSA is discussed and the power of the European Commission to take decisions. Chapter 15 specifically addresses the powers of law enforcement and incident management.

8.7 Choice

The framework of analysis set out in this chapter is leading in the organisation of this book, and in particular of this second part of the book. It provides the logic in the way issues have been chosen as topics or as sub-topics. This has led us to certain choices regarding the analysis in this book that arguably could have been and by others probably would have been and actually are made differently.

In our book, for example, genetically modified organisms are a sub-topic to three different topics: authorisation (in Chapter 10), traceability (in Chapter 13) and labelling (in Chapter 14). If the analysis had been based on separate pieces of legislation (such as Regulation 1829/2003) or on political attention, a separate chapter would definitely have been dedicated to GMOs where these three aspects would have been discussed together.[273]

Another consequence, for example, of the choice to organise provisions on the basis of whom they address is that incident management obligations on food business operators and incident management powers for public authorities are treated in different places in this book (Chapters 13 and 15 respectively). If an incident occurs, businesses and authorities usually work closely together and implement the applicable rules in close connection with one-another.

The reader is invited to use the framework wisely as a tool, but not as a cage.

[273] As it is, for example, in: E. Sirsi, GM food and feed, in: L. Costate and F. Albisinni (eds.) European Food Law, CEDAM Wolters Kluwer, Padua, Italy, 2012, pp. 337-350; C. MacMaolain, EU Food Law: Protecting Consumers and Health in a Common Market, Hart Publishing Oxford, UK, 2007, pp. 241-263; D. Holland and H. Pope, EU Food Law and Policy, Kluwer Law International, 2004, pp. 110-120, and; R. O'Rourke, European Food Law. 3rd Edition, Sweet & Maxwell, London, UK, 2005, pp. 149-166.

9. The General Food Law: general provisions of food law

Bernd van der Meulen and Anna Szajkowska

9.1 The General Food Law

The first step in the realisation of the reform of food law as planned in the White Paper on Food Safety was the passage of *Regulation (EC) No 178/2002 of the European Parliament and of the Council of 28 January 2002 laying down the general principles and requirements of food law, establishing the European Food Safety Authority and laying down procedures in matters of food safety*. The popular name of this regulation is the General Food Law (GFL). As is apparent from the regulation's official title, the General Food Law seeks to accomplish three objectives:[274]
1. To lay down the principles on which modern food legislation should be based in the European Union as well as in the Member States.
2. To establish the European Food Safety Authority.
3. To establish procedures for food safety crises, including the so-called Rapid Alert System.

To avoid confusion, it should be noted that the General Food Law is not a code encompassing all food legislation. It is the fundament to a general part of food law.[275] Next to it many (dozens, even hundreds) of other European and national rules and regulations continue to play their role.

This chapter treats the general provisions, laid down in the first two chapters of the GFL. The rest of the GFL, the institutional aspects and incident management, are covered in Chapter 5 Sections 5.9-5.11 and Chapter 15.

9.2 Aim and scope

The General Food Law provides a general framework for both national and European Union food law in the EU. In Article 5 – discussed here below – the GFL gives the objectives of these national and European provisions of food law. In Article 1 the GFL gives its own aim and scope (Law text box 9.1).

[274] These are the first three of the eighty-four legislative initiatives mentioned in the Action Plan on Food Safety, the Annex to the White Paper on Food Safety.
[275] The role the General Food Law plays with regard to food law is comparable to the role a constitution plays in a legal order. It must be stressed, however, that, technically speaking, the General Food Law is a regulation, and it is not given a higher status than other EU regulations or directives. Only by its purpose and wide scope of application does the General Food Law acquire its specific position in the food law system.

Law text box 9.1. Article 1 GFL on aim and scope of the GFL.

Article 1 GFL
Aim and scope

1. This Regulation provides the basis for the assurance of a high level of protection of human health and consumers' interest in relation to food, taking into account in particular the diversity in the supply of food including traditional products, whilst ensuring the effective functioning of the internal market.
   It establishes common principles and responsibilities, the means to provide a strong science base, efficient organisational arrangements and procedures to underpin decision-making in matters of food and feed safety.
2. For the purposes of paragraph 1, this Regulation lays down the general principles governing food and feed in general, and food and feed safety in particular, at Community and national level.
   It establishes the European Food Safety Authority.
   It lays down procedures for matters with a direct or indirect impact on food and feed safety.
3. This Regulation shall apply to all stages of production, processing and distribution of food and feed. It shall not apply to primary production for private domestic use or to the domestic preparation, handling or storage of food for private domestic consumption.

There are some basic notions in this article that have already been mentioned:

- The principles and responsibilities are common to all food law in the EU.
- The principles and responsibilities are common to all stages of food and feed production and distribution ('from farm to fork').[276] This is sometimes called the 'holistic' approach to food law. Only primary production for private domestic use or the domestic preparation, handling or storage of food for private domestic consumption is exempted from the scope of application of the General Food Law.

9.3 Definitions

The first step that the GFL takes in integrating the whole body of national and Union food law in the EU is the development of a common 'language', by providing definitions for the most important notions. It is important to note that the definitions are given 'for the purpose of this Regulation'. As a consequence the GFL definitions do not apply automatically in the legislation of the Member States – although the GFL is a regulation – nor in other pieces of Union legislation. The legislators will have to ensure that the GFL definitions are introduced in all relevant legislative acts. An example can be found in the Hygiene Regulation for food of animal origin (853/2004) (Law text box 9.2).

[276] This farm to fork continuum is also referred to as a 'stable to table' or 'plough to plate' approach.

Law text box 9.2. Inclusion of GFL definitions in Regulation 853/2004.

Article 2
Definitions
The following definitions shall apply for the purposes of this Regulation:
1. the definitions laid down in Regulation (EC) No 178/2002;
2. the definitions laid down in Regulation (EC) No 852/2004;
3. the definitions laid down in Annex I;
4. any technical definitions contained in Annexes II and III.

Let food be thy medicine
and let thy medicine be food.
Hippocrates[277]

9.3.1 Food

The definition of 'food' provided in the second article of the General Food Law is essential (Law text box 9.3).[278] Its fulfilment is a precondition for the applicability of the GFL. If a product meets this definition, it is a food in the sense of the GFL and the GFL applies to it. The same holds true for all the other laws and regulations that use this definition. In due course that should be the whole body of food law in the European Union and its Member States.

[277] As will be shown in this section, EU food law applies a strict separation of food from medicine. This is at odds with other approaches found in several countries, for example in Asia where the concepts overlap. Also Hippocrates, the semi-mythical father of medical sciences, is quoted as representing an opinion very different from current European food law. Interestingly, it has been argued that Hippocrates never actually said such a thing. See: D. Cardenas, Let not thy food be confused with thy medicine: Hippocratic misquotation, e-SPEN Journal 8 (2013), pp. e260-e262. In the current chapter, it is shown that the legal definition of food strictly separates it from medicine. As is discussed in Chapter 14, this separation is further reinforced by behavioural requirements. It is strictly forbidden to claim that a food can cure a certain disease. No exemption from this ban is made in situations where the curative properties of a food are a scientifically established fact. In this way, the delineation of food law from pharmaceutical law is upheld with a rigorous restriction on free speech. The discussion, if this restriction is justified, is outside the scope of this book. For the role of the human right to free speech in food law, see: B.M.J. van der Meulen and E.L. van der Zee, 'Through the wine gate'. First steps towards human rights awareness in EU food (labelling) law, EFFL 2013, pp. 41–52.
[278] Surprisingly, although the European legislator had been very active in the field of food law, the term 'food' was defined for the first time in the 2002 General Food Law. The GFL does not distinguish between food and food ingredients as some older legislation does. Ingredients fulfil the definition of food and are (therefore) subject to the same safety rules. Only in labelling legislation does the distinction still have significance. See Chapter 14.

Law text box 9.3. Definition of 'food' in Article 2 General Food Law.

Article 2
Definition of 'food'

For the purposes of this Regulation, 'food' (or 'foodstuff') means any substance or product, whether processed, partially processed or unprocessed, intended to be, or reasonably expected to be ingested by humans.

'Food' includes drink, chewing gum and any substance, including water, intentionally incorporated into the food during its manufacture, preparation or treatment. It includes water after the point of compliance as defined in Article 6 of Directive 98/83/EC and without prejudice to the requirements of Directives 80/778/EEC and 98/83/EC.

'Food' shall not include:

(a) feed;
(b) live animals, unless they are prepared for placing on the market for human consumption;
(c) plants prior to harvesting;
(d) medicinal products within the meaning of Council Directives 65/65/EEC and 92/73/EEC;
(e) cosmetics within the meaning of Council Directive 76/768/EEC;
(f) tobacco and tobacco products within the meaning of Council Directive 89/622/EEC;
(g) narcotic or psychotropic substances within the meaning of the United Nations Single Convention on Narcotic Drugs, 1961, and the United Nations Convention on Psychotropic Substances, 1971;
(h) residues and contaminants.

A product that may be used as a foodstuff, e.g. palm oil, has to meet the food law requirements, until it is made clear that it will be used for other purposes. This category of substances is included in the definition as a substance that is 'reasonably expected' to be digested by humans.

Live animals included in the definition of food, are – in the European culture – mainly oysters. All other animals are generally slaughtered (killed) before they are eaten.

Case law that has evolved so far mainly focuses on the delineation of food and pharmaceuticals (medicinal products). The directive currently applicable to medicinal products is Directive 2001/83.[279] This directive gives two definitions of medicinal product, one 'by presentation' and one 'by function'. Under the first subparagraph of Article 1(2) of Directive 2001/83, a medicinal product is '[a]ny substance or combination of substances presented for treating or preventing disease in human beings', and according to the second subparagraph thereof, '[a]ny substance or combination of substances which may be administered to human

[279] Directive 2001/83 repeals these directives, and states in Article 128: 'References to the repealed Directives shall be construed as references to this Directive'. It is unfortunate that the legislator has not bothered to update the text of the GFL in this regard.

beings with a view to making a medical diagnosis or to restoring, correcting or modifying physiological functions in human beings' is likewise to be considered a medicinal product.

A product is a medicinal product if it falls within either of those definitions. If it is a medicinal product, it drops out of the definition of food.[280] Unfortunately, in a case concerning the import into Germany of a product sold as food supplement in the Member State of origin but classified as medicine in Germany, the Court had to decide that harmonisation of pharmaceutical law is not sufficiently complete to avoid the existence of differences in the classification of products as medicinal products or foodstuffs between Member States. Thus, the fact that a product is classified as a foodstuff in another Member State cannot always prevent it from being classified as a medicinal product in the Member State of importation, if it displays the characteristics of such a product.[281] For food law, this has the unfortunate consequence that, notwithstanding the common definition of food, it remains possible that a product is considered a food in one Member State and a medicine in another.

In a second case, also concerning Germany, the Court took the view that the German government had overstepped its margin of discretion by classifying garlic concentrates as medicines on the grounds that garlic has beneficial effects on health. The Court held that the product concerned, whose effect on physiological functions is no more than the effects a foodstuff consumed in a reasonable quantity may have on those functions, does not have a significant effect on the metabolism. Therefore, it cannot be classified as a product capable of restoring, correcting or modifying physiological functions within the meaning of the second subparagraph of Article 1(2) of Directive 2001/83.[282] According to the Court, this qualification of a food as a medicine requiring pre-market approval constituted an infringement of the free movement of goods as protected in (now) Article 34 TFEU.

Food businesses have it within their own power to present their product as having properties to cure (etc.) bringing it within the ambit of pharmaceutical law. As regards the actual properties of the product, Member States enjoy some discretion in classifying products as medicines. However, the beneficial effects of the products must be above the threshold of the effects a normal food can generate (Diagram 9.1). While the yardstick the Court applies is not very precise, at least it becomes

[280] The definition in the Codex Alimentarius (see Chapter 3) applies a less strict separation of food from medicines. According to the Codex 'food' means 'any substance (...) intended for human consumption (...) but does not include (...) substances used *only* as drugs'. (Italics added).
[281] Joined Cases C-211/03, C-299/03 and C-316/03 to C-318/03, HLH Warenvertrieb and Orthica.
[282] CJEU 15 November 2007, Case C-319/05, Commission v. Federal Republic of Germany, paragraph 68.

clear that there is a yardstick and that the interpretation of the concepts of medicine and food may not vary too much between the Member States.[283]

Diagram 9.1. Medicine.

Medicine	By presentation	By function
	Business decides	MS has discretion. Effect should be above normal food consumed in reasonable quantity.

9.3.2 Food law

The General Food Law defines 'food law'. With this definition, it sets the scope for the general principles it contains (Law text box 9.4).[284]

Law text box 9.4. Definition of 'food law' Article 3(1) GFL.

Article 3(1) GFL
Definition of food law
'food law' means the laws, regulations and administrative provisions governing food in general, and food safety in particular, whether at Community or national level; it covers any stage of production, processing and distribution of food, and also of feed produced for, or fed to, food producing animals;

[283] Much has been written on the delineation of the concept of food, in particular from medicines. See for example: B. Klaus, Der gemeinschaftsrechtliche Lebensmittelbegriff. Inhalt und Konsequenzen für die Praxis insbesondere im Hinblick auf die Abgrenzung von Lebensmitteln und Arzneimitteln, Verlag P.C.O. Bayreuth, Bayreuth, Germany (diss.), 2005; A. Titz, The borderline between medicinal products and food supplements, Pharmaceuticals Policy and Law 8 (2005, 2006) pp. 37-49; P. Coppens, The Use of Botanicals in Food Supplements and Medicinal Products: The Co-existence of two Legal Frameworks, EFFL 2008, pp. 93-100; S.R. Melchor and L. Timmermans, 'It's the Dosage, stupid': The ECJ clarifies the Border between Medicines and Botanical Food Supplements, EFFL 2009, pp. 185-191; F. Capelli and B. Klaus, Is Garlic a Food or a Drug? EFFL 2009, pp. 390-399; M. Korzycka-Iwanow and M. Zboralska, Never-ending Debate on Food Supplements: Harmonisation or Disharmonisation of the Law? EFFL 2010, pp. 124-135.

[284] The wording 'food law' is not the same in all language versions of the GFL. The French and the Dutch texts apply expressions best translated into English as 'food legislation' (*législation alimentaire* and *levensmiddelenwetgeving*, respectively). The German version (*Lebensmittelrecht*) is similar to the English text. The difference in meaning is if the word covers the entire system of law, including for example case law or only prescriptive texts. What follows in the GFL definition of food law is an enumeration of such texts. It is a broad enumeration including provisions originating from the legislature and from the executive, from the Community and from the Member States. This may indicate that the French and Dutch wording are closer to the mark than the English and German wording. This book, however, and the underlying academic activity are not bound to any legal interpretation of a definition of food law. It includes case law and administrative practices in its area of interest.

9.3.3 Food business

Other important definitions are those of 'food business' and 'food business operator'. These definitions are important because many provisions address these businesses (operators). If someone meets the definition, s/he has to comply with food law. The definitions are given in Law text box 9.5.

Law text box 9.5. Definitions of 'food business' and 'food business operator', Article 3(2/3) General Food Law.

Article 3(2/3) GFL
Definitions of food business and food business operator.
'food business' means any undertaking, whether for profit or not and whether public or private, carrying out any of the activities related to any stage of production, processing and distribution of food;
'food business operator' means the natural or legal persons responsible for ensuring that the requirements of food law are met within the food business under their control;

The definition of food business is very broad. In the European Parliament the question[285] has been raised as to whether the requirements of traceability (Article 18 GFL) apply to charities that collect and distribute food. From the response of the European Commission[286] it follows that such charities are indeed food businesses to which the requirements of food law apply (national agencies can however be lenient in the enforcement of compliance by charities).

Unfortunately, the definition of food business operator is imprecise. The main function of this definition is to identify the natural or legal person who is responsible for complying with food law. To this effect, see Article 17(1) GFL. By including in the definition that the operator is the one who is responsible, we end up with a perfect circle where the GFL says that responsibility for compliance with food law rests with the one who is responsible.

Fortunately, the intention of the legislator is not as enigmatic as the words. Food business operator is the one who is in charge of the business, usually that would be the owner. This is the person who has the power to take the decisions that are decisive for compliance or non-compliance.

[285] E-2704/04.
[286] E-2704/04DE.

The Hygiene package – in particular Article 6 of Regulation 852/2004 – imposes a general obligation on food business operators to notify the national competent authorities and to register each establishment under its control that carries out any of the stages of production, processing and distribution of food.[287] See Law text box 9.6.

On top of the registration requirement, establishments of food businesses to which Regulation 853/2004 applies (and some others), that are establishments working with food of animal origin, need an approval after an on-site inspection (Diagram 9.2).

Law text box 9.6. Registration and approval of food businesses.

Article 6 Regulation 852/2004
Official controls, registration and approval

1. Food business operators shall cooperate with the competent authorities in accordance with other applicable Community legislation or, if it does not exist, with national law.
2. In particular, every food business operator shall notify the appropriate competent authority, in the manner that the latter requires, of each establishment under its control that carries out any of the stages of production, processing and distribution of food, with a view to the registration of each such establishment.
   Food business operators shall also ensure that the competent authority always has up-to-date information on establishments, including by notifying any significant change in activities and any closure of an existing establishment.
3. However, food business operators shall ensure that establishments are approved by the competent authority, following at least one on-site visit, when approval is required:
   (a) under the national law of the Member State in which the establishment is located;
   (b) under Regulation (EC) No 853/2004; or
   (c) by a decision adopted in accordance with the procedure referred to in Article 14(2).
   Any Member State requiring the approval of certain establishments located on its territory under national law, as provided for in subparagraph (a), shall inform the Commission and other Member States of the relevant national rules.

Diagram 9.2. Registration obligations of food business operators.

Notify	Each establishment for registration
Notify	Significant changes
Ensure approval following on-site visit	Establishments handling food of animal origin

[287] See also Article 31 of Regulation 882/2004. See in more detail on this subject Chapter 15, Section 15.2.2.

9.3.4 Placing on the market

Many provision of food law apply to the 'placing on the market' of food by the food business operator. This concept is used in a rather wide meaning defined in Article 3(8) of the General Food Law (Law text box 9.7).

Law text box 9.7. Definition of 'placing on the market' in Article 3(8) GFL.

Article 3(8) GFL
Definition of placing on the market
'placing on the market' means the holding of food or feed for the purpose of sale, including offering for sale or any other form of transfer, whether free of charge or not, and the sale, distribution, and other forms of transfer themselves;

9.3.5 Competent authority

Part of the sovereignty of the Member States is the so-called 'principle of institutional autonomy'. EU law has little to say about the organisation of the public sector in the Member States. Usually, obligations are conferred to the national 'competent authority'. It is for the national legislature to decide which state organ is its 'competent authority' in any given matter and to endow it with the powers necessary to fulfil its obligations under EU law. In most Member States, food law is in the domain of either the Minister of Agriculture or the Minister of Public Health. Most Member States also have a more-or-less independent food safety authority.

Next to the European Commission, the competent authority in the Member States is the prime addressee of food law at the side of the public authorities. The concept has not been defined in the General Food Law. Regulation 852/2004 does however give a definition that is suitable to be applied in food law in general. See Law text box 9.8.

Law text box 9.8. Definition of 'competent authority' in Article 2(1)(d) Regulation 852/2004.

Article 2(1)(d) Regulation 852/2004
Definition of competent authority
'competent authority' means the central authority of a Member State competent to ensure compliance with the requirements of this Regulation or any other authority to which that central authority has delegated that competence; it shall also include, where appropriate, the corresponding authority of a third country;

In the General Food Law, the word 'Authority' refers to the European Food Safety Authority. In general, the word authority is used in a broader meaning.[288]

9.3.6 Food safety

Even though food safety is the main objective of food law, the GFL does not give a definition of food safety. As we will see here below it does provide for a ban on *un*safe food and explains in that context when food is deemed to be unsafe. This does not mean, however, that we can simply strike the negatives to come to a definition of food safety – as some authors do.[289] Legal texts often apply double negations. This indicates that the legislator does not have a bipolar situation in mind, but a continuum with a grey area between the opposites. In this line of thinking 'not black' does not equal 'white', but leaves space for all kind of shades of grey, red, yellow and blue. All are ok, as long as it is not black. The GFL does not define food safety. It only says that food must not be unsafe.

In contrast to the General Food Law, the Codex Alimentarius[290] does define the concept of food safety (Law text box 9.9).[291] It refers to not causing harm. There is no reason to presume that the notion of food safety is understood differently in EU food law.

Law text box 9.9. Definition of 'food safety' in Codex Alimentarius.

Food safety – assurance that food will not cause harm to the consumer when it is prepared and/or eaten according to its intended use.

What the GFL does define, are the concepts of 'risk' and 'hazard' (Law text box 9.10).

[288] Indeed, in administrative law the word 'authority' is usually reserved for bodies that have been endowed by law with the power to take binding decision. EFSA does not have such power and thus is not an authority in the administrative law sense of the word.
[289] See for example R. Riedl and C. Riedl, Shortcomings of the New European Food Hygiene Legislation from the Viewpoint of a Competent Authority, EFFL, 2008, pp. 64-83.
[290] On the Codex Alimentarius see Chapter 3.
[291] Recommended International Code of Practice General Principles of Food Hygiene, CAC/RCP 1-1969, rev. 4 (2003), p. 7.

Law text box 9.10. Definition of 'risk' and 'hazard' Article 3(9) and 3(14) GFL.

Article 3(9) and 3(14) GFL
Definitions of risk and hazard
'risk' means a function of the probability of an adverse health effect and the severity of that effect, consequential to a hazard;
'hazard' means a biological, chemical or physical agent in, or condition of, food or feed with the potential to cause an adverse health effect;

These concepts help to identify in which situations a food may cause harm and thus is no longer safe. What can go wrong in food is probably in all situations of a biological, chemical or physical nature.

9.4 General principles

The General Food Law is not entirely clear on the meaning it attaches to the word 'principle'. In its second chapter, the General Food Law identifies six 'principles' of food law; four of these address food law as such (i.e. legislation on food) and two address transparency. These principles are followed by three 'general obligations' of food trade. The second chapter ends with eight Articles containing 'general requirements' of food law.

The main distinction between the principles and the general requirements is that the principles mainly address the legislator and the requirements mainly the businesses. The general obligations of food trade address all actors engaged in international trade, businesses and authorities.

In our view the 'general obligations' and the 'general requirements' are of the same basic nature as the 'principles'. For this reason we will call them all 'principles'.[292] These principles are presented in Diagram 9.3. We will discuss some of these principles briefly and elaborate a little bit more on the objectives of food law, risk analysis, the precautionary principle and the food safety principle (or duty of care) because these are the most important for food business operators. The principle on presentation (labelling) will be discussed in Chapter 14 Section 14.5. The responsibilities of food business operators for traceability, withdrawal and recall will be discussed in Chapter 13, Section 13.3.

[292] In the initial Commission proposal, they were included in one section called 'Principles and requirements of food law'.

Diagram 9.3. Principles of food law (Chapter II GFL).

Principle	Content Details
General principles (section 1)	
General objectives (or: Focus) (Art. 5)	Protection of: • human life & health; • consumers interests; • fair practices in food trade; • animal health & welfare, plant health, environment; • free movement of feed & food products; • application of international standards.
Science based (Art. 6)	Based on risk analysis. Risk assessment based on all scientific evidence, independent, objective, transparent. Risk management takes into account results of risk assessment, other legitimate factors, precautionary principle.
Precautionary principle (Art. 7)	If the possibility of harmful effects on health is identified, but scientific uncertainty persists: • provisional risk management measures may be taken; • pending further research; • proportionate; • no more restrictive to trade than required; • with regard to technical and economic feasibility and other legitimate factors; • measures must be reviewed within a reasonable time.
Consumer protection (Art. 8)	• Provide a basis for informed choice; • Prevent fraudulent or deceptive practices; • Prevent adulteration of food; • Prevent other misleading practices.
Principles of transparency (section 2)	
Public consultation (Art. 9)	• The public must be consulted on food legislation.
Public information (Art. 10)	• Risk communication; • The general public has to be informed of suspected risks and of measures taken.
General obligations of food trade (section 3)	
Import (Art. 11)	• Imports must comply with EU food law.
Export (Art. 12)	• Exports must comply with EU food law.
International standards (Art. 13)	• Contribute to development; • Promote international organisations; • Contribute to agreements; • Special attention to needs of developing countries; • Promote consistency.

▶▶

Diagram 9.3. Continued.

Principle	Content	Details
General requirements (section 4)		
Food safety (Art. 14)	• No placing unsafe food on the market. • Applies to entire batch unless proven safe.	Food is unsafe: • injurious to health; • unfit for consumption (contamination, putrefaction, deterioration, decay). Regarding: • normal use; • information. Regarding: • long-term effects; including future generations; • cumulative toxic effects; • sensitive category of consumers. Food complying with food law is deemed safe for the aspects covered: • compliance does not bar authorities from taking measures.
Feed safety (Art. 15)	• No placing unsafe feed on the market; • No feeding unsafe feed to food producing animal.	
Presentation (Art. 16)	• No misleading consumers	
Responsibilities (Art. 17)	• Business operators from farm to fork for following food law; • Member States for enforcement from farm to fork (controls, communication & penalties).	
Traceability (Art. 18)	• From farm to fork; • One step up, one step down; • Operators have systems in place; • Information available for authorities; • Adequate identification & labelling.	
Responsibilities for food (Art. 19)	• Operator, who has reason to doubt safety, must withdraw food; • If food has reached consumer: inform consumer and if necessary recall; • Inform and co-operate with competent authorities.	
Responsibilities for feed (Art. 20)	• Operator who has reason to doubt safety, must withdraw feed; • Inform and co-operate with competent authorities.	
Liability (Art. 21)	• General provisions on product liability apply.	

9.5 Focussed objectives

The GFL provides a limited set of objectives for food legislation.[293] In other words, a principle of focus applies.

Food law shall pursue one or more of the general objectives of a high level of protection of human life and health and the protection of consumers' interests, including fair practices in food trade, taking account of, where appropriate, the protection of animal health and welfare, plant health and the environment.[294]

Next, the free movement of food products is mentioned, as is adherence to international standards.

If we take a closer look at the words connected to the objectives mentioned, we notice subtle differences. The first two are to be pursued, the next to be taken account of, then aim to be achieved and finally to be taken into consideration. See Diagram 9.4. These differences have meaning; they are not just poetic variations on the same theme.

Diagram 9.4. General objectives of food law.

Pursue general objectives (paragraph 1)	Take account of (where appropriate)	Aim to achieve (paragraph 2)	Take into consideration (paragraph 3)
High level of protection of human life and health (elaborated in Arts. 6 and 7) Protection of consumers' interests, including fair practices in food trade (elaborated in Art. 8)	Animal health Animal welfare Plant health Environment	Free movement of compliant food and feed	International standards existing or imminent (except ...)

Article 5 GFL limits the freedom of the legislator. Food law may only pursue the given objectives and no other objectives. The effect of this limitation is greatest for food law in the Member States. National law may not depart from EU law. The EU legislator himself has the option to take other decisions in later legislation. Two objectives are given: protection of human life and health and protection of (other) consumers' interests. The other principles elaborate on these objectives.

[293] Article 5 GFL.

[294] Consumer protection is considered a human right in the EU. Article 38 of the Charter of Fundamental Rights of the European Union reads: Union policies shall ensure a high level of consumer protection.

From Articles 6 and 7 GFL it follows that pursuing protection of human health is seen as a matter in which science has to be involved. Article 8 GFL focuses on the protection of other consumers' interests, for instance, protection from different types of unfair trading practices.

The objective of food (and feed!) law is to protect humans. If alternative ways to protect humans present themselves to the legislator, the interests of animals, plants and the environment can be taken into account. The method that is the most protective or least damaging to animals, plants or the environment can be chosen from different ways of protecting humans. All this should result in food and feed that can be freely traded all over the European Union. Where does the legislator find the alternatives suitable to pursue the objectives of food law? Often in international standards like the Codex Alimentarius that have already been agreed upon or are still in the pipeline. In making food law the legislator must confront himself with the question as to whether the solutions presented are suitable and then either follow them or have good reasons to depart from them.

What about interests not mentioned in Article 5 as objectives to be pursued, achieved, taken into account or taken into consideration? The most striking example of such interests is the interests of food businesses. Even though food businesses are the addressees of most provisions of food law and thus the parties most affected by these provisions, their interests are not to be pursued! The absence of food businesses as subjects to be considered in food legislation can probably be explained as an effect of the BSE crisis which prompted the new approach to food law as laid down in the GFL. As discussed in Chapter 7, Section 7.4, the European Commission was criticised for being excessively influenced by commercial interests. In drafting Article 5, the Commission seems to have gone to the other extreme.

Those interests not mentioned are protected by the principle of proportionality.[295] Stakeholders' interests may not be harmed more than is necessary to achieve the objective pursued.[296]

[295] Recognised in Protocol (2) on the application of the principles of subsidiarity and proportionality to the Lisbon Treaty.

[296] For a detailed discussion of Article 5 GFL see: Bernd van der Meulen, The function of food law. On objectives of food law, legitimate factors and interests taken into account, EFFL 2010, pp. 83-90.

9.6 Risk analysis, risk management and risk communication

9.6.1 Concepts

Food law is science-based or, in the words of the GFL, 'food law shall be based on risk analysis except where this is not appropriate to the circumstances or the nature of the measure'.[297]

This principle is at the core of the new European approach to food law as 'food safety law'. The GFL regards risk analysis as a process[298] consisting of three interconnected components: risk assessment, risk management and risk communication.[299] See Diagram 9.5.

Diagram 9.5. Components of risk analysis.

Activity	Elements	Product
Risk assessment	• Hazard identification • Hazard characterisation • Exposure assessment • Risk characterisation	Advice
Risk communication	• Interactive exchange of information and opinions	Information
Risk management	• Considering risk assessment and other legitimate factors • Including precautionary principle • Consultation with interested parties • Weighing policy alternatives • Selecting appropriate prevention and control options	Decision

The first of these three components, risk assessment, is seen as a scientifically based process consisting of four steps: hazard identification, hazard characterisation, exposure assessment, and risk characterisation.[300] The European Food Safety Authority is the key player in risk assessment. Risk assessors provide information for decision makers (public authorities) to decide whether the risks deriving from a given product or activity are acceptable, and – if need be – to take appropriate measures to mitigate these risks.

[297] Article 6 GFL. There are some areas of food regulation that do not deal with hazards of a biological, chemical or physical nature and they do not require prior scientific risk assessment. An example of such measures are provisions relating to food labelling.
[298] Or, in the words of Recital 17: 'a systematic methodology'.
[299] Article 3(10) GFL.
[300] Article 3(11) GFL.

The second, 'risk management', is a broad concept. It means the process, distinct from risk assessment, of weighing policy alternatives in consultation with interested parties, considering risk assessment and other legitimate factors, and, if need be, selecting appropriate prevention and control options.[301] The options available to risk management encompass legislation, enforcement and everything in between. To a large extent risk management is the domain of the European Commission and the national authorities. To a lesser extent, the same is true for risk communication. Risk management takes into account the results of risk assessment, however in some cases other factors relevant to the matter under consideration should also be taken into account. These so-called 'other legitimate factors' (OLFs) include societal, economic, traditional, ethical and environmental factors and the feasibility of controls. In some cases, where – after the assessment of the available information – scientific uncertainty persists, the precautionary principle may be invoked where the conditions laid down in Article 7 are met (see hereafter Section 8.7.2).

Choosing the level of protection is also an element of risk management. It is stated repeatedly that the EU chooses a high level of protection[302] and is thus likely to choose stringent measures to deal with identified risks.

The third element of risk analysis is risk communication. 'Risk communication' means the interactive exchange of information and opinions throughout the risk analysis process. This includes hazards and risks in themselves, as well as risk-related factors and perceptions of risk among assessors, managers, consumers, feed and food businesses, the academic community and other interested parties. It also includes the explanation of risk assessment findings and the basis of risk management decisions.[303] Risk communication provides the explanation for risk assessment findings and the basis of risk management decisions, and closes the gaps between experts, policy makers and other stakeholders. The BSE crisis in the UK taught a very important lesson about the importance of risk communication, its transparency and openness. Now EFSA can communicate risks on its own initiative in all fields within its mission, except for communicating risk management decisions, which falls within the competence of the Commission. Close cooperation between EFSA, the Commission and the Member States is therefore necessary to ensure coherence of the risk communication process.

While in the ideal situation risk assessment comes before risk management,[304] risk communication plays a role in all the stages. Communication is crucial both

[301] Article 3(12) GFL.

[302] For example Article 1 and 5 GFL.

[303] Article 3(13) GFL.

[304] Many publications represent the three activities in a circular way indicating that risk management decisions give rise to new risk assessment. While this may be true in many cases, risk management is not a legal precondition for risk assessment, while risk assessment is a condition for risk management.

for risk assessment and risk management to reach their intended results and then to disseminate these. Risk communication in the guise of a warning can even be a form of risk management.

9.6.2 Scope

The wording of Article 6 GFL might lead an unsuspecting reader to believe that all legislation on food or at least all legislation aiming to protect human life and health is preceded by scientific risk assessment. In so far as it gives rise to this expectation, Article 6 is a gross overstatement. In fact not one single example can be found of a regulation or directive of a general nature referring to an opinion of EFSA or another form of risk assessment that has been taken into account.

In other words: food legislation is not based on risk analysis. Risk analysis is an instrument of food law in so far as decisions are taken relating to specific foods. In other words, risk analysis is a methodology that is applied in taking decisions in this field. If we take a look at opinions issued by EFSA, they all relate to applications for approval (as discussed in Chapter 10), to the setting of food safety targets (as discussed in Chapters 11 and 12) or to the measures to be taken regarding certain risks. Such measures may be taken in the form of legislation (regulation or directive), they are however of a specific nature and thus in essence are administrative decisions.

This need not be surprising if we take another look at the definition of risk assessment. Hazard identification, hazard characterisation, exposure assessment in particular, and risk characterisation, by their very nature focus on specific products, substances or life forms.

9.7 Precautionary principle

The precautionary principle adds subtlety to the risk-analysis principle. In specific circumstances where, following an assessment of available information, the possibility of harmful effects on health is identified but scientific uncertainty persists, provisional risk management measures necessary to ensure the high level of health protection chosen in the EU may be adopted, pending further scientific information for a more comprehensive risk assessment.[305]

9.7.1 Background

The precautionary principle deals with decision making in the face of scientific uncertainty. In European Union law, the principle appeared for the first time in 1992 in the environmental title of the EC Treaty. However, its scope was not

[305] Article 7 GFL.

limited to environmental issues. The precautionary principle applied to other EU policies, including human, animal and plant health. A famous application of the precautionary measures can be found in the judgements on the validity of the EU measure imposing a worldwide embargo on British beef exports, linked to the outbreak of BSE. In the BSE judgements, the Court of Justice of the EU stated: 'Where there is uncertainty as to the existence or extent of risks to human health, the institutions may take protective measures without having to wait until the reality and seriousness of those risks become fully apparent'. This formulation was repeated in subsequent rulings.

Recourse to the precautionary principle in the area of food safety has been a subject of controversy at the international level. This is because, under the strict scientific regime of the WTO Agreement on Sanitary and Phytosanitary Measures, such a principle - allowing for the adoption of a measure in case of insufficiency of scientific evidence - becomes a very powerful instrument, which can be used as a disguised form of protectionism.

Regardless of the status of the precautionary principle in international law, the European Union has consequently advocated its application to the new EU food safety policy. In 2000, the European Commission published its Communication on the use of the precautionary principle, in order to inform all interested parties of a manner in which the principle was implemented in EU policies.[306]

The principle is now – for the first time in EU law – defined in Article 7 GFL. The definition also determines conditions for the application of the principle.

The GFL's preamble states that, in some cases, scientific risk assessment alone cannot provide all the necessary information on which the risk management should be based. The GFL invokes the precautionary principle to ensure health protection in the Union and to set a uniform basis for the use of precautionary measures within the Union, in order to avoid arbitrary decisions giving rise to unjustified barriers to the free movement of food and feed. The preamble states that in circumstances where 'a risk to life or health exists but scientific uncertainty persists, the precautionary principle provides a mechanism for determining risk management measures or other actions in order to ensure the high level of health protection chosen in the Community'.[307]

[306] Brussels 2.2.2000 COM(2000) 1 final.

[307] See Recitals 20 and 21 GFL.

9.7.2 Conditions

Article 7(1) GFL sets out three pre-requisite conditions that have to be met if the precautionary principle is to be invoked in the EU or national policy:

1. There must be an assessment of available scientific data before a decision is made.
2. The possibility of harmful effects on health is identified.
3. The risk assessment is inconclusive or inconsistent and the risk cannot be determined – scientific uncertainty persists.

9.7.3 Precaution and risk analysis

Article 6(3) mentions the precautionary principle among the elements that risk managers should take into account when taking decisions in the area of food safety. Thus, the precautionary principle is a risk management tool and it can only be used within the framework of the risk analysis methodology. Recourse to the principle presupposes a scientific risk evaluation which – due to the insufficient, inconclusive or imprecise information – cannot assess the risk with sufficient certainty. The insufficiency of available information makes it impossible to base a risk management decision on risk assessment. However, the risk evaluation may identify potential risks, which, although not fully scientifically proven, are not acceptable to society and not consistent with the high level of protection of human health chosen in the Union. In these circumstances, the precautionary principle enables public authorities to act without having to wait until the reality and seriousness of those risks become fully apparent.

9.7.4 Provisional and proportional measures

Measures taken on the basis of the precautionary principle are provisional, 'pending further scientific information for a more comprehensive risk assessment'. The measures based on the precautionary principle should be maintained as long as the scientific data are inadequate, imprecise or inconclusive and as long as the risk is considered too high to be imposed on society. The provisional character is not linked to a time factor, but to the development of scientific knowledge. The measures must be reviewed within a reasonable period of time, depending on the nature of the risk to life or health identified.

According to the Commission's Communication, measures resulting from the precautionary principle may take various forms (Diagram 9.6). Besides the adoption of legal measures, there are other possibilities of action that can be inspired by the precautionary principle, such as funding a research programme, informing the public about the adverse effects, etc. Precautionary measures must be proportional to the desired level of protection and no more restrictive of trade than it is required to achieve the high level of protection.

Diagram 9.6. Application of the precautionary principle.

Conditions	Measures
Risk assessment	Proportionate
Potential hazard	Provisional
Scientific uncertainty regarding risk	Reviewed after improved risk assessment

The definition of the precautionary principle includes a notion of 'cost-effectiveness': in considering precautionary measures regard must be had for technical and economic feasibility and other factors regarded as legitimate, even though, as set out in the case law of the CJEU, the protection of health will normally take precedence over economic considerations.[308]

9.8 Food law and science

Legal standards and science can become interrelated in different ways; static and dynamic (Diagram 9.7). A standard can be called static if the specific outcomes of risk assessment in a specific moment in time, determine the standard, for example, when a maximum residue limit or level (MRL) is set. This MRL stays the same as long as the standard is not changed regardless of developments in scientific knowledge. A standard can be called dynamic if it relates to the state of the art of scientific knowledge at the moment of application of the standard. Article 14 GFL to be discussed hereafter is an example of a dynamic standard. It sets a ban on bringing unsafe food to the market. As knowledge on food safety risks develops, the exact meaning of the ban on unsafe food develops along with it. The disadvantage of static standards is that they easily get outdated, and thus lose relevance from a scientific point of view. The advantage is legal certainty. The standard can be understood at face value. The advantage of a dynamic standard is that scientifically it is always up to date. The disadvantage is a lack of legal certainty. Legislation is disclosed in a specific way at a specific place.[309] This makes it possible to inform oneself of changes in the text of the law. Scientific knowledge is disseminated in defuse ways. The state of the art cannot be established with the same certitude. Food business operators can never be entirely sure if their scientific knowledge is up to date. If it is not, they are not compliant with food law.

[308] On the precautionary principle, see in more detail: A. Szajkowska, The impact of the definition of the precautionary principle in EU food law, Common Market Law Review 2010, pp. 173-196. On risk analysis see A. Szajkowska, Regulating food law: Risk analysis and the precautionary principle as general principles of EU food law (diss. Wageningen University), Wageningen Academic Publishers, Wageningen, the Netherlands, 2012.
[309] For European legislation, this specific place is the Official Journal.

Diagram 9.7. Food law & science.

	Static	**Dynamic**	**Zero tolerance**
Standard	represents scientific opinion at the time of drafting	meaning evolves with scientific progress	meaning evolves with technological progress
Legal certainty	+	–	–
Scientific relevance	–	+	–

An especially intriguing standard is the contamination limit of zero. While this standard is static in its appearance, the meaning of zero (detection limit) evolves not with scientific opinion but with technological capacity. The smaller the quantities that can be measured, the more strict the norm 'zero' becomes. However, it may also become scientifically meaningless if the traces measured are no longer relevant from a food safety point of view.[310] Current technology can measure contamination in ppbs (parts per billion). The consequence is that if the sample is large enough anything can be found in it..

9.9 Other consumer interests

Article 8 GFL further details the other interests of consumers indicated in Article 5. It states that food law shall aim at the protection of the interests of consumers and shall provide a basis for consumers to make informed choices in relation to the foods they consume. It shall aim to prevent:

a. fraudulent or deceptive practices;
b. the adulteration of food; and
c. any other practices which may mislead the consumer.

Diagram 9.8 graphically represents this interpretation of consumer protection.

Diagram 9.8. Consumer protection.

Protect consumer interests	
Provide	**Prevent**
Basis for informed choice	a. fraudulent or deceptive practices
	b. the adulteration of food
	c. any other misleading practices

[310] See for this opinion: H. Lelieveld and L. Keener, Global harmonization of food regulations and legislation: the Global Harmonization Initiative, Trends in Food Science & Technology, 2007, pp. S15-S19.

9.10 Transparency

9.10.1 Public consultation

Two principles fall under the heading 'transparency'. The first of them is that the public must be consulted in the development of food law.[311] The European Commission publishes proposals for new legislation on the Internet and invites comments from the public. This principle is in line with the reforms in European governance, launched by the Commission in the White Paper on European Governance in 2001.[312] The White Paper highlighted the need for opening up the policy-making process and getting citizens involved in shaping EU policies, under five guiding principles of EU governance: openness, participation, accountability, effectiveness, and coherence.

9.10.2 Public information

The authorities have a duty to inform the public of food-borne health risks and the measures that are taken to counteract them.[313] This is part of the concept of 'risk communication'.

9.11 International trade

9.11.1 Import

European food safety requirements apply to imported food.[314] As we will see hereafter, these requirements on food safety not only address the condition of the product when it arrives at the European border, but also the way it has been handled in processing and trade and even in the choice of raw materials. This principle therefore implies considerable extra-territorial ambitions of EU food law.

9.11.2 Export

Exported food must also comply with the European food safety standards.[315] This principle on the one hand makes the European origin of a food product a quality guarantee. On the other hand, it facilitates controls and enforcement as all production in the EU in principle has to meet the same safety standards, regardless of the market for which they are producing.

[311] Article 9 GFL.
[312] COM(2001) 428 final.
[313] Article 10 GFL.
[314] Article 11 GFL. See also Chapter 18.
[315] Article 12 GFL.

An exception applies if the authorities or the legislation in the country of destination request differently. In other circumstances, food not complying with EU food law may only be (re)exported if the competent authorities in the country of destination have been informed of this non-compliance with EU law and have expressly agreed. Food that is injurious to health may never be (re)exported, regardless of the opinion of the authorities in the country of destination.

9.11.3 International standards

The EU and the Member States contribute to the development and application of international standards.[316] As discussed above, these standards in turn must be taken into account when developing food law in the EU (Article 5(3) GFL). This principle was addressed in Chapter 3.

9.12 Food safety: a duty of care

9.12.1 Responsibility

The most important notion in the General Food Law is the central place it accords to safety. The GFL imposes on food business operators the responsibility for the safety of the food they bring to the market. Article 17, which bears the title 'Responsibilities'[317], states in its first paragraph that food business operators must ensure compliance with food law (Law text box 9.11).

Law text box 9.11. Article 17 (1) GFL on compliance with food law.

Article 17 (1) GFL
Food and feed business operators at all stages of production, processing and distribution within the businesses under their control shall ensure that foods or feeds satisfy the requirements of food law which are relevant to their activities and shall verify that such requirements are met.

This sounds like stating the obvious: 'one should comply with the law', but there is more to it than this. Article 14 GFL forbids bringing food to the market if it is unsafe. Food is deemed to be unsafe if it is injurious to health or unfit for human consumption. This imposes a general responsibility to ensure the safety of any food brought to the market. Furthermore, this implies a general risk for the food business operator in case a food that has been brought to the market turns out to have been injurious to health. Even the effects on the health of subsequent generations are taken into consideration. In other words, adverse effects of food may constitute an offence committed by the food business operator.

[316] Article 13 GFL.
[317] The German version uses a heading 'Zuständigkeiten' best translated as 'competences'.

Article 17(1) GFL is also of relevance for the legislator. If food business operators are to be responsible, food legislation should be drafted in such a way that they can indeed take responsibility. In the explanatory memorandum to the initial proposal[318] for what became the General Food Law, the European Commission put it this way:

> ***Responsibilities***
>
> *In some areas of European food law, notably in hygiene legislation, the primary responsibility for ensuring compliance with food law, and in particular the safety of the food, rests with the food business. To complement and support this principle there must be adequate and effective controls organised by the competent authorities of the Member States. In other areas of food law this principle is not so widely applicable. This proposal will extend this principle to all food law, and lead to a general review of food law to establish if this principle is respected or whether there are rules where Community legislation has unnecessarily taken responsibility away from the feed or food business by prescribing how a given objective has to be achieved instead of fixing the objective.*

In other words, given the choice, the legislator should formulate norms in terms of results to achieve rather than in terms of ways to behave.

9.12.2 Ban on unsafe food

Undoubtedly, the single most important provision in EU food law is Article 14 GFL (Law text box 9.12).[319]

Article 14(1) GFL puts a general ban on marketing unsafe food. The remainder of Article 14 GFL elaborates when and how food is deemed unsafe for the application of this ban. It brings together substantive and formal approaches. That is to say, first it addresses the factual effects of the food and then the conformity with legal requirements is taken into account.

In addressing the issue of unsafety, normal conditions of use are taken as a starting point, including the information provided to the consumer. The presence or absence of proper instructions of use or warnings (allergen labelling!) may determine the safety or unsafety of a food.

---

[318] Proposal for a Regulation of the European Parliament and of the Council laying down the general principles and requirements of food law, establishing the European Food Authority, and laying down procedures in matters of food, 8.11.2000, COM(2000) 716 final, pp. 10-11.

[319] For a detailed analysis of this provision, see: B.M.J. van der Meulen, The Core of Food Law: A Critical Reflection on the Single Most Important Provision in All of EU Food Law, EFFL 2012, pp. 117-125.

Law text box 9.12. Article 14 GFL ban on unsafe food.

Article 14
Food safety requirements

1. Food shall not be placed on the market if it is unsafe.
2. Food shall be deemed to be unsafe if it is considered to be:
   (a) injurious to health;
   (b) unfit for human consumption.
3. In determining whether any food is unsafe, regard shall be had:
   (a) to the normal conditions of use of the food by the consumer and at each stage of production, processing and distribution, and
   (b) to the information provided to the consumer, including information on the label, or other information generally available to the consumer concerning the avoidance of specific adverse health effects from a particular food or category of foods.
4. In determining whether any food is injurious to health, regard shall be had:
   (a) not only to the probable immediate and/or short-term and/or long-term effects of that food on the health of a person consuming it, but also on subsequent generations;
   (b) to the probable cumulative toxic effects;
   (c) to the particular health sensitivities of a specific category of consumers where the food is intended for that category of consumers.
5. In determining whether any food is unfit for human consumption, regard shall be had to whether the food is unacceptable for human consumption according to its intended use, for reasons of contamination, whether by extraneous matter or otherwise, or through putrefaction, deterioration or decay.
6. Where any food which is unsafe is part of a batch, lot or consignment of food of the same class or description, it shall be presumed that all the food in that batch, lot or consignment is also unsafe, unless following a detailed assessment there is no evidence that the rest of the batch, lot or consignment is unsafe.
7. Food that complies with specific Community provisions governing food safety shall be deemed to be safe insofar as the aspects covered by the specific Community provisions are concerned.
8. Conformity of a food with specific provisions applicable to that food shall not bar the competent authorities from taking appropriate measures to impose restrictions on it being placed on the market or to require its withdrawal from the market where there are reasons to suspect that, despite such conformity, the food is unsafe.
9. Where there are no specific Community provisions, food shall be deemed to be safe when it conforms to the specific provisions of national food law of the Member State in whose territory the food is marketed, such provisions being drawn up and applied without prejudice to the Treaty, in particular Articles 28 and 30 thereof.

9.12.3 Health effects

Food is deemed unsafe if it is injurious to health.[320] The concept of unsafe food is based first and foremost on the effects that its consumption may have on the health of the consumer. According to paragraph 4 of Article 14 this includes immediate, short-term and long-term effects and even effects on subsequent generations. As for this last category: experience in pharmacy has shown that consumption of the artificial hormone DES (diethylstilbestrol) is harmless to the consumer (patient) but does great damage to the child of which the consumer is pregnant at the time of consumption. DES was once the most notorious illegal hormone used in cattle rearing. It would therefore be conceivable that it is present in a food. If this is the case, it renders the food unsafe under this provision.

In determining the unsafety of the food, cumulative toxic effects have to be taken into account, e.g. probable negative effects of consumption in combination with other foods.

In determining whether a food is unsafe, an *ex-ante* position is taken. Regard is taken of the probable effects. For an effect to be probable, the relevant knowledge must exist at the time of placing on the market.

Finally, the particular health sensitivities of a specific category of consumers must be considered where the food is intended for that category of consumers. For example, baby food should be free from adverse effects on babies. Food intended for the general public is not unsafe solely because of the fact that it may have adverse effects on babies. If, however, a specific category is part of the general public as is the case with people suffering from certain food allergies, food presented to the general public must be considered to be intended for this category as well and thus needs the appropriate warnings to be considered safe.

9.12.4 Unfit for consumption

Food is also deemed unsafe if it is unfit for human consumption (Article 14(2)(b) GFL). In determining whether any food is unfit for human consumption, one should consider whether the food is unacceptable for human consumption according to its intended use, for reasons of contamination, whether by extraneous matter or otherwise, or through putrefaction, deterioration or decay (Article 14(5) GFL). Some scholars criticise this provision. They understand it to mean that it applies to foods that are *not* injurious to health but that are nevertheless labelled 'unsafe'. We prefer another reading. It is mainly a matter of burden of proof. If a food is deteriorated to such an extent that it is unfit for human consumption, no separate information is required regarding its probable effects on health in case of consumption. The

[320] Article 14(2)(a) GFL.

nature of the microbes responsible for the deterioration (pathogenic or not) does not need to be established. If the food is unfit for consumption it is considered unsafe without further ado and should be treated as such.[321]

9.12.5 Conclusions from sampling

Paragraph 6 of Article 14 addresses the consequences of the finding of an unsafe food. Where any food which is unsafe is part of a batch, lot or consignment of food of the same class or description, it must be presumed that all the food in that batch, lot or consignment is also unsafe, unless, following a detailed assessment, there is no evidence that the rest of the batch, lot or consignment is unsafe. Findings regarding unsafety apply to all foods regarded as 'the same' under this provision until counter proof has been provided.

9.12.6 Foods complying with food safety requirements

What level of contamination with toxins or pathogens renders a food unsafe? In general, it has been left to science to answer this question. In many situations, however, a generic answer has been laid down in legislation setting levels. Paragraphs 7 (EU provisions) and 9 (Member State provisions) address this issue. A food in compliance with legal requirements is considered safe insofar as the aspects covered by these requirements are concerned. In other aspects, they may still be unsafe. For example, a food is not unsafe due to pesticide residues if these residues are below the statutory maximum residue limit (MRL). If the same food is however contaminated by other toxins or pathogens it is still unsafe. New scientific data may even indicate that the food in compliance is unsafe because the MRL has been set too high.

To deal with such situations, paragraph 8 indicates that the presumption of safety based on compliance is open to counterproof. Competent authorities may still take measures if they have reason to suspect that, despite the conformity, the food is unsafe.

9.12.7 Foods not complying with food safety requirements

Article 14 GFL basically takes a substantive approach to unsafety. Unsafety is determined on the basis of the probable effect. As we have seen above, this substantive approach is somewhat limited by formal statements of safety. Article 14 does not however take its consideration with legal provisions so far as to define foods unsafe on the basis that they are not in conformity with food safety requirements. Article 14 does not say that a food is unsafe if it surpasses an MRL

[321] We may need to reconsider this position. The CJEU seems to accept that a product can be unsafe without being injurious to health. See CJEU 11 April 2013 Case C-636/11, Karl Berger v. Freistaat Bayern.

or if it has not passed a required pre-market approval procedure! The ban of Article 14(1) to bring an unsafe food to the market does not apply to such 'illegal' foods if no adverse effect on health is probable.[322] To remain with the example of MRLs, usually in setting these, such a wide safety margin is taken into account that indeed a food slightly surpassing the limit can still be safe in the sense that no adverse health effects are probable.

The fact that the ban of Article 14(1) GFL does not apply, does not mean, however, that food businesses are free to bring noncompliant foods to the market. Usually the specific legal framework applying to specific food safety requirements includes a provision banning noncompliant products from the market. Indeed Article 19 GFL on withdrawal and recall applies to all foods not in compliance with food safety requirements. So unlike Article 14 GFL, Article 19 GFL takes a formal rather than a substantive approach to food safety. As a consequence the scope of Article 19 GFL includes all foods not in compliance with Article 14 GFL and all foods not in compliance with other food safety requirements.

Diagram 9.9 sums up the above. The Diagram can be read like a decision tree if the aspects in the left-hand column are seen as questions. An affirmative answer takes the reader to the right; a negative answer takes the reader the next step down.

Note that foods qualified as unsafe have not been marked in the column representing Article 14(8) GFL. Member States need to have enforcement powers to demand withdrawal or recall of unsafe products.[323] Article 14 only addresses the fact that compliance with specific food safety requirements in itself does not provide full protection against the exercise of such powers. It does not grant those powers.

9.12.8 Unsafety and risk assessment

Article 14 does not distinguish between products that are inherently unsafe and products that are categorically not unsafe but have been rendered so by their specific condition. Many aspects of unsafety are known. The presence of pathogens, toxins or foreign objects can be established by visual inspection or by testing. Such inspection and testing is done by businesses as part of food hygiene (see Chapter 13) and by inspectors as part of official controls (see Chapter 15). Also with regard to safety, scientific questions may arise requiring risk assessment. If this risk assessment reveals scientific uncertainty as to the safety of a product, authorities can base safeguard measures on the precautionary principle.

[322] Unless one argues that the mere fact of surpassing an MRL constitutes a contamination rendering the food unfit for human consumption in the sense of Article 14(2)(b) GFL.
[323] See Article 54 Regulation 882/2004.

Diagram 9.9. Categories of unsafe food.

Condition	Qualification		Consequence of Article GFL			
			14(3)(b) remedy by information	14(8) authority measures	14(1) ban	19 withdrawal recall
Compliance with specific Community food safety requirements						
• Yes	Compliant			X		
• No	Non-compliant					X
• No, not applicable and other aspects				X		
Compliance with specific national food safety requirements						
• Yes	Compliant			X		
• No	Non-compliant					X
• No, not applicable and other aspects				X		
Contamination	Unfit for human consumption	Unsafe (distinguish all consumers and specific category in particular in applying information remedy)				
• Yes					X	X
Putrefaction						
• Yes					X	X
Deterioration						
• Yes					X	X
Decay						
• Yes					X	X
• No						
Immediate adverse health effects	Injurious to health					
• Yes			X		X	X
Short term adverse health effects						
• Yes			X		X	X
Long term adverse health effects						
• Yes			X		X	X
Adverse effects on subsequent generations						
• Yes			X		X	X
Cumulative toxic effects						
• Yes			X		X	X

9.13 Implementation of the General Food Law

The General Food Law was adopted on 28 January 2002. It was published in the Official Journal on 1 February 2002. Article 4 GFL states in its third paragraph that existing food law principles and procedures shall be adapted as soon as possible and by 1 January 2007 at the latest in order to comply with Articles 5 to 10.

It is not very common for a regulation to address the legislators in the Member States the way the GFL does in Article 4. The form of EU legislation usually used to harmonise national legislation in the Member States is the directive. For this reason one can say that a directive lays hidden within the General Food Law. If one looks again at the Action Plan on Food Safety,[324] in action 3 the initial plan was to propose a General Food Law Directive. At the moment of publication of this Action Plan a proposal for this directive must already have been under construction. This proposal for a directive was then turned into a proposal for a regulation retaining much of its directive kind of content.

For some Member States this has been confusing. By general theory, the obligations of Member States with regard to regulations are to abolish conflicting national legislation, and to create the necessary structures for compliance and enforcement. Transposition of the regulation into national law is deemed forbidden.[325] In the case of the General Food Law, this general theory has been overruled by the explicit command in Article 4 to the legislators to implement. A regulation is binding on all addressees including, if they are addressed, legislators.

Also by general theory, European law can impose obligations on national authorities, but whether these authorities acquire the powers needed to fulfil these obligations remains a matter of national law. The preliminary question has been raised with the European Court of Justice as to whether under circumstances national authorities can derive powers directly from a European regulation. The Court ruled on 13 March 2008.[326] Strictly speaking, the Court did not answer the question. It does however strongly reiterate the obligation of Member States to comply with EU law, regardless of attribution of the power under national law. The message seems to be: better to act *ultra vires* (beyond the limit of legal powers) infringing national law than to infringe European law by not acting. The question as to whether an obligation implies a power probably remains a matter of national constitutional law.

---

[324] Annex to the White Paper on Food Safety.

[325] This understanding is based on the definition of regulation and on case law such as CJEU 7 February 1972, Case C-39/72, *Commission v. Italy*.

[326] CJEU 13 March 2008, Joined Cases 383/06 to 385/06.

In food law, this issue may be of particular relevance in situations where action is due underArticle 7 of the General Food Law (the precautionary principle), but conditions to act under national law are not met.[327]

[327] For example, in many cases enforcement measures rely on the existence of an infringement of the law, or at least a serious suspicion that one has been committed. This is distinct from scientific uncertainty relating to the safety of a food.

10. Authorisation requirements

Dominique Sinopoli, Jaap Kluifhooft and Bernd van der Meulen

10.1 Introduction

10.1.1 Legislation addressing the product

As indicated in Chapter 8, the major instruments addressing food businesses are rules regarding the food (product) as such, rules regarding the process (the handling of the product) and rules regarding communication about food. This chapter and the next address the first category: rules regarding the product.

As discussed in Chapter 9, food may only be brought to the market if it is not unsafe. This principle applies both when a food is unsafe due to its condition (deterioration, contamination, decay) and when a food is unsafe due to its inherent properties. If, for example, a product is toxic by nature it is an unsafe food. This chapter addresses the safety of a food as such. On a generic basis certain ingredients, organisms or other materials are judged to be marketable or not – mainly for reasons of unsafety.

German scholars distinguish two approaches to food law which they call the 'principle of abuse' and the 'prohibition principle with reservation of permission'.[328] If the former principle applies, food businesses are free in their actions but will be held responsible if they infringe the general norm of food safety. In other words the food is considered not to be categorically unsafe. If the latter principle applies, putting the food on the market is forbidden unless an express permission has been obtained. The food is considered categorically unsafe (until authorities decide otherwise).[329]

If the law does not explicitly state otherwise, the former principle applies. In other words in the absence of a prohibition freedom of choice applies. The number of prohibitions is however continuously increasing.

While we do not oppose the German approach, in this book we distinguish not two but four legislative approaches to food (components) as such. The use and/or presence of ingredients, organisms or other materials in food can be free, conditional, restricted or banned (Diagram 10.1). Food businesses are free to use

---

[328] Translation by Margret Will and Doris Guenther, Food Quality and Safety Standards, 2nd edition GTZ 2007, p. 16.

[329] In the USA scholars call this approach 'precautionary'. In the transatlantic debate this leads to some confusion. The precautionary principle as defined in Article 7 GFL deals with scientific uncertainty (which may warrant banning a product from the market). This precautionary approach requires scientific certainty (before the product may come to the market).

any conventional ingredients having a history of safe use; for several innovative foods a requirement applies that their safety has to be approved before they may be brought to the market; for many substances or organisms known to pose food safety risks safety limits apply like the maximum residue limits of pesticides and veterinary drugs. Finally, for some materials the food safety risk is judged to be such that they may not be used for and/or be present in food products at all.

Diagram 10.1. Product legislation.

Legislative approach	Example
Free	• conventional ingredients with a history of safe use in the EU
Conditional	• additives • supplements • genetically modified foods • (other) novel foods
Restricted	• residues of pesticides • residues of veterinary drugs • (other) contaminants
Banned	• BSE risk material • zero tolerance residues/contaminants

This chapter addresses the first two categories indicated in Diagram 10.1 with an emphasis on the second. The next two chapters address the last two categories indicated in Diagram 10.1 with an emphasis on the third.

10.1.2 Pre-market approval

The European legislator seems to work from the assumption that conventional foods, that is to say foods that have a tradition of use in the EU, can be considered safe unless new scientific findings indicate otherwise. In that case it is up to the legislator or, in urgent situations, the executive to take action to protect safety. For other products, that is products that are in some way artificial or new, the safety must be proven before they may come to the market. The proof of safety is a matter of science. The assumption of safety of conventional foods may be understood as a mixture of experience as a knowledge base and of political realism. If a product has been consumed by many people for a significant period of time, apparently without doing them harm, this may be a reasonable basis on which to consider it safe. On the other hand the legislator would probably overstep the limits of democratic legitimacy if s/he tried to take off the market products known to create food safety risks but that many consumers were unwilling to give up (like alcohol and coffee). In this situation authorities have no alternative but to provide the consumer with the relevant information and leave it to the consumer to decide about consumption.

10.1.3 Positive lists

Some of the earliest EEC legislation on food pertains to colorants. 'EEC Council Directive on the approximation of the rules of the Member States concerning the colouring matters authorised for use in foodstuffs intended for human consumption',[330] set out to establish a single list of colouring matters whose use is authorised for colouring foodstuffs and to lay down criteria of purity which those colouring matters must satisfy. Thus the first list was created of food products of a certain type that could be used, while products not on the list could not be used. This is a so-called positive list. The law does not say what is forbidden leaving the rest free as it usually does, but says what is allowed making the rest forbidden. The list is a part of the law (in this case an annex to the regulation). To later include a product in the list (delete a product from the list or adapt the list in some other way) the law must be changed by the applicable procedure.

While the details differ greatly, this system of positive lists set by the law is still the core mechanism of pre-market approval schemes in EU food law. Below we discuss the major schemes: additives (Section 10.2), food supplements (Section 10.3), novel foods (Section 10.4), genetically modified foods (Section 10.5), functional foods as far as relevant are also discussed (Section 10.6). A few other approval schemes are mentioned but not discussed (Section 10.7). The chapter ends with some general observations on pre-market approval schemes (Section 10.8) and an overview of developments regarding proposals for new legislation (Section 10.9).

10.2 Food Improvement Agents Package

In 2008, the Food Improvement Agents Package (FIAP) was passed, consisting of four new regulations to govern substances used in foods. The legislation includes Regulation 1331/2008 establishing a common authorisation procedure for food additives, food enzymes and food flavourings,[331] Regulation 1332/2008 on food enzymes,[332] Regulation 1333/2008 on food additives[333] and Regulation 1334/2008 on flavourings.[334]

---

[330] The uniform numbering system of European legislation did not yet exist. The directive was published in OJ 115, 11 November 1962, pp. 2645-2654.

[331] Regulation 1331/2008 of the European Parliament and of the Council of 16 December 2008 establishing a common authorisation procedure for food additives, food enzymes and food flavourings.

[332] Regulation 1332/2008 of the European Parliament and of the Council of 16 December 2008 on food enzymes and amending Council Directive 83/417/EEC, Council Regulation (EC) No 1493/1999, Directive 2000/13/EC, Council Directive 2001/112/EC and Regulation (EC) No 258/97.

[333] Regulation 1333/2008 of the European Parliament and of the Council of 16 December 2008 on food additives.

[334] Regulation 1334/2008 of the European Parliament and of the Council of 16 December 2008 on flavourings and certain food ingredients with flavouring properties for use in and on foods and amending Council Regulation (EEC) No 1601/91, Regulations (EC) No 2232/96 and (EC) No 110/2008 and Directive 2000/13/EC.

10.2.1 Food additives

At present, the basic provision setting the rules and procedures on additives is Regulation 1333/2008 of the European Parliament and of the Council of 16 December 2008 on food additives. There the definition of additives is given (Law text box 10.1).

Law text box 10.1. Definition of 'additive'.

Article 3(2)(2) Regulation 1333/2008
Definition of additive
any substance not normally consumed as a food in itself and not normally used as a characteristic ingredient of food whether or not it has nutritive value, the intentional addition of which to food for a technological purpose in the manufacture, processing, preparation, treatment, packaging, transport or storage of such food results, or may be reasonably expected to result, in it or its by-products becoming directly or indirectly a component of such foods.

Paraphrasing the definition one could call an additive a substance added to food for a technological purpose. In its Annex I the Additives Regulation sets out a comprehensive list of 27 functional classes of food additives.[335]

Some examples of coined names of additives categories are:
- sweeteners;
- colours;
- anti-oxidants;
- preservatives;
- emulsifiers;
- gelling agents; and
- anti-caking agents.

The functional class descriptors refer to the technological function that the members of that category generally exert in food.

For food additives to be approved it must be demonstrated that on the basis of the scientific evidence available, the additive does not pose a safety concern to the health of the consumer at the level of use proposed; there is a reasonable technological need that cannot be achieved by other economically and technologically practicable means; and that it does not mislead the consumer (Diagram 10.2).[336]

[335] It shall be noted that the concept of '*additive*' is being used for other products, e.g. for feed. Yet, for feed, the meaning of '*additive*' is quite different, and – at variance with food additives – includes a great many enzymes, micro-organisms, and coccidiostats. On feed see Chapter 21.
[336] Article 6(1) of Regulation 1333/2008.

Diagram 10.2. Criteria for the authorisation of food additives.

Authorisation criteria	
Presence	Technological need
Absence	Safety concern Misleading

Sweeteners and colours must meet additional conditions to be included in the list of approval food additives. In addition to the three requirements discussed above, sweeteners must serve one or more of the following purposes:

a. replacing sugars for the production of energy-reduced food, non-cariogenic food or food with no added sugar; or
b. replacing sugars where this permits an increase in the shelf-life of the food; or
c. producing food intended for particular nutritional uses as defined as Article 1(2)(a) of Directive 89/398.[337]

Colours must serve one of the following additional purposes:

a. restoring the original appearance of food of which the colour has been affected by processing, storage, packing and distribution, whereby visual acceptability may have been impaired;
b. making food more visually appealing;
c. giving colour to food otherwise colourless.

There are also additional restrictions for the use of food additives in foods for infants and young children.[338]

10.2.2 Processing aids

Outside the scope of the concept of additive and therefore outside the scope of the regulation, are processing aids.[339] 'Processing aid' has been defined in Article 3(2)(b) of Regulation 1333/2008. See Law text box 10.2.

The distinction between additive and processing aid is in the presence or absence of a technological effect (function) in the final product. Processing aids are outside the scope of both the authorisation requirement that applies to additives and the labelling requirement.

[337] On food for special purposes, see Chapter 17.
[338] As stated in Article 16 Regulation 1333/2008; see Directive 89/398 for details.
[339] Article 2(2)(a) Regulation 1333/2008.

Law text box 10.2 Definition of processing aid in Article 3(2)(b) Regulation 133/2008.

'processing aid' shall mean any substance which:
(i) is not consumed as a food by itself;
(ii) is intentionally used in the processing of raw materials, foods or their ingredients, to fulfil a certain technological purpose during treatment or processing; and
(iii) may result in the unintentional but technically unavoidable presence in the final product of residues of the substance or its derivatives provided they do not present any health risk and do not have any technological effect on the final product

10.2.3 Authorisation of additives

Regulation 1331/2008 provides a common authorisation procedure for food additives, food enzymes and food flavourings to test for the safety of the substances before they can be placed on the market for human consumption. The establishment of a harmonised authorisation procedure which is effective, time-limited and transparent aims to facilitate their free movement within the European Union market, and thus has a beneficial effect on the health, well-being, social interests and economic interests of European citizens.

Food additives must be approved before they can be placed on the European market. The business wishing to market the additive must send an application to the Commission. EFSA performs a risk assessment and forms an opinion about the safety of the additive, after which the Commission and the SCFCAH make the final decision regarding its approval.

The legislation on additives is built on positive lists. As explained above, these are exhaustive lists that indicate which additives may be used in which food product and – many times – at which maximum level.[340] Any material not listed as authorised in the Food Additive Regulation is prohibited as a food additive. The positive list for food additives consists of two parts: Annex II of Regulation 1333/2008 is for food additives that may be placed on the market and used in foods, and Annex III which is for food additives that may be used in food additives, food enzymes and food flavourings. Additives in the latter category are typically carriers[341] and additives used in nutrients.

[340] See, however, the concept of QS hereafter.

[341] Carriers are substances used to dissolve, dilute, disperse or otherwise physically modify a food additive or a flavouring, food enzyme, nutrient and/or other substance added for nutritional or physiological purposes to a food without altering its function (and without exerting any technological effect themselves) in order to facilitate its handling, application or use (defined in Annex I of Regulation 1333/2008).

Food additives listed in these Annexes must include the following:

a. The name of the additive and its E number
E numbers are assigned to all approved food additives, which are used for identification purposes. If a substance is approved it is assigned an E number. The E number can be used to draw up the ingredient list for the label on food products that contain the additive. However, the full name may be used as well instead of the E number.[342]

b. The foods to which it may be added
Food additives are listed in groups of food categories in which they may be used. This food categorization system (FCS) consists of 18 food categories,[343] which are further divided into 153 subcategories. The number of additives which may be used in a food product varies widely. Zero food additives are allowed in unprocessed foodstuffs,[344] honey, butter, pasteurised and sterilised milk, natural mineral and spring water, coffee (excluding flavoured instant coffee), unflavoured leaf tea, sugars, dry pasta (excluding gluten-free and others for special diets) and plain unflavoured buttermilk. Colour additives are also not allowed in various other food products. On the other hand, more food additives are authorised for use in more processed foods, such as confectionary, savoury snacks and flavoured beverages. As an example, more than 250 additives are allowed to be used in the food category of edible ices.[345]
Additionally, legislation[346] has allowed Member States to prohibit the use of certain food additives in foods that were considered to be traditional. For a prohibition to be allowed to continue, it had to exist before 1 January 1992. Such an exception was requested by Germany with a view to prevent the use of intense sweeteners in beers brewed according to the *Reinheitsgebot*.[347] Vigorous opposition from many sides led finally to the permission for all Member States

[342] See also Chapter 14 on food information. For background information on each E-number see www.food-info.net.

[343] These categories are dairy products and analogues; fats and oils and fat and oil emulsions; edible ices; fruit and vegetables; confectionary; cereals and cereal products; bakery wares; meat; fish and fisheries products; eggs and egg products; sugars, syrups, honey and table-top sweeteners; salts, spices, soups, sauces, salads and protein products; foods intended for particular nutritional uses as defined by Directive 2009/39; beverages; ready-to-eat savouries and snacks; desserts excluding products covered in earlier categories; food supplements; processed foods not covered in earlier categories, excluding foods for infants and young children. The count does not include category '0' which is for additives that can be used in all foods.

[344] Defined in Article 3 of Regulation 1333/2008 as 'a food which has not undergone any treatment resulting in a substantial change in the original state of the food, for which purpose the following in particular are not regarded as resulting in substantial change: dividing, parting, severing, boning, mincing, skinning, paring, peeling, grinding, cutting, cleaning, trimming, deep-freezing, freezing, chilling, milling, husking, packing or unpacking'.

[345] Food Safety Authority of Ireland. 2010. Guidance on Food Additives, available at: http://tinyurl.com/k3o89km.

[346] Food additives were previously regulated under Council Directive 89/107 on the approximation of the laws of the Member States concerning food additives authorised for used in foodstuffs intended for human consumption. This Directive was amended by European Parliament and Council Directive 94/34.

[347] See Chapter 7.

to make exemptions for 'traditional' foodstuffs from the horizontal legislation on additives in general; not just on sweeteners. In order to cope with the problem that EU law did not define 'traditional', the Commission published a Decision (292/97) listing once and for all which foodstuffs could claim the exception 'traditional' from the additive rules.[348] Member States had to have their traditional foods approved under the European Parliament and Council Directive 94/34. The foods which are prohibited from containing certain or all food additives are beer (Germany), feta cheese (Greece), 'traditional French bread' (France), preserved truffles (France), preserved snails (France), goose (France), duck (France), turkey preserves (France), 'Bergkäse' (Austria) and mämmi (Finland).[349]

c. The conditions under which it may be used
The conditions refer to the allowed level of use. The level should be set at the lowest level necessary to achieve the desired effect, taking into account any acceptable daily intake (ADI) or equivalent assessment and the estimated daily intake (EDI),[350] including situations in which the food additive is to be used in foods expected to be eaten by certain groups of consumers.
In some situations, no maximum level of use is set for a food additive (*quantum satis*).[351] *Quantum satis* literally means 'the amount that satisfies' and essentially means the amount that is necessary to achieve the desired result, but not any more than that.
Additives approved to be used under *quantum satis* conditions are those of minor concern, such as calcium carbonate (E 170), lactic acid (E 270), citric acid (E 330), pectins (E 440), fatty acids (E 570) and nitrogen (E 941). However, some additives are allowed only under restricted conditions, for example, natamycin (E 235), erythorbic acid (E 315) and sodium ferrocyanids (E 535). Natamycin is approved only as a preservative for the surface treatment of cheese and dried sausages, erythorbic acid is approved only as an antioxidant in some meat and fish food products and sodium ferrocyanids are approved only as anti-caking agents in salts and salt substitutes.[352]
Authorisations of additives are generic. This means that food legislation does not limit the use of approved additives to the business that applied for approval.[353] All food businesses may use approved additives under the same conditions. The focus of the authorisation is on the product, not on the business.

---

348 A consolidated version of the Additives Framework Directive can be found in Eur-Lex. The Decision with the list of 'traditional' foodstuffs however (292/97) is NOT part of that file.

349 R. O'Rourke, European Food Law, 3rd Edition. Sweet & Maxwell, London, UK, 2005.

350 The ADI and EDI are established, when possible, by EFSA during the safety evaluation. The ADI is the amount of a substance that people can consume daily for a lifetime, usually expressed in mg/kg body weight/day. Available at: http://www.efsa.europa.eu/en/topics/topic/additives.htm.

351 Article 11 Regulation 1333/2008.

352 DG Sanco Questions and Answers on Food Additives, available at: http://tinyurl.com/mok7h2e.

353 If the additive is also patented, the owner may bar other businesses from using the additive on the basis of patent law.

10.2.4 Specifications: criteria of purity and identity

The EU has issued specifications (criteria of purity) for most additives authorised, in the annexes to Regulation 1333/2008.

10.2.5 Enzymes

Food enzymes not used in the production of food additives are covered under Regulation 1332/2008 on food enzymes.[354] Food enzymes are defined as a product obtained from plants, animals or microorganisms which contains enzymes capable of catalysing a specific biochemical reaction and which is added to food for a technological purpose during manufacturing, processing, preparation, treatment, packaging, transport or storage. They are different from food additives because they have specific biochemical actions which serve technological purposes, but they do not typically become components of food and instead are used as processing aids.

The Regulation does not apply to enzymes that are used for a function that is not technological, such as those added for nutritional or digestive reasons. Additionally, microbial cultures which may produce enzymes, such as those used in the production of cheese and wine, are not included.

Food enzymes must meet three conditions to be placed on the market: not pose a safety concern, meet a technological need and not mislead the consumer. All enzymes must undergo a premarket authorisation procedure before they can be used in foods, under the approval process established by Regulation 1331/2008.

10.2.6 Flavourings

Flavourings are covered under Regulation 1334/2008.[355] A 'flavouring' is defined as a product which is added to foods in order to impart or modify odour and/or taste. It consists of various categories, such as flavouring substances, flavouring preparations, thermal process flavourings, smoke flavourings and flavour precursors.

Flavourings under the meaning of this Regulation include flavourings used in foods, food ingredients with flavouring properties, and source materials for and foods containing the former two substances (Article 2). The Regulation does not apply to substances which have exclusively a sweet, sour or salty taste; raw foods; and non-compound foods and mixtures such as spices, herbs and teas (Article 2).

---

[354] Regulation No 1332/2008 of the European Parliament and of the Council of 16 December 2008 on food enzymes and amending Council Directive 83/417/EEC, Council Regulation (EC) No 1493/1999, Directive 2000/13/EC, Council Directive 2001/112/EC and Regulation (EC) No 258/97.

[355] Regulation 1334/2008 of the European Parliament and of the Council of 16 December 2008 on flavourings and certain food ingredients with flavouring properties for use in and on foods and amending Council Regulation (EEC) No 1601/91, Regulations (EC) No 2232/96 and (EC) No 110/2008 and Directive 2000/13/EC.

It also does not apply to smoke flavourings (which are covered by Regulation 2065/2003).

In order to be approved for use, flavourings and food ingredients with flavouring properties must meet two requirements: not unsafe and not misleading to the consumer (Article 4). The approval procedure is the same as that for food additives as it is also covered in Regulation 1331/2008.

10.3 Food supplements

10.3.1 General

Another category to which a system of positive lists applies is food supplements. Food supplements are governed by Directive 2002/46 of the European Parliament and of the Council of 10 June 2002 on the approximation of the laws of the Member States relating to food supplements, or more precisely by the national legislation implementing this directive.[356] In Article 2 the directive defines food supplements. See Law text box 10.3.

Law text box 10.3. Definitions in Article 2 Directive 2002/46.

Article 2 Directive 2002/46

For the purposes of this Directive:

(a) 'food supplements' means foodstuffs the purpose of which is to supplement the normal diet and which are concentrated sources of nutrients or other substances with a nutritional or physiological effect, alone or in combination, marketed in dose form, namely forms such as capsules, pastilles, tablets, pills and other similar forms, sachets of powder, ampoules of liquids, drop dispensing bottles, and other similar forms of liquids and powders designed to be taken in measured small unit quantities.

(b) 'nutrients' means the following substances:
   (i) vitamins,
   (ii) minerals.

In short, food supplements are additional doses of vitamins, minerals and other substances. The directive has two annexes. One listing the vitamins and minerals that may be used in the production of food supplements and one giving the form in which they may be used. The lists are generic. If a vitamin or mineral is included in the lists everyone is entitled to include them in food supplements.

[356] The country reports in EFFL/1/2006 are dedicated to the implementation of this directive in the Member States.

10.3.2 Authorisation

The procedure for including other vitamins or minerals in the list is easier than the one that applies to additives. The lists can be modified through comitology. Also applying this procedure, minimum and maximum amounts can be set per daily portion as recommended by the manufacturer.[357] The directive does not give any criteria for authorisation of vitamins or minerals. Annexes I and II of the Directive contain the lists of permitted vitamins, minerals and vitamin and mineral substances which may be used in the manufacture of food supplements.

10.3.3 Monitoring

To facilitate efficient monitoring of food supplements, Member States may require food businesses to notify the competent authorities of the placing on the market by sending them a mock-up of the label used for the product.[358]

10.4 Novel foods

10.4.1 Concept

Food products and ingredients[359] that have not been used to a significant degree for human consumption within the EU prior to passage of the Novel Foods Regulation 258/97[360] (15 May 1997), are so-called novel foods. They have to pass a safety assessment before they may be brought to the market. The Novel Foods Regulation signifies an important step in the development of pre-market approval schemes in EU food law. The scheme is not limited to foods with a certain function but potentially covers a wide spectrum of products. Article 1 of the Novel Foods Regulation specifies four categories of novel foods, see Law text box 10.4 for the specifications.

[357] Article 5 Directive 2002/46.

[358] Article 10 Directive 2002/46.

[359] In the Novel Foods Regulation ingredients are separately mentioned. This is explained by the fact that this regulation pre-dates the General Food Law and its definition of food. An ingredient meets the definition of food and therefore would no longer need to be mentioned separately.

[360] Regulation 258/97 of the European Parliament and of the Council of 27 January 1997 concerning novel foods and novel food ingredients, OJ L 43, 14 February1997, p. 1. Article 15: entry into force 90 days after its publication in the Official Journal.

Law text box 10.4. Definition of 'novel foods', Article 1(2) Novel Foods Regulation 258/97.

Article 1(2) Novel Foods Regulation 258/97

This Regulation shall apply to the placing on the market within the Community of foods and food ingredients which have not hitherto been used for human consumption to a significant degree within the Community and which fall under the following categories:

(c) foods and food ingredients with a new or intentionally modified primary molecular structure;

(d) foods and food ingredients consisting of or isolated from micro-organisms, fungi or algae;

(e) foods and food ingredients consisting of or isolated from plants and food ingredients isolated from animals, except for foods and food ingredients obtained by traditional propagating or breeding practices and having a history of safe food use;

(f) foods and food ingredients to which has been applied a production process not currently used, where that process gives rise to significant changes in the composition or structure of the foods or food ingredients which affect their nutritional value, metabolism or level of undesirable substances.

Until 18 April 2004, genetically modified foods were mentioned as (a) and (b). As from this date the lemmas (a) and (b) are deleted and a separate regime applies for GM foods.

This definition seems to have aimed initially to cover all new food products, whether originating from animals, plants or micro-organisms, but only those new processing methods that have the effects mentioned in subpoints c and f. Foods can be considered new in two different ways. They may be the result of technical innovation and may not have existed before anywhere in the world or they may be known as food elsewhere in the world and be new only to the EU. The latter novel foods are often referred to as 'exotic'. New breeds of plants or animals are exempt if they are propagated/bred in a normal way from 'conventional' parents. Innovations consisting of new combinations of conventional (or approved) ingredients are outside the scope of the regulation.

Article 2(1) of the Novel Foods Regulation further exempts from its scope additives, flavourings and extraction solvents. These products have their own regulatory frameworks as discussed above.

As from 2004 genetically modified organisms are also outside the scope of the Novel Foods Regulation. Even though one could argue that after the deletion of lemmas (a) and (b) in Article 1 of the Novel Foods Regulation, they still fall within the scope of the other lemmas, the specific regulations on GM food take precedence over the more general Novel Foods Regulation.[361]

[361] In legalese: *lex specialis derogat legi generali*. In English: the specific law takes precedence over the general law.

For foods that existed before 15 May 1997, it is crucial to establish if they have been consumed within the EU to a significant degree before that date. The criterion as worded in the regulation is however rather vague.

In a case where among other things the novelty of a food was at issue, the CJEU has been asked to clarify the concept. On 3 February 2005, the Advocate-General delivered an interesting opinion on the condition set in Article 1(2) of the Novel Foods Regulation:

> *This condition consists of two elements: a time element and a quantitative element. As regards the time element, the parties in the main proceedings, the Member States that have made observations and the Commission all agree that it refers to the date of entry into force of the regulation, that is, 15 May 1997. I share this view. As regards the quantitative element 'not ... used ... to a significant degree within the Community', opinions differ somewhat. In my view, in order to interpret this element reference should be made to the purpose of Article 1(2) of the regulation. This provision is aimed at restricting the substantive scope of the regulation to 'novel' products. In fact, a product which, when the regulation entered into force, was already being marketed in one or more Member States and thus was available to the consumer was being used for human consumption to a significant degree and thus could not be novel. It therefore seems to me that the test should be the product's being on the market, a requirement which has the additional advantage of being simple and objectively verifiable. This leads me to the following answer:*
> *Foods, within the meaning of Article 1(2) of Regulation No 258/97, are not used to a significant degree within the Community, if upon the entry into force of that regulation they were not on the market in one or more Member States. The reference date for determining the degree of significance of human consumption of the food in question is 15 May 1997.*

This interpretation clarifies an important concept in the definition of novel food. Unfortunately, the Court in its ruling did not, at least not explicitly, adopt the interpretation the Advocate General proposed. Instead it gave a more ambiguous interpretation.

The Court ruled:[362]

> *Article 1(2) of Regulation (EC) No 258/97 of the European Parliament and of the Council of 27 January 1997 concerning novel foods and novel food ingredients is to be interpreted as meaning that a food or a food ingredient has not been used for human consumption to a significant degree within the Community if, when all the circumstances of the case are taken into account, it is established that that food or that food ingredient has not*

[362] CJEU 9 June 2005, Joined Cases C-211/03, C-299/03, C-316/03, C-317/03 and C-318/03. HLH Warenvertriebs GmbH and Orthica BV v. Bundesrepublik Deutschland.

> *been consumed in a significant quantity by humans in any of the Member States before the reference date. 15 May 1997 is the reference date for the purpose of determining the extent of human consumption of that food or food ingredient.*

The Court explains the vague term 'significant degree' by substituting it for another vague term 'significant quantity'. This interpretation is closer to the text than the one proposed by the Advocate General, but less clarifying.[363]

10.4.2 Authorisation

Before being placed on the market, novel foods must undergo a safety assessment after which an authorisation decision may be taken.[364]

The criteria to judge the marketability of novel foods are laid down in Article 3(1) of the Novel Foods Regulation (Diagram 10.3):

Foods and food ingredients falling within the scope of this Regulation must not:
- present a danger for the consumer;
- mislead the consumer;
- differ from foods or food ingredients which they are intended to replace to such an extent that their normal consumption would be nutritionally disadvantageous for the consumer.

Diagram 10.3. Criteria for the authorisation of novel foods.

Authorisation criteria	
Absence	Danger for consumer Misleading Nutritional disadvantage

The assessment procedure is divided into two stages: an initial stage at Member State level, in most cases followed by an additional stage at EU level. Under the assessment procedure, the competent body of the Member State, which receives an application for pre-market approval, must make an initial assessment and

[363] On the need for clarification, see: H. Verhagen *et al.* Novel Foods: an explorative study into their grey area. British Journal of Nutrition 2009, pp. 270-1277; and, C. Sprong *et al.* Grey area Novel Foods: an investigation into criteria with clear boundaries, European Journal of Nutrition & Food Safety 2014, pp. 342-363.

[364] Guidance for the submission of an application is given in: Commission Recommendation 97/618 of 29 July 1997 concerning the scientific aspects and the presentation of information necessary to support applications for the placing on the market of novel foods and novel food ingredients and the preparation of initial assessment reports under Regulation (EC) No 258/97 of the European Parliament and of the Council.

determine whether an additional assessment is required. If it is found that no additional assessment is required, and neither the Commission nor a Member State raises an objection, the applicant will be informed that s/he may place the product on the market. In other cases[365] the Commission must take an authorisation decision with the assistance of the Standing Committee on the Food Chain and Animal Health.

The authorisation decision defines the scope of the authorisation and specifies, as appropriate, the conditions of use, the designation of the food or food ingredient, its specification and the specific labelling requirements. The authorisation is specific. It gives only the applicant the right to market the novel food concerned. If another business wants to market the same food it also has to apply for an authorisation. In this situation a simplified procedure applies. An example of a novel food that has been approved repeatedly using the simplified procedure, is noni juice. Noni juice is an exotic fruit juice originating in Southeast Asia and new to the EU market.[366]

The regulation lays down specific requirements concerning the labelling of these foodstuffs. The following must be mentioned:

- any characteristics such as composition, nutritional value or the intended use of the new foodstuff which renders it no longer equivalent to an existing food;
- the presence of materials which may have implications for the health of some individuals;
- the presence of materials which give rise to ethical concerns.

This is in addition to the general requirements on labelling discussed in Chapter 14.

Example

A well-known example of an approved novel food is Becel Pro Activ; a spreadable fat with a blood cholesterol lowering[1] effect. See Commission Decision 2000/500 of 24 July 2000.

[1] Hence bcl, the abbreviation that turned into a name.

10.5 Genetically modified foods

10.5.1 Introduction

A special category of novel foods, which since 2004 is subject to a separate regulatory framework, is genetically modified organisms (GMOs) (including microorganisms) used for human consumption (GM food). GM foods need an authorisation on the basis of a double safety assessment before they may be brought to the market.

[365] Experience shows that this constitutes virtually all cases.

[366] The applications and the list of approved (or rejected) novel foods are available on the Internet.

They need an authorisation for the deliberate release into the environment, under the criteria laid down in Directive 2001/18 and an authorisation for use in food and/or feed under the criteria laid down in Regulation 1829/2003.

10.5.2 The GM package

On 18 April 2004 the core provisions of the current EU regulatory framework for genetically modified food products came into force. They supplemented some existing provisions and were fleshed out by some further texts. The EU regulatory framework on GM food now consists of several regulations of the Council and the European Parliament, regulations of the Commission and directives (Diagram 10.4).

Diagram 10.4. The GM package.

EU legislation on GM food

The most important text surviving from the situation prior to 2004, is *Directive 2001/18* on the deliberate release into the environment of genetically modified organisms. It regulates experimental releases and the placing on the market of genetically modified organisms. This directive has been amended by Decision 2002/623; Regulation 1829/2003 and Regulation 1830/2003.

The core of the regulatory framework is *Regulation 1829/2003* on GM food and feed. It regulates the placing on the market of food and feed products containing or consisting of GMOs and provides for the labelling of such products to the final consumer.

Regulation 1830/2003 on traceability and labelling of GMOs and the traceability of food and feed products from GMOs further elaborates the standards set in *Regulation 1829/2003* on labelling and adds a specific regime on traceability for GMOs, departing from the general traceability regime in the GFL.[1]

Commission *Regulation 65/2004* establishes a system for the development and assignment of unique identifiers for genetically modified organisms. This system is used in the traceability regime introduced in Regulation 1830/2003.

Commission *Regulation 641/2004* on the detailed rules for the implementation of Regulation 1829/2003, elaborates the authorisation procedure.

For the sake of completeness one more directive, one more regulation and one recommendation have to be mentioned that are relevant for GMOs, but fall outside the scope of this book. Directive 90/219/EEC, as amended by Directive 98/81/EC, on the contained use of genetically modified micro-organisms (GMMs). This Directive regulates research and industrial work activities involving GMMs (such as genetically modified viruses or bacteria) under conditions of containment, i.e. in a closed environment in which contact with the population and the environment is avoided. This includes work activities in laboratories. Regulation 1946/2003 on transboundary movements of genetically modified organisms implements the Cartagena Protocol on Biosafety.

Commission Recommendation C (2003) 2624 (OJ 2003 L 189/36) on guidelines for national strategies and best practices to ensure coexistence provides Member States with policy options to protect conventional agriculture from GM admixture.

[1] See Chapter 13.

10.5.3 Novel Foods and GMOs

The legislation discussed in this section takes GM foods outside the scope of the Novel Foods Regulation and provides them with their own regulatory framework. This legislation is *lex specialis*, so, as far as it applies, the Novel Foods Regulation no longer applies unless this Regulation applies for another reason.[367] To this effect, recital 11 of Regulation 1829/2003 states, *inter alia*:

> *(F)oods covered by an authorisation granted under this Regulation will be exempted from the requirements of Regulation (EC) No 258/97 concerning novel foods and novel food ingredients, except where they fall under one or more of the categories referred to in Article 1(2)(a)*[368] *of Regulation (EC) No 258/97 in respect of a characteristic which has not been considered for the purpose of the authorisation granted under this Regulation.*

10.5.4 Environmental approval

History has shown that the balance in the natural environment from co-evolution is vulnerable to alien species. When man started to roam the globe, in his wake he wrought environmental disaster by accidental or intentional introduction of foreign plants and animals into each new environment. Today, protective (sanitary and phytosanitary) measures try to avoid further damage.

It is conceivable that genetic engineering results in new species to which the environment proves vulnerable.[369] Protective measures have been agreed upon at the global level (Cartagena Protocol[370]) and have also been taken in the EU. On 8 May 1990 the Directive of 23 April 1990 on the deliberate release into the environment of genetically modified organisms (90/220) was published in its Official Journal.[371] This Directive was replaced by the current Directive 2001/18.[372]

This, the latter Directive is subdivided in four parts: A through D. Part A contains general provisions. Part B addresses the deliberate release of GMOs for any other purpose than for placing on the market. Part C deals with placing GMOs as such

[367] For example, if an exotic food (i.e. a food having a history of use only outside the EU) were to be genetically modified, or if a novel technical process (e.g. high pressure processing the foodstuff) were to be applied to a GMO, then both sets of rules would apply simultaneously.

[368] The reference to lemma (a) in Article 1(2) is obviously a mistake and should be read as (c)-(f).

[369] It would be conceivable, for example, that resistance to certain natural enemies ('pests') or herbicides provides GMOs with an edge over natural species or human attempts to redress an unwanted situation.

[370] OJ L 201, 2002, p. 50.

[371] Council Directive 90/220 of 23 April 1990 on the deliberate release into the environment of genetically modified organisms.

[372] Directive 2001/18 of the European Parliament and of the Council of March 12, 2001 on the deliberate release into the environment of genetically modified organisms and repealing Council Directive 90/220/EEC; see also D. Lawrence, *et al.*, New Controls on the Deliberate Release of GMOs, European Environmental Law Review 2002, pp. 51-56.

or in products on the market. Part D contains final provisions. The Directive takes a 'no unless' approach to GMOs. Releases into the environment are prohibited unless specifically approved, under procedures outlined in Part B (commonly used with a view to field trials). Part C of the Directive is the most important part with regard to GM foods.

The regular procedure first requires notification of the competent authority in the Member State where the GMO is to be placed on the market for the first time.[373] This authority informs the Commission and the other Member States. The competent authority assesses the notification.[374] At this point, the subsequent procedure depends upon whether the assessment report is favourable or not and whether other Member States file objections or not. If the competent authority concludes that the GMO must not be placed on the market, the notification is rejected. If the report is favourable and no objections are made, the competent authority gives consent for a renewable maximum period of ten years.

In case of objections, the decision shall be taken in comitology.[375] Member States may not prohibit, restrict, or impede placement of GMOs on the market – as or in products – which comply with the requirements of Directive 2001/18.[376]

10.5.5 Food approval

Article 4(2) of Regulation 1829/2003 gives a general prohibition on GMOs on the market for food use unless the particular GMO is covered by an authorisation and the conditions to this authorisation are satisfied. The whole regulatory framework for pre-market approval for GM food is built as a set of exceptions to this prohibition.

Scope

The authorisation requirements of Regulation 1829/2003 apply to: (a) GMOs for food use; (b) food containing or consisting of GMOs; and (c) food produced from or containing ingredients produced from GMOs.[377]

[373] Article 13 Directive 2001/18.
[374] Article 14 Directive 2001/18.
[375] Articles 18 and 30 Directive 2001/18.
[376] Article 22 Directive 2001/18 states that Member States may not prohibit, restrict or impede the placing on the market of GMOs which comply with the requirements of this directive. In several instances the CJEU had to intervene to uphold this provisions against non-compliant Member States. See for example: CJEU 16 July 2009 Case C-165/08 Commission v. Poland; CJEU 9 December 2008 Case C-121/07 Commission v. France; CJEU 13 September 2007 Cases C-439/05 and C-454/05 Commission v. Austria. See also Case C-313/11 Commission v. Poland regarding a ban on GM animal feed.
[377] Article 3 Regulation 1829/2003.

The first two are straightforward; however, the concept 'produced from' may raise some questions. This concept can be understood to mean highly refined food (ingredients) made from GMOs, but no longer containing proteins or DNA (like soy oil or maize oil).[378] In theory, processing aids fall outside the scope of the Regulation. The Regulation does not apply to food produced with GM enzymes. Recital 16 of Regulation 1829/2003 reads:

> *This Regulation should cover food and feed produced 'from' a GMO but not food and feed [produced] 'with' a GMO. The determining criterion is whether or not material derived from the genetically modified source material is present in the food or in the feed. Processing aids which are only used during the food or feed production process are not covered by the definition[379] of food or feed and, therefore, are not included in the scope of this Regulation. Nor are food and feed which are manufactured with the help of a genetically modified processing aid included in the scope of this Regulation. Thus, products obtained from animals fed with genetically modified feed or treated with genetically modified medicinal products will be subject neither to the authorisation requirements nor to the labelling requirements referred to in this Regulation.*

10.5.6 Application procedure

According to Article 5 of Regulation 1829/2003, an authorisation is granted exclusively on the basis of an application. Regulation 1829/2003 gives some requirements for the application procedure, which have been further elaborated in Commission Regulation 641/2004.

Criteria

The Regulation stipulates that GM food/feed must not: 'have adverse effects on human health, animal health, or the environment; mislead the consumer; or differ from the food/feed it is intended to replace to such an extent that its normal consumption would be nutritionally disadvantageous for the consumer/animals'[380] (Diagram 10.5). The burden of proof is on the applicant.

---

[378] This requirement raises some discussion on the question of what a GMO is and whether processing can remove this quality from a product. In particular in the USA it seems to have been argued that a GMO is no longer a GMO if the modified genes are no longer present.

[379] This assumption is debatable. Remember the definition of food in Article 2 of the GFL discussed in Chapter 2: 'For the purposes of this Regulation, 'food' (or 'foodstuff') means any substance or product, whether processed, partially processed or unprocessed, intended to be, or reasonably expected to be ingested by humans'. Processing aids are not intended to be ingested by humans, but depending upon the specific processing aid in question, traces or residues may be expected to be ingested. This does not, however, alter the scope of Regulation 1829/2003.

[380] Article 4 Regulation 1829/2003.

Diagram 10.5. Criteria for the authorisation of GM foods.

Authorisation criteria	
Absence	Adverse effects on health
	Adverse effects on the environment
	Misleading
	Nutritional disadvantage

The applicant

The applicant must define the scope of the application, indicate which parts are confidential,[381] and must include a monitoring plan, a labelling proposal and a detection method for the new GM food or feed.[382] S/he must present copies of available studies that have been carried out and any other available material demonstrating that the GM food complies with the mentioned criteria.[383] Applications are to be submitted to the competent authority of the Member State where the GM food product will be marketed first.[384]

'One door one key'

The GM regulatory framework is based on the 'one door one key' principle. This phrase contains several elements. First, authorisation for GM foods is valid throughout the Community. Unlike, for instance, pharmaceutical products, there is no need to acquire authorisation from each Member State where the product is brought to market. Second, Regulation 1829/2003 makes it possible to file a single application for obtaining both the authorisation under Directive 2001/18 for release into the environment and the authorisation under Regulation 1829/2003 for placement on the market as a food or feed.[385] However, the applicant may also choose to follow two separate procedures. S/he may, for instance, want to perform field tests long before an authorisation for food use is relevant. Third, this single application is followed by a single risk assessment process, for which EFSA is responsible, and a single risk management process, involving both the Commission and the Member States through a regulatory committee procedure.[386]

Finally, if a product is likely to be used as both a food and a feed, it must be authorised for both or not at all.[387] A single application shall be submitted and

[381] Article 30 Regulation 1829/2003.
[382] Articles 2 and 3 Commission Regulation 641/2004.
[383] Article 5(3) Regulation 1829/2003.
[384] Article 5 Regulation 1829/2003.
[385] Articles 5 and 7 Regulation 1829/2003.
[386] On the concepts risk assessment and risk management, see Chapter 9, Section 9.6.
[387] Articles 15-26 Regulation 1829/2003. This book does not go into the details of the prescriptions that apply to feed in particular.

shall give rise to a single opinion from EFSA and a single EU decision.[388] A single authorisation is given for a GMO and all its possible uses, thus insuring against a repeat of the US experience with Starlink maize.[389]

The national competent authority

Compared to the Novel Foods Regulation, the role of the national authorities with regard to these applications is very limited. It does not go much beyond that of a mailbox. They receive applications[390] and must both acknowledge receipt of applications in writing within fourteen days and inform EFSA.[391] The application, and any supplementary information supplied by the applicant, must be made available to EFSA, which is responsible for a scientific risk assessment covering both environmental risks and a human and animal health safety assessment.

EFSA may ask national authorities to carry out risk assessments.[392] EFSA may also ask 'a competent authority designated in accordance with Article 4 of Directive 2001/18/EC' to carry out an environmental risk assessment. EFSA is even under an obligation to do so 'if the application concerns GMOs to be used as seeds or other plant-propagating material'.

Opinion of EFSA

EFSA receives the application and any supplementary information supplied by the applicant from the national authorities. EFSA is responsible for a scientific risk assessment, covering both environmental risks and a human and animal health safety assessment. EFSA can conduct the risk assessment itself, or it can ask a national food assessment body to perform this task. EFSA's assessments are subject to a six month time limit, although this may be extended if EFSA requests further information from the applicant.

Community Reference Laboratory

In Article 6(3)(d) of Regulation 1829/2003, EFSA is required to forward to the 'Community Reference Laboratory'[393] the particulars necessary to test and validate the method of detection and identification of the GMO proposed by the applicant.[394]

---

[388] Article 27 Regulation 1829/2003.

[389] 'Starlink' maize (corn) was a GM maize which was only authorised for feed but turned up in food.

[390] Article 5(2) Regulation 1829/2003.

[391] Idem; see also Article 5(2)(a) Commission Regulation 641/2004.

[392] Article 6 Regulation 1829/2003.

[393] The Community Reference Laboratory is also referred to as the Community's 'Joint Research Centre'.

[394] Regulation 1829/2003 institutes the 'Community Reference Laboratory' in Article 32 and in its Annex.

It is assisted by a consortium of national reference Laboratories – the 'European Network of GMO laboratories'.

The Community reference laboratory is responsible for:
- reception, preparation, storage, maintenance and distribution to national reference laboratories of the appropriate positive and negative control samples;
- testing and validation of the method for detection, including sampling and identification of the transformation event and, where applicable, for the detection and identification of the transformation event in the food or feed;
- evaluating the data provided by the applicant for authorisation for placing the food or feed on the market, for the purpose of testing and validation of the method for sampling and detection;
- submitting full evaluation reports to EFSA.[395]

Publication

EFSA makes a public version[396] of its opinion available to the public. The public is allowed to make comments to the Commission within thirty days of the publication.[397]

Commission

Within three months of receiving EFSA's opinion, the Commission will draft a proposal for granting or refusing authorisation on the basis of that opinion.[398] The proposal must be approved by a qualified majority[399] of the Member States within the Standing Committee on the Food Chain and Animal Health,[400] which is composed of representatives of the Member States. If the Committee gives a favourable opinion, the Commission adopts the Decision. If not, the draft Decision is submitted to the Council of Ministers for adoption or rejection by a qualified majority. If the Council does not act within three months, the Commission shall adopt the decision.[401]

---

[395] Annex of Regulation 1829/2003.

[396] That is to say, EFSA provides the public with a text from which confidential information has been deleted.

[397] Articles 2; 5(2)(b), 6(7), 29-31 Regulation 1829/2003. Article 38(1) GFL. See also Article 38(1) Commission Regulation 65/2004. The comments received by EFSA are published on the Internet.

[398] Article 7(1) Regulation 1829/2003.

[399] See Chapter 5.

[400] Article 58 GFL; Article 35 Regulation 1829/2003.

[401] These shifts in competence, depending upon the content of the decision and the amount of consent, seem deplorable from an accountability point of view.

Time involved

From the above it is obvious that the authorisation procedure for GM food is lengthy. Diagram 10.6 gives the official timescale. There is no mechanism, however, to ensure that deadlines are actually respected.

Diagram 10.6. Timeline GM food authorisation procedure.

14 days	MS to acknowledge and notify EFSA
6 months	EFSA risk assessment
PM[1]	EFSA extension additional info
3 months	Commission draft proposal
PM	SCFCAH opinion
3 months	Council if SCFCAH unfavourable
PM	Commission if Council misses deadline
54 weeks + PM	Total

[1] PM = *pro memoria*, i.e. to be decided.

10.5.7 Authorisation

Scope

The authorisation lifts the prohibition against bringing a GM food onto the market. The decision is addressed to the applicant. As a result, the applicant receives a *de facto* monopoly to bring the authorised GM product to the market until other applicants acquire authorisation as well.

Once granted, market authorisations for GM foods are valid for ten years throughout the EU.[402] Conditions and restrictions may be connected to authorisations, and the applicant may be obliged to implement a monitoring plan.[403]

[402] Article 7(5) Regulation 1829/2003. National legislation seeking to impose on supermarkets an obligation to place genetically modified foods in a place specially designated for them on separate shelves from non-genetically modified goods, is declared non-admissible by the Commission: Commission Decision (2006/255) of 14 March 2006 concerning national provisions imposing on supermarkets an obligation to place genetically modified foods on separate shelves from non-genetically modified foods, notified by Cyprus pursuant to Article 95(5) of the EC Treaty.
[403] Articles 5(3)(k), 5(5)(b), 9(1) Regulation 1829/2003.

Publication and registration

The applicant is informed without delay of a decision by the Commission. The details of the decision are published in the Official Journal of the European Union.[404] Products authorised shall be entered into a public register of GM food and feed.[405]

10.5.8 Identification

In Regulation 65/2004, the Commission devised a system of unique identifiers that are to be assigned to each GMO. The Annex to this regulation gives the format. The Commission follows the formats for unique identifiers that have been established by the Organization for Economic Cooperation and Development for use both in the context of its BioTrack product database and in the context of the Biosafety Clearing-House,[406] established by the Cartagena Protocol on Biosafety to the Convention on Biological Diversity.[407]

The applicant for an authorisation must develop the unique identifier for each GMO concerned.[408] The Commission specifies the identifier in the authorisation decision,[409] records it in the relevant registers[410] and ensures that it is communicated to the Biosafety Clearing-House as soon as possible.[411]

The unique identifiers consist of 9 alphanumeric digits. The first three indicate the business concerned. On 19 May 2004 for example, Syngenta Seeds BV received authorisation for Sweet maize (fresh or canned) with Unique ID: SYN-BT Ø 11-1. The transformation event (Bt11) is recognisable in the ID.

10.5.9 Liability

Article 7(7) Regulation 1829/2003 states, 'The granting of authorisation shall not lessen the general civil and criminal liability of any food operator in respect of the food concerned'. Notwithstanding this provision, an authorisation certainly has influence on liability. Part of general civil liability law is product liability, as harmonised by Directive 85/374.[412] An important defence in European product liability law is the 'development risk' defence: the producer can disclaim liability if he proves that the state of scientific and technological knowledge at the time

[404] Article 7(4) Regulation 1829/2003.
[405] Article 28 Regulation 1829/2003; Community Register of GM Food and Feed.
[406] See Biosafety Clearing House on the website of the OECD.
[407] Recital 6 of Regulation 65/2004.
[408] Article 2 Regulation 65/2004.
[409] Article 3(a) Regulation 65/2004.
[410] Articles 3(c) and 5(3) Regulation 65/2004.
[411] Article 5(4) Regulation 65/2004.
[412] See *Chapter* 16 Section 16.7.

the product was put into circulation did not allow the existence of the defect to be discovered.[413] A favourable assessment by EFSA followed by an authorisation would seem to constitute considerable evidence that would sustain this defence.

10.5.10 Modification and renewal

The authorisation-holder can propose to modify the terms of the authorisation, by application *mutatis mutandis* of the procedure to the proposal.[414] Authorisations are renewable for ten year periods.[415] Applications must be sent to the Commission one year before the expiration date of the authorisation at the latest.

This system of authorisations limited in time and renewals requiring application, seems to imply that the use of specific GM foods is phased out when a business is no longer willing to support it through application for renewal.[416]

10.5.11 Suspension and revocation

An authorisation, once granted, is not untouchable during its ten year period of validity. EFSA may, on its own initiative or on request, issue an opinion as to whether an authorisation for a product still meets the conditions. The Commission may then decide whether the authorisation shall be modified, suspended or revoked.

10.6 Functional foods

Functional foods is not a legal category. According to Claire M. Hasler,[417] the term functional foods was first introduced in Japan in the mid-1980s and refers to processed foods containing ingredients that aid specific bodily functions in addition to being nutritious.

According to the International Food Information Council (IFIC), functional foods are foods or dietary components that may provide a health benefit beyond basic nutrition.[418]

[413] Article 7(e) Directive 85/374.
[414] Article 9(2) Regulation 1829/2003.
[415] Article 11(1) Regulation 1829/2003.
[416] See for example: Commission Decision of 25 April 2007 on the withdrawal from the market of products derived from GA21xMON810 (MON-ØØØ21-9xMON-ØØ81Ø-6) maize (notified under document number C(2007) 1810) (2007/308).
[417] C.M. Hasler, Functional foods: their role in disease prevention and health promotion: a publication of the Institute of Food Technologists Expert Panel on Food Safety and Nutrition, Food Technology 1998, pp. 57-62, available at: http://www.nutriwatch.org/04Foods/ff.html.
[418] See: http://tinyurl.com/mfqojxx.

As functional foods is not as such a legal category, no pre-market approval requirement exists. If these foods are safe and if they do not fall within another category for which approval is required, they can be brought to the market.

From the food law point of view the most important aspect of functional foods arises if the beneficial properties of functional foods are communicated to consumers. Such communication will constitute a nutrition or health claim that may only be used if it is in accordance with Regulation 1924/2006. This regulation is treated in Chapter 14 on labelling.

The authorisation is different from the safety assessment of the products discussed in this chapter (Diagram 10.7). Central in safety assessment is what the product does not do (cause harm), central in the assessment of health claims is what the food does do (provide the promised health effects).

Diagram 10.7. Criteria for the authorisation of health claims.

Authorisation criteria	
Presence	Scientific substantiation of claimed effect
Absence	Misleading Raise doubt about other foods Encourage excessive consumption

10.7 Other approval schemes

Outside the domain of food law approval schemes that may be of relevance for the food sector are pharmaceuticals, veterinary drugs, pesticides and chemicals (REACH), but they cannot be discussed here.[419]

Also within the domain of food law further approval procedures exist for example for extraction solvents,[420] infant formulae,[421] some other foods for particular nutritional

[419] For sources on veterinary drugs and pesticides see Chapter 10. For REACH see Regulation 1907/2006.
[420] Council Directive 88/344 of 13 June 1988 on the approximation of the laws of the Member States on extraction solvents used in the production of foodstuffs and food ingredients.
[421] Directive 91/321 on infant formulae and follow-on formulae. On special foods and the replacement of this directive, see Chapter 17.

uses,[422] novel food contact materials,[423] and decontaminants.[424] In addition the use of certain messages must be approved in advance: protected designations of origin, protected geographical indications[425] and traditional specialties.[426] Even voluntary labelling of beef and veal must be approved.[427]

10.8 Pre-market approval schemes

10.8.1 Precautionary approach

In this chapter we have encountered an expanding EU policy of requiring safety assessment of certain types of foods before they may enter the market. This policy applies positive lists containing the products that may be brought to the market and banning those that have not been included in the list. This policy is sometimes referred to as 'precautionary'. Scientific certainty about safety is required upfront. The burden of proof to provide this certainty is on the businesses that wish to bring a product to the market. Risk assessment (by EFSA) consists mainly of checking the homework done by the applicant. Risk management takes the form of deciding whether or not to grant approval.

The European Commission sees this approach as an application of the precautionary principle. In its Communication on the precautionary principle,[428] the Commission states:

> *Community rules and those of many third countries enshrine the principle of prior approval (positive list) before the placing on the market of certain products, such as drugs, pesticides or food additives. This is one way of applying the precautionary principle, by shifting responsibility for producing scientific evidence. This applies in particular to substances deemed 'a priori' hazardous or which are potentially hazardous at a certain level of absorption. In this case the legislator, by way of precaution, has clearly reversed the burden of proof by requiring that the substances be deemed hazardous until proven otherwise. Hence it is up to the business community to carry out the scientific work needed to evaluate the risk. As long as the human health risk cannot be evaluated with sufficient certainty,*

[422] Directive 89/398; Directive 2001/15 on substances that may be added for specific nutritional purposes in foods for particular nutritional uses. On special foods and the replacement of these directives, see Chapter 17.
[423] See Regulation 1935/2004.
[424] Article 3(2) first sentence Regulation 853/2004: Food business operators shall not use any substance other than potable water – or, when Regulation (EC) No 852/2004 or this Regulation permits its use, clean water – to remove surface contamination from products of animal origin, unless use of the substance has been approved in accordance with the procedure referred to in Article 12(2).
[425] See Regulation 1151/2012.
[426] See Regulation 1151/2012.
[427] See Regulation 1760/2000.
[428] Brussels 2.2.2000 COM(2000) 1 final.

> *the legislator is not legally entitled to authorise use of the substance, unless exceptionally for test purposes.*

This interpretation of the precautionary principle goes beyond Article 7 GFL.[429] This Article limits the scope of the precautionary principle to a situation where a risk assessment has already taken place and scientific uncertainty persists. It is not about shifting the burden of proof of risk assessment. Generally speaking, in pre-market approval schemes there will be little room left for applying the precautionary principle as defined in Article 7 GFL. If the dossier provided by the applicant leaves scientific uncertainty as to the safety of the product, the standard of proof for authorisation has not been met and authorisation will be denied.[430]

So far no firm structure for pre-market approval has been established. In each instance where a new requirement is introduced, a new procedure has been developed. The results might seem messy, but at least it shows an interesting diversity.

10.8.2 Criteria

The criteria applied in pre-market approval schemes are not identical but similar, usually requiring that the product is safe, the claims are not misleading and there is no nutritional disadvantage.

Diagram 10.8 gives an overview of the criteria as worded in the different provisions. If we assume that the legislator has similar notions in mind when referring to safety hazards, danger for the consumer and adverse effects on health, the core issues are protecting the consumer from products that pose a risk to health, that appear to be something which they are not and that are nutritionally of lower quality than their conventional counterparts. An additional justification is only required for additives (technological need).

The three main criteria are all negative, e.g. they are about what the food must not do. The text of the law seems to leave little leeway for balancing risks and benefits. Possible benefits are not mentioned as criteria, therefore it seems unlikely that benefits can outweigh disadvantages. For example, a calorie-free sweetener with beneficial effects for diabetics may not find its way to the EU market if it is likely to be allergenic to some other consumers.

---

[429] On this issue, see: B.M.J. van der Meulen, H.J. Bremmers, J.H.M. Wijnands and J. Poppe, Structural Precaution: The Application of Premarket Approval Schemes in EU Food Legislation, Food and Drug Law Journal, 2012, pp. 453-473.

[430] This can only be different if authorities see reasons to bring forward counter-proof against the favourable outcome of the risk assessment or if they want to impose conditions along with the approval.

Diagram 10.8. Criteria for market authorisation.

		Additives	Flavourings	Enzymes	Food supplements	Novel foods	GM-food	Functional foods
Presence	Technological need	X		X				
	Scientific substantiation of claimed effect							X
Absence	Safety hazard	X	X	X				
	Danger for consumer					X		
	Adverse effects on health				?		X	
	Adverse effects on the environment						X	
	Misleading consumer	X	X	X	?	X	X	X
	Nutritional disadvantage					X	X	
	Raise doubt about other foods							X
	Encourage excessive consumption							X

10.8.3 Competences

The competences to deal with and decide on an application for pre-market approval vary significantly from scheme to scheme. The competence to receive an application is either with the national competent authority or the Commission, but never with EFSA who in many cases will actually be the first to work on the case. Risk assessment is the competence of either EFSA or a national body. As discussed above, if EFSA is competent it may pass on the task to a colleague at national level. The biggest diversity, however, concerns the competence to actually decide on the application, this may even vary within one scheme. Diagram 10.9 gives an overview of the various competences.

10.8.4 Authorisations

EU pre-market approval schemes on food differ in the legal effect of authorisation, e.g. inclusion of a product in the positive list or register. Sometimes the authorisation is generic, and sometimes it is exclusively for the applicant. Generic authorisations address the product. The product acquires the status 'fit for the EU market'. Exclusive authorisations address the applicant. S/he acquires the privilege to bring the product to the market. Diagram 10.10 gives an overview.

Diagram 10.9. Competences in authorisation schemes.

	Receive application	Risk assessment	Decision		
Additives	Commission	EFSA	Commission & SCFCAH		
Flavourings	Commission	EFSA	Commission & SCFCAH		
Enzymes	Commission	EFSA	Commission & SCFCAH		
Food supplements	Commission	EFSA	Commission & SCFCAH		
Novel foods initial assessment	National competent authority	National risk assessment body	National competent authority		
Novel foods additional assessment		EFSA	Commission & SCFCAH		
GM environmental approval	National competent authority	National risk assessment body	Assessment is negative: National competent authority	Assessment favourable no objections: National competent authority	Assessment favourable objections: Commission & SCFCAH
GM food approval	National competent authority	EFSA	SCFCAH favourable: Commission	SCFCAH not favourable: Council	SCFCAH not favourable, but Council misses 3 month deadline: Commission

Diagram 10.10. Characteristics of authorisations.

	Generic	Exclusive
Additives	X	
Flavourings	X	
Enzymes	X	
Food supplements	X	
Novel foods		X
GM food		X
Functional foods	X	X
Protected designations	X	

As will be discussed in Chapter 14 on food information, different authorisation schemes apply to different types of nutrition and health claims.

From the point of view of consumers generic authorisations have the advantage that once a product is approved it can come to the market easily. For some businesses generic authorisations have the advantage that once the safety of a product has been established no more resources have to be invested to acquire the right to access the market. Exclusive authorisations provide a (temporary) advantage to businesses who invest resources in fulfilling the procedure. As a reward they receive a monopoly in marketing the product concerned. This monopoly is somewhat comparable to a patent. The major difference between a patent and a market authorisation, however, is that enforcement of a patent is a private law matter for which the patent holder is responsible. Keeping unauthorised foods from the market is a public law food safety issue for which the enforcement authorities of the Member States are responsible.

In practical terms, exclusive authorisation is to the advantage of businesses that are first to go through the procedure, while generic authorisations are to the advantage of businesses that follow after another has dealt with the procedure.

10.8.5 Road map

Because the procedures differ it is all the more important for businesses in need of market clearance to find the right procedure. To this end the decision tree set out in Diagram 10.11 may be helpful. The value of this decision tree should not be over-estimated, however. In practice the decision as to the applicable procedure is not always taken on the basis of interpretation of the law, but on the basis of a compromise between Member State representatives involved in comitology. Also, there is a certain tendency to simply follow the road the applicant has chosen.

10.8.6 Patentability

Pre-market approval schemes are the domain par excellence of the innovative players in the food industry. A food not yet approved for the EU market suggests a discovery either in science or in some other geographical part of the world or an invention either regarding a product or a process.

As discussed in Chapter 22 inventions that exhibit an inventive step which is of practical use and has not been disclosed before may be eligible for protection by a patent. While 'newness' is not understood in the same way in patent law as in pre-market approval schemes, there is considerable overlap.

Diagram 10.11. Decision tree on authorisation schemes.

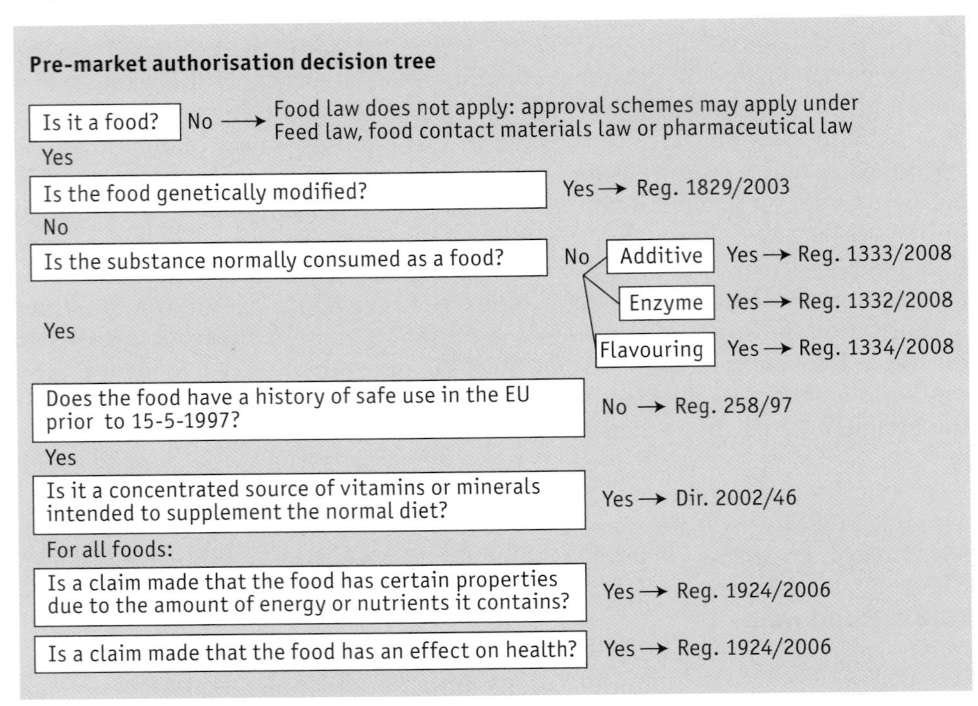

There has been some debate as to whether innovations in the field of biotechnology can be protected by patents.[431] Some argue that living and self-reproducing matter cannot be the subject of inventions as understood in patent law. For the EU this issue has been resolved by Directive 98/44 on the legal protection of biotechnological inventions,[432] extending patent protection to such inventions.

10.9 Developments

10.9.1 Novel Foods

In January 2008 the Commission adopted a proposal for a Regulation of the European Parliament and of the Council on Novel Food.[433] The proposed changes were to streamline the authorisation procedure, develop a modified system to assess the safety of traditional foods from other countries (which are considered

[431] On this debate see: H. Somsen (ed.) The regulatory challenge of biotechnology. Human genetics, food and patents, biotechnology regulation series, Edward Elgar, Cheltenham, UK, 2007.
[432] Directive 98/44 of the European Parliament and of the Council of 6 July 1998 on the legal protection of biotechnological inventions.
[433] Proposal for a regulation of the European Parliament and of the Council on novel foods and amending Regulation (EC) No XXX/XXXX [common procedure]. Brussels, 14.1.2008. COM(2007) 872 final.

novel under Regulation 258/97), and clarify the definition of 'novel food', including new technologies which have an impact on food, and the scope of the Regulation. However, in 2011, after years of discussions, the proposal was not adopted largely due to disagreements on how to regulate the cloning of farm animals.

On 18 December 2013 a new proposal for the regulation of novel foods was issued.[434] This new proposal addresses only the safety of novel foods. It proposes a streamlined authorisation procedure (18 months instead of 3 years) and a simplified procedure for the placing on the market of traditional foods from third countries, but does not address new technologies which may have an impact on food. A generic authorisation scheme is proposed to avoid the resubmission of applications, with a 5-year 'data protection' period as an incentive to develop innovative products.

10.9.2 Cultivation of GMOs

A group of thirteen Member States asked the European Commission to limit the scope of the authorisation for cultivation of GMOs in the EU and to return to them the freedom to decide whether or not they wish to cultivate GM crops on their territory. In response, the European Commission submitted a proposal to the European Parliament to amend Directive 2001/18 in such a way that Member States may adopt 'measures restricting or prohibiting the cultivation of all or particular GMOs authorised'. Such measures must be 'based on grounds other than those related to the assessment of the adverse effect on health and environment which might arise from the deliberate release or the placing on the market of GMOs'.[435] The Commission has in mind 'grounds relating to the public interest other than those already addressed by the harmonised set of EU rules which already provide for procedures to take into account the risks that a GMO for cultivation may pose to health and the environment'.[436] In the public debate, these grounds are loosely referred to as 'socio-economic'.

The European Parliament, however, has proposed to connect the competence of Member States to restrict cultivation to 'duly justified grounds relating to local or regional environmental impacts'.[437] The opinions of the European Commission and the European Parliament seem to differ fundamentally. For this reason it does seem likely that a lengthy procedure will precede any actual amendment.

[434] Proposal for a regulation of the European Parliament and of the Council on novel foods. Brussels, 18.12.2013. COM (2013) 894 final.
[435] European Commission, Proposal for a Regulation of the European Parliament and of the Council amending Directive 2001/18/EC as regards the possibility for the Member States to restrict or prohibit the cultivation of GMOs in their territory, COM(2010) 375 final.
[436] Draft recital 8.
[437] Position of the European Parliament of 5.7.2011, COD(2010) 208.

10.9.3 Implementation

Will the new legislation come into force any time soon? This is difficult to predict. The feeling of urgency to improve food legislation that was felt after the BSE crisis is waning. The European Parliament and the Member States seem to take a more critical position towards Commission proposals to change food legislation. It remains to be seen whether the European Parliament is willing to agree to the proposals that are on the table at the time of writing.

The authors of this book are of the opinion that political decisions should focus on formulating the general requirements for the approval of food products. With regard to specific applications the decision should be limited to applying these requirements in a legal technical manner on the basis of scientific advice. Political discussion and compromise should concern the principles, not the specific products.

11. Contaminants and restricted substances

Ans Punt, Dasep Wahidin and Bernd van der Meulen[438]

11.1 Introduction

Beside raw materials that the producer intentionally includes in a food product, all kinds of compounds may find their way into the final product unintentionally before it reaches the consumer. They may reach the food from agricultural production, pollution of the environment, storage or transport, or can even be formed during heating of foods. Contaminants can be separated into those that are avoidable, including veterinary drugs, pesticides, food contact materials, and those that are unavoidable including environmental chemicals (e.g. PCBs, dioxins, and heavy metals), radioactivity, natural toxins, and chemicals formed during food processing. In the previous chapter we introduced four legislative approaches to food as such. These approaches are summarised in Diagram 11.1. The present chapter deals with the legislation related to these restricted contaminants (both avoidable and unavoidable) as well as banned substances. Substances that are intentionally added to foods (free and conditional substances) have been discussed in the previous chapter. Microbiological contamination will be discussed in Chapter 12.

Diagram 11.1. Product legislation.

Legislative approach	Example
Free	• conventional ingredients with a history of safe use in the EU
Conditional	• additives • supplements • genetically modified foods • (other) novel foods
Restricted	• residues of pesticides • residues of veterinary drugs • unavoidable contaminants • radioactivity
Banned	• hormones • prohibited substances (e.g. nitrofurans, chloramphenicol, etc.)

[438] Ans Punt rewrote, updated and expanded the chapter originally written by Bernd van der Meulen. Dasep Wahidin added the section on radioactive contamination. Many thanks to Rozita Spirovska Vaskoska for her support and suggestions.

11.2 Legislation

For various reasons criteria for the presence of contaminants need to be set, specifically to minimise exposure of consumers to contaminants and to permit free circulation of products. Within the EU, the basic principles of controlling chemical food contaminants have been laid down in the Framework Regulation 315/93. The regulation opens in the first paragraph of Article 1 with a definition of 'contaminant' (Law text box 11.1). The same definition is used within the Codex general standard for contaminants and toxins in food and feed.[439]

Law text box 11.1. Definition of 'a contaminant'.

Article 1(1) Framework Regulation 315/93
Definition of a contaminant
Contaminant means any substance not intentionally added to food which is present in such food as a result of the production (including operations carried out in crop husbandry, animal husbandry and veterinary medicine), manufacture, processing, preparation, treatment, packing, packaging, transport or holding of such food, or as a result of environmental contamination. Extraneous matter, such as, for example, insect fragments, animal hair, etc., is not covered by this definition.[1]

[1] Such extraneous matter may, however, render a food unfit for human consumption. See Article 14(5) GFL.

The application of this definition shows that the concept of contaminant covers two of the three hazards usually distinguished in food law:[440] chemical and certain biological hazards (i.e. chemicals produced by microorganisms). Physical hazards and biological hazards from microorganisms as such are not included in the definition of contaminants. Within the Framework Regulation 915/93 it is specified that the regulation does not apply to contaminants which are the subject of more specific Union rules.[441] These include, for example, pesticides, veterinary drugs and food contact materials which are covered within specific regulations.

The basic principles of Framework Regulation 315/93 include:

- food containing a contaminant to an amount unacceptable from the public health viewpoint and in particular at a toxicological level, shall not be placed on the market;
- contaminant levels shall be kept as low as can reasonably be achieved following recommended good working practices;
- the Commission may, where necessary, establish the maximum tolerances for specific contaminants.

[439] CODEX STAN 193-1995.
[440] See Article 3(14) GFL.
[441] Article 1(2) Regulation 315/93; refers to these more specific rules.

The tolerances shall be adopted in the form of a non-exhaustive[442] Union list and may include:

- limits for the same contaminant in different foods;
- analytical detection limits;
- a reference to the sampling and analysis methods to be used.

Commission Regulation 1881/2006 implements the Framework Regulation, by setting maximum levels for certain contaminants. It further elaborates on the principles. In the case of contaminants which are considered to be genotoxic carcinogens or in cases where current exposure of the population or of vulnerable groups in the population is close to or exceeds the tolerable intake, maximum levels should be set at a level which is as low as reasonably achievable (ALARA).[443]

Furthermore, it contains a prohibition on decontamination through mixing contaminated foods with less or non-contaminated ones[444] (to acquire a product that no longer surpasses the limits). Foods contaminated with mycotoxins shall not be decontaminated by chemical treatments.[445] However, sorting as a method to separate contaminated from non-contaminated foods, as well as other physical treatments are acceptable for groundnuts, other oilseeds, tree nuts, dried fruit, rice and maize.[446]

11.3 Setting legal limits

Legal limits go by different names; at product level terms include 'maximum residue levels' and/or just 'maximum levels', often expressed as maximum amount per gram or litre of product. In setting specific contaminant levels in legislation, two parameters are applied; necessity and safety. The level of necessity is based on what is unavoidable when applying good practices. What is avoidable, should be avoided and is not accepted on the market. Safety on the other hand is based on risk assessment in which toxicology plays an important role. The lower of the two levels defines acceptability.

The evaluation of both necessity and safety in setting legal limits can be seen from the definition of maximum residue levels (MRLs) for pesticides in Regulation 396/2005. MRL means the upper legal level of a concentration for a pesticide residue in or on food or feed based on good agricultural practice and the lowest consumer exposure necessary to protect vulnerable consumers.[447] The MRLs for

[442] This means that the list has no effect for situations not included in the list, that is to say not mentioned contaminant-food combinations. For these the general safety requirement applies (Article 14 GFL and Article 2(1) Regulation 315/93).
[443] Recital 4 Regulation 1881/2006.
[444] Article 3(3) Regulation 1881/2006.
[445] Article 3(4) Regulation 1881/2006.
[446] Article 4 Regulation 1881/2006.
[447] Article 3(2) Regulation 396/2005.

pesticides are thus primary set based on necessity, to achieve sufficient effects of the pesticide in the field but resulting in the lowest consumer exposure possible. Nevertheless, MRLs are also intended to be toxicologically acceptable.

Within the EFSA scientific opinions on the safety of substances are carried out by different panels. For food contaminants there is the Panel on Contaminants in the Food Chain (CONTAM), for pesticides there is the Panel on Plan Protection Products and their Residues (PPR). For veterinary drugs scientific opinions are not carried out by any of the EFSA committees, but by the European Medicines Agency (EMA). Overall, the risk assessment of the different chemicals in foods relies on the comparison of two components: daily human exposure to the compound via food and other routes, and the dose that that has been established not to cause adverse health effects.[448] Whereas most panels can perform their risk assessments based on the applications and data presented to EFSA, the CONTAM Panel primary relies on scientific information that is in the public domain. To complement these open data sources, the Data Collection and Monitoring (DCM) Unit of EFSA regularly launches a call for data on occurrence of the substance(s) of interest and collects food consumption data.

11.4 Contaminants within Framework Regulation 315/93

The annex to Regulation 1881/2006 sets limits for nitrates in spinach, lettuce and cereal-based baby foods; for mycotoxins (in particular aflatoxin, ochratoxin A, patulin, deoxynivalenol, zearalenone, fumonisins, T-2 and Ht-2 toxin), heavy metals (lead, cadmium, mercury, tin), 3-MCPDs, dioxins and PCBs, PAHs (benzo(a) pyrene) (Diagram 11.2).

11.5 Acrylamide

Acrylamide is a specific case in the approach to contaminants. Acrylamide forms in certain food as a result of cooking practices and is commonly found in starchy foods, such as potato and cereal products which have been deep-fried, roasted or baked at temperatures above 120 °C. This substance is known to have potential carcinogenic effects. Research is being undertaken to better understand the process of its formation and to find ways to reduce its presence in food. A careful selection of raw material as well as certain cooking practices are known to help limit the formation of acrylamide in potato products and bread. In close co-operation with the European Commission the food and drink industry (FDE) has developed a toolbox to highlight ways in which to lower levels of acrylamide in food. Based on this, the Commission issued a Recommendation on the monitoring of acrylamide levels in food whereby the Member States are invited to perform such monitoring

[448] EFSA Journal 2012;10(10):s1004.

Diagram 11.2. Format of the annex of Regulation 1881/2006.

ANNEX

Maximum levels for certain contaminants in foodstuffs

Section 1: Nitrate

	Foodstuffs	Maximum levels (mg NO_3/kg)	
1.1	Fresh spinach (*Spinacia oleracea*)	Harvested 1 October to 31 March	3 000
		Harvested 1 April to 30 September 2	500
1.2	Preserved, deep-frozen or frozen spinach		2 000
1.3	Fresh Lettuce (*Lactuca sativa* L.) (protected and open-grown lettuce) excluding lettuce listed in point 1.4	Harvested 1 October to 31 March:	
		• lettuce grown under cover	4 500
		• lettuce grown in the open air	4 000
		Harvested 1 April to 30 September:	
		• lettuce grown under cover	3 500
		• lettuce grown in the open air	2 500
1.4	Iceberg-type lettuce	Lettuce grown under cover	2 500
		Lettuce grown in the open air	2 000
1.5	Processed cereal-based foods and baby foods for infants and young children		200

in selected food categories.[449] In accordance with the monitoring results obtained, the Commission produced another non-binding document: a Recommendation on the investigation into the levels of acrylamides in food.[450] This Recommendation provides indicative levels of acrylamide in foods. They are not to be treated as safety limits, because when they are surpassed they do not call for enforcement actions but rather for further investigations into the production and processing methods by the food producers. These recommendations are continuously updated depending on the monitoring results.

11.6 Pesticide residues

The legislation on pesticide residues is a good example of the EU approach to the presence of undesired substances in food. The regulatory framework on pesticides consists of two parts: the pre-market approval of pesticides for use in agriculture and the establishment of maximum residue limits in food commodities.

[449] Latest version: Commission Recommendation 2010/307 of 2 June 2010 on the monitoring of acrylamide levels in food.

[450] Latest version: Commission Recommendation 2013/647 of 8 November 2013 on investigations into the levels of acrylamide in food.

11.6.1 Pre-market approval of pesticides

Pre-market approval of pesticides is arranged in Regulation 1107/2009. This regulation lays down the rules for the authorisation of plant protection products in commercial form and for their placing on the market, use and control within the European Union.[451] The criteria for approval are provided in the annex of the regulation and include an evaluation of the impact of the compound on human health and the environment.[452] The approval of products (as opposed to active substances) is left to the Member States. Substances excluded from authorisation include substances classified as category 1A or 1B mutagenic, carcinogenic or toxic for reproduction, or when the substance is considered to have endocrine disrupting properties.[453] The setting of MRLs is very closely coordinated with the parallel activities on this evaluation of pesticides active ingredients.

11.6.2 Pesticide MRLs

The setting of MRLs has been codified in Regulation 396/2005, which came into force on 2 September 2008.[454] MRLs are not maximum toxicological limits as such. They are based on good agricultural practice and they represent the maximum amount of residue that might be expected on a commodity if good agricultural practice was adhered to during the use of a pesticide. Nonetheless, when MRLs are set, care is taken to ensure that the maximum levels do not give rise to toxicological concerns.

In principle, MRLs are set on the basis of the following:

a. Supervised agricultural residue trials establish the residue level in or on an agricultural crop treated with a pesticide under specified use conditions (Good Agricultural Practice = GAP). This results in the Supervised Trial Median Residue Level (STMR) and the Highest Residue Level (HR).

---

[451] Article 1(1) Regulation 1107/2009. See also: I. Kireeva and R. Black, Chemical Safety of Food: Setting of MRLs for Plant Protection Products ('Pesticides') in the European Union and in the Russian Federation, EFFL 2011, pp. 174-186.

[452] Previous pesticide MRL legislation was derived from/based on four Council Directives: Directive 76/895 establishing MRLs for selected fruits and vegetables; Directive 86/362 establishing MRLs for cereals and cereal products; Directive 86/363 establishing MRLs in products of animal origin; and Directive 90/642 establishing MRLs in products of plant origin, including fruits and vegetables. Until the entry into force, Chapters II, III and V to the Regulation are not applicable; and the national MRL of each Member State for each pesticide/crop combination remains in force.

[453] See the sources mentioned in I. Kireeva and R. Black, Chemical Safety of Food: Setting of MRLs for Plant Protection Products ('Pesticides') in the European Union and in the Russian Federation, EFFL 2011, pp. 174-186.

[454] See the sources mentioned in I. Kireeva and R. Black, Chemical Safety of Food: Setting of MRLs for Plant Protection Products ('Pesticides') in the European Union and in the Russian Federation, EFFL 2011, pp. 174-186.

b. Using appropriate consumer intake models, the daily residue intake under normal and worst-case conditions can be estimated for the European population and for national populations and sub-populations (e.g. infants).
c. Data from toxicological tests on the pesticide allow for the fixing of an 'acceptable daily intake' (ADI). Usually this involves finding the highest dose that would produce no adverse effects over a (simulated) lifetime (chronic) exposure period and then applying appropriate safety factors. From animal testing the NOAEL is derived; the 'No observed adverse effects level'. The quantity of pesticide residue represented by the NOAEL is two times divided by 10; once to account for the difference in sensitivity between species and once to account for the difference in sensitivity between individuals. As a consequence the ADI generally is 1/100 of the NOAEL. The ADI is set for chronic exposure.
d. In addition, with a view to an extreme diet the Acute Reference Dose (ARfD) is fixed for substances that may be acutely toxic.
e. If the estimated daily consumer intake for all commodities calculated under (b) is lower than the ADI calculated under (c) and the HR for extreme daily intake is lower than the ARfD calculated under (d), then the STMR under (a) is set as the MRL. In cases where the calculated intake is higher, the conditions of use described in (a) need to be modified to reduce the residue level in the commodity. If this is not possible the use of that pesticide on that crop cannot be tolerated and the MRL is set at the limit of determination[455] (LOD; effectively zero). This zero level does not stipulate that even minute traces of the substance are unsafe, but that the substance may not be used.

MRLs for processed products and composite foodstuffs are normally calculated on the basis of the MRL set for the agricultural commodity by application of an appropriate dilution or concentration factor. For composite foodstuffs MRLs are calculated taking into account the relative concentrations of the ingredients in the composite foodstuff. Only exceptionally may specific MRLs be determined for certain processed products or certain composite foodstuffs. Some substances are excluded from authorisation (like DDT, Aldrin, HCH).[456]

11.6.3 Default MRL

Regulation 396/2005 sets a general maximum pesticide residue level in foodstuffs of 0.01 mg/kg. This general level is applicable 'by default', i.e. in all cases where an MRL has not been specifically set for a product or product type. In this way, all foods intended for human (or animal) consumption in the European Union are now subject to a maximum residue level of pesticides in their composition.

[455] The limit of detection (LOD) is the lowest concentration of a pesticide residue that can be measured using routine analysis.
[456] Directive 79/117.

If a specific MRL has been set (for a food/pesticide combination), this takes precedence over the general MRL. The products to which MRLs apply are listed in Annex I of Regulation 396/2005.[457] Specific MRLs are listed in Annex II.[458]. Annex III (part A and part B)[459] holds temporary MRLs.[460] These are national MRLs still to be harmonised at EU level. In Annex IV[461] active substances shall be listed that need no MRL because their residues cannot be distinguished from levels arising naturally (e.g. pepper, garlic extract). An EU pesticide database is available listing all current MRLs.[462]

11.7 Veterinary drugs

A comparable structure to that for pesticide residues is applied to residues of veterinary drugs, requiring a double procedure. One for approval as a veterinary drug based on the effectiveness and safety of the product for the animal concerned,[463] and one setting the maximum residue limit of the product in the food of animal origin (mainly meat and dairy).

11.7.1 Residues of veterinary drugs

The framework regulation for veterinary drugs is Regulation 470/2009. This Regulation lays down a Community procedure for the establishment of maximum residue limits of veterinary medicinal products in foodstuffs of animal origin.[464]

The operative part of the regulation is in its annex, which sets out the classification regarding the maximum residue limits. Table 1 of the Annex contains the classification regarding MRLs in alphabetic order of the substance. For some of the compounds no MRLs are required, for example for homeopathic drugs. Table 2 of the Annex shows the list of prohibited substances. Some of these are for example known to be genotoxic carcinogens and are therefore banned. A zero tolerance applies to substances not listed in Table 1. A default level as described for pesticides does not apply. However, some reference points for action have been set. In 2013, the CONTAM panel provided a guideline for establishing these reference points of action.[465] They take into account the lowest residue

[457] This annex has been drawn up by Regulation 178/2006.
[458] This annex has been drawn up by Regulation 178/2006.
[459] This annex has been drawn up by Regulation 178/2006.
[460] See regulation 149/2008 for the latest update of these annexes.
[461] This annex has been drawn up by Regulation 178/2006.
[462] See: http://ec.europa.eu/sanco_pesticides/public/?event=homepage.
[463] On the basis of Directive 2001/82.
[464] Regulation 470/2009 of the European Parliament and of the Council of 6 May 2009 laying down Community procedures for the establishment of residue limits of pharmacologically active substances in foodstuffs of animal origin.
[465] EFSA Journal 2013;11(4):3195.

concentration which can be quantified with an analytical method validated in accordance with EU requirements.

11.7.2 Hormones

Directive 96/22[466] prohibits the use in stock farming of certain substances having a hormonal or thyrostatic action and of β-agonists. Or, to put it more precisely, it requires the Member States to prohibit the placing on the market of these products[467] and their administration to farm or aquaculture animals.[468] Some exceptions can be made for therapeutic purposes.[469]

The placing on the market[470] for human consumption and the import[471] of aquaculture animals to which such substances have been administered and of processed products derived from such animals shall also be prohibited, even if the exception for therapeutic purposes applies.[472]

11.8 Food contact materials

Legal limits are also set for contaminants that may merge into products from food contact materials. Framework regulation 1935/2004 lays down the general requirements for all food contact materials. This regulation requires that food contact materials do not transfer their constituents to food in quantities which could endanger human health, or bring about an unacceptable change in the composition of the food or a deterioration in the organoleptic characteristics.[473] Specific migration limits for different materials are laid down in Directive 84/500 (ceramics), Directive 2007/42 (regenerated cellulose film), Regulation 282/2008 (recycled plastic materials), Regulation 450/2009 (active and intelligent materials and articles), and Regulation 10/2011 (plastics). Examples of compounds of concern in plastics include phthalates and bisphenol A. These are plasticizers that may leak into the product and potentially cause endocrine disruption. Bisphenol A has been banned in Regulation 321/2011 from use in plastic infant feeding bottles.

[466] Council Directive 96/22 of 29 April 1996 concerning the prohibition on the use in stock farming of certain substances having a hormonal or thyrostatic action and of β-agonists. Official Journal L 125, 23 May 1996.
[467] Article 2 Directive 96/22.
[468] Article 3 Directive 96/22.
[469] Article 4 Directive 96/22.
[470] Article 3(c)(d) and (e) Directive 96/22.
[471] Article 11(2) Directive 96/22.
[472] Article 7(2) Directive 96/22.
[473] Article 3(1) Regulation 1935/2004.

11.9 Radioactive contamination

Radioactivity is part and parcel of the living environment. It was not considered a food safety risk in need of regulation, until the catastrophe that struck the Chernobyl Nuclear Power Plant (Ukraine – at that time USSR) on 26 April 1986. In reaction the EU has set limits to the presence of radioactive contamination in food.

Unlike the other pieces of EU food law discussed in this book, most of the provisions relating to radioactive contamination have not been based on the TFEU (or its predecessors) but on the EURATOM Treaty. Also the framework for risk analysis is different. Risk assessment is not performed by EFSA, but by the so-called Article 31 group of experts.[474]

Article 2 of the EURATOM Treaty requires that 'in order to perform its task, the Community shall, as provided for in this Treaty (...) establish uniform safety standards to protect the health of workers and of the general public and ensure that they are applied'.[475] Protection of the general public includes protection from exposure to radiation vià the intake of contaminated food. The current body of legislation consists of three parts. The first relates to foods in general, the second relates specifically to imports from Japan. This second group of requirements was put in place after the nuclear accident at the Fukushima Daini Nuclear Power Plant on 11 March 2011. Both groups are intended for nuclear emergency situations. The third group relates to long-term exposure situations. The latter was put in place for imports to the European Union from specified countries which were directly affected by the Chernobyl disaster.

11.9.1 General limits to radioactive contamination in foods in emergency situations

The legislation concerning radioactive contamination in general consists of the following directive and regulations. Council Directive (EURATOM) 29/1996 laying down basic safety standards for the health protection of the general public and workers against the dangers of ionising radiation, sets out the so-called Basic Safety Standards (BSS). It is also known as the BSS Directive. On the basis of the BSS including reference levels and an ingestion dose, specific limits have been set in regulations for the various radioactive contaminations on foodstuffs and feedingstuffs. These regulations are: Council Regulation (EURATOM) 3954/87 laying down maximum permitted levels of radioactive contamination of foodstuffs and of feedingstuffs following a nuclear accident or any other case of radiological

[474] The name refers to its legal basis: Article 31 of the EURATOM treaty. See generally: http://ec.europa.eu/energy/nuclear/radiation_protection/article_31_en.htm. Also the international framework is different. Risk assessment is performed not by JECFA, but by the International Committee on Radiation Protection (ICRP). Standards, however, can be found in the Codex Alimentarius.
[475] See: http://tinyurl.com/nqmnxrh.

emergency,[476] Council Regulation (EURATOM) 944/89 laying down maximum permitted levels of radioactive contamination in minor foodstuffs following a nuclear accident or any other case of radiological emergency; and Council Regulation (EURATOM) 770/90 laying down maximum permitted levels of radioactive contamination of feedingstuffs following a nuclear accident or any other case of radiological emergency. At the time of writing, a proposal is pending to consolidate these three regulations into one single regulation.[477]

The annexes to these regulations set limits for diverse radionuclide isotopes notably strontium (Sr-90); iodine (I-131); plutonium and transplutonium (Pu-239, Am-241); and cesium (Cs-134, Cs-137) for baby foods, dairy produce, other foodstuffs except minor foodstuffs; liquid foodstuffs; minor foodstuffs and feedingstuffs. The limits are expressed in Becquerel (Bq) per kilogram for solid foods and per litre for liquid foods. Bq is a unit of radioactivity. One Bq is defined as the activity of a quantity of radioactive material in which one nucleus decays per second.[478]

Diagram 11.3. Format of the annex to Regulation 3954/1987.

ANNEX

Maximum permitted levels for foodstuffs and feedingstuffs (Bq/kg or Bq/l)

	Baby foods	Dairy Produce	Other foodstuffs except minor foodstuffs	Liquid foodstuffs
Isotopes of strontium, notably Sr-90	75	125	750	125
Isotopes of iodine, notably I-131	150	500	2,000	500
Alpha-emitting isotopes of plutonium and transplutonium elements, notably Pu-239, Am-241	1	20	80	20
All other nuclides of Half-life greater than 10 days, notably Cs-134, Cs-137, except C-14 and H-3	400	1000	1,250	1000

[476] As amended by Council Regulation (EURATOM) 2218/89.

[477] COM(2010) 184 final.

[478] The EU legislature determined these maximum permitted levels by using the following formula: $MPL = E/(f \times D \times I \times C)$

Where: MPL is the activity concentration limit specified in the Regulation for a given radionuclide category and food group; E is the reference individual effective dose arising from consumption of contaminated foods in a year subsequent to the accident; f is the factor which reflects a judgment that the average annual concentration in food actually consumed by the individual is a fraction of the activity concentration limit; D is the ingestion dose coefficient in Sv/Bq; I is the annual consumption rate of the relevant food in kg; and C is a correction factor to allow for the additivity of foods within the category other foods, taken as 5 for all radionuclide with physical half-life greater than a few weeks, and as 1 for radionuclide, e.g. iodine-131, with half-life of days or shorter.

Law text box 11.2. The maximum permitted levels for minor foodstuffs in Regulation 944/1989.

Article 2
For the minor foodstuffs given in the Annex, the maximum permitted levels to be applied are 10 times those applicable to 'other foodstuffs except minor foodstuffs' fixed in the Annex of Regulation (Euratom) No 3954/87 or pursuant to Regulations adopted on the basis of Article 3 of that Regulation.

11.9.2 Special limits for imports from Japan after the Fukushima accident

In the aftermath of the Fukushima accident in 2011, initially the general limits as expressed in the annex of Council Regulation (EURATOM) 3954/87 also applied to imports from Japan.[479] Later specific limits were formulated for imports from Japan first in Regulation 351/2011, then in Commission Implementing Regulation (EU) No 284/2012, followed by Regulation (EU) 996/2012 and lastly in Regulation 322/2014. These limits are based on the action levels applied in Japan. As the percentage of foods contaminated foods[480] due to the Fukushima accident is much larger in Japan than in the EU, Japanese standards need to be stricter to ensure food safety. As a result of adjusting limits for imported foods from Japan to these Japanese standards, in the EU a higher level of protection applies with regard to Japanese imports than the levels set in the general regulation (Regulation 3954/1987).

Diagram 11.4. Format of the annex to Regulation 322/2014.

ANNEX
Maximum levels for food (Bq/kg) as provided in the Japanese legislation

	Foods for infants and young children	Milk and milk-based drinks	Other foodstuffs, with the exception of - mineral water and similar drinks - tea brewed from unfermented leaves	Mineral water and similar drinks and tea brewed from unfermented leaves
Sum of Cs-134 and Cs-137	50	50	100	10

[479] Article 7 Regulation 297/2011.
[480] F in the formula set out in footnote 479.

11.9.3 Limits in case long-term exposure persists

The EU established levels for long-term exposure in Council Regulation 733/2008 to counteract radiation contamination in agricultural products imported from third countries. Legislation such as this and its predecessors is only kept in force in case there is a need due to persisting radioactive contamination.

Council Regulation 733/2008 was scheduled to expire on 31 March 2010; however it was amended by Council Regulation 1048/2009 to ensure its extension until 31 March 2020. This was done for two main reasons:

a. According to scientific reports the level of radioactive contaminant Cs-137 in particular foodstuffs originating from third countries is still above the maximum permitted levels of Regulation 733/2008.
b. The radioactive contaminant Cs-137 following the Chernobyl accident still exists in the environment due to its long physical half-life which is approximately 30 years.

According to Council Regulation 733/2008 the accumulated maximum radioactive level for caesium-134 and -137 shall be:

a. 370 Bq/kg for milk and milk products listed in Annex II of the regulation and for foodstuffs intended for the special feeding of infants during the first four to six months of life, which meet, in themselves, the nutritional requirements of this category of persons and are put up for retail sale in packages which are clearly identified and labelled 'food preparation for infants'.
b. 600 Bq/kg for all other products concerned.

11.10 Overview

The legislative approach to undesired substances in food adds specific limits for certain food/substance combinations to the general ban on unsafe food. Often sampling methods are regulated as well. While these have not been discussed in this chapter, they have been included in Diagram 11.5. Diagram 11.5 gives an overview of substances discussed in this chapter and the legislation that applies to their presence. In addition to this legislation, some examples are given of legislation dealing with sampling and analysis and some guidance documents available on the EU website to help businesses and inspectors deal with their legal obligations.

Diagram 11.5. Overview of contaminants legislation.

Substance	Limit	Sampling method	Guidance
Food contact materials	Reg. 1935/2004	Reg. 1935/2004	EFSA guidelines on submission of a dossier for safety evaluation by the EFSA of active or intelligent substances present in active and intelligent materials and articles intended to come into contact with food Union guidelines on Regulation (EU) No 10/2011 on plastic materials and articles intended to come into contact with food EU guidance to the Commission Regulation (EC) No 450/2009 of 29 May 2009 on active and intelligent materials and articles intended to come into contact with food EU guidelines on conditions and procedures for the import of polyamide and melamine kitchenware originating in or consigned from People's Republic of China and Hong Kong Special Administrative Region, China Union guidance on Regulation (EU) No 10/2011 on plastic materials and articles intended to come into contact with food as regards information in the supply chain
Mycotoxins	Reg. 1881/2006	Reg. 401/2006	
• Patulin	Reg. 1881/2006	Reg. 401/2006	Rec. 2003/598
• Aflatoxin	Reg. 1881/2006	Reg. 401/2006	Guidance document for Competent Authorities for the control of compliance with EU legislation on aflatoxins (no n°2010)
Pesticides	Reg. 396/2005	Reg. 396/2005	
Veterinary drugs	Reg. 470/2009		Notice to applicants F2/AW D(2003)
Hormones	Dir. 96/22 (ban)		
Radioactive contamination	Council Directive (EURATOM) 29/1996		

▶▶

Diagram 11.5. Continued.

Substance	Limit	Sampling method	Guidance
Chemicals	Reg. 1881/2006	Reg. 1881/2006	
• Lead, cadmium, mercury, inorganic tin, 3-MCPD, benzo(a)pyrene	Reg. 1881/2006		
• Methyl mercury	Reg. 1881/2006		D/530286
• Dioxin	Reg. 1881/2006	Reg. 1883/2006 Guidance of sampling of whole fishes of different size and/or weight (no n°)	Guidelines; Rec. 2006/88; COM(2001) 593 Guidelines for the enforcement of provisions on dioxins in the event non-compliance with the maximum levels for dioxins in food (no n°)
• PCBs	Reg. 1881/2006	Reg. 1883/2006	Rec. 2006/88
• PAHs	Reg. 1881/2006		Rec. 2005/108
• Furans	Reg. 1881/2006		Rec. 2006/88; COM(2001) 593
• Polychlorinated biphenyls			COM(2001) 593

12. Biological hazards

Rozita Spirovska Vaskoska

12.1 Introduction

The food on our plates not only provides us with the nutrients we need and the sensory experiences we long for, it also often introduces into our bodies organisms invisible to our eyes without a microscope. These so-called microorganisms (for instance, probiotics[481]) can bring health benefits, can be harmless, but can also cause adverse health effects ranging from diarrhoea to severe deadly diseases. Hence, microorganisms and other biological agents with the potential to cause an adverse health effect represent biological hazards. Microorganisms appear in various forms: bacteria, viruses, yeasts, moulds, algae, parasitic protozoa, microscopic parasitic helminths, and their toxins and metabolites.[482] To prevent the presence completely or at levels dangerous to human health, if possible, legal limits are being established in the EU and worldwide. Legal limits can only be established if scientists have developed adequate methods to verify compliance to those limits. Next to legal limits, other legislative forms are used to control the spread of biological hazards, such as causative agents of zoonoses,[483] disease-causing agents that are not microorganisms (prions), as well as animal-derived materials that can pose a risk to human and animal health. [484]

12.2 Microbiological criteria

Within the EU, most legal limits on the presence of microorganisms in foods are introduced with Regulation 2073/2005.[485] Most of what is understood under biological hazards in the EU and subsequently more regulated is represented by bacteria, because of their ubiquity in the food chain, pathogenicity[486] and more developed methods for their detection, identification and enumeration in food. Nevertheless, there are also bacterial strains such as *Vibrio vulnificus* and *Vibrio parahaemolyticus,* for which the latter methods are not sufficiently developed to subject these microorganisms to a legal limit.[487] For this same reason the intention to set up a limit for Norwalk-like viruses and viruses in live bivalve molluscs, which represent viruses of public health importance, has been thwarted.[488] Yeasts

[481] According to the World Health Organization: live micro-organisms which, when administered in adequate amounts, confer a health benefit on the host.
[482] Article 2(b) Regulation 2073/2005.
[483] A disease transmissible from animal to human.
[484] The author would like to thank Martine Reij, from the European Chair in Food Safety Microbiology at Wageningen University, for her valuable comments.
[485] Commission Regulation 2073/2005 of 15 November 2005 on microbiological criteria for foodstuffs.
[486] Ability to cause a disease.
[487] Commonly associated with seafood.
[488] Norwalk-like viruses are the most common cause of what is known as 'winter diarrhoea'.

and moulds,[489] in their vegetative forms, mostly render food unfit for human consumption (by spoiling the food), rather than injurious to health,[490] and thus they don't have to be regulated. From the parasites, *Trichinella*, a worm that can be present in domestic animals and game, is subject to legal provisions. Microorganisms, as living organisms, produce metabolites and toxins that can lead to allergic reactions, intoxication or mutagenicity (leading to an increase in the cancer risk). Mycotoxins, toxins produced by moulds, are treated as contaminants in EU food law.[491] In contrast, histamine, an allergenic substance created by enzymatic conversion of histidine (a compound naturally present in fish) by bacteria is regulated in the frame of microbiological limits.[492]

Regulation 2073/2005 on microbiological criteria for foodstuff lays down the microbiological criteria for certain microorganisms and the implementing rules to be complied with by food business operators when implementing the general and specific hygiene measures referred to in Article 4 of Regulation 852/2004,[493] or the so-called Hygiene 1 Regulation. Microbiological criteria are listed in the most important part of Regulation 2073/2005, Annex I of Chapter I. The Hygiene 1 Regulation refers to the usage of the comitology procedure[494] and consultation of EFSA when criteria are being set.[495]

12.2.1 Definition and types of microbiological criteria

A microbiological criterion means a criterion defining the acceptability of a product, a batch of foodstuffs or a process, based on the absence, presence or number of microorganisms, and/or on the quantity of their toxins/metabolites, per unit(s) of mass, volume, area or batch.[496] There are two types of microbiological criteria used in the EU: food safety criteria and process criteria.

A 'food safety criterion' means a criterion defining the acceptability of a product or a batch of foodstuff applicable to products placed on the market.[497] Food safety criteria apply at three main points: when the product is ready to be placed on the

[489] Yeast and moulds are different from bacteria, since they have cells more similar to cells of plants and humans. Yeasts are composed of one cell, and moulds of multiple cells.
[490] The concepts have been explained in Chapter 9 Section 9.12.
[491] See Chapter 11.
[492] Following the classification of mycotoxins.
[493] This follows from Article 1 of Regulation 2073/2005. Regulation 852/2004 of the European Parliament and of the Council of 29 April 2004 on the hygiene of foodstuffs, also known as 'Hygiene 1' discussed in Chapter 13.
[494] Article 4(4) Regulation 2073/2005.
[495] Article 15 Regulation 2073/2005.
[496] Article 2(b) Regulation 2073/2005.
[497] Article 2(c) Regulation 2073/2005.

market but still in the premises of the food business,[498] when the product is in the distribution and retail chain, and when a product imported from a third country is at the point of entry to the EU. Food safety criteria are mandatory and in case of non-compliance the procedures described in Article 19 GFL are likely to apply. However, the regulation is not clear cut on how to carry out these procedures and allows for two alternative options once a non-compliant food is discovered. One option is to perform further processing that will eliminate the presence of the microorganisms in food that is not yet at the retail level.[499] The other option is to change the purpose of the batch from that originally intended, as long as there is no risk to public health and as long as it is in conformity with the Hazard Analysis and Critical Control Points (HACCP) procedures and approved by the competent authorities.[500] On the format of food safety criteria, see Diagram 12.1.

A 'process hygiene criterion' is a criterion indicating the acceptable functioning of the production process.[501] These criteria are listed in Annex I Chapter II of Regulation 2073/2005. The process hygiene criteria are more like a guideline; they represent indicative levels that when surpassed require a corrective action to be taken by the food business operator. Tailor-made corrective actions are described in conjunction with the criterion. Corrective actions mainly include improvement of production hygiene, and more specifically: improvement in slaughter hygiene; review of process control, origin of animals and of the biosecurity measures in the farms of origin; improvements in selection and/or origin of raw materials; check on the efficiency of heat treatment, prevention of recontamination and additional tests. On the format of process hygiene criteria, see Diagram 12.1.

[498] In Annex I of Regulation 2073/2005 it is written that the food safety criteria apply to products placed on the market, and since placing on the market includes the holding for sale (Article 3(8) GFL), the stage of the ready products held in the premises of the food business is included here.

[499] Article 7(2) Regulation 2073/2005.

[500] See Chapter 13 for more details.

[501] Article 2(d) Regulation 2073/2005.

Diagram 12.1. Format of food safety criteria in Regulation 2073/2005.

Food category	Micro-organisms/ their toxins, metabolites	Sampling plan		Limits		Analytical reference method	Stage where the criterion applies
		n	c	m	M		
1.1 Ready-to-eat foods intended for infants and ready-to-eat foods for special medical purposes	*Listeria monocytogenes*	10	0	Absence in 25 g		EN/ISO 11290-1	Products placed on the market during their shelf-life
1.2 Ready-to-eat foods able to support the growth of *L. monocytogenes*, other than those intended for infants and for special medical purposes	*Listeria monocytogenes*	5	0	100 cfu/g		EN/ISO 11290-2	Products placed on the market during their shelf-life
		5	0	Absence in 25 g		EN/ISO 11290-1	Before the food has left the immediate control of the food business operator, who has produced it
(...)							
1.23 Dried follow-on formulae	*Salmonella*	30	0	Absence in 25 g		EN/ISO 6579	Products placed on the market during their shelf-life
1.24 Dried infant formulae and dried dietary foods for special medical purposes intended for infants below 6 months of age	*Cronobacter* spp. (*Enterobacter sakazakii*)	30	0	Absence in 10 g		ISO/TS 22964	Products placed on the market during their shelf-life
(...)							
1.29 Sprouts	Shiga toxin producing *E. coli* (STEC) O157, O26, O111, O103, O145 and O104:H4	5	0	Absence in 25 g		CEN/ISO TS 13136	Products placed on the market during their shelf-life
(...)							
2.2.11 Dried infant formulae and dried dietary foods for special medical purposes intended for infants below six months of age	Presumptive *Bacillus cereus*	5	1	50 cfu/g	500 cfu/g	EN/ISO 7932	End of the manufacturing process

12.2.2 Setting up microbiological criteria

Principles in the setting up of microbiological criteria

The issue of scientific opinions needed in the setting up as well as the evaluation of microbiological criteria was the task of the Scientific Committee on Veterinary measures relating to Public Health (SCVPH) and the SCF. It has now been transferred to the Biological Hazards (BIOHAZ) panel of EFSA. The work of the BIOHAZ Panel is supported administratively and scientifically by the work of the Biological Hazards and Contaminants Unit. Before Regulation 2073/2005 was adopted, vertical directives were the legal instruments introducing microbiological criteria separately for specified categories of foods.[502] The first attempt of the Commission to define general conditions on establishing microbiological criteria in the EU was not finalised.[503] In accordance with the previous Hygiene Directive 93/43, a good explanatory guideline on the development of microbiological criteria was created by the Commission.[504] However, the setting of microbiological criteria in the EU, in their first form, did not follow a structured approach. The SCVPH in 1999 evaluated the old criteria and reported that the microbiological criteria at the time (concerning fresh meat, poultry, milk, fish, eggs) were outdated (5-10 years old), were not based on the principles of the Codex Alimentarius, were not meaningful for consumer health protection, did not provide for corrective actions and were mainly focused on the production site.[505] At the time, foodborne pathogens of great importance such as *Campylobacter* spp., and enterohaemorrhagic *Escherichia coli*, particularly serogroup O157, were just emerging and the SCVPH proposed the possibility to dedicated microbiological criteria. Corrections were proposed and they were taken into account when the strategy for the current criteria was formulated, as explained below.

In the meantime, as for other legislative topics, the CAC came up with an international standard for the establishment and application of microbiological criteria related to foods.[506] The Codex Alimentarius general principles for the establishment and application of microbiological criteria are shown in Law text box 12.1.[507]

---

[502] See Chapter 7.

[503] Commission proposal for a Council Decision on general conditions to be followed for establishing microbiological criteria for foodstuffs and feedingstuffs, including the conditions for their preparation, in the veterinary, foodstuffs and animal nutrition sectors. OJ C 252/7. 02.10.1981.

[504] European Commission. Principles for the development of microbiological criteria for animal products and products for animal origin for human consumption, September 1997.

[505] DG SANCO, Scientific committee on veterinary measures relating to public health. The evaluation of microbiological criteria for food products of animal origin for human consumption, 23 September 1999.

[506] Codex Alimentarius Commission. Principles and Guidelines for the establishment and application of microbiological criteria related to foods, CAC/GL 21-1997.

[507] The CAC and its standards are elaborated in Chapter 3 Section 3.6.

Law text box 12.1. General principles for the establishment and application of microbiological criteria according to Codex Alimentarius.

A microbiological criterion should be appropriate to protect the health of the consumer and where appropriate, also ensure fair practices in food trade.

- A microbiological criterion should be practical and feasible and established only when necessary.
- The purpose of establishing and applying a microbiological criterion should be clearly articulated.
- The establishment of microbiological criteria should be based on scientific information and analysis and follow a structured and transparent approach.
- Microbiological criteria should be established based on knowledge of the microorganisms and their occurrence and behaviour along the food chain.
- The intended as well as the actual use of the final product by consumers needs to be considered when setting a microbiological criterion.
- The required stringency of a microbiological criterion used should be appropriate to its intended purpose.
- Periodic reviews of microbiological criteria should be conducted, as appropriate, in order to ensure that microbiological criteria continue to be relevant to the stated purpose under current conditions and practices.

In 2005, just after the adoption of the Hygiene 1 Regulation, the Commission created a strategy for setting up microbiological criteria for foodstuffs.[508] Limitations in the availability of finalised risk assessments were recognised and alternative risk profiles were suggested.[509] It has been admitted that the establishment of microbiological criteria has not faithfully followed the prescribed structured approach of the Codex Alimentarius or the International Commission on Microbiological Specifications for Food (ICMSF), but it has been based on experience of food production and processing, research and expert opinions of what was necessary to ensure food safety on the one hand, and what was considered achievable in relation to the application of good hygienic practices on the other.[510] In that way, with microbiological criteria we deal with the same concepts as with chemical contaminants: necessity and safety, but with different methods to establish them.

[508] DG SANCO, Discussion paper on strategy for setting microbiological criteria for foodstuffs in Community legislation, 2005.

[509] DG SANCO, Discussion paper on strategy for setting microbiological criteria for foodstuffs in Community legislation, 2005. Risk profiles provide for a description of a food safety problem and its context, and are developed to identify those elements of a hazard or risk that are relevant to risk management decisions.

[510] EFSA, Opinion of the scientific panel on biological hazards on microbiological criteria and targets based on risk analysis. The EFSA Journal (2007) 462, pp. 1-29.

Choosing the elements of a microbiological criterion

A microbiological criterion comprises six basic elements: the food category, the microorganism (toxin or metabolite), the sampling plan, the limit, the analytical reference method and the stage to which it applies:

a. Selecting the pathogen, food category and their association

   The guidelines of the Commission in accordance with the old Hygiene directive, next to the Codex Alimentarius principles, describe most accurately the considerations that should be given so a microbiological criterion fulfils its purpose in general, or the considerations to be taken into account in the selection of the food category, microorganism and their association in particular. The Commission listed the following considerations:
   - evidence of risk to health;
   - microbiological status of the raw material (s);
   - the effect of processing on the microbiological status of the food:
     - the likelihood and consequences of microbial contamination and/or growth during subsequent handling, storage and use;
     - the categories of consumers concerned; and
     - the cost/benefit of applying such a criterion.

   The risk to health is essential for deciding which microorganisms in which foods should be limited and normally this is established through epidemiological evidence. Epidemiological evidence means that a specific microorganism in a specific food category is a proven cause of some common and dangerous disease. For instance, *Cronobacter* spp., a more recently discovered microorganism epidemiologically linked to severe diseases and death in infants, has been listed in Regulation 2073/2005.

   As for the microbiological status and the effect of processing, it is important to realise that microorganisms are living organisms; once they are present in food, they multiply when there are favourable conditions for their growth (e.g. temperature of storage) and they are inactivated (killed) when the food is treated (e.g. cooked, baked, irradiated). The ability of microorganisms to multiply is taken into account in the set-up of the criterion on *Listeria monocytogenes* in ready-to-eat foods.[511] Ready-to-eat food means food intended by the producer or the manufacturer for direct human consumption without the need for cooking or other processing that would effectively eliminate or reduce microorganisms of concern to an acceptable level.[512] Since the number of microorganisms on the plate may be higher than the number in the food that left the food business because of the favourable growth conditions during the storage of the food, two categories of ready-to-eat foods are distinguished: ready-to-eat foods that allow growth and ready-to-eat foods that do not allow growth. The former category is bound to a stricter criterion for the absence of *L. monocytogenes,* to account

[511] *Listeria monocytogenes* is a microorganism that may cause meningitis and other severe infections in elderly and people with underlying disease, and may lead to abortion in pregnant women.
[512] Article 2(g) Regulation 2073/2005.

for the possible increase up to the level of 100 cfu/g for the latter.[513] Annex II of Regulation 2073/2005 includes requirements for the shelf-life studies; food business operators need to verify compliance with this criterion for ready-to-eat foods. The ability of microorganisms to be inactivated is taken into consideration when setting up criteria on staphylococci[514] and their toxins in milk-derived products. An assessment by the SCVPH in 2003[515] split the criteria that were initially only tailored to the microorganism into two categories: where the enterotoxin is the only indicator (mostly when the microorganism has been inactivated by a treatment) and where the staphylococci are first assessed and if they are at the toxin-producing level,[516] a testing for the toxin is also done. The target consumer of the food category is also a highly relevant factor; microbiological criteria on foods for infants and special medical needs are much stricter than for food intended for the healthy adult population.[517]

b. Setting up of the microbiological limit
The microbiological limits for food safety criteria should be selected based on a risk assessment. The dose response of the microorganism/toxin, where the number of people becoming ill is a function of the number of microorganisms ingested, gives a rough indication of the threshold below which the contaminated food is safe to consume. Basically, there is always some probability that people will get ill when a certain amount of food contaminated with microorganisms is ingested and a choice has to be made about which level is acceptable. In Regulation 2073/2005 there are not many dubious cases; two pathogens *Salmonella* and *Cronobacter* have 'zero tolerance' criteria (absence is demanded), because small numbers of cells of the microorganisms are reported or believed to cause a disease. Distinctively, *Listeria monocytogenes* is considered to cause a disease when present at higher numbers and as recommended by the SCVPH, SCF, and Codex Alimentarius, it should not exceed 100 cfu/g at the point of consumption.[518] At the beginning of Chapter 11, biological hazards were classified in the group of restricted substances. One could argue that some microbiological criteria like those for *Salmonella* in various foods, Shiga-toxin producing *E. coli* (STEC) strains in sprouts, *L. monocytogenes* in ready-to-eat foods intended for infants and not only for infants before they leave the producer and *Enterobacteriaceae* in dried infant formulae have a banning approach rather than a restrictive one. That is due to the fact that the microbiological criterion

[513] Cfu/g means colony forming units/gram.
[514] Microorganisms naturally present on human skin, but also in milk and meat.
[515] DG SANCO, Opinion of the scientific committee on veterinary measures relating to public health on staphylococcal enterotoxins in milk products, particularly cheeses. 26-27.03.2003.
[516] 100,000 microorganisms/g.
[517] The same amount of some microorganism may cause a deadly disease in infants, but no symptoms at all in healthy adults.
[518] This controlled level in a risk assessment is predicted to lead to 1 in 100,000 infections with *Listeria monocytogenes*, according to Nørrung (2000).

is absence of these organisms in 10 or 25 gram[519] of food. Statistically, absence cannot be guaranteed with a sampling plan. There is always a probability, predetermined by the selection of the criterion, that a 'defective' lot will be accepted.[520] Therefore, in their practical application, these kinds of 'zero tolerance' criteria are restrictive rather than banning by the law of statistics. The limits chosen for process hygiene criteria are generally not based on the health effects, but rather on what is attainable in the meaning of necessity; which at the same time indirectly can lead to safety. Indirectly because indicator microorganisms are being used primarily; i.e. not all *E. coli* or *Enterobacteriaceae* are dangerous to human health, but their high numbers indicate faecal contamination, an unhygienic process and/or the presence of pathogenic species among them.

c. Selecting a sampling plan
To check the compliance of the food with the criterion, the business or the competent authority must rely on the results from the analysis of selected samples since it is impossible to test everything. A sample is a set composed of several units or portions of matter that should provide information in this case of the presence and/or number of selected microorganisms.[521] The samples are selected based on a predetermined plan and the sampling plans to verify the microbiological criteria in Regulation 2073/2005 can be either two-class of three-class plans. A two-class sampling plan distinguishes between two categories: an unsatisfactory sample and a satisfactory sample. The decision yardstick for an unsatisfactory microbiological result will be the presence of some microorganisms or a number above a defined level of others. In Diagram 12.1 the criteria on *Salmonella* are based on a two-class sampling plan, so the detection of any *Salmonella* in the selected foods will be an unsatisfactory result. Three-class sampling plans allow an intermediate category. The yardsticks here are: a lower limit 'm', below which the results of the sample are satisfactory; an intermediate level between 'm' and 'M' where a limited number of samples 'c' out of the total 'n' are allowed in order to declare the presence-level as acceptable; and higher level 'M' that differentiates acceptable from unsatisfactory. In Diagram 12.1, the criterion for *Bacillus cereus* is an example of a criterion based on a three-class sampling plan, where five samples of dried infant formula should be taken, of which one can contain between 50 and 500 cfu/g (acceptable), and the other four should contain less than 50 cfu/g (satisfactory). None of the samples may have counts over 500 cfu/g (unsatisfactory). Overall, a compliance with a criterion means obtaining

[519] This absence can also be written as 0.1 cfu/g or 0.04 cfu/g, respectively. The bigger the lot to be tested the higher the likelihood of finding something. This makes absence in 25 grams a stricter criterion than absence in 10 grams.
[520] This is so-called consumer risk, see: M. van Schothorst, M.H. Zwietering, T. Ross, R.L. Buchanan and M.B. Cole, (ICMSF), Relating microbiological criteria to food safety objectives and performance objectives, Food Control 2009, pp. 967-979.
[521] Article 2(j) Regulation 2073/2005.

satisfactory or acceptable results, as described above, when testing against the values set in Annex I of Regulation 2073/2005 through the taking of samples, the conduct of analyses and the implementation of corrective action, in accordance with food law and the instructions given by the competent authority.[522]

d. Selecting a testing method and corrective actions
The analytical methods proposed for testing against microbiological criteria are internationally used, validated reference methods of the International Organization for Standardization (ISO).[523] Corrective actions for process hygiene criteria are selected based on the purpose of the criterion.

12.2.3 Application of microbiological criteria

The determination of acceptability by microbiological criteria, according to the CAC, can be applied in various ways that we can classify as: food business related (such as verification of prerequisite programs or HACCP; verification of specifications between food business operators; and providing information about what can be achieved when best practices are applied) and government-related (which includes verification as to whether the measures will meet public health goals (performance objectives (PO) and food safety objectives (FSO)[524]). Part of the latter group, of government-related purposes, is also the verification as part of official controls.

Business related application

The Hygiene 1 Regulation asks food business operators to be compliant with microbiological criteria.[525] The obligation of food business operators to comply with the relevant criteria applies in all stages of the food chain, from raw materials to finished product, and should be fulfilled in the context of the HACCP.[526] See Law text box 12.2. HACCP is a system for food safety management for food businesses.[527] As part of the implementation of the HACCP system, critical control points are chosen based on the severity and probability of a hazard and the ability of the control measures to effectively control the hazard. When the hazard is biological, it is underlined with some microbiological criteria.[528] However, Regulation 2073/2005 does not ask for regular testing for confirmation of compliance to criteria, neither

[522] Article 2(l) Regulation 2073/2005.
[523] ISO standards are strictly speaking private standards. On private standards, see Chapter 23. We see here an interesting example where public law is required to follow private standards. One could call this a form of 'hybrid' regulation.
[524] Explained below.
[525] Article 4(3)(a) Regulation 2073/2005.
[526] Article 3(1) Regulation 2073/2005.
[527] See Chapter 13 for further details.
[528] Literature: R.L. Buchanan, The role of microbiological criteria and risk assessment in HACCP, Food Microbiology, 1995, 421-424.

a regular end point testing for release on the market (positive release).[529] It only asks for defined frequency of sampling and testing for selected foods listed in Annex I, Chapter 3b. Food business operators have the freedom to select their criterion by defining a different limit (lower), sampling plan and analytical method.[530] As for the analytical methods, if companies decide to use an alternative to the one given in Annex I for the purpose, then the method should be validated against the reference method. Because of the irregular nature of the criteria and the fact that microbiological tests are impractical for routine monitoring,[531] the monitoring step of the HACCP system, explained further in Chapter 13, uses mostly other procedures that can give the same safety guarantee as a microbiological test would.[532] Following that, microbiological criteria mainly serve as a verification tool for the HACCP-based procedures or HACCP plans.

As part of their contacts with other food businesses, companies develop product specifications as a private law agreement. Microbiological criteria also serve as a means of verifying such specification in business-to-business relations.

Law text box 12.2. Article 3 of Regulation 2073/2005 on compliance with microbiological criteria.

General requirements

1. Food business operators shall ensure that foodstuffs comply with the relevant microbiological criteria set out in Annex I. To this end the food business operators at each stage of food production, processing and distribution, including retail, shall take measures, as part of their procedures based on HACCP principles together with the implementation of good hygiene practice, to ensure the following:
   (a) that the supply, handling and processing of raw materials and foodstuffs under their control are carried out in such a way that the process hygiene criteria are met,
   (b) that the food safety criteria applicable throughout the shelf-life of the products can be met under reasonably foreseeable conditions of distribution, storage and use.
2. As necessary, the food business operators responsible for the manufacture of the product shall conduct studies in accordance with Annex II in order to investigate compliance with the criteria throughout the shelf-life. In particular, this applies to ready-to-eat foods that are able to support the growth of *Listeria monocytogenes* and that may pose a *Listeria monocytogenes* risk for public health.
   Food businesses may collaborate in conducting those studies.
   Guidelines for conducting those studies may be included in the guides to good practice referred to in Article 7 of Regulation (EC) No 852/2004.

[529] Food Standards Agency, General guidance for food business operators EC Regulation No. 2073/2005 on microbiological criteria for foodstuffs.
[530] Article 5(5) Regulation 2073/2005.
[531] Test results are issued in a few days with classical microbiological methods.
[532] Such examples are: time-temperature combinations that lead to inactivation of the microorganisms, pH values that prevent growth, controlled rapid cooling and cold storage, etc.

Government-related application

Food law should be based on risk analysis.[533] In an attempt to create a science-based policy, numerical benchmark expressions of risk management goals are proposed to be set at different points of the food chain. They should all contribute to a defined level of people getting ill that is acceptable for the society, or the so-called Appropriate Level of Protection (ALOP). On the international scene, the CAC has defined additional risk management metrics that together with microbiological criteria can be used to achieve a public health goal such as reduction of an incidence of a foodborne illness (see Law text box 12.3). Theoretically, these definitions also apply to chemical hazards.[534]

Law text box 12.3. Definitions of Codex Alimentarius on FSO, PC and PO.[535]

Definitions of risk analysis terms related to food safety

Food Safety Objective (FSO)

The maximum frequency and/or concentration of a hazard in a food at the time of consumption that provides or contributes to the appropriate level of protection (ALOP).

Performance Criterion (PC)

The effect in frequency and/or concentration of a hazard in a food that must be achieved by the application of one or more control measures to provide or contribute to a PO or an FSO.

Performance Objective (PO)

The maximum frequency and/or concentration of a hazard in a food at a specified step in the food chain before the time of consumption that provides or contributes to an FSO or ALOP, as applicable.

These measures are not yet part of EU food law, but more and more research[536] is being done on their incorporation in policy-making, for instance by determining a microbiological criterion based on a defined ALOP and FSO. Basically, if this became common practice, governments could select the ALOP or the acceptable number of people getting ill with a foodborne illness, derive from the ALOP the FSO or the number of microorganisms in the plate of the consumer and from there determine the maximum number of microorganisms that the food business operator will allow to be present in the food going out of some equipment or out of its premises. This is exactly where microbiological criteria meet with these new metrics and can be linked to the public health goal. Care should be taken to not

[533] Article 6(1) GFL.

[534] In this direction see: E.G. Garcia-Cela, A.J. Ramos, V.S. Sanchis, S.M. Marin, Emerging risk management metrics in food safety: FSO, PO. How do they apply to the mycotoxin hazard?, Food Control, 2012, 797-808; and C.S de Swarte and R.A. Donker, Towards an FSO/ALOP based food safety policy, Food Control, 2005, 825-830.

[535] Definitions of Risk Analysis Terms Related to Food Safety, in: Codex Alimentarius Commission, Procedural Manual, 21st edition, Rome 2013, p. 115.

[536] A lot of work has been done by the ICMSF.

mix food safety criteria with FSO, or process hygiene criteria with PO, because the criteria are composed of multiple elements, while FSO and PO are only a microbiological limit.

Microbiological criteria change over time and their dynamics is dependent on scientific developments,[537] but also on risk management decisions.

Preventive risk management: control of Salmonella and other zoonotic agents

Microbiological criteria have been applied as a preventive risk management measure in the reduction of prevalence of *Salmonella* in defined species of poultry and pigs. Regulation 2160/2003 on the control of *Salmonella* and other zoonotic agents,[538] whose provisions will be elaborated hereafter, has been the basis for setting a target of $<1\%$ prevalence of *Salmonella* in the selected species of animals.[539] EFSA, as a reference point of scientific opinion, has issued a positive assessment of the benefits of such a management measure based on a modelling approach. In the pursuit of this EU target established under Regulation 2160/2003, dynamic process hygiene criteria on *Salmonella* in carcasses of broilers and turkey moving in time from less to more strict ones (with less samples allowed as unsatisfactory) have been enacted in Regulation 2073/2005.[540] In addition, a food safety criterion on *Salmonella* Enteritidis and *Salmonella* Typhimurium was established in fresh poultry meat for the same goal. [541]

Reactive risk management: STEC in sprouts[542]

The case of STEC in sprouts is an example of a criterion that became a necessity after a public health scare in the form of reactive risk management. The SCVPH has issued an opinion on STEC in foodstuffs, where it was concluded that meaningful risk reductions are unlikely to be delivered by a criterion where the microorganism is of sporadic occurrence and low prevalence.[543] That would not be the case if prior evidence indicated faecal contamination or high prevalence of STEC, where microbiological criteria would be considered as risk management measures.[544]

---

[537] On food law and science interaction, see Chapter 9 Section 9.8.
[538] Regulation 2160/2003 of the European Parliament and of the Council of 17 November 2003 on the control of salmonella and other specified food-borne zoonotic agents.
[539] Covered species are breeding flocks of *Gallus gallus*, laying hens, broilers, turkeys, herds of slaughter pigs and breeding herds of pigs.
[540] Recital 14 Regulation 1086/2011.
[541] These two microorganisms are selected as the serotypes that contribute most to human salmonellosis.
[542] STEC is also referred to as VTEC and EHEC.
[543] DG SANCO, Opinion of the scientific committee on veterinary measures related to public health on verotoxicogenic *E. coli* (VTEC) in foodstuff. 21-22 January 2003.
[544] DG SANCO, Opinion of the scientific committee on veterinary measures related to public health on verotoxicogenic *E. coli* (VTEC) in foodstuff. 21-22 January 2003.

However, although the SCVPH found that STEC is a hazard to public health in several categories of foods,[545] it was not subject to a microbiological criterion. This situation remained until 2013, when in the wake of the STEC outbreak in the EU, and particularly in Germany, a criterion on STEC in sprouts was established. A scientific opinion was issued by EFSA that thoroughly considered at which point in the food chain a criterion would be useful and which sampling plans would be most effective. Subsequently, six serotypes of STEC were subjected to a new two-class microbiological criterion.

12.3 Other microbiological limits

Regulation 853/2004 laying down specific hygiene rules for food of animal origin, sets up a limit (not strictly a criterion in the definition of Regulation 2073/2005) on the plate count which is the total number of microorganisms in raw milk and colostrum.

12.4 Control of zoonoses

The control of zoonoses, caused by zoonotic microorganisms, is not exclusively related to food safety, but also with public and animal health. Directive 2003/99[546] lays down rules to ensure monitoring of zoonoses and causative agents, monitoring of antimicrobial resistance, proper epidemiological investigations of outbreaks and collection of relevant information.[547]

In the frame of the monitoring and control of zoonoses, especially those caused by *Salmonella*, another form of legal limits is established in Regulation 2160/2003. Regulation 2160/2003 is based on Article 168(4b) TFEU on the protection of public health through veterinary measures and is devoted to the control of *Salmonella* and other zoonotic diseases. The purpose of this regulation is to ensure that proper and effective measures are taken to detect and control *Salmonella* and other zoonotic agents at all relevant stages of production, processing and distribution, particularly at the level of primary production, including in feed, in order to reduce their prevalence and the risk they pose to public health.[548] Four aspects are covered within this Regulation: the adoption of targets for reduction of prevalence of *Salmonella* and other zoonotic microorganisms (Chapter II), the approval of national control programs within the Member States (Chapter III), adoption of specific rules on the control methods in the reduction of prevalence (Chapter IV) and the adoption of the rules concerning trade in the Union and imports from

[545] Raw or undercooked beef and maybe meat from other ruminants, minced meat and fermented beef and products thereof, raw milk and raw milk products, fresh produce, in particular sprouted seeds, and unpasteurised fruit and vegetable juices.
[546] Directive 2003/99 of the European Parliament and of the Council of 17 November 2003 on the monitoring of zoonoses and zoonotic agents.
[547] Article 2 Directive 2003/99.
[548] Article 1(1) Regulation 2160/2003.

third countries (Chapter V). The targets represent a numerical expression and the elements are shown in Law text box 12.4. Targets are established with consideration given to the experience gained from national measures and information forwarded to the Commission and the Member States under previous requirements.[549]

Law text box 12.4. Article 2 Regulation 2160/2003.

The targets referred to in paragraph 1 shall consist at least of:
(a) a numerical expression of:
  (i) the maximum percentage of epidemiological units remaining positive; and/or
  (ii) the minimum percentage of reduction in the number of epidemiological units remaining positive;
(b) the maximum time limit within which the target must be achieved;
(c) the definition of the epidemiological units referred to in (a);
(d) the definition of the testing schemes necessary to verify the achievement of the target; and
(e) the definition, where relevant, of serotypes with public health significance or of other subtypes of zoonoses or zoonotic agents listed in Annex I, column 1, having regard to the general criteria listed in paragraph 6(c) and any specific criteria laid down in Annex III.

The targets are to be achieved with the establishment of national control programs for the selected zoonotic agents and zoonoses. National control programs need to be approved by the Commission.[550] Even the food business operators may establish their own control programs that should be approved by the competent authority of the Member State.[551] The regulation also provides for rules on the control methods to be used for the achievement of the target, including detailed rules on procedures and implementation.[552] As for trade among the EU Member States, the conditions of the national program of the country of destination should be fulfilled if the country is authorised for the purpose. Third countries that export the specified animals and foodstuffs into the EU should also have an approved national control program equivalent to the EU-based programs.

12.5 Specific hazards

Regulation 854/2004[553] names the following specific hazards of biological origin: transmissible spongiform encephalopathies (TSEs), cystercosis, trichinosis, glanders, tuberculosis and brucellosis.[554] The meat from animals infected with these diseases

[549] Article 4(1) Regulation 2160/2003.
[550] Article 6 Regulation 2160/2003.
[551] Article 7(1) and (2) Regulation 2160/2003.
[552] Article 8(1) Regulation 2160/2003.
[553] Regulation 854/2004 of the European Parliament and of the Council of 29 April 2004 laying down specific rules for the organisation of official controls on products of animal origin intended for human consumption.
[554] Annex I Section IV, Chapter IX, C. Regulation 854/2004.

is declared unfit for human consumption. Regulation 2075/2005[555] lays down the specific rules for the official control of *Trichinella* in meat. The meat of horses, wild boar and other farmed and wild animal species susceptible to trichinosis shall be systematically sampled as part of their post-mortem examination. The Annexes of Regulation 2075/2005 describe in detail the detection methods, freezing methods, examination of animals other than swine and the detailed conditions for *Trichinella*-free holdings and regions with a negligible *Trichinella* risk. The latter holdings and regions are exempted from *Trichinella* testing.

12.6 TSE risk material

The first TSE disease was discovered in 1986 as bovine spongiform encephalopathy (BSE), better known as 'mad cow disease'. Regulation 999/2001[556] lays down rules for the prevention, control and eradication of transmissible spongiform encephalopathies (TSEs) in animals.[557] It applies to the production and placing on the market of live animals and products of animal origin and in certain specific cases to exports thereof.[558] It demands separation of live animals from animal-derived products to prevent the risk of cross contamination.[559] A classification of the Member States and third countries, based on defined criteria and risk analysis[560] is provided based on the BSE status in countries with: negligible BSE risk, controlled BSE risk and undetermined BSE risk.[561] The status is obtained after an application for its determination to the Commission.[562] The Member States are obliged to carry out annual monitoring programs,[563] according to the Annex V of the Regulation. The risk material that should be removed and disposed of is listed in Annex V of the Regulation and also in the Regulation on animal by-products (explained hereafter).[564] Member States that have suspected or infected animals should notify the Commission and should restrict the movement of the animal or kill and dispose of the animal.[565] Export, import requirements related to this Regulation and reference laboratories are listed in its Annexes VIII, IX and X, respectively.

[555] Commission Regulation 2075/2005 of 5 December 2005 laying down specific rules on official controls for *Trichinella* in meat.
[556] Regulation 999/2001 of the European Parliament and of the Council of 22 May 2001 laying down rules for the prevention, control and eradication of certain transmissible spongiform encephalopathies.
[557] Article 1(1) Regulation 999/2001.
[558] Article 1(1) Regulation 999/2001.
[559] Article 2 Regulation 999/2001.
[560] Annex II Regulation 999/2001.
[561] Article 5(1) Regulation 999/2001.
[562] Article 5(1) Regulation 999/2001.
[563] Article 6 Regulation 999/2001.
[564] Article 8 Regulation 999/2001.
[565] Article 11 Regulation 999/2001.

12.7 Animal by-products

In the past, animals that died on the farm and were unfit for human consumption could enter the animal feed chain. This practice of recycling cadavers and material unfit for human consumption into the feed chain was the main factor in the spreading of the BSE epidemic, but also of other food safety and animal health scares, such as the dioxin crises and foot and mouth disease. As a consequence, these practices are now heavily regulated.

Regulation 1069/2009[566] lays down the rules on animal by-products and derived products. Animal by-products encompass entire bodies or parts of animals, products of animal origin or other products obtained from animals, which are not intended for human consumption, including oocytes, embryos and semen.[567] Examples are animal feed, organic fertilisers and soil improvers, technical products (pet food, wool, hides and skins for leather, fishmeal, processed animal protein, blood for diagnostic purposes).[568] They were either initially not intended for human consumption, or irreversibly grouped as animal by-products by the operator.[569] Three categories of animal by-products are differentiated based on the risk to human and animal health.

Category 1 is the most risky category and it includes, for instance, animals that have been infected with TSE, by-products contaminated with PCBs and mycotoxins. Category 2 materials include manure, products containing veterinary drugs above the legal limits, products unfit for human consumption because of foreign bodies, imported food incompliant with veterinary legislation, etc. Category 3 materials include foods that pose the least risk, such as those fit for human consumption but rejected for commercial reasons or foods unfit for human consumption but not showing signs of a communicable disease. Category 1 materials are treated more severely, for example by incineration, co-incineration, landfilling after processing, or used as fuel by combustion. By-products of category 2 can be used as raw materials for biogas, organic fertilisers, soil improvers and even applied directly on the soil for products such as manure not considered as a risk by the competent authorities. Finally, by-products of category 3, next to the purposes listed for category 2, can also be used as raw materials for manufacturing animal or pet feed.

[566] Regulation 1069/2009 of the European Parliament and of the Council of 21 October 2009 laying down health rules as regards animal by-products and derived products not intended for human consumption and repealing Regulation 1774/2002 (Animal by-products Regulation).
[567] Article 3(1) Regulation 1069/2009.
[568] Http://ec.europa.eu/food/food/biosafety/animalbyproducts/index_en.htm.
[569] Article 2 Regulation 1069/2009.

Registered operators conduct the collection, identification and transport of the by-products, while keeping traceability records and maintaining HACCP-based procedures.[570] Traceability is important to prevent diversion of by-products in the food chain. Further provisions on by-products and the procedures of their disposal and usage are provided in Regulation 124/2011 implementing Regulation 1069/2009.[571]

[570] Articles 21-29 Regulation 1069/2009.

[571] Commission Regulation 142/2011 of 25 February 2011 implementing Regulation (EC) No 1069/2009 of the European Parliament and of the Council laying down health rules as regards animal by-products and derived products not intended for human consumption and implementing Council Directive 97/78/EC as regards certain samples and items exempt from veterinary checks at the border under that Directive.

13. Process: hygiene, traceability and recall

Rozita Spirovska Vaskoska, Bernd van der Meulen and Menno van der Velde

13.1 Introduction

Without a doubt the rules on the process, i.e. what businesses have to do, represent the hard core of food safety law addressing food business operators. In this chapter the subject matter is subdivided into three parts. Requirements aimed at actions to be taken to prevent food safety problems, requirements regarding actions that must be taken to be prepared for food safety problems and requirements for actions to be taken to deal with food safety problems. The latter two issues are each the subject of – essentially – one provision in the General Food Law. A whole package of regulations and directives applies to the former. This difference in complexity is reflected in the attention given to it in this chapter. See Diagram 13.1 on food handling.

Diagram 13.1. Food handling.

	Internally	Externally
Prevention (Section 13.2)	Hygiene/HACCP (Section 13.2.8)	
Preparedness (Section 13.3)	(HACCP) (Section 13.3.1)	Tracking & tracing (Sections 13.3.2-13.3.5)
Response (Section 13.4)	(corrective action under HACCP)	Withdrawal & recall (Sections 13.4.1-13.4.4)

When a food business operator starts a new production process, or engages in new trade relations, s/he has to ascertain that both the handling of the food by the company internally and between companies externally is in agreement with the applicable provisions of food law. According to Article 17 of the General Food Law, food business operators are responsible for meeting the relevant (food safety) requirements at all stages of production, processing and distribution of food. Under the Hygiene Regulation 852/2004 they are even held responsible for formulating these requirements.

13.2 Prevention

13.2.1 Food hygiene

In Greek mythology, *Hygieia* – from whose name the word hygiene has been derived – daughter of Asclepius the demi-god of medicine and healing (Aesculapius in Latin) and sister to Panacea who represented the cure to sickness, was the goddess for maintenance of good health and prevention of disease. Safe food is probably the first prerequisite for maintaining good health. It is therefore fitting that in food law the word hygiene is used in a broad sense for all measures to ensure food safety even though in common speech the word has the more narrow meaning of cleanliness. See Law text box 13.1 for the definition of food hygiene in EU law.

Law text box 13.1. The definition of 'food hygiene' in Regulation 852/2004.

Article 2 Definitions

1. For the purposes of this Regulation:
   (a) 'food hygiene', hereinafter called 'hygiene', means the measures and conditions necessary to control hazards and to ensure fitness for human consumption of a foodstuff taking into account its intended use;

While this is not fully apparent from the wording of the definition, the measures referred to are measures to be taken by food businesses. Measures taken by public authorities are not included.

The safety of food products on a consumer's plate depends largely on the way they have been produced. For this reason rules have been made to ensure that safe methods of production are used. The legislation on the safe production of food is distributed over four EU regulations, 852/2004, 853/2004, 854/2004, and 882/2004, that came into force in 2006. They replaced the national food hygiene laws of the Member States that had been harmonised on the basis of one general EU Directive (93/43), and 17 vertical (product specific) directives for food products of animal origin (meat, fish, eggs, etc.).

13.2.2 Overview

In this section food hygiene law is set out in its complexity. It is organised as follows. First an overview is given of the package of legislation (Section 13.2.3-13.2.5). Next the requirement for all food businesses to be registered is addressed and for businesses working with products of animal origin to be approved as well (Section 13.2.6). Hygiene legislation contains requirements on the premises where food is handled (Section 13.2.7) and on the way this is done (Section 13.2.7-13.2.9). To a large extent businesses are themselves responsible for making the rules that apply to their processes either individually (HACCP) (Section 13.2.8) or collectively (hygiene guides) (Section 13.2.9).

13.2.3 EU food hygiene legislation

Regulation 852/2004 on the hygiene of foodstuffs is at the centre of EU food hygiene legislation. This regulation provides the principles, rules, requirements and instruments for food hygiene law, and addresses the food business operators of all foodstuffs of every stage of the food production chain.

The scope of the EU hygiene legislation

The scope of a regulation is the collection of activities and situations to which the rules and determinations of the regulation must be applied. Article 1 of Regulation 852/2004 is written to state the scope, at least that is what the title of that article suggests.[572] Article 1 comes closest to the description of its scope, the activities that the hygiene regulation is made to regulate, in the second sentence of its first paragraph:

> *This Regulation shall apply to all stages of production, processing and distribution of food and to exports, and without prejudice to more specific requirements relating to food hygiene.*

In this scope we recognise the holistic approach to food law discussed in Chapter 8. This description of the scope is somewhat more at a distance from the food production chain than the immediate predecessor of Regulation 852/2004, Directive 93/43, in its description of its objective. Article 3(1) of that directive prescribed:

> *The preparation, processing, manufacturing, packaging, storing, transportation, distribution, handling and offering for sale or supply of foodstuffs shall be carried out in a hygienic way.*[573]

Legislation is often more to the point when it excludes particular activities or situations from its scope. The four categories of matters that remain outside the rules of Regulation 852/2004 are:

1. primary production for private domestic use;
2. domestic preparation, handling or storage of food for private domestic consumption;
3. direct supply, by the producer, of small quantities of primary products to the final consumer or to local retail establishments directly supplying the final consumer;
4. collection centres and tanneries which fall within the definition of food business only because they handle raw material for the production of gelatine or collagen.[574]

---

[572] However, it is mainly a list of policy preferences that the legislator wants to achieve and his leading principles. Article 1 is the list of objectives of the regulation in the area and for the activities that fall under its sway. That is not the same as the extent of the area where the powers can or have to be exercised.

[573] Article 3 Directive 93/43.

[574] Article 1(2) Regulation 852/2004.

The first two exclusions deal with two types of food production where the produce is not put on the market. In category 1 the food business operator, who is active in the primary stage of the food production chain, consumes the produce privately.

In category 2 a person who is not a food business operator at all, performs a series of actions that belong to the core activities of a group of food business operators but are in this case part of self-provision outside the commercial food chain.[575]

The third category excludes direct local supply from the scope of Regulation 852/2004. This exclusion aims to prevent the full weight of EU food hygiene law applying to a small local direct supply of primary products to the consumer, and the local supply of local retailers that stay away from the internal market by keeping their transactions with primary producers and local consumers close to home.[576]

The last category relates to the activities of businesses that supply by-products to food producers. Further processing is such that the former businesses do not contribute in any relevant way to the safety of the ultimate product.

The hygiene package

The scope of Regulation 852/2004 (this regulation is sometimes referred to as 'Hygiene 1') determines the outer reaches of the scopes of the other hygiene regulations because these all add more specific legislation to the basic rules of Regulation 852/2004 and therefore remain within its bounds.

Regulation 853/2004 (sometimes referred to as Hygiene 2') lays down rules on the hygiene of food of animal origin. These rules supplement those laid down by Regulation 852/2004.[577]

Regulation 854/2004 (sometimes referred to as Hygiene 3') adds the rules that are necessary to organise official controls on compliance with Regulation 853/2004.[578]

---

[575] Article 1(2)(a) and (b) Regulation 852/2004.

[576] Article 1(2)(c) Regulation 852/2004. The Member States are given the assignment in Article 1(3) to make national law to govern this situation. The freedom of the Member States to make the rules that fit local-regional situations best is circumscribed by the condition that this national law must 'ensure the achievement of the objectives of this Regulation'. In effect this is a mini-directive surfacing in the midst of a regulation. This additional legislative task is akin to, but not mentioned in, Article 13 where the Member States are given other assignments or freedoms to legislate in order to safeguard traditional methods or to assist food business operators in areas with geographic constraints (Article 13(4)(a) Regulation 852/2004). The construction that allows national deviations from EC food hygiene law, provided they are so small and local that they do not disturb the internal market, resembles the '*de minimis*' or bagatelle rule in EC competition law.

[577] Article 1 Regulation 853/2004.

[578] Article 1(1) Regulation 854/2004.

The rules of Regulation 853/2004 and Regulation 854/2004 often deal with the same situation from opposite perspectives. Where, for example, Regulation 853/2004 commands the food business operator to cooperate with the competent authorities,[579] Regulation 854/2004 directs the Member States to see to it that food business operators offer all assistance needed to ensure that official controls carried out by the competent authority can be performed effectively.[580]

Regulation 882/2004 adds controls to prevent, eliminate or handle risks to humans and animals and guarantee fair practices in feed and food trade to protect consumer interests.[581]

The four regulations form a stack of rules addressed mainly to food business operators, but also to the Commission, to the Standing Committee on the Food Chain and Animal Health (SCFCAH), and to the Member States. The rules are organised according to the definitions, principles and rules on food safety of the General Food Law and on the hygiene of foodstuffs of Regulation 852/2004. They become more specialised with Regulation 853/2004 that deals first with food of animal origin in general, and then with food of specific animal origins like fish, game, meat and milk, and continues with subdivisions that address food business operators responsible for ever more specialised foodstuffs of animal origin.

Those who follow the trail will encounter rules like:

> *the temperature of the meat of lagomorphs is maintained at not more than 4 °C during cutting, boning, trimming, slicing, dicing, wrapping and packaging.*[582]

Regulations 854/2004 and 882/2004 cover the rules on how food hygiene is controlled and enforced. The combination of rules on hygiene and rules on control creates new rules that are directly relevant for the food business operators.

The four regulations call for implementing regulations, made by the Commission in cooperation with the representatives of the Member States in the Standing Committee on the Food Chain and Animal Health. To this category belong Regulations 2073/2005 on microbiological criteria for foodstuffs,[583] 2074/2005 implementing certain rules on food of animal origin, 2075/2005 on official controls for *Trichinella* in meat, and 2076/2005 with transitional arrangements. They entered into force together with the four hygiene regulations. Amidst the regulations one

---

[579] Article 4(4) Regulation 853/2004.
[580] Article 4(1) Regulation 854/2004.
[581] On official controls see also Chapter 15.
[582] Regulation 853/2004, Annex III Specific requirements, Section II: Meat from poultry and lagomorphs, Chapter V: Hygiene during and after cutting and boning.
[583] Article 1 Regulation 882/2004. On this regulation see also Chapter 15.

Directive 2004/41 was made to repeal most of the old hygiene directives and provide for transitory measures – this directive is also referred to as 'Hygiene 4'.

The stack of rules contains not only regulations but also explanations of the regulations in guidance documents usually written by civil servants from the Commission's Directorate-General Health & Consumer Protection. One of these guidance documents is presented by the SCFCAH. These documents offer a cautious type of interpretation: on the one hand they present the opinions of the public servants who are more involved with these matters and command greater expertise than anyone else, and present these opinions to the stakeholders in the light of transparency, on the other hand they caution every reader that these opinions carry no weight and are used by each at his own peril, because only the Court of Justice has the power to present the final interpretation. While it is proper that the ultimate power of interpretation of the Court of Justice is respected, the non-binding character of the guidance documents does not go so far that the Commission can hold it against food businesses if they act as suggested in the guidelines.

Law text box 13.2 presents a simplified[584] overview of the regulations, directive, and guidance documents of the EU food hygiene legislation. Diagram 13.2 gives a schematic representation of the relations between the regulations. The guidance documents are attachments to the regulations that they explain.

Law text box 13.2. A simplified overview of the regulations, directive, and guidance documents of the EU food hygiene legislation: the 'hygiene package'.

0. Regulation (EC) No 178/2002 of the European Parliament and of the Council of 28 January 2002 laying down the general principles and requirements of food law, establishing the European Food Safety Authority and laying down procedures in matters of food safety.

0a. Standing Committee on the Food Chain and Animal Health, Guidance on the Implementation of Articles 11, 12, 16, 17, 18, 19 and 20 of Regulation (EC) N° 178/2002 on General Food Law, Conclusions of the Standing Committee on the Food Chain and Animal Health, 26 January 2010.

00. Regulation 882/2004 of the European Parliament and of the Council of 29 April 2004 on official controls performed to ensure the verification of compliance with feed and food law, animal health and animal welfare rules.

[584] One could also argue that the following also belong to the Hygiene package: Directive 2002/99 laying down the animal health rules governing the production, processing, distribution and introduction of products of animal origin for human consumption; with the connected General guidance on EU import and transit rules for live animals and animal products from third countries, DG Sanco Directorate D – Animal health and welfare 30 April 2006; Regulation (EC) No 183/2005 laying down requirements for feed hygiene and the Opinion of the European Economic and Social Committee on 'Hygiene rules and artisanal food processors' (2006/C 65/25; OJ C 65, 17 March 2006, 141-148).

Law text box 13.2. Continued.

1. Regulation (EC) No 852/2004 of the European Parliament and of the Council of 29 April 2004 on the hygiene of foodstuffs.

1a. DG Sanco, Guidance document on the implementation of certain provisions of Regulation (EC) No 852/2004 on the hygiene of foodstuffs.

1b. DG Sanco, Guidance document on the implementation of procedures based on the HACCP principles, and on the facilitation of the implementation of the HACCP principles in certain food businesses.

1c. DG Sanco, Hygiene and control measures. Technical Specifications in Relation to the Master List and the Lists of EU Approved Food Establishments.

1d. DG Sanco, Guidance document on certain key questions related to import requirements and the new rules on food hygiene and on official food controls.

2. Regulation 853/2004 of the European Parliament and of the Council of 29 April 2004 laying down specific hygiene rules for food of animal origin.

2a. DG Sanco, Guidance document on the implementation of certain provisions of Regulation (EC) No 853/2004 on the hygiene of food of animal origin.

3. Regulation 854/2004 of the European Parliament and of the Council of 29 April 2004 laying down specific rules for the organisation of official controls on products of animal origin intended for human consumption.
4. Directive 2004/41 of the European Parliament and of the Council of 21 April 2004 repealing certain Directives concerning food hygiene and health conditions for the production and placing on the market of certain products of animal origin intended for human consumption and amending Council Directives 89/662/EEC and 92/118/EEC and Council Decision 95/408/EC.
5. Commission Regulation 2073/2005 of 15 November 2005 on microbiological criteria for foodstuffs.
6. Commission Regulation 2074/2005 of 5 December 2005 laying down implementing measures for certain products under Regulation (EC) No 853/2004 of the European Parliament and of the Council and for the organisation of official controls under Regulation (EC) No 854/2004 of the European Parliament and of the Council and Regulation (EC) No 882/2004 of the European Parliament and of the Council, derogating from Regulation (EC) No 852/2004 of the European Parliament and of the Council and amending Regulations (EC) No 853/2004 and (EC) No 854/2004.
7. Commission Regulation 2075/2005 of 5 December 2005 laying down specific rules on official controls for *Trichinella* in meat.
8. Commission Regulation 2076/2005 of 5 December 2005 laying down transitional arrangements for the implementation of Regulations (EC) No 853/2004, (EC) No 854/2004 and (EC) No 882/2004 of the European Parliament and of the Council and amending Regulations (EC) No 853/2004 and (EC) No 854/2004.

Diagram 13.2. The hygiene package and related texts.

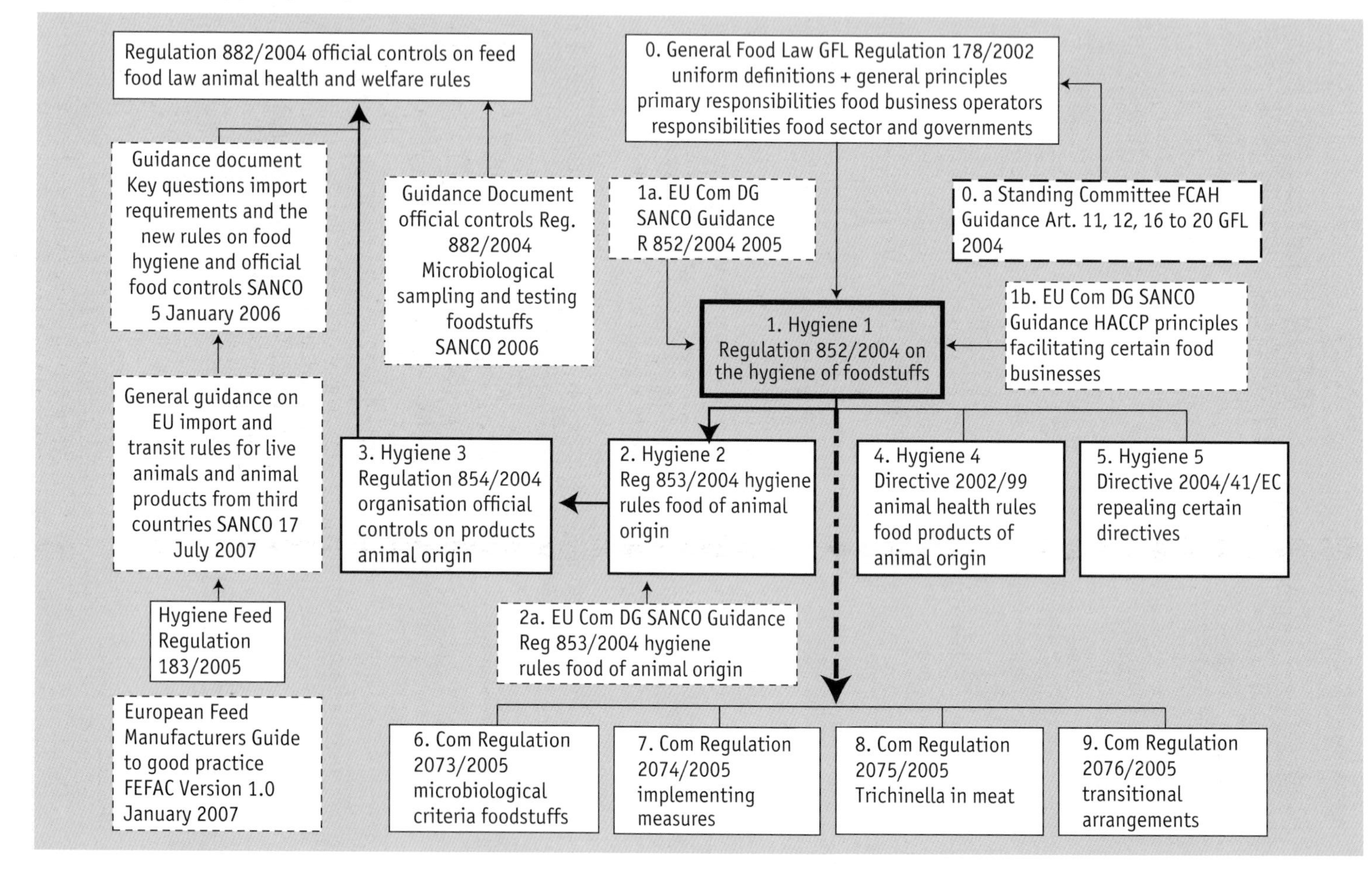

13.2.4 Instruments of food hygiene law

Regulation 852/2004 uses mainly three types of instruments to regulate the hygiene of foodstuffs:

1. prescriptive rules,
2. procedures based on the Hazard Analysis and Critical Control Point (HACCP) system, and
3. guides to good hygiene practice.

EU regulations often contain a mix of rules and decisions addressed to a mix of EU institutions, the Member States, the citizens/businesses and their organisations. Regulation 852/2004 is directed mainly at the food business operators, but also at the food business sectors, consumer organisations and authorities on the national level[585] and on the Union level,[586] to the Commission,[587] to the SCFCAH[588] and to the Member States.[589]

The ultimate aim of all these rules, invitations, exceptions, delegations of power and assignment of duties is to influence the activities and positions of the people who are involved with the hygiene aspects of food to secure a high level of food safety.

Rules as an instrument for food hygiene

Among the rules directed at the different persons and organisations mentioned above, the first group addresses the food business operators directly. To distinguish these rules from the other instruments, we will call them 'direct legislation'. These are the examples of legislation in the classic sense of rules laid down for persons and situations outside the organisation of government to be applied time and again,

[585] For example, in Article 8(1) Regulation 852/2004 where the national food sector organisations are asked to make guides to good hygiene practices.

[586] In Article 9(2)(a) where the Commission has to consult them on European guides to good practice.

[587] In Article 15 where the Commission is charged to consult with EFSA on hygiene matters that could have a significant impact on public health.

[588] In its Comitology Article 14.

[589] In Article 13(3) and 13(4)(a)(i) where Member States get the power to take national measures to enable the continued use of traditional methods, at any of the stages of production, processing or distribution of food. The special position of the EU as a half-way house between an international organisation and a (federal) state gives its rules and decisions a hybrid character. Rules directed to the authorities of the Member States are applied by them to take decisions that are necessary to give effect to EU law. When a system of EU hygiene law is introduced replacing directives by regulations it operates directly for the persons or organisations that are addressed by the articles. That does not mean that the rules and decisions of the regulation take care of themselves, or that the authorities of the Member States have to wait and see what happens. When the regulations mention a duty to approve establishments where food of animal origin is processed, arrangements have to be made that some authority will issue approvals. In most cases the EU institutions do not set up an EU system. Authorities of the Member States will then create national systems of approval that will issue national approvals based on EU criteria.

as long as the conditions set by the legislation for its application are fulfilled. The direct legislation can be found in a few articles of the main body of Regulation 852/2004, in its two Annexes and in many articles of Regulation 853/2004.

The second and most important type of instrument is the creation of the obligation for food business operators to apply a procedure based on the HACCP system. This system of food safety management was developed in the USA and has been promoted since the late 1960s[590] by the Codex Alimentarius Commission.

The third type is the creation of guides to good hygiene practice, developed by food business sectors, accepted voluntarily by food business operators, promoted and approved by the Member States and registered by the European Commission.

The layout of Regulations 852/2004 and 853/2004

The food hygiene legislation of the EU is a system of stacked regulations where each regulation has a systematic exposition of its contents. This layout is presented in Diagrams 13.3 and 13.4 for Regulation 852/2004 and Regulation 853/2004.

13.2.5 Direct legislation

Article 3 of Regulation 852/2004 reiterates the responsibility of food business operators expressed in Article 17(1) GFL. Food business operators shall ensure that all stages of production, processing and distribution food under their control satisfy the relevant hygiene requirements of this regulation.

Regulation 852/2004 deals in Article 4 with the complexity of EU hygiene legislation. This article elaborates which provisions apply to whom:
- businesses in primary production apply Annex I A of Regulation 852/2004 and any specific requirements of Regulation 853/2004;
- other businesses apply Annex II of Regulation 852/2004 and any specific requirements of Regulation 853/2004;
- all food businesses shall comply with hygiene measures set by the European Commission (on microbiology, temperature control and the like);
- all food business may use guides to Good Hygiene Practice to comply with their obligations.

Article 4 connects the rules of Regulation 852/2004 with the rules of Regulation 853/2004 explicitly. It underlines the binding force and the unity of the articles in the main body of Regulation 852/2004 with its two Annexes and the corresponding parts of Regulation 853/2004. This is especially important for the unity of the direct legislation directed at the food business operators.

[590] CAC/RCP 1-1969.

Diagram 13.3. The internal organisation of Regulation 852/2004 on the hygiene of foodstuffs.

Chapter I General provisions	**Chapter II Food business operators' obligations**	**Chapter III Guides to Good Practice**	**Chapter IV Imports and Exports**	**Chapter V Final provisions**	**Annex I Primary Production**	**Annex II General hygiene requirements for all food business operators (Except when Annex I applies) Chapter I-XII**
				Art. 12 Implementing measures and transitional arrangements	Part A: General Hygiene Provisions for Primary Production and Associated Operations	I General requirements for food premises (other than those specified in Chapter III) II Specific requirements in rooms where foodstuffs are prepared, treated or processed (excluding dining areas and those premises specified in Chapter III) III Requirements for movable and/or temporary premises (such as marquees, market stalls, mobile sales vehicles), premises used primarily as a private dwelling-house but where foods are regularly prepared for placing on the market and vending machines
				Art. 13 Amendment and adaptation of Annexes I and II		
	Art. 3 General obligation					
Art. 1 Scope	Art. 4 General and specific hygiene requirements	Art. 7 Development, dissemination and use of guides	Art. 10 Imports	Art. 14 Committee procedure		IV Transport
Art. 2 Definitions	Art. 5 Hazard analysis and critical control points	Art. 8 National guides	Art. 11 Exports	Art. 15 Consultation of the European Food Safety Authority		V Equipment requirements VI Food waste VII Water supply VIII Personal hygiene
	Art. 6 Official controls, registration and approval	Art. 9 Community guides		Art. 16 Report to the European Parliament and to the Council	Part B: Recommendations for Guides to Good Hygiene Practice	IX Provisions applicable to foodstuffs X Provisions applicable to the wrapping and packaging of foodstuffs XI Heat treatment XII Training
				Art. 17 Repeals		
				Art. 18 Entry into force		
What is this regulation about?	*Introduction of obligations, direct legislation, and alternative system of hazard management*	*Instrument for voluntary cooperation and preparation*	*Art. 3-6 < Art. 11 and 12 GFL*	*Internal maintenance of legislation*	*A: direct legislation, B: advise*	*Detailed direct legislation for food business operators active after primary production and dealing with food of vegetal origin, animal origin or both.*

Diagram 13.4. The internal organisation of Regulation 853/2004 laying down specific hygiene rules for food of animal origin.

Chapter I General provisions	**Chapter II Food business operators' obligations**	**Chapter III Trade**	**Chapter IV Final provisions**	**Annex I Definitions**	**Annex II Requirements concerning several products of animal origin**	**Annex III Specific requirements**
				1. Meat		I: Meat of domestic ungulates
			Art. 9 Implementing measures and transitional measures			II: Meat from poultry and lagomorphs III: Meat of farmed game IV: Wild game meat
			Art. 10 Amendment and adaptation of Annexes II and III		Section I Identification marking A. Application of the identification mark B. Form of the identification mark C. Method of marking Section II: Objectives of HACCP-based procedures	V: Minced meat, meat preparations and mechanically separated meat (msm) VI: Meat products
	Art. 3 General obligations		Art. 11 Specific decisions			
Art. 1 Scope	Art. 4 Registration and approval of establishments	Art. 7 Documents	Art. 12 Committee procedure	2. Live bivalve molluscs		VII: Live bivalve molluscs
Art. 2 Definitions	Art. 5 Health and identification marking	Art. 8 Special guarantees on negative outcome of microbiological tests and elimination of salmonella for meat and eggs	Art. 13 Consultation of the European Food Safety Authority	3. Fishery products		VIII: Fishery products
	Art. 6 Products of animal origin from outside the Community		Art. 14 Report to the European Parliament and to the Council	4. Milk 5. Eggs 6. Frogs' legs and snails	Section III: Food chain information	IX: Raw milk and dairy products X: Eggs and egg products XI: Frogs' legs and snails XII: Rendered animal fats and greaves
			Art. 15 Entry into force	7. Processed products		XIII: Treated stomachs, bladders and intestines XIV: Gelatine XV: Collagen
				8. Other definitions		Appendix to Annex III
What is this regulation about?	*Introduction of general obligation, announcement of instruments*	*Special guarantees for Sweden and Finland*	*Internal provisions for the application and maintenance of the regulation*	*Concepts determining application of regulation*	*Direct legislation for identification of products*	*Direct legislation for foodstuff indicated by Section headings*

The fact that the most detailed rules, prescribing directly what behaviour is expected from the food business operators, are part of the annexes does not mean that these rules have a lower ranking than the rules in the main body of the regulations. They are an integral part of the regulations, as Article 4 shows for Regulation 852/2004.

These rules of a technical nature are placed in the annexes because the annexes can be changed by the comitology procedure. Normally, the 'ordinary legislative procedure'[591] where the full power of the European Parliament is brought to bear, applies to changes (in the main body) in the regulation. In comitology the representatives of the Member States take decisions with the weighted votes of the Member States, under the chairmanship of a representative of the Commission.[592] The European Parliament agreed to give up some of its power to facilitate changes needed to keep up with technical developments.

13.2.6 Registration and approval

As has been mentioned in Chapter 9, Section 9.3.3 establishments used for any stage in the food chain must be registered and in many cases approved. This requirement is one of the examples where the complexity of the hygiene package is felt. Provisions dealing with this issue are: Article 6 of Regulation 852/2004, Article 4 of Regulation 853/2004, Article 3 of Regulation 854/2004, and Article 31 of Regulation 882/2004.

Registration of establishments where food is handled

Article 6 of Regulation 852/2004 has the heading 'Official controls, registration and approval'. It addresses food business operators. It is the basis of the obligation to have each food establishment registered, but not the basis of the EU approval system.

Every food business operator has to apply for registration of each establishment under her/his control where any of the stages of production, processing and distribution of food is carried out. The registration has to contain the essential data on the address, and the food hygiene related activities at that address.[593] The address of each premise has to be registered if the establishment has different premises.[594]

The food business operator has to see to it that the competent authority has up-to-date information on establishments at all times. Significant changes in activities and closure of an existing establishment have to be reported to the authorities.

[591] See Chapter 5.
[592] On comitology see Chapter 5, Section 5.5.
[593] Article 6 Regulation 852/2004.
[594] Article 6 in combination with Chapter I of Annex II Regulation 852/2004.

The authorities in the Member States have to draw up the procedures on how to obtain registration.[595]

Approval of establishments

Every establishment in the EU where food of animal origin is produced or handled has to be approved by the authorities of the Member States applying EU legislation. What is the basis of this approval system? It takes some close reading to find that basis in Regulation 854/2004, the regulation on hygiene controls.[596]

In fact, four regulations deal with the subject of approval of establishments. Article 6 of Regulation 852/2004 prepares the approval systems by making the food business operator, in defined instances, responsible for the approval by the competent authorities and even for the fact that this approval is given only after the authority has paid at least one on-site visit. This is clearly more than an obligation to cooperate with the authorities.

However, these responsibilities of the food business operator do not create the approval system. Article 6(3) merely prepares the food business operator in case an approval system is part of national law, Regulation 853/2004 or one of the comitology decisions.

Regulation 853/2004 and the consequences

Article 4 of Regulation 853/2004 has the heading 'registration and approval'. It supports the requirements of registration and approval by dealing with its consequences. Food business operators handling food of animal[597] origin are not allowed to place their products on the market if they have been made in, or are otherwise related to, an establishment that has not been approved.[598] Likewise operators have to cease to operate when the approval of their establishment has been withdrawn.[599] The granting and withdrawal of approvals are based on the next food hygiene regulation.[600]

---

[595] Article 31(1)(a) Regulation 882/2004.
[596] It is unclear why registration and approval are based on different regulations. The obvious place would have been the General Food Law rather than the hygiene package.
[597] The following categories of food of animal origin are covered: meat of domestic ungulates, meat from poultry and lagomorphs, meat of farmed game and wild game meat, meat products, minced meat, meat preparations and mechanically separated meat (MSM), live bivalve molluscs, fishery products, raw milk, colostrum, dairy products and colostrum-based products, egg and egg products, frogs' legs and snails, rendered animal fats and greaves, treated stomachs, bladders and intestines, gelatine, collagen.
[598] Article 4(2) of Regulation 853/2004.
[599] Article 4(4) of Regulation 853/2004.
[600] Article 4(3) of Regulation 853/2004.

Regulation 854/2004 approval and approval number

Article 3(5) of Regulation 854/2004 creates the EU approval system for establishments that begin placing products of animal origin on the market. It prescribes that the competent authority has to make an on-site visit and will approve an establishment only if the food business operator has demonstrated that s/he meets the relevant requirements of Regulations 852/2004 and 853/2004 and others of food law.[601] The competent authority gives each approved establishment an approval number with a code that indicates the type of product of animal origin that is manufactured in that establishment.[602]

Regulation 882/2004 and the list of all approved establishments in the EU

Regulation 882/2004, finally, is involved because it orders the competent authorities of the Member States to draw up the procedures that food business operators have to follow when they apply for approval of their establishment.[603] The Member States have to maintain up-to-date lists of approved establishments, with their respective approval numbers and other relevant information, and make them available to other Member States and to the public.[604]

The Commission has created a website that provides access to the national lists of the EU Member States, the other EEA countries and other countries with special agreements (Switzerland, Føroyar, Greenland and San Marino).[605]

13.2.7 Prerequisites

The annexes to Regulation 852/2004 contain direct legislation of a technical and detailed nature. Annex I is made for the primary sector. The special position of primary production is discussed at the end of this section.

Annex II is called General Hygiene Requirements for All Food Business Operators (Except When Annex I Applies). It contains direct legislation addressed to all food business operators outside primary production and its associated activities. The rules in this annex are often referred to as the 'prerequisites'. This term expresses that it only makes sense to apply HACCP (discussed hereafter) if basic requirements are fulfilled. The rules are organised into twelve chapters by applying

601 Article 3(1)(a) Regulation 854/2004 that provides the procedure.
602 Article 3(3) Regulation 854/2004.
603 Article 31(2)(a) Regulation 882/2004.
604 Article 3(6) Regulation 854/2004 and Article 31(2)(a) Regulation 882/2004. The fact that even the Commission's civil servants occasionally get lost in the forest of seemingly redundant articles of the hygiene complex, is shown on the website of the list of approved food establishments, where the obligation to approve establishments is attributed to Regulation 882/2004 in general. It is however very particularly Article 3(5) of Regulation 854/2004 that provides the procedures.
605 The Community Register of approved food establishments is available on the website of DG Sanco.

two principles. First, the locality where the activities of the food business take place (Chapters I-III). Then the aspects of hygiene that are relevant in all stages of the production chain. These aspects are the production inputs, equipment, clean working circumstances and water supply (Chapters V-VII), the human factor in personal hygiene and training (Chapters VIII and XII) and the protection, wrapping and packaging of the cause of it all, the product food (Chapters IX and X). Transport holds the position between the localised factors and the mobilised factors in Chapter IV.

Food premises

Annex II opens with a puzzle about the location where food hygiene legislation is to be applied. The convoluted sequence of the first three chapters give the distinct impression that the legislator is trying to get a grip on the many possible establishments, places, rooms, tents and other things that have to be captured under the law that is laid down for the operator who is active in them. The legislator uses the term 'premises' to differentiate between the relevant places. There is no definition here and no use of the term 'establishment' that is used to register or approve a food business. Regulation 852/2004 does give the definition of the term establishment, namely 'any unit of a food business', but does not use it when localisation is the subject of the legislation.[606] The combination with the GFL definition of food business makes clear that an establishment is a unit of a concerted activity where both place and activity are at issue but the activity dominates the concept. One thing seems clear however: when food business operators have to provide information on the whereabouts of their business units they have to be specific about the premises level.[607] The presentation of the direct legislation in Annex II is organised according to premises that are immovable and a recognisable part of the means used in the food production chain dedicated to food or movable and equally dedicated such as fishing ships in Chapter I. They become detailed at the room level and in close contact with spaces used for other activities from which they have to be separated in Chapter II, and pass into the impermanent, temporarily built structures in Chapter III to become fully mobile in Chapter IV, but then they are no longer business units, they have become means of transport only.

[606] Article 2(1)(c) Regulation 852/2004.

[607] The legislation made to assist the implementation of the EU food hygiene regulations in the part of the UK after devolution called England provides the following definition of premises: 'premises' includes any establishment, any place, vehicle, stall or moveable structure and any ship or aircraft. Article 2(1) Interpretation, Statutory Instruments 2005 No. 2059 Food, England. The Food Hygiene (England) Regulations 2005, Made 21st July 2005, Laid before Parliament 10th August 2005, Coming into force 1st January 2006.

Rules addressed to a food business operator deal with an operator who is either stationary or on the move. When s/he is stationary s/he has a location that is temporarily or more or less permanent.[608]

Food handling

The function of direct legislation in the annexes is to provide a guaranteed minimum.

Article 4 performs a core function for the larger set of direct rules on food hygiene. Two of the legislator's instruments, HACCP procedures and voluntary guides, are based on a high degree of autonomy for the food business operators. They will use these instruments according to their knowledge and experience .

But to leave it at that would make the objective of secured food hygiene dependent on the uncertain success of 'procedures based on' HACCP, something that is a set of procedures by itself, and on the voluntary adoption of good practices. A degree of dependence and uncertainty that is unacceptable for a legislator who wants to maintain a high level of food safety, after earlier failures. So, in short, the direct legislation in the annexes acts as the available reserve legislation that is made to guarantee the objective of a high level of food hygiene as best as legislation can. The other two instruments can improve that result or compensate for the weaknesses of direct regulation.

The direct legislation represents the old style food hygiene legislation: the legislator prescribes the rules that have to be applied by all those involved with food hygiene. A control system is set up to guard over the implementation of the legislation and enforcement measures are available. In relation to the other elements of Regulation 852/2004 it can be said that the rules provide the clearest information for food business operators on what they have to do to meet the requirements of the law.

Set against these other approaches, the binding rules set the stage for the other instruments of Regulation 852/2004 by providing the floor on which they have to perform their part. If that floor is set high enough it can spur a dynamic mix of instruments to compensate for each other's weak points. Recital 12 of Regulation 852/2004 wishes just that:

> *Food safety is a result of several factors: legislation should lay down minimum hygiene requirements; ...*

[608] Reality is a flux. Persons are even more flexible as no one would imagine a person to be permanently located on some spot. It is not the position of the food business operator as such that is of interest here; we are not looking for the operator in or after business hours, but the position of the relevant activity or inactivity.

Law text box 13.3 gives the chapter headings each with one article to illustrate the contents and organisation of Annex II to Regulation 852/2004. It gives an impression of the direct legislation in Annex II as it is not possible to discuss these very detailed rules at this point. The examples are representative for the rules in Annex II of Regulation 852/2004. They also indicate the character of the annexes with direct legislation of the other three major food hygiene regulations.

Law text box 13.3. Annex II to Regulation 852/2004: chapter headings with one representative article per chapter.

Annex II General Hygiene Requirements for All Food Business Operators (Except When Annex I Applies)

I Food premises

2. The layout, design, construction, siting and size of food premises are to: (a) permit adequate maintenance, cleaning and/or disinfection, avoid or minimise air-borne contamination, and provide adequate working space to allow for the hygienic performance of all operations;

II Rooms

1. In rooms where food is prepared, treated or processed (excluding dining areas and those premises specified in Chapter III, but including rooms contained in means of transport) the design and layout are to permit good food hygiene practices, including protection against contamination between and during operations. In particular:
   (b) wall surfaces are to be maintained in a sound condition and be easy to clean and, where necessary, to disinfect. This will require the use of impervious, non-absorbent, washable and non-toxic materials and require a smooth surface up to a height appropriate for the operations unless food business operators can satisfy the competent authority that other materials used are appropriate;

III Special types of food premises: not permanent, not fixed, nor exclusive

Requirements for movable and/or temporary premises (such as marquees, market stalls, mobile sales vehicles), premises used primarily as a private dwelling-house but where foods are regularly prepared for placing on the market and vending machines

1. Premises and vending machines are, so far as is reasonably practicable, to be so sited, designed, constructed and kept clean and maintained in good repair and condition as to avoid the risk of contamination, in particular by animals and pests.
2. In particular, where necessary:
   (a) appropriate facilities are to be available to maintain adequate personal hygiene (including facilities for the hygienic washing and drying of hands, hygienic sanitary arrangements and changing facilities);

IV Transport

1. Conveyances and/or containers used for transporting foodstuffs are to be kept clean and maintained in good repair and condition to protect foodstuffs from contamination and are, where necessary, to be designed and constructed to permit adequate cleaning and/or disinfection.
5. Where conveyances and/or containers have been used for transporting anything other than foodstuffs or for transporting different foodstuffs, there is to be effective cleaning between loads to avoid the risk of contamination. ►►

Law text box 13.3. Continued.

V Equipment

1. All articles, fittings and equipment with which food comes into contact are to:
   (a) be effectively cleaned and, where necessary, disinfected. Cleaning and disinfection are to take place at a frequency sufficient to avoid any risk of contamination;

VI Food waste

1. Food waste, non-edible by-products and other refuse are to be removed from rooms where food is present as quickly as possible, so as to avoid their accumulation.

VII Water supply

1. (a) There is to be an adequate supply of potable water, which is to be used whenever necessary to ensure that foodstuffs are not contaminated;
   (b) Clean water may be used with whole fishery products. Clean seawater may be used with live bivalve molluscs, echinoderms, tunicates and marine gastropods; clean water may also be used for external washing. When clean water is used, adequate facilities and procedures are to be available for its supply to ensure that such use is not a source of contamination.

VIII Personal hygiene

1. Every person working in a food-handling area is to maintain a high degree of personal cleanliness and is to wear suitable, clean and, where necessary, protective clothing.
2. No person suffering from, or being a carrier of a disease likely to be transmitted through food or afflicted, for example, with infected wounds, skin infections, sores or diarrhoea is to be permitted to handle food or enter any food-handling area in any capacity if there is any likelihood of direct or indirect contamination. Any person so affected and employed in a food business and who is likely to come into contact with food is to report immediately the illness or symptoms, and if possible their causes, to the food business operator.

IX Foodstuffs

3. At all stages of production, processing and distribution, food is to be protected against any contamination likely to render the food unfit for human consumption, injurious to health or contaminated in such a way that it would be unreasonable to expect it to be consumed in that state.

X Provisions applicable to the wrapping and packaging of foodstuffs

3. Wrapping and packaging operations are to be carried out so as to avoid contamination of the products. Where appropriate and in particular in the case of cans and glass jars, the integrity of the container's construction and its cleanliness is to be assured.

XI Heat treatment

2. to ensure that the process employed achieves the desired objectives, food business operators are to check regularly the main relevant parameters (particularly temperature, pressure, sealing and microbiology), including by the use of automatic devices;

XII Training

Food business operators are to ensure:

1. that food handlers are supervised and instructed and/or trained in food hygiene matters commensurate with their work activity;
2. that those responsible for the development and maintenance of the procedure referred to in Article 5(1) of this Regulation or for the operation of relevant guides have received adequate training in the application of the HACCP principles;

13.2.8 HACCP

Space age experience has shown that the testing of food products does not provide sufficient certainty that similar products in the same lot are safe. To ensure that astronauts will not suffer from diarrhoea or food poisoning during their stay in space, in 1959 the US space agency NASA developed a method of food production aimed at eliminating all possible hazards from the production process. This is the so-called HACCP system (Hazard Analysis and Critical Control Points).[609] Within this system the production process is analysed to establish what kind of hazards may enter the product in which part of the process. Procedures have been developed to prevent hazards or to deal with the consequences. Meticulous application of such a system in combination with testing of products achieves the highest possible level of food safety. The World Health Organisation (WHO) and the Food and Agriculture Organization (FAO) have recommended the worldwide use of HACCP-based systems to ensure safe production of food (e.g. food hygiene).[610] This recommendation has been implemented in EU legislation. The application of HACCP is obligatory for virtually[611] all food businesses in the EU. As a consequence food businesses have to analyse their processes, establish procedures to ensure hygiene and control the functioning of these systems themselves. In other words: the businesses have to formulate and uphold the rules that apply to their processes.

The 'Hazard analysis and critical control point', system is the second type of instrument of Regulation 852/2004. It holds a prominent place in EU food hygiene law. According to Article 5(1) Regulation 852/2004, food business operators shall put in place, implement and maintain a permanent procedure or procedures based on the HACCP principles. The EU definition of the HACCP principles is in line with the international standard. The HACCP system is well established. It has been adopted by the Codex Alimentarius Commission in 1997. The present Codex version is the fourth version with its latest revision in 2003. See Law text box 13.4 for the position of the HACCP system in EU hygiene law.

[609] See on HACCP and other quality assurance systems: M. Fogden, Hygiene, in: K. Goodburn (ed.) EU food law, Woodhead Publishing, Cambridge, UK, 2001; P.A. Luning, W.J. Marcelis and W.M.F. Jongen, Food quality management. a techno-managerial approach, Wageningen Academic Publishers, Wageningen, the Netherlands 2002; and M. van der Spiegel, Measuring effectiveness of food quality management (diss.) Wageningen University, Wageningen, the Netherlands, 2004.

[610] See: for instance CAC/RCP 1-1969, rev. 4 (2003) Recommended International Code of Practice General Principles of Food Hygiene; CAC/GL 21 – 1997; Principles for the Establishment and Application of Microbiological Criteria for Foods; CAC/GL-30 (1999) Principles and Guidelines for the Conduct of Microbiological Risk Assessment. These texts have been published in: Codex Alimentarius, Basis texts on Food Hygiene, 3rd edition, FAO, Rome, Italy, 2003. For a hyperlink, see the reference list.

[611] There are some exceptions. E.g. a somewhat less stringent regime of food hygiene applies to primary production.

Law text box 13.4. Definition of HACCP principles in Article 5, Regulation 852/2004.

Article 5
Hazard analysis and critical control points

1. Food business operators shall put in place, implement and maintain a permanent procedure or procedures based on the HACCP principles.
2. The HACCP principles referred to in paragraph 1 consist of the following:
   (a) identifying any hazards that must be prevented, eliminated or reduced to acceptable levels;
   (b) identifying the critical control points at the step or steps at which control is essential to prevent or eliminate a hazard or to reduce it to acceptable levels;
   (c) establishing critical limits at critical control points which separate acceptability from unacceptability for the prevention, elimination or reduction of identified hazards;
   (d) establishing and implementing effective monitoring procedures at critical control points;
   (e) establishing corrective actions when monitoring indicates that a critical control point is not under control;
   (f) establishing procedures, which shall be carried out regularly, to verify that the measures outlined in subparagraphs (a) to (e) are working effectively; and
   (g) establishing documents and records commensurate with the nature and size of the food business to demonstrate the effective application of the measures outlined in subparagraphs (a) to (f).

   When any modification is made in the product, process, or any step, food business operators shall review the procedure and make the necessary changes to it.
3. Paragraph 1 shall apply only to food business operators carrying out any stage of production, processing and distribution of food after primary production [...].
4. Food business operators shall:
   (a) provide the competent authority with evidence of their compliance with paragraph 1 in the manner that the competent authority requires, taking account of the nature and size of the food business;
   (b) ensure that any documents describing the procedures developed in accordance with this Article are up-to-date at all times;
   (c) retain any other documents and records for an appropriate period.
5. Detailed arrangements for the implementation of this Article may be laid down in accordance with the procedure referred to in Article 14(2). Such arrangements may facilitate the implementation of this Article by certain food business operators, in particular by providing for the use of procedures set out in guides for the application of HACCP principles, in order to comply with paragraph 1. Such arrangements may also specify the period during which food business operators shall retain documents and records in accordance with paragraph 4(c).

The HACCP system

With regard to the HACCP approach, the food business operator is set on a course that will give her/him control over the processes s/he is responsible for, more information on what goes on and what goes wrong in her/his business and hence the tools to guarantee food hygiene and to improve on standards already achieved. The application of HACCP is a continuing process that is founded on the efforts already made, on changes in the food business internally, and in the outside world. The responsible food business operator is approached in a way that is very different from a set of rules specified in advance and applied to situations that must be isolated to determine whether a certain rule has been applied correctly. The HACCP approach offers not a specified set of rules on what is allowed but a method to achieve what is allowed, and an incentive to do even better.

The food business operator applying HACCP has to analyse his or her own processes; has to identify the hazards and has to decide on the measures to deal with them. For this reason HACCP is often referred to as enforced (or imposed) self-regulation. 'Self-regulation' because ultimately the rules applying to the production process have to be formulated by the food business operator. 'Imposed', because unlike other types of self-regulation, there is no choice.

The EU applications of the HACCP system

The requirement for documents and record keeping, the obligation to keep them up-to-date at all time and to retain them for an appropriate period have the added effect that official controls have extensive inside information (Article 5(2)(a-f)).

13.2.9 Guides to Good Hygiene Practice

The third type of instrument of Regulation 852/2004 for food hygiene (after direct legislation and procedures based on HACCP), the guides to good hygiene practice, provides an alternative to the individual application of HACCP. See Law text box 13.5.

Law text box 13.5. Article 4(6) Regulation 852/2004 on Guides to Good Hygiene Practice.

Article 4(6) Regulation 852/2004
Food business operators may use the guides provided for in Articles 7, 8 and 9 as an aid to compliance with their obligations under this Regulation.

Compliance with the applicable guide means compliance with the HACCP obligation. The HACCP obligations come first; the guides are a way (an alternative) to do it. The national guides are made by the food business sectors on their own initiative. The legislation for the European guides makes it possible for the Commission to take the initiative on the EU level, if the Member States allow it. But it can hardly be more than a spark that ignites a burning desire by food producers to make a European guide, there is no sense in forcing voluntary participation.

Chapter III of Regulation 852/2004 contains rules that have to be applied when a Guide is made, that instruct the Member States to assess the national guides and to forward the approved guides to the Commission. Guides can also be made on the EU level.

National guides

The national guides will be developed by the food business sectors themselves. The persons or organisations in the food business sectors who take the initiative to develop a guide have to meet three conditions:

- They have to consult representatives of the persons and organisations whose interest may be substantially affected. The regulation mentions competent authorities and consumer groups explicitly;[612]
- The drafters have to use the relevant codes of practice of the Codex Alimentarius;
- They have to use the recommendations set out in Annex I Part B of Regulation 852/2004 when primary production is involved.[613]

The legislator recommends that each initiative is supported by the national standards institute to benefit from its authority.

Approval by the Member State

The Member States have to assess the national guides to ensure that the three conditions set for their development have been observed.[614] In addition the Member States have to test the guides on two other aspects, their practicality and potential to help the food business operator to fulfil a part of her/his food hygiene obligations. Regulation 852/2004 specifies the parts where good hygiene practices can be most effective: the general obligation of Article 3, the rules and the two Annexes based on Article 4, the permanent procedures based on the HACCP principles of Article 5 and the specific obligations under Regulation 853/2004. The Member States have to send the guides that have passed the test to the Commission who will place them in a register without further testing.

EU register of national guides

The Commission has to set up a register of national guides and make that available to all Member States.[615] The Commission made the EU Register of national guides to good hygiene practice' publicly available on a webpage of DG Sanco.[616] The Register began with a list of 415 national guides and gives for each one the following

[612] Article 8(1)(a) Regulation 852/2004.
[613] Article 8(1)(a) – (c) Regulation 852/2004.
[614] Article 8(3)(a) Regulation 852/2004.
[615] Article 8(4) Regulation 852/2004.
[616] The Register's internet address can be found in the reference list.

information: title in the original language, title in English, country, language, author, edition, and ISBN or ISSN. It presents a list of keywords on good hygiene practices with an internet address or other contact in the Member States.

European Guides

The Commission takes a more active part in the creation of guides at the EU level than the governments do at the Member State level. The Commission has to use the Comitology procedure to decide with the Member States whether European guides will be made, and what scope and subject matter they will have. Article 9 of Regulation 852/2004 mentions explicitly European Community guides to good hygiene practice and guides for the application of the HACCP principles.

When the Member States have decided to make a European guide, it is for the Commission to ensure that they will be made. The Commission will perform tasks that are comparable with the tasks of the Member States in the development process of the national guides, but are markedly more active. European guides are published in the Official Journal of the EU. The Commission has published guidelines for the development of EU guides to good practice for hygiene and for the application of the HACCP principles.[617] At the time of writing 5 EU guides to good practice have been published.[618]

National and European guides

When the law for EU voluntary guides to good hygiene practice is compared with the law on the national guides there are a number of marked differences. Two main points are the dominance of the EU in the procedure to develop European guides, and the position of the Member States. Subsidiarity explains part of the elaborate testing that has to be done before guides on the EU level are even contemplated.

On the other hand the more active position of the Commission can be explained by a shortage of representative partners over the full range of affected interests due to the weaker civil society on the EU level when compared with the national level. The Commission is more active on the EU level to compensate for the missing parts of the civil society. Law text box 13.6 on the role of governments in the promotion of guides to good hygiene practice compares the different rules for the development of these guides on the national and EU levels.

[617] Guidelines for the development of community guides to good practice for hygiene and for the application of the HACCP principles, in accordance with Article 9 of Regulation (EC) No 852/2004 on the hygiene of foodstuffs and Article 22 of Regulation (EC) No 183/2005 laying down requirements for feed hygiene.
[618] See http://tinyurl.com/poze8kt.

Law text box 13.6. The role of governments in the promotion of guides to good hygiene practice.

Task in the development of guides to good hygiene practice	Performed by the government of the Member States for national guides Art. 8 Reg. 852/2004	Performed by the Commission for EU guides. Art. 9 Reg. 852/2004
Initiative	No government decisions on desirability or necessity and contents	
Functions of public law authorities	National governments encourage national guide. Art. 7	Member States decide first in Comitology procedure whether European guides are made, their scope and subject matter: Art. 9(1)
	National governments make information available on how to make guides (No Article)	
Development and dissemination	By food business sectors: Art. 8(1)	'By' appropriate representatives of European food business sectors, including SMEs and other interested parties like consumer groups: Art. 9(2)(a) 'by'.
Public law authorities do more than encouraging and take the initiative	No	'or in' *consultation with* same groups: Art. 9(2)(a)
	Commission can take the initiative. Art. 9(2)(a) 'or in'	
Parties whose interests may be substantially affected	such as competent authorities and consumer groups; Art. 8(1)(a)	including competent authorities: Article 9(2)(b)
Contribution of parties with substantially affected interests to the development of guides	In consultation with their representatives Art. 8(1)(a)	Collaboration of those parties: Article 9(2)(b)
Relevant codes of practice of the Codex Alimentarius	Having regard to Codex Art. 8(1)(b)	Having regard to Codex Art. 9(2)(c)
Guides for primary production and those associated operations listed in Annex I, having regard to the recommendations set out in Part B of Annex I.	Yes, Art. 8(1)(c)	Yes, Art. 9(2)(d)

►►

Law text box 13.6. Continued.

Task in the development of guides to good hygiene practice	Performed by the government of the Member States for national guides Art. 8 Reg. 852/2004	Performed by the Commission for EU guides. Art. 9 Reg. 852/2004
Role for standards institute in development of guides	Developed under the aegis of a national standards institute explicitly mentioned. Refer to institutes in Annex II to Directive 98/34/EC Art. 8(2)	Existing European institutes not mentioned
Who makes assessment?	National government of the Member States Art. 8(3)	Member States in SCFCAH. The Committee referred to in Article 14). Art. 9(3)(a)
Standards	Developed	
Assessment object	National guides	*Draft* EU guides
Developed in accordance with demands of Regulation 852/2004	Art. 8(3)(a)	paragraph 2; Art. 9(3)(a)
Practicable for the sectors to which they refer throughout the EU	Art. 8(3)(b)	Art. 9(3)(b)
Suitable as guides to compliance with Articles 3, 4 and 5, Regulation 852/2004	Art. 8(3)(c)	Art. 9(3)(c)
Who reviews?	No review	Commission shall invite the SCFCAH of Article 14. Art. 9(4)
Review	No review	Periodically to review EU guides prepared in accordance with this Article, in cooperation with the bodies mentioned in paragraph 2. Art. 9(4)
Aim of the review	No review	To ensure that the guides remain practicable and to take account of technological and scientific developments. Art. 9(4)(par. 2)
Central collection of guides	Forward national guides that comply with Art. 8(3)(a-c) to the Commission. Art. 8(4)	

►►

Law text box 13.6. Continued.

Task in the development of guides to good hygiene practice	Performed by the government of the Member States for national guides Art. 8 Reg. 852/2004	Performed by the Commission for EU guides. Art. 9 Reg. 852/2004
Public information on guides	Commission runs a register with the titles and references of the national guides and makes it available for the Member States. Art. 8(4)	The titles and references of EU guides prepared in accordance with this Article shall be published in the C series of the Official Journal of the European Union. Art. 9(5)
Guides made according to preceding Directive 93/43/EEC	Continue to apply when compatible with Regulation 852/2004 Art. 8(5)	No mention of European guides to good hygiene practice made according to preceding Directive 93/43/EEC

13.2.10 Primary production

Primary production holds a special position in the food hygiene regulations. It has its own Annex I in Regulation 852/2004 with direct legislation in Part A and forceful advice on how to make guides to good hygiene practice in Part B.[619] It is exempted from the HACCP obligation,[620] but encouraged to experiment with it.[621] It is cautioned that it may be subjected to the HACCP system in the future.

The borderline between primary production and the next production stages

The article of Annex I called Scope is of special importance because it determines the dividing line between primary production and the next stages of the food production chain. Article I of Annex I defines the associated operations, the operations so closely connected to primary production that they are categorised with primary production for legislative purposes. Regulation 852/2004 does not present a definition of primary production itself. Article 3(17) GFL states: primary production means the production, rearing or growing of primary products including harvesting, milking and farmed animal production prior to slaughter. It also includes hunting and fishing and the harvesting of wild products. Article 2 of Regulation 852/2004 gives the following definition: 'primary products' means products of primary production including products of the soil, of stock farming, of hunting

[619] Articles 8(1)(c) and 9(2)(d) Regulation 852/2004.
[620] Recital 11 and Article 5(3) Regulation 852/2004.
[621] Recital 14 Regulation 852/2004. As discussed in Chapter 23, private standards like GlobalGAP include obligations to apply HACCP at the farm level.

and fishing'.[622] Therefore primary production is the combination of this definition and the associated operations. Diagram 13.5 presents these operations.

The article on hygiene provisions contains two general duties that can be found in paragraphs 2 and 6 and a repetition of already existing obligations based on other legislation in paragraph 3. These paragraphs for all operators are separated by paragraph 4 that addresses operators dealing with products of animal origin, and paragraph 5 written for operators dealing with plant products. This pattern of a common paragraph for the vegetal and animal branches of food production, followed by a separate paragraph for each of them, is also applied in Article A (III) on record-keeping.

13.2.11 Private standards

The HACCP-based system and hygiene guides are true 'self-regulation' in the sense that the businesses and the business associations set rules to be applied by themselves and their members. It has, however, increasingly been realised that the safety of a business' products depends not only on the business having its own processes under control, but also on the reliability of the processes applied by the businesses upstream (the suppliers and their suppliers). To ensure themselves of the reliability of the processes applied by other businesses, players downstream – retailers in particular – increasingly regulate the behaviour of upstream players through so-called private standards. These standards set out in great detail how the chain partners should conduct their business. To be eligible to supply, these partners have to provide evidence of compliance through certification issued by an independent third party. This system is discussed in some detail in Chapter 23.

13.3 Preparedness

13.3.1 HACCP

The HACCP system discussed above is not only relevant for the prevention of food safety problems. The principle mentioned in Article 5(1)(e) of Regulation 852/2004 requires that the HACCP plan includes corrective action to be taken in case of problems. The legislator leaves it largely to the food business operator to deal with problems that occur within the business. However, for problems that may affect other stages in the food chain, some mandatory requirements are in place.

[622] Article 2(1)(b) Regulation 852/2004.

13.3.2 Traceability

The General Food Law requires that food and food ingredients be traceable. Recitals 28 and 29 to the General Food Law read:

> *Experience has shown that the functioning of the internal market in food or feed can be jeopardised where it is impossible to trace food and feed. It is therefore necessary to establish a comprehensive system of traceability within food and feed businesses so that targeted and accurate withdrawals can be undertaken or information given to consumers or control officials, thereby avoiding the potential for unnecessary wider disruption in the event of food safety problems.*
>
> *It is necessary to ensure that a food or feed business including an importer can identify at least the business from which the food, feed, animal or substance that may be incorporated into a food or feed has been supplied, to ensure that on investigation, traceability can be assured at all stages.*

The intention of the traceability system is therefore to enable food safety problems to be identified at the source, and across the food chain. To this end food business operators must keep comprehensive records of exactly where their food material originated and where it went.

For food products of animal origin traceability is nothing new. The General Food Law, however, broadens the scope of traceability to all foods. In this section we will discuss the general provisions in the General Food Law, requirements for traceability of food producing animals and traceability requirements for genetically modified foods. The concept of traceability is defined in Article 3(15) GFL. See Law text box 13.8.

Law text box 13.8. Article 3(15) GFL on traceability.

Article 3(15) Regulation 178/2002
'traceability' means the ability to trace and follow a food, feed, food-producing animal or substance intended to be, or expected to be incorporated into a food or feed, through all stages of production, processing and distribution.

The main aim of traceability is to be able to quickly identify the source of a food safety problem and to conduct well-aimed recalls to take affected products from the market.[623] In practice, traceability schemes are used for other purposes as well. They play a role, for example, in assuring consumers that products possess certain invisible qualities relating to their origin and the way they have been handled (Kosher, Hallal, organic, etc.). The regulation regarding food contact materials has a similar traceability scheme.[624] This regulation expresses what had been suspected all along: traceability is also meant to facilitate attribution of liability.[625] There is no reason to suppose that this notion applies to food contact materials exclusively.

Whether the possibility of using traceability to establish liability of the business that caused a food safety problem is considered an advantage or a disadvantage depends upon one's point of view. For businesses at the end of the food chain, it seems advantageous for liability to be passed on to the companies upstream where the problem originated. Businesses in the latter position would probably see this as a disadvantage.

If no other, more specific requirements apply based on the General Food Law, businesses must be able to trace their inputs and outputs one step up and one step down.[626] For the majority of farm animals more specific requirements do apply.[627]

13.3.3 General Food Law

The General Food Law requires in Article 18 that food, feed, food-producing animals, and any other substance intended or expected to be incorporated into a food or feed shall be traceable at all stages of production, processing and distribution (see Law text box 13.9).

To this end, food and feed business operators must be able to identify any person who has supplied them with a food, feed, food-producing animal, or any substance intended or expected to be incorporated into a food or feed. They must also be able to identify all businesses to which their products have been supplied.

---

[623] The SCFCAH has identified the following areas in which traceability records help: facilitating targeted withdrawal and recall of food, thereby avoiding unnecessary disruption of trade; enabling consumers to be provided with accurate information concerning implicated products, thereby helping to maintain consumer confidence; facilitating risk assessment by control authorities. Guidance on the Implementation of Articles 11, 12, 16, 17, 18, 19 and 20 of Regulation (EC) n° 178/2002 on General Food Law, Conclusions of the Standing Committee on the Food Chain and Animal Health (2004). For hyperlink see reference list.

[624] Regulation 1935/2004 on materials and articles intended to come into contact with food.

[625] Article 17 Regulation 1935/2004.

[626] Recital 29 GFL.

[627] See also: S. Ammendrup and A.E. Füssell, Legislative Requirements for the Identification and Traceability of Farm Animals within the European Union, Revue Scientifique et Technique (International Office of Epizootics), 2001, p. 437.

Law text box 13.9. Article 18 GFL on traceability.

Article 18 Regulation 178/2002
Traceability

1. The traceability of food, feed, food-producing animals, and any other substance intended to be, or expected to be, incorporated into a food or feed shall be established at all stages of production, processing and distribution.
2. Food and feed business operators shall be able to identify any person from whom they have been supplied with a food, a feed, a food-producing animal, or any substance intended to be, or expected to be, incorporated into a food or feed.
   To this end, such operators shall have in place systems and procedures which allow for this information to be made available to the competent authorities on demand.
3. Food and feed business operators shall have in place systems and procedures to identify the other businesses to which their products have been supplied. This information shall be made available to the competent authorities on demand.
4. Food or feed which is placed on the market or is likely to be placed on the market in the Community shall be adequately labelled or identified to facilitate its traceability, through relevant documentation or information in accordance with the relevant requirements of more specific provisions.
5. Provisions for the purpose of applying the requirements of this Article in respect of specific sectors may be adopted in accordance with the procedure laid down in Article 58(2).

Business operators shall have in place systems and procedures that allow this information to be made available to competent authorities on demand. The law has little to say on the makeup of the traceability system. This is left to the discretion of the businesses concerned. An important decision they have to make is on the batch size they apply in their system. Does it cover, for example, one hour of production, a day, a week? The bigger the batch, the easier it is to manage the system. The smaller the batch, the smaller the damage if a product recall has to be undertaken.[628]

Food or feed which is or is likely to be placed on the market in the European Union must be adequately labelled or identified to facilitate its traceability through relevant documentation or information in accordance with the relevant requirements of more specific provisions.[629] Often the date marking is used for this purpose.[630]

[628] However, the aforementioned guidance document of the SCFCAH says more on the information to be kept by proposing the type of traceability information companies need to record and also the time of keeping the respective documents.
[629] See also Chapter 14.
[630] This explains why sometimes on the label the durability of a product is not only expressed in a date but in hours and minutes as well.

The requirement includes only businesses; it does not extend to the consumer. In practice this means that traceability terminates at the retail end. It begins before primary production as it includes the inputs of the primary sector.[631] In a guidance document that provides an interpretation of the most important provisions in the GFL,[632] the Standing Committee on the Food Chain and Animal Health took the position that traceability requirements apply only from entry past the EU border onwards; that is to say, authorities will not demand information regarding the origin of the product in a third country from which it was imported.[633] It is unclear what the basis is for this limited interpretation – the text of the GFL does not provide a foothold.[634]

This same document also concludes that businesses are not obligated to ensure internal traceability. In other words, they have to know where the ingredients came from and where the products went but not necessarily which ingredients went into which products. This interpretation seems to expose a major flaw in the system. In cases where internal traceability is not assured, it seems highly problematic to reconstruct the entire chain to trace the origins and consequences of a food safety problem.[635]

The burden to reconstruct the whole food chain when an incident occurs is on the authorities and to that end traceability information has to be made available to those authorities on demand.

13.3.4 Traceability of food-producing animals

Traceability legislation on living animals goes well beyond the general requirements of food law as it includes veterinary purposes in their framework.[636] That is to say, infectious diseases are not only being controlled for food safety reasons in a strict sense, but also for economic reasons including preventing residues of veterinary products entering the food chain, the protection of healthy animals and the protection of the reputation of the EU and its Member States of safe

[631] According to the interpretation of the SCFCAH on GFL, natural persons supplying food at the primary production (like mushroom collectors and hunters) should also be identified as part of the backward traceability.

[632] Guidance on the Implementation of Articles 11, 12, 16, 17, 18, 19 and 20 of Regulation (EC) n° 178/2002 on General Food Law, Conclusions of the Standing Committee on the Food Chain and Animal Health (2004). For hyperlink see reference list.

[633] Practice shows that traceability can be applied even in third countries, further than the importer; in the case of the EHEC outbreak in 2011 in Germany, EFSA and ECDC took a tracing action supported with the official controls by the FVO up to the farm level.

[634] Article 11 GFL rather suggests the contrary.

[635] As is discussed in Chapter 23, this flow is remedied in several private standards that explicitly do require internal traceability.

[636] For a detailed account see: B.M.J. van der Meulen and A.A. Freriks, 'Beastly Bureaucracy' Animal Traceability, Identification and Labeling in EU Law, Journal of Food Law & Policy 2(2), pp. 317-359.

agricultural products of high quality. Identification of animals also plays a role in the common agricultural policy – in particular, for the supervision of premiums.

An identification system must be regarded as a prerequisite for effective traceability. When discussing traceability of animals and animal products it is important to note that although of primary importance, identification is only one of the issues at stake. The basic requirements in European legislation to provide for an adequate system of identification of animals predate the General Food Law. They are laid down in two Directives. Article 3(1)(c) of Directive 90/425 states that animals for intra-EU trade must be identified in accordance with the requirements of EU rules and be registered in such a way that the original or transit holding, centre or organisation can be traced. These identification and registration systems are to be extended to the movements of animals within the territory of each Member State.[637] In addition, Article 14 of Directive 91/496[638] states that the identification and registration as provided for in Article 3(1)(c) of Directive 90/425 of imported animals must be carried out before area-specific checks have been made, except in the case of animals for slaughter and registered equidae (donkeys and horses).[639]

The Member States must collect all information in a database. Originally the implementing rules concerning the identification and registration of animals in aforementioned Directives were laid down in Directive 92/102 on the identification and registration of animals. Following several crises (discussed in Chapter 7), specific requirements were drafted for bovine, ovine, and caprine animals. However, the original Directive from 1992 is still in force for porcine animals (pigs). On the basis of Directive 92/102 Member States must have in place systems for the identification and registration of groups of pigs, including ear tags, registers per holding and a computerised database at national level.[640]

Before the BSE crisis, the rules concerning the identification and the registration of bovine animals (cattle and buffaloes) were also laid down in Directive 92/102. In order to re-establish market stability after the BSE crisis, the European legislature held that the transparency of the conditions for the production and marketing of the products concerned, particularly in regard to traceability, had to be improved. This led to the establishment of, on the one hand, a more efficient system for the identification and registration of bovine animals at the production stage and, on the other hand, a specific EU labelling system in the beef sector. The new system was laid down in Regulation 820/97 establishing a system for the identification and registration of bovine animals and regarding the labelling of beef and beef

[637] Article 3(1)(c) Directive 90/425. See also Directive 2002/33 (amending directive 90/425).
[638] See also Directive 96/43 (amending Directive 91/496).
[639] Directive 91/496 laying down the principles governing the organization of veterinary checks on animals entering the Community from third countries and amending Directives 89/662/EEC, 90/425/EEC and 90/675/EEC.
[640] See also Commission Decision 2000/678.

products. Apart from identification requirements, the Regulation introduced a labelling system that was optional for operators and organisations marketing beef until 1 January 2000 in the sense that operators and organisations wishing to label their beef had to do so in accordance with the Regulation. A compulsory beef-labelling system for all the Member States had to be introduced after 1 January 2000.

The improvements in the regulatory system brought about by this Regulation, exerted a positive influence on consumption of beef. 'In order to maintain and strengthen the confidence of consumers in beef and to avoid misleading them, it was deemed necessary to further develop the framework in which the information was made available to consumers by sufficient and clear labelling of the product'. This led to a mandatory labelling system that is laid down in Regulation 1760/2000 that replaces the former Regulation entirely. Since 2000, both the identification and labelling requirements are therefore set out in Regulation 1760/2000. Although the system has been set out in a regulation, to a large extent, it addresses the national legislatures in the Member States. Member States must set up a cattle identification and registration system. The system for the identification and registration of bovine animals must be comprised of ear tags to identify animals individually, computerised databases, animal passports, and individual registers on each holding.[641]

The basic objectives of the requirements set out in Regulation 1760/2000 are:
- the localisation and tracing of animals for veterinary purposes, which is of crucial importance for the control of infectious diseases;
- the traceability of beef for public health reasons; and
- the management and supervision of livestock premiums as part of the reform of management of the common agricultural policy.[642]

Ovine and caprine animals (sheep and goats) were excluded from the scope of Directive 92/102 in 2004. Based upon Article 3(1) of Regulation 21/2004 the system for the identification and registration of animals comprises the means of identification to identify each animal, up-to-date registers kept on each holding, movement documents and a central register or a computer database. For traceability requirements regarding fish see Regulation 2065/2001.

[641] Specific requirements on ear tags, passports and holding registers are laid down in Commission Regulation 911/2004 implementing Regulation 1760/2000 of the European Parliament and of the Council as regards ear tags, passports and holding registers; and Commission Regulation 887/2004.
[642] See Report from the European Commission to the Council and the European Parliament on the Possibility of Introduction of Electronic Identification for Bovine Animals, COM(2005) 9 final, 25 January 2005, p. 5.

13.3.5 GM traceability

Aim of GM traceability

Article 1 of Regulation 1830/2003 states that traceability of GM foods has as its objectives:
- facilitating accurate labelling;
- monitoring the effects on the environment and, where appropriate, on health; and
- implementing appropriate risk management measures including, if necessary, withdrawal of products.[643]

Content of GM traceability

Regulation 1830/2003 defines traceability as 'the ability to trace GMOs and products produced from GMOs at all stages of their placing on the market through the production and distribution chains'. Unlike the GFL, Regulation 1830/2003 requires a paper trail to accompany GM food. This paper trail ensures internal traceability. The flaw indicated above, which may exist in the general system of traceability, therefore does not occur with regard to GM food.

At the first stage of placing a product consisting of or containing GMOs on the market, including bulk quantities, operators must ensure that information: (a) that the product contains or consists of GMOs and (b) providing the unique identifier(s) assigned to those GMOs[644] is transmitted in writing to the operator receiving the product. At every following stage, the same information must be passed on for each ingredient or additive that it concerns.[645]

This seems easy on paper, but in practice, it is next to impossible to preserve the identity of each raw material through to the end-products in which they are used. Identity preservation[646] is hard to realise, for example, in bulk storage, in continuous production processes, and in cases where failed products re-enter the production chain as raw materials.[647] Additionally, all information must be

[643] Regulation 1830/2003 concerning the traceability and labelling of genetically modified organisms and the traceability of food and feed products produced from genetically modified organisms and amending Directive 2001/18/EC. On GM traceability see: Margaret Rosso Grossman, Traceability and Labeling of Genetically Modified Crops, Food and Feed in the European Union, Journal of Food Law & Policy, 1(1), 2005, pp. 43-85.

[644] On these identifiers, see Chapter 10, Section 10.5.8.

[645] Article 4(1) and (2) Regulation 1830/2003.

[646] Also called internal traceability.

[647] Also called 'rework'.

kept for five years.[648] Small traces – no more than 0.9% – are exempted from the traceability requirements if they are adventitious and unavoidable.[649]

13.4 Response

13.4.1 Withdrawal and recall

As elaborated in Chapter 8, food business operators may not bring food to the market if it is unsafe (Article 14 GFL). The product must be withdrawn from downstream businesses or recalled from the consumer if unsafe food nonetheless is discovered to have made it to market. The General Food Law deals with this situation in Article 19 shown in Law text box 13.10.

Law text box 13.10. Responsibilities for food: food business operators, Article 19 General Food Law Regulation 178/2002.

Regulation 178/2002
Article 19
Responsibilities for food: food business operators

1. If a food business operator considers or has reason to believe that a food which it has imported, produced, processed, manufactured or distributed is not in compliance with the food safety requirements, it shall immediately initiate procedures to withdraw the food in question from the market where the food has left the immediate control of that initial food business operator and inform the competent authorities thereof. Where the product may have reached the consumer, the operator shall effectively and accurately inform the consumers of the reason for its withdrawal, and if necessary, recall from consumers products already supplied to them when other measures are not sufficient to achieve a high level of health protection.
2. A food business operator responsible for retail or distribution activities which do not affect the packaging, labelling, safety or integrity of the food shall, within the limits of its respective activities, initiate procedures to withdraw from the market products not in compliance with the food-safety requirements and shall participate in contributing to the safety of the food by passing on relevant information necessary to trace a food, cooperating in the action taken by producers, processors, manufacturers and/or the competent authorities.
3. A food business operator shall immediately inform the competent authorities if it considers or has reason to believe that a food which it has placed on the market may be injurious to human health. Operators shall inform the competent authorities of the action taken to prevent risks to the final consumer and shall not prevent or discourage any person from cooperating, in accordance with national law and legal practice, with the competent authorities, where this may prevent, reduce or eliminate a risk arising from a food.
4. Food business operators shall collaborate with the competent authorities on action taken to avoid or reduce risks posed by a food which they supply or have supplied.

[648] Articles 4(5) and 5(2) Regulation 1830/2003.
[649] Article 7 Regulation 1830/2003.

The food business operator who considers or has reason to believe that a food s/he has imported, produced, processed, manufactured or distributed, is not in compliance with the food safety requirements, has at least four duties:

- First, there is the duty to immediately initiate procedures to *withdraw* the food in question from the market.
- Second, the operator must immediately *inform* the authorities that s/he has reason to believe that an unsafe food has been placed on the market. The operator must also inform the authorities of all actions taken to deal with the problem.
- Third, in case the product has already reached consumers, the operator shall effectively and accurately inform those consumers of the reason for its withdrawal, and *recall* products already supplied when other measures are deemed insufficient to achieve a high level of health protection.
- Finally, the food business operator has a duty to *collaborate* with the competent authorities on actions taken to avoid or reduce risks posed by foods, which s/he supplied.

If the food business operator in question has an adequate traceability system in place, withdrawal should not be too much of a problem. All the information on which product to withdraw from which customer should be present in the system. However, as discussed above, traceability systems are not required to include the consumer. Therefore in most cases the business operator will not have in his possession information on the identity of the consumers concerned. For this reason recall actions need to resort to publicity in the media.

13.4.2 Responsible food business operator

Who is responsible for withdrawal and recall? In its first paragraph Article 19 GFL mentions all possible types of food business that may have handled the food at issue. The only condition setting Article 19 in motion is that the business operator has reason to believe that the food is not in compliance with food safety requirements. Article 19 is not limited to the business that may have caused this non-compliance. Each player in the chain bears full responsibility. A business may not excuse itself by stating that it is not at fault. Naturally, it may prompt the business that was at fault to take the required action or hold that business liable for costs incurred. Once a business starts adequate actions, other businesses can limit themselves to supporting this action.

The responsibility is somewhat limited in section 2 for food businesses that by their very nature are unlikely to have caused the problem.

13.4.3 Non-compliance

The condition set in Article 19(1) GFL is not that the food is unsafe, but that it is not in compliance with the food safety requirements. The food safety requirements are all legal requirements placed on food with a view to ensuring food safety, including for example the food safety targets discussed in Chapter 11. As explained in that chapter, these targets often apply safety margins. The ADI for pesticide residues for example is usually set at 1/100 of the NOAEL. As a consequence of this margin a non-compliant food need not be unsafe within the meaning of article 14 GFL (injurious to health). In paragraph 3 of Article 19 GFL, the obligation to immediately inform the competent authorities is limited to the situation where the food may be injurious to health (thus unsafe).

13.4.4 Measures

Article 19 GFL explicitly mentions withdrawal (from other businesses) and recall (from consumers) as measures to be taken. Slightly hidden at the end of paragraph 1 is the sentence: 'when other measures are not sufficient to achieve a high level of health protection'. This sentence implies that the food business operator is empowered to take other measures than withdrawal or recall if these other measures are sufficient to achieve a high level of health protection. In practice we see, for example, that businesses that omitted to mention allergens on the label of a product issue a public warning without calling the product back.

The exception that 'other measures' may be taken if they are effective does not apply to the obligation to inform the authorities.

Discussion on the obligation to co-operate and the rights of the defence

It is likely that in several Member States discussion will arise as to what extent the information that business operators are obliged to provide under Article 18 or 19 GFL, may be used by the authorities when imposing sanctions on the operator for infringing on Article 14 GFL.

The German application of this provision – § 44(6) of the Lebensmittel-, Bedarfsgegenstände- und Futtermittelgesetzbuch (the Code on food, food contact materials and feed) – explicitly states that the information provided by the food business operator may not be used against him in criminal proceedings. This provision will undoubtedly stimulate operators to come forward with problems they discover within their organisation. However, there is also the risk that such a provision will be misused to escape punishment for intentional neglect.

13.5 Food safety law for businesses

EU food hygiene legislation with the central place its gives to the HACCP principles, traceability and recall requirements are tell-tale implementations of the principle underlying Article 17(1) GFL. Food business operators are responsible. EU legislation imposes upon them to achieve food safety. To a large extent is has been left to the business operators to decide how best to meet this responsibility in a specific case.

14. Food labelling and beyond

Harry Bremmers and Bernd van der Meulen

14.1 Introduction

The European Union is constantly being challenged to achieve an internal market, which implies a requirement to protect consumers against unfair information practices and enable them to make informed choices. Food information plays an intermediary role as transparency is a condition for smooth functioning of markets on the one hand, and making substantiated choices on the other. From the point of view of the business, the most important messages will usually be the identity of the brand under which the product is sold. This brand can be protected from other businesses by registering it as a trademark. This option is discussed in Chapter 22. The freedom of expression[650] that food business operators can exercise is limited in food law. Certain information is mandatory, other information is prohibited. The General Food Law lays down the principles on consumer information in Articles 8 and 16 (See Law text boxes 14.1 and 14.2). Their function is to give consumers the opportunity to make informed choices and to protect them from misleading practices.

Law text box 14.1. Protection of consumers' interests, Article 8 General Food Law.

Article 8
Protection of consumers' interests
Food law shall aim at the protection of the interests of consumers and shall provide a basis for consumers to make informed choices in relation to the foods they consume. It shall aim at the prevention of:
(a) fraudulent or deceptive practices;
(b) the adulteration of food; and
(c) any other practices which may mislead the consumer.

[650] Article 11(1) of the Charter of Fundamental Rights of the European Union reads: 'Everyone has the right to freedom of expression. This right shall include freedom to hold opinions and to receive and impart information and ideas without interference by public authority and regardless of frontiers'. In Article 52 the Charter sets the conditions for limiting the exercise of fundamental rights: 'Any limitation on the exercise of the rights and freedoms recognised by this Charter must be provided for by law and respect the essence of those rights and freedoms. Subject to the principle of proportionality, limitations may be made only if they are necessary and genuinely meet objectives of general interest recognised by the Union or the need to protect the rights and freedoms of others'. It is unfortunate that nowhere in food labelling law does the EU legislator make the effort to actually demonstrate that these conditions have been considered and that they are met.

Law text box 14.2. Presentation, Article 16 General Food Law.

Article 16
Presentation
Without prejudice to more specific provisions of food law, the labelling, advertising and presentation of food or feed, including their shape, appearance or packaging, the packaging materials used, the manner in which they are arranged and the setting in which they are displayed, and the information which is made available about them through whatever medium, shall not mislead consumers.

Article 8 GFL addresses the legislator, both at EU level and in the Member States. The food legislation they make should contribute to achieving these objectives. Article 16 GFL addresses the businesses. In their communication practices, they should comply with it. The principles expressed in these articles have been elaborated in the food information law.

14.2 Marketing

With regard to sales of food and other products, economists distinguish a marketing mix of four 'P's: product, price, promotion, and place.[651] Appealing products need to be presented at affordable prices and in accessible places and the customer has to be informed about this. Businesses give themselves and their products identity through branding. Brands are made recognisable with the help of trademarks, trade names, shapes, colours and by other measures. Their content, however, goes beyond these visible aspects and may include an entire image of quality, reliability, style, etc. Traditionally, branding has been the domain of producers. Nowadays, retail businesses develop brands of their own, which are referred to as 'private labels'. Initially they were understood to represent low quality and low price as compared to premium brands, but nowadays retail chains bring out high quality private labels as well. It should be noted that in economic literature the expression 'labelling' covers a concept of image building. As we will see below, in food law the expressions 'label' and 'labelling' are much more mundane in meaning. They do not go far beyond the piece of paper actually stuck to a food product and the message it contains. However, in the course of time the scope of food information law has expanded as well.

14.3 Food Information

From the beginning of the development of EU food law, most pieces of legislation contained provisions on labelling, indeed many still do. The foundation for the present generic food information requirements have been laid down in *Directive*

[651] See among others: Ph. Kotler and K. Keller's Marketing Management, 14th Edition. Prentice Hall, New York. 2012.

2000/13 on the approximation of the laws of the Member States relating to the labelling, presentation and advertising of foodstuffs. As of 13 December 2014 this directive will be replaced by a regulation. Here below, in discussing the different aspects of food information, reference will be made to this Food Information to Consumers Regulation (FIC).[652] The concept of food information as applied in the FIC is wider than the scope of 'labelling, advertising and presentation of foodstuffs'. It is defined as the information concerning a food and made available to the final consumer[653] by means of a label, other accompanying material, or any other means including modern technology tools or verbal communication.[654] Media for transfer of information are not only the label (that is:[655] any tag, brand, mark, pictorial or other descriptive matter, written, printed, stencilled, marked, embossed or impressed on, or attached to the packaging or container of food) or other accompanying documents, advertisements or flyers, but also information intended for the final consumer available on websites or other electronic information carriers.

The broadening of the scope may among other things have been induced by the way foods are processed and marketed nowadays (for instance by means of long, heterogeneous supply chains instead of short, homogeneous ones) and the changes in preferences and consumption patterns (convenience foods, foods with special functions, consumption in catering facilities rather than at home, etc.). The FIC is not only the follow-up to Directive 2000/13, but also most notably to the Nutrition labelling directive,[656] coming into effect from 13th December 2016. To a major extent the horizontal requirements that are laid down in the FIC are comparable to the ones in the preceding directive, but there are major exceptions, which are addressed in the remaining chapter. The FIC creates a solid foundation for informing the consumers:

- some specific directives and regulations are repealed and integrated in the FIC (the most prominent being the above indicated rules for nutrition information);
- a solid base is defined for food information, in the form of a set of conclusive principles;
- food information law, as a sub-disciplines of food law, is solidly connected to and embedded in food law, in adopting defined concepts in other acts (e.g. 'food', 'enzymes', 'additives' or 'food business'),[657] and regulating mandatory food information at the EU level instead of the level of Member States.

[652] Regulation 1169/2011 of the European Parliament and of the Council of 25 October 2011 on the provision of food information to consumers, amending Regulations 1924/2006 and 1925/2006 of the European Parliament and of the Council, and repealing Commission Directive 87/250, Council Directive 90/496, Commission Directive 1999/10, Directive 2000/13 of the European Parliament and of the Council, Commission Directives 2002/67 and 2008/5 and Commission Regulation 608/2004.
[653] Including also the mass caterer.
[654] Article 2(2)(a) of the FIC.
[655] Article 2(2)(i) of the FIC.
[656] Council Directive 90/496 of 24 September 1990 on nutrition labelling for foodstuffs.
[657] Article 2(1) of the FIC.

Next to the generic mandatory requirements in the FIC, other mandatory rules in specific directives and regulations derogate on or supplement it:

- vertical food information rules, originating from the common market organisation,[658] as for fruit and vegetables or milk and milk products, veal and poultry, or for eggs or from other, product-related directives or regulations (like for chocolate or juices);[659]
- the rules for specific foods (like GMO-based foods, foods for special purposes, or food supplements[660]); and
- requirements for substances used in or on foods, or during production processes (like additives, enzymes, flavourings, or food contact materials).[661]

Next to mandatory food information, food businesses may also provide food information on a voluntary basis. Voluntary information may be regulated, structure and/or content being specified in public rules and regulations, or non-regulated. For instance, information concerning nutrition and health claims is regulated in a Nutrition and Health Claims Regulation (NHCR).[662] A claim is any message or representation, which is not mandatory under Union or national legislation, including pictorial, graphic or symbolic representation, in any form, which states, suggests or implies that a food has particular characteristics.[663] Other voluntary information is to a large extent not specifically regulated (as is the case for private food law, also called self-regulation).[664] At the European level information provided to the consumer is governed by general principles. Food information may not be misleading, ambiguous or confusing, must (where appropriate) be substantiated with scientific facts and may not be provided to the detriment of mandatory information. The FIC starts by outlining generic information requirements (like principles or responsibilities) and moves on to

---

[658] Regulation 1308/2013 of the European Parliament and the Council of 17 December 2013 establishing a common organisation of the markets in agricultural products.

[659] Directive 2000/36 of the European Parliament and of the Council of 23 June 2000 relating to cocoa and chocolate products intended for human consumption; Council Directive 2001/112 of 20 December 2001 relating to fruit juices and certain similar products intended for human consumption.

[660] Regulation 1829/2003 of the European Parliament and of the Council of 22 September 2003 on genetically modified food and feed; Regulation 609/2013 of the European Parliament and the Council of 12 June 2013 on food intended for infants and young children, food for special medical purposes, and total diet replacement for weight control; Directive 2002/46 of the European Parliament and the Council of 10 June 2002 on the approximation of the laws of the Member States relating to food supplements.

[661] Regulation 1333/2008 of the European Parliament and the Council of 16 December 2008 on food additives; Regulation 1334/2008 of the European Parliament and of the Council of 16 December 2008 on flavourings and certain food ingredients with flavouring properties for use in and on foods; Regulation 1332/2008 of the European Parliament and of the Council of 16 December 2008 on food enzymes; Commission Regulation 10/2011 of 14 January 2011 on plastic materials and articles intended to come into contact with food.

[662] Regulation 1924/2006 of the European Parliament and of the Council of 20 December 2006 on nutrition and health claims made on foods.

[663] Article 2(2)(1) of the NHCR.

[664] See: T. Havinga, Private Regulation of Food Safety by Supermarkets. Law & Policy, 2006, pp. 515-533.

specific, mainly labelling-related, ones. Technical specificities are placed in the 15 annexes. The remainder of this chapter focuses first on generic mandatory requirements, starting with responsibilities and principles and next reviewing mandatory particulars. Then additional rules contained in adjacent or vertical legal and non-legal acts are addressed. Next, the rules and regulations on key topics of voluntary food information law will be reviewed. Finally, remarks are made on the effect of European food law on international law, and vice versa. The structure of the remaining chapter is visualised in Diagram 14.1.

Diagram 14.1. Structure of this chapter.

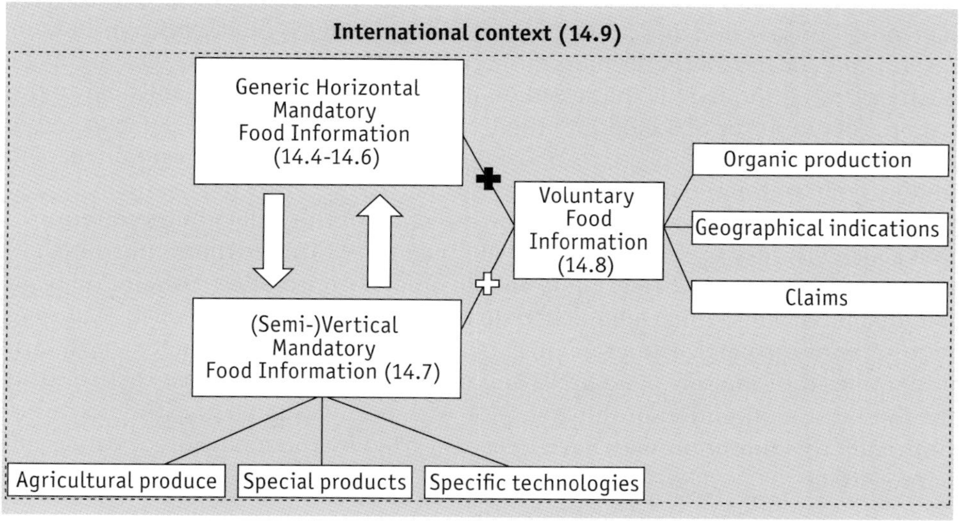

International context (14.9)
Generic Horizontal Mandatory Food Information (14.4-14.6)
(Semi-)Vertical Mandatory Food Information (14.7)
Voluntary Food Information (14.8)
Organic production
Geographical indications
Claims
Agricultural produce
Special products
Specific technologies

14.4 Responsibilities

Responsibilities can be discerned in:

- those of the retailer/caterer who sells foods to the final consumer;
- those of previous stages in the supply chain.

The general principle that food business operators are responsible across the entire food chain[665] applies also to food information. In Case C-315/05 ('Lidl Italia') the question was whether a retailer can be held responsible for incorrect information on branded liquors that it sells. The Italian authorities had imposed a fine on Lidl because an alcoholic beverage had been sold with a lower alcohol content than the manufacturer had indicated on the label. Retailer Lidl had argued to the Italian authorities that the responsibility for the labelling of brands from third parties

[665] Article 17(1) of the GFL.

applies to the brand-holder, not the retailer. The Court of Justice expressed in a preliminary ruling that (former) Directive 2000/13 must be interpreted as not precluding legislation of a Member State *'which makes it possible for an operator, established in that Member State, which distributes a pre-packaged alcoholic beverage to be delivered as such within the meaning of Article 1 of that directive, produced by an operator established in another Member State, to be held liable for an infringement of that provision, established by a public authority, resulting from the producer's inaccurate statement on the product label of the alcoholic strength by volume of the product and, consequently, to be penalised by an administrative fine, even where, as the mere distributor, it simply markets the product as delivered to it by the producer.'*[666] Article 8(1) of the FIC may be interpreted as changing the law away from the interpretation given in this ruling. Here it is stated that the food business operator responsible for the food information shall be the operator under whose name or business name the food is marketed or, if that operator is not established in the Union, the importer of a food into the Union market. This would seem to imply that the retailer is no longer responsible for the information on products sold under the brand name of a supplier to a supermarket. However, Article 8(5) of the FIC widens the responsibility in the direction of the Lidl ruling and requires that food business operators, within the businesses under their control, shall ensure compliance with the requirements of food information law and relevant national provisions which are relevant to their activities and shall verify that such requirements are met. Paragraph 8(1) places responsibility on the business that undertakes the labelling. Paragraph 8(5) places it with the business that controls the food. If the intention of the legislator with 8(1) was to change the law away from Lidl, this intention may have been thwarted by the inclusion of paragraph 5 into Article 8.

Previous stages have to provide all necessary and mandatory information to next stages, so that subsequent stages are adequately informed and finally the consumers' information needs can be satisfied.[667]

14.5 Principles and generic requirements

The principles that have been adopted via the FIC are in concordance with general aims as proclaimed in the treaties of the European Union, as well as in the General Food Law (Articles 3-5 of the FIC):

- the objective of a high level of consumer protection;
- enabling the consumer to make informed choices taking into account health, social, environmental and ethical considerations;
- fostering the free movement of goods, while considering producers' interests and the provision of quality food;

[666] CJEU 23 November 2006 Case C-315/05, Lidl Italia Srl v. Comune di Arcole (VR).
[667] Article 8(8) of the FIC.

- open and transparent communication with stakeholders, applying predefined principles of mandatory food information and – if public health is involved – consultation of EFSA before measures are adopted.

Food information practices are required to be 'fair'. What 'fair' is, is content-wise delineated in Article 7 of the FIC (Law text box 14.3.).

Law text box 14.3. Article 7 of the FIC: fair information practices.

Article 7 Fair information practices

1. Food information shall not be misleading,[1] particularly:
   (a) as to the characteristics of the food and, in particular, as to its nature, identity, properties, composition, quantity, durability, country of origin or place of provenance, method of manufacture or production;
   (b) by attributing to the food effects or properties which it does not possess;
   (c) by suggesting that the food possesses special characteristics when in fact all similar foods possess such characteristics, in particular by specifically emphasising the presence or absence of certain ingredients and/or nutrients;
   (d) by suggesting, by means of the appearance, the description or pictorial representations, the presence of a particular food or an ingredient, while in reality a component naturally present or an ingredient normally used in that food has been substituted with a different component or a different ingredient.
2. Food information shall be accurate, clear and easy to understand for the consumer.
3. Subject to derogations provided for by Union law applicable to natural mineral waters and foods for particular nutritional uses, food information shall not attribute to any food the property of preventing, treating or curing a human disease, nor refer to such properties.

[1] The addition 'to a material degree' which appeared here in Directive 2000/13 has been removed (footnote added).

The requirement 'clear and easy to understand' is further elaborated upon in Article 13. This provision rules that mandatory food information shall be put in a conspicuous place in such a way that it is easily visible, clearly legible and, where appropriate, indelible. It also sets specific formats for the information (like expression in words or numbers with a general minimum font-size for all food information of 1.2 mm,[668] or the use of approved pictograms or symbols). The requirement that food information shall not be misleading is, for instance, violated by providing a wrong durability date, by stating that it lowers cholesterol levels when it does not, by using artificial cheese on a quattro formaggi pizza without proper notification, or hog meat as an ingredient for producing goulash soup instead of beef. Substitution is a fraudulent act that seems to be ineradicable, given the 2012-2014 horsemeat scandal. This scandal fits perfectly into a series

[668] Exceptions apply with respect to very small packages; see Article 13 of the FIC.

of incidents of fraud, adulteration and deception (like dioxin and melamine).[669] The response of the European legislator is to tighten the rules and fill the gaps in food information law, such as the uncontrolled use of the designation 'natural', or the false or hidden use of health claims, or of misleading geographical indications (like 'Belgian' chocolate produced in China).[670]

14.6 Mandatory particulars

Articles 9-35 of the FIC address mandatory particulars for foodstuffs. First, a list of mandatory particulars is stated (Article 9, see Law text box 14.4.), next the particulars are addressed one by one in articles, while specific details are included in the 15 annexes to the FIC.

Law text box 14.4. Mandatory particulars in the FIC.

Article 9 List of mandatory particulars

1. In accordance with Articles 10 to 35 and subject to the exceptions contained in this Chapter, indication of the following particulars shall be mandatory:
   (a) the name of the food;
   (b) the list of ingredients;
   (c) any ingredient or processing aid listed in Annex II or derived from a substance or product listed in Annex II causing allergies or intolerances used in the manufacture or preparation of a food and still present in the finished product, even if in an altered form;
   (d) the quantity of certain ingredients or categories of ingredients;
   (e) the net quantity of the food;
   (f) the date of minimum durability or the 'use by' date;
   (g) any special storage conditions and/or conditions of use;
   (h) the name or business name and address of the food business operator referred to in Article 8(1);[1]
   (i) the country of origin or place of provenance where provided for in Article 26;
   (j) instructions for use where it would be difficult to make appropriate use of the food in the absence of such instructions;
   (k) with respect to beverages containing more than 1.2% by volume of alcohol, the actual alcoholic strength by volume;
   (l) a nutrition declaration.

[1] That is the food business operator that is responsible for the food information (footnote added).

[669] See Chapter 7.

[670] See for example: http://tinyurl.com/np656re [in Dutch].

The mandatory particulars have to be placed directly on the package or on the label,[671] with some exceptions:

- unless national law states otherwise for non-prepackaged foods only allergen information is mandatory. Member States may determine what and how other particulars mentioned in Articles 9-10 have to be provided, after consent of the Commission;[672]
- in case of distance selling the date of minimum durability or the 'use-by' date are made available ultimately at time of delivery;[673]
- for some foods, like alcoholic beverages with more than 1.2% alc./vol., the ingredient list is not mandatory and also nutrition labelling is voluntary for these products;[674]
- some categories of products have their own nutrition labelling requirements (like food supplements and natural mineral waters).

Member States may also extend the mandatory particulars for specific categories of foods for the protection of public health, of consumers, the prevention of fraud and the protection of intellectual property rights,[675] or derogate from the requirements for certain foods, like milk or milk products.[676]

For traceability purposes, lot-identification information has to be provided during all stages of the supply chain and not only on the package for the final consumer. That is why this mandatory particular is not an item in Article 9(1) of the FIC.[677]

14.6.1 Name

The name requirement is elaborated in Article 17 of the FIC. The use of the word 'name' is a little confusing. Other language versions use words that may be translated as 'designation' or 'indication'.[678] The name that must appear on the labelling is not what in common speech is understood by 'name' either: the brand name under which the product is marketed (like 'Twix'). The name in legal terms is the generic name ('subtitle') which describes the nature of the product ('milk chocolate covered with caramel and biscuit').

There are three modalities for the name of a food: the legal name available on a European or a national level, the customary name, or a descriptive name (in this order of relevance).

[671] Article 12 of the FIC.
[672] Article 44 of the FIC.
[673] Article 14 of the FIC.
[674] The Commission will probably design special rules for regulating 'alcopops'.
[675] Article 39 of the FIC.
[676] Article 40 of the FIC.
[677] It is addressed in Directive 2011/91 of the European Parliament and of the Council of 13 December 2011 on indications or marks identifying the lot to which a foodstuff belongs.
[678] In German: *Verkehrsbezeichnung*, in Dutch: *aanduiding*, in French: *dénomination de vente*.

Legal names can be found in standards for single types of foodstuffs (such as chocolate and jam), for categories of foodstuffs (like the designation 'food supplement') or jointly in framework and Commission acts (like the regulation regarding common market organisation (CMO)[679] and implementing rules based thereon (for instance for 'veal'), together with additional particulars (like the use of the expression 'free range chicken'). In the provision of legal names the impact of vertical food legislation is felt. This type of legislation connects compositional standards to legal names.

A customary name is the name that is accepted and recognised in the Member State in which a food is sold (like 'hamburger', or 'goulash soup'). If neither a legal name nor a customary name is available, then a description of the product is due (like 'spinach-cream vegetable dish with cheese topping').

Some EU Member States have extensive provisions on legal food names.[680] Surprisingly, at the EU level no reference is made to the Codex Alimentarius. The Codex contains a great many vertical standards. Given the commitment the EU professes regarding the development of international food standards,[681] it would have been fitting to recognise in the FIC the names put forward by the Codex as legal names. The situation now is that the names provided by Codex may exercise some influence through adoption in vertical legislation at EU or Member State level, by becoming customary or through case law. In deciding if vertical provisions in Member State law constitute barriers to trade by reserving a name, the CJEU sometimes takes the Codex Alimentarius into consideration.[682]

The FIC requires additional particulars[683] with the name. These particulars are given in Annex III (see Diagram 14.2).

[679] Regulation 1308/2013 of the European Parliament and of the Council of 17 December 2013 establishing a common organisation of the markets in agricultural products.

[680] Germany and Austria, for example, have their 'Lebensmittelbuch'. These books can be compared to official encyclopaedias of food.

[681] As expressed in Articles 5(3) and 13 of the General Food Law.

[682] See, for example, the Emmenthal cheese case discussed in Chapter 3, Section 3.6.6 (CJEU 5 December 2000, Case C-448/98).

[683] See Article 10 of the FIC.

Diagram 14.2. Additions to the name of a food (based on Annex III of the FIC).

Type/Category of food/Process step	Particulars
Physical condition/specific treatment	For example: powdered, quick-frozen, smoked
Frozen before sale/defrosted before sale	Defrosted
Ionisation	'irradiated', 'treated with ionising radiation'
Substitution of a normally used/naturally present ingredient	Indication of the substitute
Added proteins in meat products, meat preparations and fishery products of a different animal origin	Presence of such proteins and their origin
Added water to meat/fishery products/preparations (a cut/joint/slice/portion or carcase) > 5%	Presence of added water
Glued/joint meats/fishery products	'formed meat', 'formed fish'
Minced meat	Indication of fat and collagen content

14.6.2 Ingredients

Article 2(2) of the FIC defines 'ingredient' as any substance or product, including flavourings, food additives and food enzymes, and any constituent of a compound ingredient, used in the manufacture or preparation of a food and still present in the finished product, even if in an altered form. It states further that residues shall not be considered as 'ingredients'. The obligation to provide *a list* of ingredients has been elaborated in Article 18 of the FIC, while technical implementation rules are included in Annex VII. The list of ingredients shall include any substances, including additives, enzymes and flavourings, and any constituent of a compound ingredient, used in the manufacture or preparation of a foodstuff and still present, even if in altered form, in the finished product.[684] The rules discussed in the previous section regarding the name of a food apply to the designation of ingredients as well. Ingredients that themselves consist of several ingredients (so-called 'compound ingredients'[685]) may be mentioned as single ingredient, provided they are no more than 2% of the finished product. Otherwise the ingredients of which they consist must be mentioned separately.[686]

Several exceptions apply to the obligation to list ingredients (for example, for fresh fruit, vinegar, cheese, butter, products consisting of one single ingredient recognisable from the name, and beverages containing more than 1.2% of alcohol). The most important exception to the obligation to mention substances used in the manufacture applies to additives or other substances used as processing aids.

[684] Article 2(2)(f) of the FIC.
[685] Article 2(2)(h) of the FIC.
[686] For details see part E of Annex VII.

The FIC refers for the definitions of processing aids and additives to Regulation 1333/2008.[687] A processing aid is any substance not consumed as a food ingredient by itself, intentionally used in the processing of raw materials, foods or their ingredients, to fulfil a certain technological purpose during treatment or processing and which may result in the unintentional but technically unavoidable presence of residues of the substance or its derivatives in the final product, provided that these residues do not present any health risk and do not have any technological effect on the finished product.

Special requirements

Special requirements for information on ingredients include:
- the possibility to designate some ingredients by generic names (like 'herbs and spices', or 'cheese');[688]
- added water has only to be indicated as an ingredient if the amount exceeds 5%, but for meat, meat preparations, unprocessed fishery products and bivalve molluscs water is considered an ingredient in any case;
- ingredients are listed in descending order of weight, unless the ingredients are less than 2% of the finished product;
- the indication of nano-engineered ingredients with the addition 'nano' between brackets;
- the highlighting (italics, bold, or underline) of all allergens mentioned in Annex II.

Allergens

There is a limited set of allergens that have to be indicated on or near foods, whether they are pre-packaged or not and whether otherwise an exception to the labelling requirement applies or not. These are listed in Annex II of the FIC. This is the only EU mandatory particular on non-prepackaged foods included in the FIC.[689] If no ingredient list is given, then allergens have to be indicated using a statement like 'Contains [name allergen(s)]'.

[687] Article 3(2)(b) of Regulation 1333/2008 on food additives.
[688] Cheese can be made of cow's milk, goat, sheep or yet another mammal. If cheese as such is sold, then the consumer should be informed about the kind of cheese that is being marketed, so as not to be misled. A general definition of 'cheese' at the European level does not exist; national and international legislation fill the gap.
[689] Article 44(1)(a) of the FIC.

Diagram 14.3. Allergenic ingredients listed in Annex II of the FIC.

Category	Exceptions[1]
Cereals containing gluten	+
Crustaceans and products thereof	–
Eggs and products thereof	–
Fish and products thereof	+
Peanuts and products thereof	–
Soybeans and products thereof	+
Milk and products thereof (including lactose)	+
Nuts (almonds, hazelnuts, walnuts, cashews, pecan nuts, Brazil nuts, pistachio nuts, macadamia nuts and Queensland nuts) and products thereof	+
Celery and products thereof	–
Mustard and products thereof	–
Sesame seeds and products thereof	–
Sulphur dioxide and sulphites at concentrations of more than 10 mg/kg or 10 mg/litre expressed as SO_2	–
Lupin and products thereof	–
Molluscs and products thereof	–

[1] In case certain exceptions apply to the obligation to label (derivates of) these substances, this is indicated with a '+' in the Diagram. For details see the Annex II to the FIC.

Quantity of Ingredients Declaration (QUID)

The ingredients must be listed in descending order of weight as added to a food during processing. For some ingredients a *QU*antity of *I*ngredient *D*eclaration (QUID) has to be given.[690] This is the case if the ingredient is in the name of the food (like 'tomato soup') or is associated with that food (e.g. in 'goulash soup': the amount of meat), is depicted on the labelling in words, pictures or graphics (for instance, a picture of an orange on a 'mixed juice drink') or is essential to characterise a food and distinguish it from other products (e.g. for 'Spicy Mexican Vegetables Mix' the amount of spicy ingredients, like pepper).

Specific substances and ingredients

For some substances, used as ingredients and/or added to foods for technological purposes, specific requirements exist. This is for instance the case with respect to

[690] See Article 22 of the FIC; exceptions are summed up in Annex VIII.

food improvement agents, like enzymes,[691] additives (including colours),[692] and flavourings.[693] Labelling of food additives is an issue of great importance, to the consumer as well as to the user industry. Additives and enzymes are ingredients and must be included in the ingredient list. There are, however, some exemptions. The most notable is drinks with an alcohol content of more than 1.2%. So far, these need not carry ingredient labelling. Consequently, alcohol-free beers will routinely show a list of ingredients. Alcoholic drinks often do not show an ingredients list, except in countries where national law requires it, such as in Germany.

Additives and enzymes whose presence in a foodstuff is merely due to carry-over[694] and that have no technological function in the final foodstuff need not be indicated in the ingredient list. Carriers (and other substances which perform the same function as carriers) need not be indicated either.[695] Additives in the ingredient list must be preceded by the name of their additive category indicating its function (colour, preservative, glazing agent, etc.) followed by their specific name, or their E-number.[696] The E-number (E is an abbreviation for EU) serves to identify the approved additives.[697]

Allergenic additives have to be highlighted. The only additives to which this requirement applies are the sulphites. Irrespective of whether it is for the purpose of intolerance warnings or for ingredient listing, sulphites are considered not present if their total quantity from all sources does not reach 10 mg/kg of product.

In addition to the rules on labelling for additives on 'pre-packaged food', there are also rules for the labelling of 'food additives sold as such', to manufacturers (additives as ingredient raw materials by themselves), as well as to the ultimate consumer (in baking powder, or in table-top sweeteners).[698] Note that table-top sweeteners are a specific type of foodstuff mainly consisting of (mixtures of) additives. Similar requirements to those for additives have been put in place for enzymes and flavourings. For flavourings, additional mandatory particulars may

---

[691] Regulation 1332/2008 of the European Parliament and of the Council of 16 December 2008 on food enzymes and amending Council Directive 83/417, Council Regulation 1493/1999, Directive 2000/13, Council Directive 2001/112 and Regulation 258/97.

[692] Regulation 1333/2008 of the European Parliament and of the Council of 16 December 2008 on food additives.

[693] Regulation 1334/2008 of the European Parliament and of the Council of 16 December 2008 on flavourings and certain food ingredients with flavouring properties for use in and on foods and amending Council Regulation (EEC)1601/91, Regulations 2232/96 and 110/2008 and Directive 2000/13.

[694] Carry over means that they have been used in an ingredient and with this ingredient came into the food but no longer serve a purpose in this food.

[695] Article 20 of the FIC.

[696] See for additives in foods Annexes I and II of Regulation 1333/2008.

[697] See Chapter 10.

[698] See Article 23 of Regulation 1333/2008 on food additives.

apply, for instance when using the term 'natural',[699] or the application of smoke flavours.[700]

14.6.3 Net quantity

On the labelling the net quantity has to be indicated.[701] In certain cases the net quantity does not have to be provided (for instance, if a food loses a considerable amount of volume or mass and they are sold by the number).[702] It should be expressed in metric units of mass or, for liquid foods, volume (i.e. g or kg and ml or litre respectively). The net quantity is often followed by the symbol 'e' (Diagram 14.4).

Diagram 14.4. E-mark.

This symbol indicates that the manufacturer has carried out a statistical control according to an EU-approved method, with the guarantee that the net weight is not below the declared value. This e-symbol has been put forward by Directive 2007/45[703] for pre-packaged liquids and in Directive 76/211[704] for other products. When the pre-packaged item contains two or more individual pre-packaged items containing the same quantity of the same product, the net quantity in each individual package should be indicated, together with the total number of packages (unless this can be clearly observed from outside). For foods normally sold by number, the total number should be indicated rather than the net quantity. For solid foods sold in liquid media, the drained net weight must also be indicated.[705]

[699] Article 16 of the FIC.
[700] Regulation 20165/2003; Commission Implementing Regulation 1321/2013.
[701] Article 23 of the FIC.
[702] See Annex IX of the FIC.
[703] Directive 2007/45 Of the European Parliament and of the Council of 5 September 2007 laying down rules on nominal quantities for pre-packaged products.
[704] Council Directive of 20 January 1976 on the approximation of the laws of the Member States relating to the making-up by weight or by volume of certain prepackaged products
[705] J. Claude Cheftel, Food and Nutrition Labelling in the European Union, Food Chemistry, 2005, pp. 531-550.

Technical requirements for the net quantity declaration are contained in Annex IX to the FIC.

The indication of the net quantity is important for the consumer's informed choice regarding value for money. In this context Directive 98/6[706] is relevant. This directive requires indication of the selling price and the price per unit of measurement of products offered by traders to consumers.

16.4.4 Durability

Most[707] products must carry a 'minimum durability date' (that is: a 'best before' indication), or a 'use-by' date.[708] The 'use-by' date is mandatory if, from a microbiological point of view, a food is highly perishable and therefore likely after a short period to constitute an immediate danger to human health.[709] In the latter case, a food is considered unsafe, in accordance with Article 14(2) to (5) of the General Food law. The food business operator that attaches his/her name to the food is responsible for setting these dates. The business has the discretion to choose the date more or less conservatively. That is to say, s/he is under no obligation to choose the latest possible date.

A date of minimum durability is regarded as a quality indication (i.e. the food keeps its specific properties when properly stored, see Article 2(2)(r) of the FIC) but is defined as alternative for the use-by date. Logically it therefore also indicates a period in which the food is considered safe. But it is not forbidden to sell a food beyond the 'best-before' date. In that case, often a discount is given. After expiry of the 'best-before' date consumers and retailers will have to assess whether a food is still fit for consumption by studying its external characteristics (like colour and smell).

14.6.5 Origin information

In the EU, significant requirements have been put forward with respect to the rules on country of origin indications or place of provenance (further jointly named 'origin'). In the FIC, 'place of provenance' is defined[710] as any place (so whether it be for instance a region, city or rural area) where a food is indicated to come from, not being the 'country of origin' as determined in Articles 23-26 of former

[706] Directive 98/6 of the European Parliament and of the Council of 16 February 1998 on consumer protection in the indication of the prices of products offered to consumers.
[707] Some exceptions remain, like for salt.
[708] Article 24 of the FIC.
[709] Article 24 of the FIC.
[710] In Article 2 of the FIC.

Regulation 2913/92,[711] which has been replaced by Regulation 952/2013.[712] The latter determines in Article 60 that:

- goods which are wholly obtained in a single country or territory will have as origin that country or territory;
- goods, the production of which involves more than one country or territory, shall be deemed to originate in the country or territory where they underwent their last, substantial, economically-justified processing or working, in an undertaking equipped for that purpose, resulting in the manufacture of a new product or representing an important stage of manufacture.

In correspondence with the General Food Law, the FIC[713] adopts as a general principle that food information shall not be misleading as to the characteristics of a food.[714] The indication of the country of origin or place of provenance is mandatory where failure to indicate this might mislead the consumer as to the true country of origin or place of provenance of the food.[715]

For certain agricultural products, to which vertical legislation applies, mandatory origin information is applicable. For (unprocessed) beef and veal, on the basis of Regulation 1760/2000, three indications as to the origin have to be stated on the package (that is: the place of birth, of rearing and the place of slaughter of the animal). Under the legislation before the FIC only poultry that was imported from a third country the origin had to be indicated. The FIC expands the scope of origin labelling to meat of swine, sheep, goat and all poultry, but not to the same extent as for beef due to – among other things – reasons of cost and identification (see Article 26 j° Annex XI of the FIC). The Commission has been granted the power to create a further extension of origin labelling regarding primary ingredients (that is, ingredients that represent more than 50% of a food, or that are usually associated by the consumer with the name of that food). An even further enlargement of scope that the Commission can put in place regards other meats than beef/veal, goat, sheep, poultry and swine, for milk and milk as ingredient, for unprocessed foods, single ingredient products and ingredients which take in more than 50% of a food. Moreover, one of the instruments proposed to address the 2012-2014 horse-meat crisis is to make origin labelling mandatory for all processed meats and meat ingredients.

[711] Council Regulation 2913/92 of 12 October 1992 establishing the Community Customs Code.

[712] Regulation 952/2013 of the European Parliament and of the Council of 9 October 2013 laying down the Union Customs Code. According to Article 286(3) of Regulation 952/2013, references to the repealed regulations – including Regulation 2913/92 – shall be construed as references to Regulation 952/2013.

[713] Article 26(2)(a) of the FIC.

[714] Article 7(1)(a) of the FIC.

[715] Article 9(1)(i) of the FIC.

14.6.6 Conditions and instructions

Article 25 of the FIC provides that special storage conditions have to be given (for instance, 'store at a temperature between 2 and 6 °C') if a food requires this. The remaining time before consumption after opening the package has to be indicated as well, if necessary. Article 25 focuses primarily on safeguarding the quality and safety of a food before consumption. Instructions for use safeguard the proper preparation and consumption of a food (for instance, 'cook in 8 minutes'). They are mandatory according to Article 27 of the FIC. The Commission can adopt implementing acts in this respect for certain foodstuffs.

14.6.7 Nutrition Information

Nutrition information is mandatory for most foods as from 13 of December 2016. The previous Directive 90/496[716] ruled that the provision of nutrition information was voluntary, with the exception that it becomes mandatory if a claim is made. Nevertheless, under these previous rules voluntary nutrition information had to be provided in legally prescribed formats. The requirements in the FIC (i.e. in Section 23 of the FIC, that is Articles 29-35) are not applicable to food supplements (Directive 2002/46[717]) and the exploitation and marketing of natural mineral waters (governed by Directive 2009/54[718]). The nutrition information for these foods is restricted to those nutrients which are actually present in the foodstuff.

The nutrition information section of the FIC rules in:
- the content and format of the nutrition declaration;
- and the calculation, expression and presentation of energy and nutrients.

As to the content, the mandatory nutrition declaration includes the energy value and the amounts of fat, saturates, carbohydrate, sugars, protein and salt.[719] New compared to the previous directive, is: the absence of fibres in the format; the fixed and closed[720] format of the nutrition table; the use of simpler expressions (like 'salt' instead of 'sodium'); the fixed basic units of expression (per 100 g or per 100 ml, with the possibility of supplementary expression per portion and/or as a percentage of recommended daily intake); the possibilities of complementary forms of expression (like 'traffic-lights' or a 'checkmark'); as well the possibility

[716] Council Directive 90/496 of 24 September 1990 on nutrition labelling for foodstuffs.
[717] Directive 2002/46 of the European Parliament and of the Council of 10 June 2002 on the approximation of the laws of the Member States relating to food supplements.
[718] Directive 2009/54 of the European Parliament and of the Council of 18 June 2009 on the exploitation and marketing of natural mineral waters.
[719] Article 30 of the FIC.
[720] This means that no other items can be included than those indicated by law. All other information has to be provided near or under the nutrition information table. For instance, when a claim with respect to vitamin b6 is made, the amount of this vitamin has to be expressed. But the designated place for it is under the table (Annex XV) in units specified in Annex XIII, Part A.

of repetition of nutritional information in the principal field of vision (a field of vision is all surfaces of a package that can be read from a single viewing point; the principal field of vision is the field of vision of a package which is most likely to be seen at first glance by the consumer at the time of purchase[721]). The standard table, as well as the units of measurement, are given in Annex XV (Diagram 14.5).

Diagram 14.5. Standard table, based on Annex XV of the FIC.

Energy	kJ/kcal
Fat	g
Of which:	
- saturates,	g
- mono-unsaturate[1],	g
- poly-unsaturates[1],	g
Carbohydrate	g
Of which:	
- sugars,	g
- polyols[1],	g
- starch[1]	g
Fibre[1]	g
Protein	g
Salt	g
Vitamins and minerals[2]	(in measuring units μg/mg and nutrient reference values)[3]

[1] Not part of the mandatory items, but can be included in the table.
[2] Must be indicated if a claim is made, but outside the main table.
[3] As provided in Annex XIII.

14.7 (Semi-)vertical food information requirements

The FIC refers in Article 2(2)(b) to horizontal as well as vertical rules as components of food information law. Vertical legislation has been designed for a wide range of primary products (in or based on market organisation for agricultural products (CMO),[722] for fishery and aquaculture products)[723] and some specific products (like honey, chocolate, juices and fruit jams).[724]

[721] Article 2(2)(k) and (l) of the FIC.

[722] Regulation 1308/2013 of the European Parliament and of the Council of 17 December 2013 establishing a common organisation of the markets in agricultural products.

[723] Regulation 1379/2013 of the European Parliament and of the Council of 11 December 2013 on the common organisation of the markets in fishery and aquaculture products.

[724] Directive 2000/36 of the European Parliament and of the Council of 23 June 2000 relating to cocoa and chocolate products intended for human consumption; Council Directive 2001/110 of 20 December 2001 relating to honey; Council Directive 2001/113 of 20 December 2001 relating to fruit jams, jellies and marmalades and sweetened chestnut purée intended for human consumption.

14.7.1 Products derived from primary produce

Already very early in the development of the European Economic Community, a common agricultural policy was concluded upon and codified in the Treaty of Rome. In a Europe that was in a state of devastation after the Second World War, security of agricultural and food production was of primary importance. So the Treaty, first and foremost aiming at free exchange between the Member States, also laid the foundation for a common policy for primary produce. The Common Market Organisation defines product-oriented processing and quality standards, market organisation instruments (like licences to produce), as well as (supplementary) labelling requirements for several products indicated in Annex I to the TFEU. For instance, marketing standards for fresh and processed fruits and vegetables may apply according to Article 75 of the CMO. Specific standards have been put in place, among others based on Commission Regulation 543/2011.[725]

Additional requirements are put forward in Article 76 of the CMO (derogations may apply, like for almonds, coconuts and sweet corn).[726] Fresh fruit and vegetables require information on the country of origin, the quality class and commercial type of product, as well as the net weight, if appropriate. Likewise, specific rules have for instance been concluded for bananas, for beef, pork and poultry meat, sheep- and goat-meats. Next to this, food information rules for (unprocessed)[727] beef can be found in Regulation 1760/2000[728] and Commission Regulation 1825/2000, which, as a response to the BSE crisis, impose origin (next to traceability) labelling for bovine animals aged more than 12 months. With respect to poultry, next to generic marketing standards, implementing rules[729] as well as generic food information requirements have to be met,[730] while for some treatments like ionisation and/or freezing, additional requirements have been decided upon.[731]

Vertical food information may look scattered, and this impression is right, considering that for many products other than those mentioned here specific rules have been put in place. For some processed foods that are mainly derived from agricultural materials vertical rules specifying recipes have been concluded upon (like for fruit jams, honey, or chocolate). These can be regarded as remnants

[725] Commission Implementing Regulation 543/2011 of 7 June 2011 laying down detailed rules for the application of Council Regulation 1234/2007 in respect of the fruit and vegetables and processed fruit and vegetables sectors.
[726] See Article 4(6) of Regulation 543/2011.
[727] For processed beef, the generic labelling requirements apply; for minced meat specific requirements for the designation 'minced meat' have been concluded upon (Part B of Annex VI to the FIC).
[728] Regulation 1760/2000 of the European Parliament and of the Council of 17 July 2000 establishing a system for the identification and registration of bovine animals and regarding the labelling of beef and beef products and repealing Council Regulation 820/97.
[729] Commission Regulation 543/2008.
[730] See Article 26 of the FIC.
[731] In the FIR (Annex VI), in Directive 1999/2, and in Council Directive 89/108, respectively.

of a period of product-specific standard setting. As already addressed in Chapter 7, the Cassis de Dijon ruling made national product standards based on European directives outdated.

14.7.2 Foods derived using specific technologies

The application of specific technologies may require additional information. If products are derived by means of genetic modification or contain genetically modified organisms (GMOs), are derived from genetically modified organisms or from ingredients thereof[732] the labelling has to provide information to the consumers in this respect. As an exception a threshold level of 0.9% has been set, under the condition that presence of GMO is adventitious or technically unavoidable.

Materials in or on a package that can migrate into a food (i.e. active contact materials) are governed by a specific regulation.[733] Some are especially designed to preserve the food or indicate the physical condition of the content (so-called 'intelligent contact materials'). Special indications during their journey along the supply chain apply (like the indication 'for food use').[734]

Some products, like poultry meat, may be irradiated to kill off microorganisms that could be dangerous to the consumer's health. In that case the consumer has to be informed that the food has undergone such a treatment.[735]

Foods may contain nano-engineered ingredients. Apart from the fact that in general novel foods must carry additional information on differences as to composition, ingredients, nutritional value or nutritional effects, or intended use of the food if these are different from equivalent existing foods, nano-engineered materials have to be indicated in the ingredients list (see Section 14.6.2).

As already indicated (Section 14.6.2), some mandatory particulars (included in Annex VI) referring to special treatments accompany the name of a food (for instance 'smoked', 'irradiated', or 'formed'; see Diagram 14.2).

[732] See Regulation 1829/2003 of the European Parliament and of the Council of 22 September 2003 on genetically modified food and feed, especially Article 12.
[733] Regulation 1935/2004 of the European Parliament and of the Council of 27 October 2004 on materials and articles intended to come into contact with food and repealing Directives 80/590 and 89/109.
[734] See Article 15 of this Regulation.
[735] Directive 1999/2 of the European Parliament and of the Council of 22 February 1999 on the approximation of the laws of the Member States concerning foods and food ingredients treated with ionising radiation.

14.7.3 Foods with specific functions

Foods with specific functions[736] (consumed 'as such') are among others:
- food supplements (regulated in Directive 2002/46);
- foods for special categories of consumers (governed by Regulation 609/2013).

With some exceptions, the generic food information requirements also apply to food supplements. These may only be marketed as pre-packaged foods. Their legal name is 'food supplement'. They may contain vitamins and/or minerals; for these ingredients, minimum and/or maximum levels may be set. As these products are governed by a directive, Member States have discretionary opportunities to implement the directive in individual legal codes. Also, Member States have been administered the authority to prevent a food supplement from entering their market motivated by substantiated health risks for their inhabitants.

Finally, Regulation 609/2013[737] addressing foods for special categories of consumers, overhauls the 'parnuts'[738] directive. Additional rules have been set, dependent on the specific category of a foodstuff (see Articles 9-10 of Regulation 609/2013[739]).

14.8 Regulated voluntary food information

Voluntary food information encompasses – among other things – indications of organic production, identification marking and the use of claims. Despite the fact that voluntary information (by definition) does not have to be provided, rules with respect to content and/or format may be applicable in case such information is given. These rules are general and/or specific in nature. General rules for voluntary food information are contained in Articles 36-37 of the FIC. In Article 36(2) of the FIC it is stated that the information may not mislead the consumer, it may not be ambiguous or confusing, and shall – if appropriate – be underpinned with relevant scientific data. In case of limited space on the package, mandatory information has priority.[740]

[736] Additives also perform specific functions, but are not consumed as such, but always in combination with other foods.
[737] Regulation 609/2013 of the European Parliament and of the Council of 12 June 2013 on food intended for infants and young children, food for special medical purposes, and total diet replacement for weight control. See Chapter 17.
[738] Foods for special nutritional purposes; Directive 2009/39 of the European Parliament and of the Council of 6 May 2009 on foodstuffs intended for particular nutritional uses.
[739] For instance, in the case of infant formula and the labelling of follow-on formula, it may not include pictures of infants, or other pictures or text which may idealise the use of such formulae.
[740] Article 37 of the FIC.

14.8.1 Organic production

Regulation 834/2007[741] and Commission Implementing Regulation 889/2008[742] establish minimum conditions[743] for the use of terms referring to the organic'[744] production method (such as 'eco' and 'bio'). Such terms may only be used in for instance the labelling, advertisements or in trademarks if they satisfy the legal requirements. In case of processed food, at least 95% of the ingredients (by weight) must be organically derived. Genetic modification and ionisation of foodstuffs are not allowed. Additional information has to be supplied as well, such as the number of the (national) control authority that checks on compliance. The regulation sets out the objectives and principles of organic production (Law text boxes 14.5 and 14.6).

Law text box 14.5. Objectives of organic production.

Article 3 Regulation 834/2007 on organic production and labelling of organic products

Organic production shall pursue the following general objectives:

(a) establish a sustainable management system for agriculture that:
   - (i) respects nature's systems and cycles and sustains and enhances the health of soil, water, plants and animals and the balance between them;
   - (ii) contributes to a high level of biological diversity;
   - (iii) makes responsible use of energy and the natural resources, such as water, soil, organic matter and air;
   - (iv) respects high animal welfare standards and in particular meets animals' species-specific behavioural needs;

(b) aim at producing products of high quality;

(c) aim at producing a wide variety of foods and other agricultural products that respond to consumers' demand for goods produced by the use of processes that do not harm the environment, human health, plant health or animal health and welfare.

---

[741] Council Regulation 834/2007 of 28 June 2007 on organic production and labelling of organic products and repealing Regulation (EEC) 2092/91.

[742] Commission Regulation 889/2008 of 5 September 2008 laying down detailed rules for the implementation of Council Regulation 834/2007 on organic production and labelling of organic products with regard to organic production, labelling and control.

[743] National or private authorities/organisations may require additional conditions to be fulfilled, connected to a logo or brand.

[744] The term 'organic', as originally conceived, derived from a belief in the farm as a self-sustaining, living organism. The English word was made popular by Lord Northbourn in his book Look to the Land (1940). Today, in its broader meaning, organic refers to the importance of the interactions between the living organisms that inhabit the farm. Many English speakers, having first come across the term in chemistry lessons where organic refers to anything derived from carbon-containing compounds, mistakenly assume that the word has a similar meaning when applied to farming and food; R. MacRae, 'A history of sustainable agriculture, Ecological Agricultural Projects', McGill University, Montréal, Canada, 1990.

Law text box 14.6. Principles of organic production.

Article 4 Regulation 834/2007 on organic production and labelling of organic products

Organic production shall be based on the following principles:

(a) the appropriate design and management of biological processes based on ecological systems using natural resources which are internal to the system by methods that:
   (i) use living organisms and mechanical production methods;
   (ii) practice land-related crop cultivation and livestock production or practice aquaculture which complies with the principle of sustainable exploitation of fisheries;
   (iii) exclude the use of GMOs and products produced from or by GMOs with the exception of veterinary medicinal products;
   (iv) are based on risk assessment, and the use of precautionary and preventive measures, when appropriate;

(b) the restriction of the use of external inputs. Where external inputs are required or the appropriate management practices and methods referred to in paragraph (a) do not exist, these shall be limited to:
   (i) inputs from organic production;
   (ii) natural or naturally-derived substances;
   (iii) low solubility mineral fertilisers;

(c) the strict limitation of the use of chemically synthesised inputs to exceptional cases these being:
   (i) where the appropriate management practices do not exist; and
   (ii) the external inputs referred to in paragraph (b) are not available on the market; or
   (iii) where the use of external inputs referred to in paragraph (b) contributes to unacceptable environmental impacts;

(d) the adaptation, where necessary, and within the framework of this Regulation, of the rules of organic production taking account of sanitary status, regional differences in climate and local conditions, stages of development and specific husbandry practices.

If a food business operator meets the requirements of organic production, the product will have to carry the European logo on the food for it as a sign of certification.[745] The place where the agricultural materials used have been farmed has to be indicated in the same field of vision as the logo.[746]

14.8.2 Voluntary identification information: PDO, PGI and TSG

Voluntary identification information, e.g. via protected designations of origin (PDO) or protected geographical indications (PGI), refers to the identification of products and their origin related to a specific place or region, which gives it special quality-related value-adding characteristics that need protection. The rules for the

[745] Article 24 of Regulation 834/2007.
[746] Article 24 of Regulation 834/2007.

voluntary protected identification of certain products are included in Regulation 1151/2012.[747] This regulation aims to help communicate valuable characteristics and farming attributes to ensure fair competition, inform consumers in a reliable way, respect intellectual property rights[748] and assure the integrity of the market.[749] The scope of the regulation is agricultural products mentioned in Annex I to the EU Treaty and other agricultural products in Annex I to Regulation 1151/2012.

For PDO, PGI as well as traditional specialties (see at the end of this section) a group application has to be made in which, among other things, the main points of the product specification are given. For PDOs and PGIs the link between product and geographical environment has to be specified.[750] Producers can file an application through their national authorities. The Commission examines requests sent via Member States, including the main elements of the specification and assesses whether they meet the requirements under the relevant regulation. It can call on independent scientific advice. If an application satisfies the rules and is registered, it may be used by any producer that comes up to the fixed specificities.[751]

New trademarks may come into conflict with applications for PDO or PGI. Trade marks submitted after submission of the registration will be refused. However, the general rule is that trademarks that already exist before the application of a PDO or PGI and conflict with it may continue to be used.[752]

Specifics of Protected Designations of Origin (PDO)

A PDO (e.g. 'Parma Ham' or 'Camembert de Normandie') is the name of a region or specific place that is used to describe an agricultural product or foodstuff originating in that region or place (or exceptionally, a certain country), the quality or characteristics of which are essentially or exclusively due to that geographical environment and of which the production steps all take place in the geographical area.[753] Hogs that are born and reared in the Netherlands, can only become 'Parma ham' if they are first transported to Italy alive where they are slaughtered and then processed. If the specific processing requirements are met in the designated region, the Commission can authorise the exclusive use of it. To obtain a Protected Designation of Origin the area must be precisely defined,

[747] See Regulation 1151/2012 of the European Parliament and the Council of 21 November 2012 on quality schemes for agricultural products and foodstuffs.
[748] See also Chapter 22.
[749] Article 1 of Regulation 1151/2012.
[750] Article 8 of Regulation 1151/2012.
[751] Article 7 of Regulation 1151/2012.
[752] Article 14 of Regulation 1151/2012.
[753] Article 5 of Regulation 1151/2012. See in this respect Consortio del Prociutto di Parma, Salumificio S. Rita SpA v Asda Stores Ltd, Hygrade Foods Ltd, Case C-108/01. It was ruled that the requirement that slicing and packaging take place in the region of production constitutes a measure having equivalent effect to a quantitative restriction on exports [...], but may be regarded as justified [...]. In this case, the processing steps are part of the product specification.

and all stages of production, processing and preparation, from the raw materials to the finished product, must take place in the area that lends its name to the product. The characteristics of the product must be essentially or exclusively due to the origin.

Specifics of Geographical Indications (PGI)

A PGI, 'protected geographical indication' (e.g. 'Ardennen Ham' or 'Danablu Cheese'), is a name that indicates that a product originates in a specific place, region or country, whose given quality, reputation or other characteristic is essentially attributable to its geographical origin; and at least one of the production steps of which take place in the defined geographical area.[754]

In contrast to the PDO, the ties between area and product are much less strict for a PGI. The characteristics of the product do not have to be essentially or exclusively due to the indicated area. For a PGI it is sufficient that only one characteristic of the product can be attributed to the area, for instance its reputation.

Traditional Specialties Guaranteed (TSG)

Title III of Regulation 1151/2012 prescribes a register of recognised traditional specialties. To obtain the TSG status a product must possess features that distinguish it from other products, and it must be traditional. The specific features required for TSG recognition are:[755]
- the product is the result of a traditional practice of production, processing or composition, or
- it is produced from raw materials or ingredients with traditional origin.

'Traditional' in this context[756] means proven usage on the EU market for a time period showing transmission between generations. This usually means a minimum of 30 years.

An application for a TSG designation must be accompanied by specifications enabling compliance with registration conditions, including a product specification (i.e. the name, product description, production method and other key elements that establish the traditional character of the product).[757]

[754] Article 5(2) of Regulation 1151/2012.
[755] Article 18(1) of Regulation 1151/2012.
[756] Hygiene legislation applies its own definition of traditional.
[757] Article 19 of Regulation 1151/2012.

Pro and contra protection

The indication of origin could hamper the exchange of goods in European markets. In the political arena of the EU, the advantages of protection of regional and traditional craftsmanship have been considered more important than free competition. Food businesses are in principle free to label the origin of their products. Member States, however, are not at liberty to introduce mandatory or even voluntary schemes. Case law of the CJEU consistently holds origin labelling legislation to constitute trade barriers not compatible with Article 34 of the TFEU. The following two cases may stand as example.

The first case[758] deals with national legislation prohibiting the retail sale of certain products imported from other Member States, unless they bear or are accompanied by an indication of origin. The Court is highly critical. For several reasons it concludes that the measure at issue is a barrier to trade for which no justification can be found in EU law. The Court argues that a requirement to label the origin in another Member State would necessarily increase the production costs of the imported article and make it more expensive. Furthermore, it has to be recognised that the purpose of indications of origin or origin-marking is to enable consumers to distinguish between domestic and imported products and that this enables them to assert any prejudices which they may have against foreign products. In short, the provisions in question are liable to have the effect of increasing the production costs of imported goods, making it more difficult to sell them. The Court also notes that if the national origin of goods brings certain qualities to the minds of consumers, it is in manufacturers' interests to indicate it themselves on the goods or on their packaging and it is not necessary to compel them to do so. In that case, the protection of consumers is sufficiently guaranteed by rules which enable the use of false indications of origin to be prohibited.

The second case[759] is about an origin-related quality label. By awarding the quality label *'Markenqualität aus deutschen Landen'* (quality label for produce made in Germany) to finished products of a certain quality produced in Germany, the Federal Republic of Germany has failed to fulfil its obligations under (what is now) Article 34 of the TFEU. The contested scheme has, at least potentially, restrictive effects on the free movement of goods between Member States. Such a scheme, set up to promote the distribution of agricultural and food products made in Germany and for which the advertising message underlines the German origin of the relevant products, may encourage consumers to buy the product with the label to the exclusion of imported products. The fact that the use of that quality label is optional does not mean that it ceases to be an unjustified obstacle to trade if the use of that designation promotes or is likely to promote the marketing of the product concerned as compared with products which do not benefit from its use.

[758] CJEU 25 April 1985, Commission vs UK, Case 207/83.
[759] CJEU 5 November 2002, Commission vs Germany, Case C-325/00.

14.8.3 Nutrition and health claims

With some exceptions,[760] information by means of claims on products is governed by Regulation 1924/2006.[761] 'Claim' means any message or representation, which is not mandatory under EU national legislation, including pictorial, graphic or symbolic representation, in any form, which states, suggests or implies that a food has particular characteristics (Article 2(2) NHCR). The regulation addresses nutrition and health claims made in commercial communications. The scope therefore not only includes messages stated on the labelling, but also commercials in public media, in advertisements or on websites. The rules in Regulation 1924/2006 are without prejudice the requirements for special foods governed by Regulation 609/2013. Often claims are connected to what in practice are called 'functional foods'[762] – a term that has not been defined or used in European law.

To the concerns which led to the design of this regulation belong possible adverse effects of wrong claims on the functioning of food markets. Unauthorised false and opportunistic food claims may distort the efficient exchange of products, for instance because the innovation advantage of one food business operator is neutralised through false or misleading information by another. A concern is also the adverse effect of wrong or false claims on the ability of consumers to make informed choices. Moreover, a situation should be avoided in which a claim is made on a food that does not fit within a nutrient profile that is to be fixed by the Commission.[763] The protection of the use of claims may stimulate innovation in food businesses, as unfair business practices are counteracted. Therefore claims have to go through an authorisation procedure, in which all scientific evidence is evaluated. Once authorised, in principle a claim may be used by any food business operator, unless exemptions apply.[764]

The NHCR states general principles which are applicable to the use of all types of claims. They may not be false, ambiguous or misleading. Also, they may not give rise to doubt about the safety and/or the nutritional adequacy of other foods. It is forbidden that a claim encourages or condones excess consumption of a food. Next it may not be stated, suggested or implied that a balanced and varied diet cannot provide appropriate quantities of nutrients in general. And the consumer may not be frightened by reference to changes in bodily functions through textual,

---

760 Like claims on spreadable fats (governed by Council Regulation 2991/94), or the claim 'gluten free'.
761 Regulation 1924/2006 of the European Parliament and of the Council of 20 December 2006 on nutrition and health claims made on foods.
762 To some authors, a functional food is a food to which a claim is attached. However, special foods and food supplements may also be ranked under the concept.
763 Article 4 of the NHCR. The situation could otherwise occur that a positive claim is made on vitamin content while at the same time the respective food contains a higher than average amount of trans-fats. Such profiles are mandatory from 2009 on. However, in 2014 they have still not been put in place.
764 See in this respect: Article 13(5) of the NHCR.

pictorial, graphic or symbolic representations.[765] On top of this, the NHCR states in Article 5 and 6 general conditions for the use of claims, among other:

- as stated, the substantiation of the claim on the basis of generally accepted scientific evidence;
- the presence, absence or reduced content of a nutrient or other substance for which the claim is made or the scientific proof that the nutrient or other substance is not present or present in reduced quantity to bring about the nutritional or physiological effect as proposed;
- that the nutrient or other substance is in a form that can be used by the body;
- that the quantity that can reasonably be expected to be consumed is provided in a significant amount;
- the 'average consumer'[766] has to be able to understand the beneficial effects which the claim intends to express.

A distinction is made between nutrition and health claims. A nutrition claim states what a product is, a health claim makes sure what it does. A nutrition claim is a claim that states, suggests or implies that a food has particular nutritional properties because of the provision, provision in reduced quantities or absence of energy and/or its nutrients or other substances. Nutrition claims that are allowed by a Commission's decision are included in a positive list (this list is included in the Annex to the regulation). An example is the claim 'light' (or 'lite'), or 'reduced'.[767]

A 'health claim' is a message that states, suggests or implies that a relationship exists between a food category, a food or one of its constituents and health (see Article 3(5) of the NHCR). An example of a health claim is a message like 'calcium is needed for the maintenance of normal teeth' or 'plant stanol esters have been shown to lower/reduce blood cholesterol. High cholesterol is a risk factor in the development of coronary heart disease'. If a health claim is stated, the labelling has to include extra information, as provided for in Article 10 of the NHCR:

- a statement indicating the importance of a varied and balanced diet and a healthy lifestyle;
- the quantity of the food and pattern of consumption required to obtain the claimed beneficial effect;
- where appropriate, a statement addressed to persons who should avoid using the food; and
- an appropriate warning for products that are likely to present a health risk if consumed to excess.

[765] See Article 3 of the NHCR.

[766] Gut Springenheide case, C-210/96; Mars case, C-470/93: Verein gegen Unwesen in Handel und Gewerbe Köln eV – Mars GmbH.

[767] It may be made if 'the reduction in content is at least 30% compared to a similar product, except for micronutrients, where a 10% difference in the reference values as set in Directive 90/496 shall be acceptable, and for sodium, or the equivalent value for salt, where a 25% difference shall be acceptable' (Annex to the Regulation).

A health claim may not suggest that a human disease is treated, cured or prevented. Except in case of special foods and natural mineral waters[768] food information in general may not attribute to any food the property of preventing, treating or curing a human disease, nor refer to such properties. While the marketing of food is in principle free, medicines have to be authorised in advance. In many countries, unlawful sales of medicine is penalised more severely than the mere abuse of the NHCR. In this context it should be noted that a medicine[769] can be recognised by the functional (medicinal) properties of a product, but also by its presentation on the labelling, in advertisement or on websites, even if in reality any functional properties are absent. The statement of a claim connected to a food may lead competent authorities to think that a medicine has been brought to the market, instead of a food or food supplement.[770]

Health claims can be further discerned in generic claims (governed by Article 13 of the regulation), disease risk reduction claims (Article 14(1)(a) NHCR) and claims referring to children's development and health (Article 14(1)(b) of the NHCR). Generic health claims are those referring to:[771]

- the role of a nutrient or substance in the growth, development or functions of the body;
- psychological and behavioural functions; or
- slimming and weight-control functions, or products that reduce the feeling of hunger (or increase the feeling of satiety).[772]

An example of an authorised generic health claim is contained in Commission Regulation 432/2012: 'Biotin contributes to normal energy-yielding metabolism'.[773] So the claim refers to the normal functioning of the body and not to anomalies.

When the NHCR was initiated, Member States were given the opportunity to submit generic health claims already used in their markets for authorisation. Initially, about 44,000 of such generic health claims were put forward by the Member States. After a review of the scientific evidence by EFSA and a comitology procedure involving the SCFCAH the Commission authorised in 2012[774] in a single regulation about 220 generic health claims. These were added to a positive list,[775] which can

[768] See Article 7(3) of the FIC.
[769] Directive 2001/83 relating to medicinal products for human use.
[770] See in this respect the 'garlic' case, C-319/05.
[771] Article 13(1) of the NHCR.
[772] Without prejudice to Commission Directive 96/8 of 26 February 1996 on foods intended for use in energy-restricted diets for weight reduction (subsequently repealed by Regulation 609/2013).
[773] The conditions for use of the claim are stated in the same decision: 'The claim may be used only for food which is at least a source of biotin as referred to in the claim 'source of [name of vitamin/s] and/or [name of mineral/s]' as listed in the Annex to Regulation 1924/2006'.
[774] Commission Regulation 432/2012 of 16 May 2012 establishing a list of permitted health claims made on foods, other than those referring to the reduction of disease risk and to children's development and health.
[775] The list can be found on the website: http//:ec.europa.eu/nuhclaims.

subsequently be extended on the basis of new scientific evidence. However, the number of authorised claims remains limited,[776] due mainly to the difficulties in providing the scientific evidence. For instance, the scientific evidence with respect to the effect of probiotics (a term that, since 2012, is no longer allowed) remains difficult to provide.

Businesses will face even more difficulties if they wish a specific health claim (on the basis of Article 14) to be authorised. In that case, extensive procedures contained in Articles 15, 16, 17 and 19 of the NHCR apply, the most significant element being the submission of scientific proof.[777] The cause-effect relationship between a nutrient or other substance and a risk factor inducing a disease should be proven. The risk factor has to be connected to a specific disease. A main problem is that scientific evidence should be provided as to healthy humans. Evidence from animal experiments or referring to diseased humans is not preferable for EFSA.

For businesses it is not always clear that they are making a statement (for instance on the labelling or in a commercial) which implies a claim that should be authorised. Also, when stating a claim, it is not always clear whether the 'food plus claim' and medicine boundary has been crossed.

The first situation is exemplified with the 'Weintor' case,[778] in which the issue was whether a description like 'easily digestible', together with a reference to the limited inclusion of substances which consumers perceive as harmful (acid), represents a health claim that is not allowed. In a preliminary ruling it was determined by the CJEU that this is indeed the case. It should be noted in this context, that alcoholic beverages containing more than 1.2% alcohol may not bear health claims at all. Nutrition claims, if authorised, are allowed but only if they stress low or reduced alcohol content or low energy level.[779]

An example of the second kind is the Article 13(1) claim 'Alpha lipoic acid helps to protect the nervous system'.[780] It suggests that a disease is prevented, which is forbidden. If, following from its suggested function or presentation (see Chapter 9 Section 9.3.1),[781] a product can be categorised as food and as medicine, the law on

[776] As per 2nd of July 2014. The regulation is regularly updated. An example of an authorised claim is: 'Consumption of alpha-cyclodextrin as part of a starch-containing meal contributes to the reduction of the blood glucose rise after that meal'.
[777] Application sent to the national competent authority which informs EFSA. EFSA informs the Member States and the Commission. On the basis of submitted evidence, EFSA will form an opinion on the validity of the claim (its content and proposed wording), submit it to the Commission and publish it. A draft decision is made by the Commission, which is then submitted to the SCFCAH. A final decision is published and the list with approved health claims is supplemented.
[778] C-544/10, Deutsches Weintor eG v. Land Rheinland-Pfalz.
[779] Article 4(3) of the NHCR.
[780] A submitted claim based on Article 13(1) of the NHCR.
[781] See Article 1(2) of Directive 2001/83 (definition of medicinal product).

medicines takes precedence over food law.[782] If this happens, legal consequences may be imminent under national law in the Member States. Depending on the national system at issue, these may include the following measures:

- the business operator is fined for infringing the NHCR;
- the responsible business is punished for infringing medicinal law (the implementation of Directive 2001/83[783]);
- the product has to be taken off the market.

However, other claims referring directly to a disease may be allowed, for instance on the basis of Article 14(1) of the NHCR: 'Vitamin C prevents the occurrence of scorbut'. In this instance, the risk factor that is affected coincides with the lack of the working substance (absence of vitamin C).

The applied wording of a claim may differ from the formulation as authorised by the Commission. It is up to the food business responsible for the food information to see that the 'commercial' expression of a claim is in line with the intentions of the claim as authorised. National and European authorities may provide guidelines for the acceptable wording of claims. However, remaining differences in interpretation between countries of the official formulation can distort the exchange of goods within the European Union.

The strict conditions for authorisation and application increase the risks food businesses are exposed to when using claims. That is why some try to circumvent the NHCR, for instance by replacing a non-authorised claim with an allowed one.[784]

14.9 International context of food information

Mandatory food information may stimulate as well as obstruct international trade. It stimulates trade by closing the information gap between the businesses' knowledge of the true characteristics of a foodstuff and the consumers' perception of its intrinsic characteristics. It may impede international trade, as food information requirements could be upheld opportunistically to hinder the entrance of foreign food products into the domestic market (content-wise, or by raising the costs of compliance).[785] For instance, protection of geographical indications can prevent

[782] Article 2 of Directive 2001/83.

[783] Directive 2001/83 of the European Parliament and of the Council of 6 November 2001 on the Community code relating to medicinal products for human use. See in this context: Damgaard (Case C-421/07), as analysed in B.M.J. van der Meulen and E.L. van der Zee, 'Through the Wine Gate' First Steps towards Human Rights Awareness in EU Food (Labelling) Law, EFFL 2013, pp. 41-52.

[784] In the case of 'probiotics', for instance, by replacing a claim with respect to the effects of a bacteria strain with the effect of an added vitamin, the vitamin being added with the only purpose to be able to make a claim. See in this context: H.J. Bremmers, B.M.J. van der Meulen and K.P. Purnhagen, Multistakeholder responses to the European health claims requirements, Journal on Chain and Network Science 2013, pp.161-172.

[785] T. Josling, The War on Terroir: Geographical Indications as a Transatlantic Trade Conflict, Journal of Agricultural Economics, 2006, pp. 337-363.

products from entering the market if they use the protected designation without conforming to the local conditions for using it.[786]

Sources for labelling provisions in an international context are the Codex Alimentarius (for labelling in general,[787] claims,[788] foods for special dietary uses[789] and production methods like halal,[790] etc.) as well as references in treaties, including conventions (like TRIPs for conflicts with intellectual property rights regarding brand names and protected designations of origin, etc.). In broad terms, the European food information provisions follow the Codex, but exceptions may apply. For instance, the number of allergens to be revealed according to EU legislation is much higher than depicted in the Codex. Authorities should, while designing new information law, restrict their efforts in two ways. One is realising that the space on a package or container of food is limited, so that trade is hampered due to lack of opportunities to commercially inform the consumer.[791] Second, the extensive set of requirements should – for as far as possible – be harmonised with adjacent sets of rules of different regions and/or nations, that struggle with the same food information issues.

[786] For instance, in 1971 a Canadian business ('Parma Foods') obtained a Canadian trade-mark ('Parma'). In 1996 'Parma' was accepted as a protected designation of origin in the European Union. See: http://tinyurl.com/nst9wuj. See: EUCJ Ravil SARL v Bellon Import SARL and Biraghi SpA, Cases-469/00 and -108/01, on the legitimacy of the requirement that Grana Padano and Pruciutto di Parma should be granted and packaged in the region of production, thus hindering trade. Although considered a quantitative barrier to trade this was deemed to be allowed, as the conditions under which these PDOs were authorised required such processing steps, under the condition that adequate publicity is given to the restrictive requirements.

[787] Codex STAN 1-1985.

[788] CAC/GL 23-1997.

[789] Codex STAN 146-1985.

[790] CAC/GL 24-1997.

[791] As already stated, when presenting food information, mandatory particulars take priority over voluntary information (Article 37 of the FIC).

15. Public powers: official controls, enforcement and incident management

Frank Andriessen, Anna Szajkowska and Bernd van der Meulen

15.1 Introduction

A typical feature of law is that people can be forced to comply with it. Sanctions and coercive measures can be applied to those who do not.

It is the responsibility of the Member States to enforce food law, and to monitor and verify that the relevant requirements of food law are fulfilled by food and feed business operators at all stages of production, processing and distribution. For that purpose, they have to maintain a system of official controls and other activities appropriate to the circumstances, including public communication on food and feed safety and risks, food and feed safety surveillance and other monitoring activities covering all stages of production, processing and distribution (Law text box 15.1).

Law text box 15.1. Article 17(2) GFL on the Member States' responsibility to enforce food law.

Article 17(2) Regulation 178/2002
Member States shall enforce food law, and monitor and verify that the relevant requirements of food law are fulfilled by food and feed business operators at all stages of production, processing and distribution.
For that purpose, they shall maintain a system of official controls and other activities as appropriate to the circumstances, including public communication on food and feed safety and risk, food and feed safety surveillance and other monitoring activities covering all stages of production, processing and distribution.
Member States shall also lay down the rules on measures and penalties applicable to infringements of food and feed law. The measures and penalties provided for shall be effective, proportionate and dissuasive.

Generally speaking, enforcement encompasses the verification of compliance with legal obligations, assurance of corrective action, and application of sanctions in case of infringements. Although Article 17 of the General Food Law holds the Member States responsible for the enforcement of food law, European food law increasingly sets standards for national enforcement and provides for supervision. On 30 April 2004 '*Regulation (EC) No 882/2004 of the European Parliament and of the Council of 29 April 2004 on official controls performed to ensure the verification of compliance with feed and food law, animal health and animal welfare rules*' was published in the Official Journal. It entered into force on 1 January 2006.

Regulation 882/2004 includes obligations for verification of food businesses' compliance by the Member States, measures to be taken in case of infringements, a framework for co-operation between national authorities and the Commission, and for the Commission to audit the performance of national authorities in the Member States and in third countries.

While the Regulation provides for a general framework for official controls, it is important to note that it does not affect ('is without prejudice to') control obligations provided for in other legal acts of the EU.

Infringements on food law may cause food safety incidents. Such incidents can however also occur for other reasons (accidents). Incident management and enforcement can be closely related, but they are not necessarily the same thing. For powers to deal with food safety incidents one has to turn to the General Food Law.

In this chapter we will discuss two enforcement issues (first official controls, then measures in case of non-compliance) and two incident management issues (rapid alert system and emergency measures).

15.2 Official controls

National inspectors supervise the application of the requirements of food law. In several Member States these inspectors work in the context of food safety authorities, which are often more or less autonomous agencies.

The national inspectors have powers under national law to inspect premises where food is handled and to report on non-compliance. Such non-compliance may result in sanctions. Some Member States use administrative law sanctions, others use criminal law. The latter usually comes into play when the non-compliance is the result of criminal behaviour (such as fraud).

15.2.1 General controls

Since the food safety scares, EU food law has increasingly been turning its attention to situations that at first sight seem to be internal matters of the Member States.

Regulation 882/2004 is concerned with food-related controls in general.[792] It distinguishes a great variety of control(-related) activities: 'official control', 'verification', 'audit', 'inspection', 'monitoring', 'surveillance', 'sampling for analysis', 'official certification', 'documentary check', 'identity check' and 'physical check'.[793] These distinctions are occasionally rather subtle and it is therefore important that they are properly defined and understood in all 28 Member States.

792 Not only in intra-Union trade.
793 Article 2 Regulation 882/2004, definitions.

Member States are responsible for ensuring that official controls are carried out regularly, with appropriate frequency, and on a risk basis.[794] Frequency depends, among other things, on identified risks and past performance of the food business operators (FBO). Good past performance by a food business may lead to a reduced frequency in inspections.

Official controls must cover the whole food chain from farm to fork'. As a rule they must be carried out without prior warning.[795] Nevertheless, the national competent authority must ensure that they carry out their activities with a high level of transparency.[796] National legislation must ensure that the control staff of the competent authorities have access to premises of and documentation kept by food business operators.[797]

The Member States may collect fees or charges to cover the costs occasioned by official controls. For some activities, they are even obliged to do so in order to avoid a distortion of intra-Union trade by different practices.[798] If non-compliance leads to extra official controls, the operators responsible will be charged.[799]

The Regulation requires Member States to prepare integrated multi-annual national control plans, to report annually on the results of their official controls and to have in place contingency plans for dealing with emergency situations.[800]

In addition to Regulation 882/2004, Regulation 854/2004 introduces specific measures to control compliance with Regulation 853/2004 (on hygiene of food of animal origin). This Regulation lays down requirements as regards, for instance, the approval of establishments, assistance in carrying out controls or presentation of documents. The controls include audits of HACCP and good hygiene practices.

15.2.2 Obligations of food business operators

Food business operators are obliged to undergo any controls and to assist the staff of the competent national authority in the accomplishment of their tasks. However, satisfactory results do not affect their duty to comply with food law, nor excuse them from future official controls. They are even under obligation to

[794] Article 3(1) Regulation 882/2004.
[795] Article 3(3) Regulation 882/2004.
[796] Article 7 Regulation 882/2004.
[797] Article 4 and 8 Regulation 882/2004.
[798] Article 27 Regulation 882/2004.
[799] Article 28 Regulation 882/2004.
[800] Articles 41-44 Regulation 882/2004.

take the initiative to inform the authorities of food safety problems that occur in their businesses.[801]

15.2.3 International trade

To ensure official controls in international trade with third countries, Member States may designate particular points of entry in their territory, which have access to appropriate control facilities.[802] Furthermore, they may require food business operators to give prior notification of imports from non-Member States.

15.2.4 Intra-EU trade

Two EU Directives concern controls of intra-Union trade in live animals and products of animal origin.[803] The Member States are under an obligation to ensure that products are accompanied by health certificates or other documents that European Union rules provide for. Products may only be intended for trade when they have been obtained, checked, marked, and labelled in accordance with EU rules. Checks must be carried out on the place of dispatch and may, under certain circumstances and without discrimination, be carried out at the place of destination. The Directives require notification to other Member States in case an animal disease occurs and adequate measures must be taken.[804]

15.2.5 Intra-EU co-operation

For people and goods, the European Union has created a common market without internal borders. The same is not true, however, for national authorities. The powers of national authorities are strictly limited to their national jurisdictions. This hampers effective surveillance of international food chains. The Regulation provides some instruments to cope with this situation.[805]

Member States must establish powers and procedures for sharing information between relevant parties. The competent authorities must provide each other with administrative assistance, which may include participation in on-the-spot controls. National authorities must provide each other with all necessary information and

[801] Article 19(3) GFL. On the staking of private and official controls see B.M.J. van der Meulen and A.A. Freriks, Millefeuille. The emergence of a multi-layered controls system in the European food sector, Utrecht Law Review, 2006, pp. 156-176.
[802] Article 17 Regulation 882/2004.
[803] Council Directive 90/425/EEC of 26 June 1990 concerning veterinary and zootechnical checks applicable in intra-Community trade in certain live animals and products with a view to the completion of the internal market. OJ L 224, 18 August 1990, pp. 29; and Council Directive 89/662/EEC concerning veterinary checks in intra-Community trade with a view to the completion of the European market. OJ L 395, 30 December 1989, corrigendum OJ L 151, 15 June 1990.
[804] Regulation 882/2004 does not introduce any changes in specific rules on controls in the field of animal health and animal welfare.
[805] Articles 34-38 Regulation 882/2004.

documents and they must arrange for the conduct of any administrative enquiries necessary to obtain such information and documents.

In cases with ramifications in several Member States or with a particular interest at EU level, the Commission is empowered to coordinate actions undertaken by the Member States. The Commission may even send an inspection team to carry out an official control on the spot.[806]

15.2.6 Second-line inspections

In 1997, the Food and Veterinary Office (FVO) was instituted. It is not an independent agency like EFSA, but a part of the European Commission's Directorate General Health and Consumers (DG Sanco). It has its headquarters in Ireland. The FVO has two main tasks. It audits the performance of national agencies in the Member States, and the performance of industry and public authorities in third countries that wish to export food products to the EU. Although the FVO is not mentioned by name, Regulation 882/2004 provides a legal basis for its activities (Law text box 15.2). The Member States must give all necessary assistance and provide all documentation that the Commission experts – the FVO – request.

Law text box 15.2. Basis for audits in Member States, Article 45(1) Regulation 882/2004.

Article 45(1) Regulation 882/2004
Commission experts shall carry out general and specific audits in Member States. The Commission may appoint experts from Member States to assist its own experts. General and specific audits shall be organised in cooperation with Member States' competent authorities. Audits shall be carried out on a regular basis. Their main purpose shall be to verify that, overall, official controls take place in Member States in accordance with the multi-annual national control plans referred to in Article 41 and in compliance with Community law. For this purpose, and in order to facilitate the efficiency and effectiveness of the audits, the Commission may, in advance of carrying out such audits, request that the Member States provide, as soon as possible, up to date copies of national control plans.

15.2.7 Controls in third countries

Although the FVO has no jurisdiction outside the EU, the Regulation indicates in Article 46 that official controls may take place in third countries.

[806] Article 40 Regulation 882/2004.

Law text box 15.3Official controls in third countries, Article 46 Regulation 882/2004.

Article 46 Regulation 882/2004
Commission experts may carry out official controls in third countries in order to verify [...] the compliance or equivalence of third country legislation and systems with Community feed and food law and Community animal health legislation. The Commission may appoint experts from Member States to assist its own experts.

These controls in third countries may only be executed if the authorities in those countries agree to them. However, as such controls may be a condition for export to the EU, these authorities often have little alternative. The inverse situation also exists, i.e. third countries carrying out inspections in the EU, and the Regulation requires the Commission to assist Member States in dealing with such situations.[807]

15.3 Measures in case of non-compliance

Generally speaking, when confronted with non-compliance, enforcement authorities have one of two options at their disposition and often both. They can take measures to remedy the problem and to ensure future compliance, or they can take measures to punish the perpetrator to deter him or her as well as others from future infringements. Both lines of action are available in the Member States.

15.3.1 Measures to remedy non-compliance

If a Member State establishes the non-compliance of a food business operator, it shall take action to ensure that the operator remedies the situation. When deciding which action to take, the competent authority has to take account of the nature of the non-compliance and the operator's past record with regard to non-compliance.[808]

Such action can include the imposition of sanitation procedures or the recall of a food product, the restriction or prohibition of placing foods on the market, or a closure of all or a part of the business concerned. In the case of imported products, these measures may include: destruction, special treatment to neutralise any potential risk associated with the non-compliance, and re-dispatch of the product to the state of origin. For the re-dispatch, a time-frame of no more than 60 days applies.

[807] Article 52 Regulation 882/2004.
[808] Article 54 Regulation 882/2004.

15.3.2 Legal protection

Article 54(3)(b) of Regulation 882/2004 states that the competent authority must provide the operator concerned with information on rights of appeal against the decision taken and on the applicable procedure and time limits, as these are provided for in the law of the Member State concerned.

15.3.3 Measures to punish non-compliance

Member States lay down the rules on sanctions applicable to infringements of feed and food law and other EU provisions relating to the protection of animal health and welfare and must take all measures necessary to ensure that they are implemented. The sanctions provided for must be effective, proportionate and dissuasive.[809]

The nature of the sanctions -administrative or criminal – is left entirely to the discretion of the Member States. The Commission in its first draft[810] intended to impose the use of criminal law on the Member States, but this intention did not survive the legislative process.

15.3.4 Public communication

Increasingly food safety inspection agencies in the Member States inform the general public about the results of their inspections.[811] The reasons for doing so seem to vary. In the case of a food safety issue, an obligation to issue a public warning follows from Article 10 GFL (Law text box 15.4).[812]

Law text box 15.4. Article 10 GFL Public information.

Article 10 Regulation 178/2002
Public information
Without prejudice to the applicable provisions of Community and national law on access to documents, where there are reasonable grounds to suspect that a food or feed may present a risk for human or animal health, then, depending on the nature, seriousness and extent of that risk, public authorities shall take appropriate steps to inform the general public of the nature of the risk to health, identifying to the fullest extent possible the food or feed, or type of food or feed, the risk that it may present, and the measures which are taken or about to be taken to prevent, reduce or eliminate that risk.

[809] Article 55 Regulation 882/2004.
[810] European Commission, Proposal for a Regulation of the European Parliament and of the Council on official feed and food controls 5.2.2003, COM(2003) 52 final, Article 55.
[811] See on this subject the special issue of the European Food & Feed Law Review 5/2007.
[812] See also Article 7 Regulation 882/2004.

Article 10 GFL belongs to the domain of risk communication. Warning the public about imminent risks is a form of remedying non-compliance. Some Member States, however, inform the public of all inspection results or of all infringements. If this is done to deter the business from infringements by threatening their reputation (often called naming and shaming), this is more a sanction of a punitive (punishing) nature.

15.3.5 Measures in case of lax enforcement

Regulation 882/2004 closely relates enforcement to incident management. In case of serious failure of a Member State's control system, the European Commission can deploy its emergency powers discussed in Section 13.5. See Law text box 15.5.

Law text box 15.5. Article 56 Regulation 882/2004 on safeguard measures.

Article 56 Regulation 882/2004
Safeguard measures

1. Measures shall be taken under the procedures provided for in Article 53 of Regulation (EC) No 178/2002 if:
   (a) the Commission has evidence of a serious failure in a Member State's control systems; and
   (b) such failure may constitute a possible and widespread risk for human health, animal health or animal welfare, either directly or through the environment.
2. Such measures shall be adopted only after:
   (a) Community controls have shown and reported non-compliance with Community legislation; and
   (b) the Member State concerned has failed to correct the situation upon request and within the time limit set by the Commission.

15.3.6 Review of the official controls Regulation

In May 2013, the Commission adopted a proposal constituting a review of Regulation 882/2004.[813] According to the Explanatory Memorandum, the proposal '... *aims to put in place a robust, transparent and sustainable regulatory framework* ...' which is 'fit for purpose'.

---

[813] Proposal for a Regulation of the European Parliament and of the Council on official controls and other official activities performed to ensure the application of food and feed law, rules on animal health and welfare, plant health, plant reproductive material, plant protection products and amending Regulations (EC) No 999/2001, 1829/2003, 1831/2003, 1/2005, 396/2005, 834/2007, 1099/2009, 1069/2009, 1107/2009, Regulations (EU) No 1151/2012, [....]/2013 of Regulation laying down provisions for the management of expenditure relating to the food chain, animal health and animal welfare, and relating to plant health and plant reproductive material], and Directives 98/58/EC, 1999/74/EC, 2007/43/EC, 2008/119/EC, 2008/120/EC and 2009/128/EC (Official controls Regulation) Document COM(2013) 265 final.

The Commission identified a number of reasons for the review. First, the review comes together with a major revision of the *acquis* in the areas of plant health, animal health, and plant reproductive material, all of which will link to/contain provisions for official controls. Second, the Commission recognised a need for a rationalisation, simplification and consolidation of provisions for official controls at EU level which are currently spread over a range of EU legal acts (e.g. mandatory controls for the presence of residues of veterinary drugs, plant health controls, controls on imported food, hygiene controls). And, third, experience with the current Regulation showed that there was scope for improvements (both in terms of wording as well as application in practice), elimination of certain redundant requirements with a view to reducing administrative burden (e.g. reporting requirements), as well as building in greater flexibility in order to deal with specific situations (e.g. in relation to requirements for laboratory accreditation).

The proposal also includes a review of the existing system of the collection of fees for official controls because that system was affected by a number of shortcomings, as a result of which the key objective of the fees – assurance of adequate resources for official controls in the Member States – was not being met throughout the EU, thus impacting on the delivery of official controls. Therefore, the proposal aims to ensure that adequate and stable resources are available, and that the financing of controls is fair, equitable, and transparent.

The overall objective of the review is to safeguard the internal market while ensuring a high level of health protection, and to ensure the proper enforcement of EU food law by providing for a set of modern and appropriate enforcement tools. The Commission also points to the benefits of a robust and EU-wide control framework in terms of building a good food safety reputation abroad, which is of key importance for EU exports of its food products.

The proposal is, at the time of writing, before the co-legislators Council and Parliament.

15.4 Incident management

The BSE crisis and other food safety incidents have demonstrated the need for improved, more comprehensive, EU-coordinated procedures for emergency and crisis management and for a system of rapid alert. The legal basis for the system and procedures is in Chapter IV of the General Food Law.

15.4.1 Rapid alert system

A system for rapid alert has existed in the field of food safety since 1979. However, the scope of the former system initially did not include feed. The Rapid Alert System for Food and Feed (RASFF) introduced by the General Food Law[814] covers food and feed, in line with the 'farm to fork' approach.

The RASFF is a network for exchanging information about direct or indirect risks to human health deriving from food or feed. The system involves the Member States, EFSA[815] and the Commission. Participation in the RASSF may be extended to third countries or international organisations, on the basis of agreements with the EU.[816]

15.4.2 Transmission of information

Where a member of the network has information about the existence of a serious direct or indirect risk relating to food or feed, it has to notify the Commission.[817] The Commission is responsible for managing the network. The Commission assesses the information received and categorises the notification under one of three categories. The information can also be rejected from transmission through the RASFF by the Commission, if the criteria for notification are not satisfied or if the information is insufficient. The notifying country is informed of this decision.

The notification is transmitted to RASFF contact points designated by all members of the network and to EFSA. Additionally, when the notification concerns an attempt to import banned products (border rejection), the information is sent to the EU Border Inspections Posts (BIPs) in order to increase the vigilance and to ensure that the rejected product does not re-enter the EU through another border post.[818] And when it is known that a product subject to a notification has been exported to a third country or when a notification concerns a product originating from a third country, the Commission also sends information to that third country.[819]

EFSA's role is to analyse the content of the notification and to supply scientific and technical information that will be helpful to Member States.[820]

[814] Article 50(1) GFL.
[815] EFSA is included to enable it to perform its task of monitoring the health and nutritional risks; Article 35 GFL.
[816] Article 50(6) GFL.
[817] Article 50(2) GFL.
[818] Article 50(3) GFL.
[819] Article 50(4) GFL.
[820] Article 50(2) GFL.

15.4.3 Notification categories

The notifications are classified in the following categories:[821]

- alert – where there is an identified direct or indirect risk to health and immediate action is required;
- information notification – relating to a food or feed that was placed on the market but the risk does not require any immediate action, or the product has not reached or is no longer present on the markets of the other members of the network; or
- border rejections – concerning consignments that have been tested and rejected at the external borders of the EU (and at the borders of the European Economic Area – EEA).

15.4.4 Public information and confidentiality rules

The Commission publishes weekly overviews of the notifications transmitted through the RASFF.[822] The public has access to information on:

- product description (e.g. 'sliced steam-cooked chicken fillets from Belgium', 'ice-cream from the UK');
- the nature of the risk (e.g. 'glass fragments', 'too high content of sulphite'); and
- the measure taken (e.g. 'product recalled from consumer', 'public warning-press release').

For the protection of commercial information, the trade names and the identity of individual companies are not made available to the public. The full information is only available to the members of the network. In the Commission's opinion, this should not influence the level of consumer protection, since a notification through the RASFF implies that all required measures have been or are in the process of being taken. However, in exceptional circumstances, where the protection of human health requires greater transparency, this information can also be made public.

15.4.5 Collateral damage

Swift action is at the heart of the RASFF. Rapidity, however, includes an inherent risk that mistakes will be made. On 10 March 2004 the General Court (at that time called: the Court of First Instance) ruled in a case initiated by Malagutti-Vezinhet SA against the European Commission.[823] Malagutti applied for compensation for the damage it allegedly suffered after the Commission issued a rapid alert message notifying the presence of pesticide residues in apples from France and giving

[821] This classification was put in place on 1 January 2008 and differs from the classification used until then. Border rejections were included in the other two while another third category – news notifications – was applied. See for example RASFF annual report 2006.

[822] On transparency and confidentiality, see; Article 52 GFL.

[823] Case T-177/02, *Malagutti-Vezinhet SA* v.*Commission*.

Malagutti's name as the exporter of the goods in question. The Court dismissed the application. Non-contractual liability on the part of the Community is subject to a number of conditions: unlawfulness of the conduct alleged against the Community institutions, actual damage and the existence of a causal link between the conduct of the institution and the damage complained of (Law text box 15.6). According to the Court, if one of those conditions is not satisfied, the entire action must be dismissed and it is not necessary to consider the other conditions.

Law text box 15.6. Article 340 TFEU on non-contractual liability.

Article 340 TFEU (previously Article 288 EC Treaty)

(...)

In the case of non-contractual liability, the Union shall, in accordance with the general principles common to the laws of the Member States, make good any damage caused by its institutions or by its servants in the performance of their duties.

(...)

Conditions elaborated by the European Court of Justice;

- unlawfulness;
- actual damage, and
- causal link between the conduct and the damage

The Court has summarised its findings as follows. The Rapid Alert System for Food and Feed confers on the national authorities only, and not on the Commission, responsibility for establishing whether there is a serious and immediate risk to the health and safety of consumers. It is thus incumbent upon the national authorities, once they have detected a serious and immediate risk the effects of which extend or could extend beyond their territory, immediately to inform the Commission and provide it with information to identify the product and the supply chain. The Commission, for its own part, confines itself to checking whether the information falls, as such, within the scope of the RASFF. The accuracy of the findings and analyses that led the national authorities to send that information does not have to be checked. As regards the prevention of risks to the health of consumers, and if any doubt remains, under the precautionary principle prevailing in the matter of the protection of public health, the competent authority may be obliged to take appropriate measures to prevent certain potential risks for public health without having to wait until the existence and seriousness of those risks has been fully demonstrated. If it was necessary to wait until all the research was completed before adopting such measures, the precautionary principle would be rendered devoid of purpose. That reasoning also applies in the case of a rapid information procedure, and, consequently, a business which is a victim of that alert system introduced in order to protect human health must accept its adverse economic consequences, since the protection of public health must take precedence over economic considerations.

In short, if mistakes are being made in applying the RASFF, there is normally no liability on the part of the European Commission.[824]

15.4.6 INFOSAN

RASFF participates in the International Food Safety Authorities Network (INFOSAN). INFOSAN is a network for the dissemination of important information about global food safety issues at world level.

In response to recommendations from several international conferences, resolutions from the World Health Assembly and guidelines from Codex Alimentarius Commission, the World Health Organization (WHO), in collaboration with the Food and Agriculture Organization of the United Nations (FAO) developed INFOSAN to promote the exchange of food safety information and to improve collaboration among food safety authorities at the national and international level. The INFOSAN network provides a mechanism for the exchange of information on both routine and emerging food safety issues.

INFOSAN Emergency is designed to provide rapid access to information during food safety emergencies. The INFOSAN network includes 181 member countries. Each member country has designated one or several INFOSAN Focal Points. Each country also has one dedicated INFOSAN emergency contact point that will be activated specifically in major international emergencies involving disease from or contamination in food.[825]

The RASFF) was nominated on 18 March 2005 as the INFOSAN emergency contact point for the transmission of INFOSAN food safety information. At the meeting of 20 September 2005 of the SCFCAH, all Member States of the EU and the EFTA (European Free Trade Association) countries agreed that the RASFF would be the single point of information exchange for the INFOSAN network.

---

[824] This ruling is in conformity with the general case law on non-contractual liability of the European institutions. Nevertheless, another approach would not have been inconceivable. In some of the Member States case law has developed that answers similar questions in a different way. The Dutch supreme court for instance judged with regard to police actions that were justified on the basis of knowledge available at the time of action, but later turned out to have been directed against an innocent person, that for reasons of fairness damages should not be borne by the innocent victim, but by society as a whole through the government. HR (Supreme Court) 2 February 1990, NJ 1990/794 (Bekkers/Staat) and HR 23 November 1990, NJ 1991/92 (Joemman/Staat). 'NJ' is the leading Dutch periodical on case law ('Nederlandse Jurisprudentie').

[825] World Health Organization in cooperation with the Food and Agriculture Organization of the United Nations, the International Food Safety Authorities Network (INFOSAN) Users Guide, available on the website of the WHO.

More close cooperation and clear procedures should be established between both systems in order to avoid overlap and misunderstanding, particularly in relation to the information transmitted to third countries.[826]

15.5 Emergencies

In the event of a serious risk to human health, animal health or the environment, urgent measures have to be taken. When there is limited time available and important values are at stake, adequate procedures allowing for effective and coordinated action to be taken are of the utmost importance.

15.5.1 Role of food business operators

The primary responsibility for ensuring that foods or feeds satisfy the requirements of food law rests with the food/feed business operators.[827] If the operators have reason to believe that their food or feed is unsafe, they shall immediately inform the competent authorities and withdraw the food or feed from the market and – if need be –recall it from the consumers. During these procedures, the operators are obliged to collaborate closely with the enforcement authorities.[828]

15.5.2 Role of national authorities

National authorities in the Member States enforce food law, monitor and verify whether food and feed business operators comply with the requirements of food law. In some cases, problems will be notified to the national authorities by food or feed business operators, who will also initiate withdrawals and recalls. There will be other instances where a problem is identified by the authorities, through inspection, outbreaks of disease, the testing of food samples or complaints by either consumers or competitors. A food alert may also result from information received through the RASFF. The authorities are the contact point for information and communication about the food or feed incident, and they coordinate investigations relating to withdrawals and recalls on a larger scale.

When the national authority adopts measures aimed at restricting the placing on the market or forcing the withdrawal or recall of food or feed, it shall immediately notify the Commission under the RASFF. It shall also inform the Commission of rejections of consignments at its Border Inspection Post, and of recommendations or agreements with food/feed business operators preventing, limiting or imposing special conditions on the placing on the market or the use of food or feed on account of a serious risk to human health requiring rapid action.

[826] RASFF annual report 2006.
[827] Article 17(1) GFL.
[828] See Chapter 14 for a more detailed description of the responsibility of food businesses.

15.5.3 Role of the European Commission

Apart from managing the RASFF and conducting the so-called second-line inspections auditing the performance of the enforcement authorities in the Member States (see Section 15.2.6), the Commission has special powers for taking emergency measures, through a single emergency procedure.

15.5.4 Powers of the European Commission

Before the General Food Law entered into force, the mechanisms for adopting emergency measures by the Commission were different in various areas of legislation. The scope of the emergency measure introduced in Article 53 GFL covers all types of food and feed, whether originating in one of the Member States or in a third country, and therefore the emergency measure ensures consistency and adequate coordination of the risks applying to different categories of foods or feeds. Where a food or feed is likely to constitute a serious risk to human health, animal health or the environment, and that – given the gravity of the situation – the risk cannot be contained satisfactorily by means of measures taken by the Member State(s)[829] concerned, the Commission shall:

- suspend the placing on the market of the food/feed in question;
- lay down special conditions for the food/feed in question;
- adopt any other appropriate interim measure.

The Commission can initiate such action at the request of a Member State, but also on its own initiative. If the Commission, following information from a Member State on the need to take emergency measures, does not initiate the procedure for the adoption of emergency measures at EU level, Article 54 GFL empowers the Member State in question to adopt interim protective measures. The Member State may maintain its national interim protective measures until an EU decision has been adopted concerning the extension, amendment or abrogation of the said measures.

In a guidance document,[830] the SCFCAH summarises the measures the European Commission can take in the scheme represented in Diagram 15.1.

[829] The subsidiarity principle applies. If action by the Member State(s) can solve the problem, the Commission should not become involved.

[830] Standing Committee on the Food Chain and Animal Health (section Toxicological Safety) Modus Operandi for the management of new food safety incidents with a potential for extension involving a chemical substance, January 2006.

Diagram 15.1. Binding measures on EU markets and/or imports.

Commission Decision	Special marketing conditions	Compulsory controls on the market
	Special import conditions	Analytical report attesting the absence of undue substance (exporters responsible) Official certificate with the analytical report Requirement of additional guarantees (sampling and analysis by the competent authorities before export) Import controls in the Member States
	Restrictions and listing	Establishment of a list of obligatory point of entry in the EU Listing of the exporters authorised to export on a positive list Listing of the third countries authorised to export on a positive list
	Suspension	Prohibition of the import of the product coming from certain countries Suspension on the placing on the market

The relation between Articles 53 and 54 GFL is subtle (Diagram 15.2). Conditions for the application by the Commission of the powers granted in Article 53 GFL are that measures by the Member State(s) cannot satisfactorily contain the risk.[831] The condition for action by Member State(s) further to Article 54 GFL is that the Commission does not act on the basis of Article 53 GFL. At first sight these conditions seem contradictory. They make sense, however, if we assume that generally the Member States in Article 53 GFL who take precedence over the Commission are not the same as the Member States in Article 54 who follow after the Commission remains passive. It seems likely that in most cases the former will be the Member States where the problem originates. They should act. If they cannot, it is up to the Commission and not to other Member States to take measures that may create barriers to trade. If, however, the Member State(s) where the problem originates nor the Commission take effective action, the threatened Member State(s) may take matters into their own hands (by closing the border to certain products).

[831] According to Article 56 Regulation 882/2004 this situation may occur if the Commission has evidence of a serious failure in a Member State's control systems.

Diagram 15.2. Emergency measures.

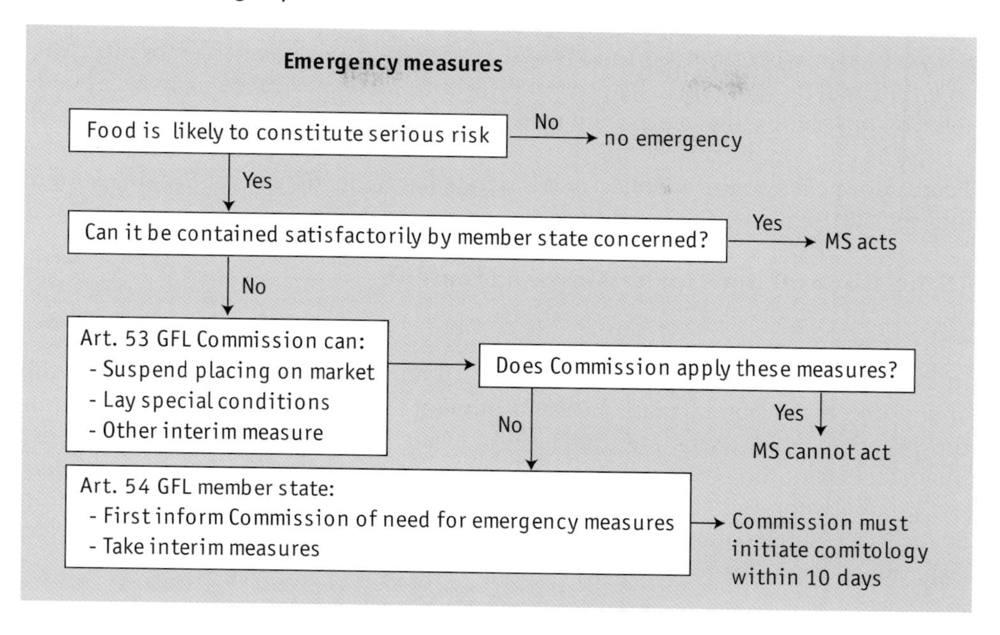

Example

Article 53 GFL empowers the Commission to adopt measures in individual cases, which are normally under the competence of the national authorities. For example, in 2006, on the basis of Article 53 GFL, the Commission took an emergency measure prohibiting the placing on the market of all curd cheese manufactured by the British company Bowland Dairy Products Ltd. During an on-site visit the Commission's inspectors from the Food and Veterinary Office found that in the production process the company used milk that did not comply with the hygiene requirements, in particular milk which had been tested positive for the presence of antibiotic residues. According to the Commission, the UK authorities did not check that the operational conditions communicated to the company had been met and therefore failed to comply with their control obligation, and – in order to address the immediate risk to human health in the Community (the products were exported to virtually all Member States) – introduced the ban on the company's products.

15.5.5 Crisis management

Some risks deriving from food and feed can be of such extreme danger or difficulty, that they will be unlikely to be prevented, eliminated or reduced to an acceptable level by provisions in place or managed solely by way of the application of Articles 53 and 54 GFL. In order to better manage these exceptional situations, the General Food Law envisaged the adoption of a general plan for food/feed crisis management and the setting up of a crisis unit.[832]

15.5.6 General plan for crisis management

The general plan for food/feed crisis management was adopted by the Commission in 2004.[833] The plan specifies: the crisis situations, the procedure leading to the application of the general plan, the establishment of a network of crisis coordinators, the practical procedures for managing a crisis, the role of the crisis unit, the practical functioning of the crisis unit (composition, means of operation, actions), the link between the crisis unit and the decision-making process, the resolution of the crisis, the management procedures in the event of a potentially serious risk, the communication strategy and the principles for transparency.

The crisis situation is characterised in particular by the following critical factors:
- the situation involves a serious direct or indirect risk to human health and/or is perceived or publicised as such or can be perceived and/or publicised as such;
- the risk is spread or could be spread to a large part of the food chain; and
- it is highly likely that the risk will spread to several Member States and/or non-EU countries.

15.5.7 Crisis unit

The crisis unit assists the European Commission in risk management, ensuring that crises are managed in a more effective, quicker and coordinated way. The crisis unit is responsible for collecting and assessing available information and identifying possible options to prevent, eliminate or reduce the risk as rapidly as possible.[834] An important task is to identify the best ways of informing the public of the risks and measures taken in a transparent way. The crisis unit consists of crisis coordinators designated by each Member State, EFSA and the Commission.

However, the crisis unit is not vested with any power to adopt decisions on the management of risk or for implementing legislation. As a consequence, crisis management will be adopted according to specific procedures already in place.

[832] Articles 55 and 56 GFL. See also Article 13 Regulation 882/2004.
[833] Commission Decision 2004/478.
[834] Article 57 GFL.

16. Consumer

Bernd van der Meulen

16.1 For you, about you, over you, without you

Even though the first and foremost objectives of EU food law are to protect consumers' health and other consumer interests, no provision can be found in food law as such actually granting the individual consumer a right s/he can uphold in a court of law.

Nevertheless, EU food law increasingly mentions 'consumers' rights'.[835] If we take a closer look at the way food law protects these rights, they turn out to be about empowerment in the marketplace, not in the courts of law. Actual legal rights for consumers have to be found outside the scope of food law.

16.2 The consumer in European case law

In case law defining the extent of Member States' powers to limit the free movement of goods with a view to protecting consumers' interests, the Court of Justice of the European Union has elaborated a consumer concept.

In European Union law, the concept of the average consumer' has become a benchmark for assessing the impact of trade practices and the responsibility of the public authorities in the field of consumer protection. A commercial practice is considered misleading if an average consumer would be misled by it. The General Food Law does not contain a definition of an 'average consumer', nor does any other piece of food legislation. The concept is interpreted in the case law of the CJEU as 'a consumer, who is reasonably well-informed and reasonably observant and circumspect, taking into account social, cultural and linguistic factors'.[836] The average consumer test is not a statistical test; it defines a way in which Union and national authorities judge particular cases, by determining the reaction in a given situation of the typical consumer who takes a serious interest in the matter at hand. Only if this consumer is unable to take care of her/his own interests are the Member States authorised to regulate on the consumer's behalf. If the consumer is unwilling or not interested for example to read the label and act on the information, this is not enough of a basis for a Member State to create legislation that may constitute a barrier to trade.

---

[835] See for the latest example (at the time of writing) the proposal (COM(2008) 40 final) that resulted in the Food information to consumers regulation discussed in Chapter 14.

[836] CJEU 16 July 1998 Case C-210/96 Gut Springenheide GmBH; CJEU 16 September 1999, Case C-220/98, ECR 2000, p. I-117, Este Lauder.

Food law also contains provisions aimed at the protection of particular groups of consumers, such as children, people with allergies, etc., who are more vulnerable to unfair trade practices. In this case, the impact of the commercial practice will be assessed from the perspective of the average consumer of that group.

In Chapter 7 we discussed the principle of mutual recognition as defined in the Cassis de Dijon ruling of the CJEU. From the point of view of the development of the concept of an 'average consumer', one consideration in the Cassis de Dijon case deserves further attention. The Court stated that consumers may be adequately protected if they are adequately informed.[837] The case is a preliminary ruling. Rewe supermarkets were denied by the German authorities the right to import Cassis de Dijon because this product's alcohol content of 20% did not conform to German product standards on fruit liquor demanding an alcohol content of at least 25%. One of the questions at issue was whether this standard was justified for reasons of consumer protection.

Law text box 16.1. The Court of Justice on the position of the consumer.

EC Court of Justice 20 February 1979, Case 120/78 (Cassis de Dijon)

'As the Commission rightly observed, the fixing of limits in relation to the alcohol content of beverages may lead to the standardisation of products placed on the market and of their designations, in the interests of a greater transparency of commercial transactions and offers for sale to the public.

However, this line of argument cannot be taken so far as to regard the mandatory fixing of minimum alcohol content as being an essential guarantee of the fairness of commercial transactions, since it is a simple matter to ensure that suitable information is conveyed to the purchaser by requiring the display of an indication of origin and of the alcohol content on the packaging of products'.

This passage illustrates the position of the consumer in European food law. Unlike German law at that time (and US law to this day), the Court will not treat European consumers as ignorant, highly vulnerable, or in need of protection by legislators who deem that they know what is best for the consumer. Rather, European food law presumes that consumers are reasonably intelligent, responsible and capable of making informed choices. As a consequence, the Court deems empowering the consumer by providing information to make informed choices, the preferable alternative to thinking on behalf of the consumer. If labelling requirements provide the consumer with sufficient protection, Member States' measures like product bans are not acceptable.

[837] See also Law text box 3.9 (Emmenthal cheese case).

This approach of the CJEU has undoubtedly influenced EU legislation. It has not, however, prescribed in EU legislation the same limits it has prescribed in national legislation in the Member States. As EU law is common to all Member States, it can hardly create trade distortions encroaching on Article 34 of the TFEU. The EU legislator therefore may well go further in consumer protection than the CJEU would accept from the Member States.

16.3 Responsibility?

The consumer is often considered to be the weak spot in EU food law. The EU Commission's Directorate General for Health and Consumers (DG Sanco) popular slogan has it that food law applies 'from farm to fork'. On closer inspection, this does not hold entirely true. Provisions of food law do not end at the moment of consumption by the final consumer ('the fork'), but at the moment of delivery to the final consumer. Food (safety) requirements do not extend into the private kitchen. If all businesses comply with food safety requirements the food may still turn unsafe if it is mishandled in the kitchen. This risk however is outside the remit of food safety law. Article 17 of the General Food Law is exemplary in this respect. It attributes responsibilities to food business operators and to the Member States, but not to consumers.

16.4 Collective rights

While the consumer is virtually absent in food law as an individual, food law does address the consumer as a species. In the Articles 9 and 10 of the General Food Law for example, food authorities are instructed to respect transparency regarding the consumer. Indeed, consumer representatives[838] are sometimes consulted in the preparation of food policy or legislation. Drafts of legislation and of certain decisions are published on DG Sanco's website to give consumers the opportunity to give their opinion.[839] What these authorities subsequently do with these opinions is, however, left to their discretion. The CJEU will not hear grievances.

16.5 Complaints

Consumers can complain to food safety inspectors if they believe they have been confronted with infringements of food law. It is a matter of national law if they hold any position in the subsequent procedure or if this procedure is considered a matter for the authority and the accused business only.

---

[838] In particular the European Consumers' Organisation BEUC Bureau Européen des Unions de Consommateurs).

[839] See for example: Article 38(1) GFL; Articles 2; 5(2)(b), 6(7), 29-31 Regulation 1829/2003; and Article 38(1) Commission Regulation 65/2004. See also Article 16 Regulation 1924/2006.

16.6 Individual consumers' rights?

While food law does not directly provide consumers with legal rights, it does influence the rights they have on other bases. By defining the legal requirements for food it gives substance to the contracts consumers conclude with food businesses. In general it can be considered fair to interpret consumer contracts as being about food in compliance with the law. If the food is not in compliance, it should not be too difficult for the consumer to get a refund. Infringements of food safety requirements may also easily constitute a basis for non-contractual (tort) liability. Consumer law has created an instrument meant to support the consumer in tort cases in their dealings with producers of defective products, called product liability law.

16.7 Product liability law

Parties who suffer damage may take legal action against operators. One could call this private law enforcement. The food business operator, who by bringing unsafe food to the market causes a customer to suffer damages, can be held liable under tort law if it can be proven that he was at fault.

Besides tort law, a more specific regime concerning defective products also exists and covers food products as well. Product liability law is not food law in a strict sense, but as it is highly relevant for the food sector it is fitting that a short description be given here.[840]

16.7.1 Strict liability

The rules on product liability have been harmonised in the European Union by Directive 85/374.[841] Directive 85/374 lays down the principle of strict liability of the producer, which means that a producer may be held responsible for a damage caused by a defective product s/he has put on the market even in the absence of fault. Diagram 16.1 compares the requirements of tort law to product liability law.

[840] In writing this section, use has been made of a paper by Nicole Coutrelis, Product Liability in the Food Sector.
[841] As amended by Directive 1999/34.

Diagram 16.1. Requirements for liability under tort law and product liability law.

Contract	Tort	Product liability
Non conformity	Unlawfulness Fault	Defective product
Damage	Damage	Personal damage
Causality	Causality	Causality

16.7.2 Producer

The producer within the meaning of the directive is not only the manufacturer of the final product, but also any other person in the chain who has produced raw materials or a component of the product, or any person who, 'by putting his name, trade mark or other distinguishing feature on the product, presents himself as its producer'.[842] All these persons bear full liability,[843] and consequently the victim may make a claim against any of these persons for complete compensation. The supplier is not responsible, except in cases when s/he does not, or cannot,[844] inform the victim of the producer's identity.

Law text box 16.2 shows side by side Article 3 of the product liability directive on the concept of producer and Article 3(2) of the General Food Law on the concept of food business. Although phrased very differently, in practice they will often refer to the same addressee.

16.7.3 Damage

Product liability law covers damages caused by death or by personal injuries and to private property of a non-commercial nature. See Law text box 16.3. Damage to property is only covered if it is more than 500 euro.[845] The text speaks of ECU. This former European Currency Unit has now been replaced by the euro at an exchange rate of 1:1.

[842] Hence the relevance of Article 9(1)(h) of Regulation 1169/2011 requiring indication of the name or business name and address of the manufacturer or packager, or of a seller established within the Community.
[843] In legalese: liability is jointly and severally. See Article 5 of Directive 85/374.
[844] If traceability requirements have been complied with, the supplier should be able to identify the producer.
[845] According to the Court of Justice of the European Union, the consequence of the choice made by the Community legislature is that, in order to avoid an excessive number of actions, in the event of minor material damage the victims of defective products cannot rely on the rules of liability laid down in the Directive but must bring an action under the ordinary law of contractual or non-contractual liability. CJEU 25 April 2002, Case C-154/00.

Law text box 16.2. Producer and food business.

Article 3 Directive 85/374	Article 3(2) Regulation 178/2002
1. 'Producer' means the manufacturer of a finished product, the producer of any raw material or the manufacturer of a component part and any person who, by putting his name, trade mark or other distinguishing feature on the product presents himself as its producer. 2. Without prejudice to the liability of the producer, any person who imports into the Community a product for sale, hire, leasing or any form of distribution in the course of his business shall be deemed to be a producer within the meaning of this Directive and shall be responsible as a producer. 3. Where the producer of the product cannot be identified, each supplier of the product shall be treated as its producer unless he informs the injured person, within a reasonable time, of the identity of the producer or of the person who supplied him with the product. The same shall apply, in the case of an imported product, if this product does not indicate the identity of the importer referred to in paragraph 2, even if the name of the producer is indicated.	'Food business' means any undertaking, whether for profit or not and whether public or private, carrying out any of the activities related to any stage of production, processing and distribution of food;

Law text box 16.3. Article 9 Directive 85/374 on damage.

Article 9

For the purpose of Article 1, 'damage' means:

(a) damage caused by death or by personal injuries;

(b) damage to, or destruction of, any item of property other than the defective product itself, with a lower threshold of 500 ECU, provided that the item of property:

   (i) is of a type ordinarily intended for private use or consumption, and

   (ii) was used by the injured person mainly for his own private use or consumption.

This Article shall be without prejudice to national provisions relating to non-material damage.

16.7.4 Conditions

The plaintiff (victim) must prove three conditions:

1. A damage to the person (death or personal injury) or to the person's private property (damages to professional property are not covered by the directive – it is about consumer protection only). For the food sector, this implies that only the final consumer can invoke product liability. Operators within the food chain can only refer to the contract provisions that they have agreed upon, or – in the absence of those – to tort law.[846]
2. A defect of the product, which is established when the product does not provide the safety that one is entitled to expect, taking into account particularly its presentation, the use which could be reasonably expected and the time at which the product was put into circulation. A mere lack of conformity to specifications (stemming from regulations or from a contract) is not sufficient to declare a product defective if safety is not at stake (Law text box 16.4).
   In the field of food, due to the number of food safety requirements, consumers are entitled to expect a high degree of safety. The producer can influence the expectations of the consumer by providing specific information, warnings, instructions on use and storage on the label. A food, for example, with allergens is not considered unsafe if the presence of these allergens is mentioned on the label.
3. A causal relationship between the defect and the damage. Consumers of unsafe food may experience some difficulty in proving the causal relationship between a specific food and the (health) damages that have been suffered. The food disappears by its very use and usually people use large varieties of food. This renders the cause of a foodborne damage difficult to determine.

Law text box 16.4. Article 6 Directive 85/374 on defective product.

Article 6

1. A product is defective when it does not provide the safety which a person is entitled to expect, taking all circumstances into account, including:
   (a) the presentation of the product;
   (b) the use to which it could reasonably be expected that the product would be put;
   (c) the time when the product was put into circulation.
2. A product shall not be considered defective for the sole reason that a better product is subsequently put into circulation.

[846] See Chapter 2 Section 2.8.5.

16.7.5 Defences

In order to disclaim liability, a producer has to prove the existence of one (or several) of the following circumstances:
- s/he did not put the product in circulation;
- the product was not defective when put into circulation;
- the product was not manufactured by the producer for sale or any form of distribution;
- the defect is due to compliance with mandatory regulations issued by public authorities;
- the state of scientific and technological knowledge at the time the product was put into circulation did not allow the existence of the defect to be discovered ('development risk');
- the defect of the final product (in case the manufacturer of a raw material or component is held liable) is attributable to the design of the final product or to the instructions of the manufacturer;
- the damage is caused, totally or in part, through the fault of the victim.

If the hygiene prescriptions have been followed and well documented, this may be of great help in building a defence that the product was not defective when put into circulation. In the case of a food that passed a pre-market approval (like novel foods and GMOs) the 'development risk' defence is worth considering, as an extensive scientific safety assessment is part of such an authorisation procedure (see Chapter 11). If the assessment did not uncover the defect, it seems probable that this defect could at that moment not be discovered.

In practice few product liability cases occur in the food sector. On the one hand it is difficult for consumers to prove their case and often damages are too low to bother trying; on the other hand food business operators do not like the bad publicity that may be involved in a court case and go to some lengths to keep their consumers happy by dealing with complaints in a generous way.

16.8 Concluding remarks

The individual consumer is not an addressee of EU food law. S/he is not held responsible for her or his contribution to food (un)safety and s/he is not endowed with legal powers under food law. Food law does, however, add to the level of protection under general consumer law by defining the level of safety a consumer may expect from food products.

III. Selected topics

17. Special foods

Foods for particular nutritional uses (PARNUTS)/Food for specific groups (FSG)

Irene Scholten-Verheijen

17.1 Introduction

Although the general approach of EU food law is to set standards and basic principles with regard to the product (quality and safety), the process and the presentation that apply to all foods, the approach to the products that are the subject of this chapter currently still has the character of product-specific legislation.[847] However, the European Commission introduced new legislation in order to, among other things, become more aligned with the general approach of food in EU legislation.[848] In this chapter we will set out the current PARNUTS legislation as well as the new FSG Regulation.

17.2 PARNUTS legislation

Foods for *par*ticular *nut*ritional uses (as indicated by the acronym PARNUTS) are foodstuffs which, owing to their special composition or manufacturing process, are clearly distinguishable from foodstuffs for normal consumption. They are suitable for their claimed nutritional purposes and are marketed in such a way as to indicate such suitability. The groups of foodstuffs for particular use are:

- infant formulae and follow-on formulae;
- processed cereal-based foods and baby foods for infants and young children;
- foods intended for use in energy-restricted diets for weight reduction;
- dietary foods for special medical purposes;
- foods intended to meet the expenditure of intense muscular effort, especially for sportsmen;
- foods for persons suffering from carbohydrate metabolism disorders (diabetes).

[847] This approach was common in the early stages of development of EU food law, but was abandoned during the 1990s, with as a result the establishment of the General Food Law; see Chapter 7. See also: B.M.J. van der Meulen, The Structure of European Food Law, Laws, 2013 (2), pp. 69-98.

[848] See press release of the European Commission: New strengthened rules for food for infants, young children and food for specific medical purpose, Brussels, 11 June 2013. available at: http://tinyurl.com/q3xnbvc.

Foodstuffs belonging to this group are expected to provide for the nutritional requirements

- of certain categories of persons whose digestive processes or metabolism are disturbed;
- of certain categories of persons who are in a special physiological condition and who are therefore able to obtain special benefit from controlled consumption of certain substances in foodstuffs;
- of infants or young children in good health.

Specific directives and regulations lay down the provisions applicable to these particular groups of foodstuffs. Such specific directives and regulations cover in particular:

- essential requirements as to the nature or composition of the products;
- provisions regarding the quality of raw materials;
- hygiene requirements;
- permitted changes;
- a list of additives;
- provisions regarding labelling, presentation and advertising;
- sampling procedures and methods of analysis needed for checking compliance with the requirements of the specific directives and regulations.

The general compositional and labelling requirements are laid down in Directive 2009/39 on foodstuffs intended for particular nutritional uses (PARNUTS Framework Directive). These requirements are complemented by a number of non-legislative Union acts,[849] which are applicable to specific categories of food. In that respect, harmonised rules are laid down in Commission Directive 2006/141 (infant formula), Commission Directive 2006/125 (processed cereal-based foods and baby foods for infants and young children), Commission Directive 96/8 (energy-restricted diets), Commission Directive 1999/21 (dietary foods for special medical purposes), Council Directive 92/52 (export of infant formula and follow-on formula), Commission Regulation 953/2009 (substances that may be added for specific nutritional purposes in foods for particular nutritional uses), and Commission Regulation 41/2009 (concerning the composition and labelling of foodstuffs suitable for people intolerant to gluten).

17.3 FSG Regulation

A new Regulation, (EU) No 609/2013, of the European Parliament and the Council on food intended for infants and young children, food for special medical purposes, and total diet replacement for weight control ('Food for Specific Groups', FSG)

[849] On the concept of non-legislative Union acts, see Chapter 5 Section 5.5.

was adopted on 12 June 2013 and applies from 20 July 2016. The new Regulation was proposed in order to:

- strengthen provisions on foods for vulnerable population groups that need particular protection, e.g. infants and children up to three years old, overweight or obese people and people with specific medical conditions;
- keep current compositional and labelling rules for infant and follow-on formulae, processed cereal-based foods and other baby foods and foods for special medical purposes;
- replace the current three lists[850] with a single EU list of substances that can be added to these foods, including minerals and vitamins;
- repeal Directive 2009/39 and abolish the concept of dietetic foods;
- reduce the administrative burden, bring clarity and consistency within the EU and flexibility in the innovative food market.

The new regulation implicates that the concept of 'foodstuffs for particular nutritional uses' should be abolished and Directive 2009/39 should be replaced.[851] The FSG Regulation sets general compositional and labelling rules for infant and follow-on formulae, processed cereal-based food and other baby foods, which will be further specified by delegated acts to be adopted by the Commission.[852]

It requires the Commission to transfer rules on gluten-free and very low gluten to Regulation 1169/2011 on Food Information to Consumers (FIC) in order to ensure clarity and consistency;[853] and to transfer rules and regulate meat replacement for weight control solely under Regulation 1924/2006 on Nutrition and Health Claims (NHCR) in order to ensure legal certainty.[854] It furthermore indicates that, for the sake of clarity and consistency, the establishment of rules on the use of statements indicating the absence or reduced presence of lactose in food should be regulated under the FIC Regulation.[855]

[850] Currently, under the PARNUTS legislation, the provisions concerning the list of the nutritional substances that may be used in the manufacture of infant formulae and follow-on formulae and of processed cereal-based foods and baby foods for infants and young children are laid down respectively in Commission Directive 2006/141 and Commission Directive 2006/125; the provisions concerning the list of the nutritional substances that may be used in the manufacture of all the other categories of foods for particular nutritional uses are laid down in Commission Regulation 953/2009.
[851] Recital 13 FSG Regulation; see also: Impact assessment, available at: http://tinyurl.com/lkvoaam.
[852] Next to Directive 2009/39 of the European Parliament and the Council, also Council Directive 92/52, Commission Directives 96/8, 1999/21, 2006/125 and 2006/141, and Commission Regulations 41/2009 and 953/2009, will be repealed.
[853] Recital 41 FSG Regulation; the Commission fulfilled this requirement by Commission Delegated Regulation (EU) No 1155/2013 of 21 August 2013, amending Regulation (EU) No 1169/2011 of the European Parliament and of the Council on the provision of food information to consumers as regards information on the absence or reduced presence of gluten in food.
[854] Recital 43 FSG Regulation.
[855] Recital 42 FSG Regulation.

It furthermore requires the Commission to prepare a report on milk-based drinks (so-called growing-up milks) and similar products intended for young children in order to analyse the necessity to establish special compositional and labelling rules for this kind of product; as well as a report on food intended for sports people in order to analyse the necessity to establish special compositional and labelling rules.[856]

The European Commission holds technical working group meetings on FSG with the Member States to discuss and tease out issues that need further consideration regarding which rules on labelling, presentation, advertising, and promotional practices should apply.

17.4 Reasons for a new Regulation

The European Commission gave several reasons for the revision of the current framework regarding PARNUTS, which led to the FSG Regulation.[857]

17.4.1 Free movement of food within the EU

The free movement of safe and wholesome food is an essential aspect of the internal market and contributes significantly to the health and well-being of citizens, and to their social and economic interests.[858] It was acknowledged that difficulties could arise from the definition of 'foodstuffs for particular nutritional uses' under the PARNUTS Framework Directive, which appeared to be open to differing interpretations by the national authorities.[859] It was indicated that an increasing number of foodstuffs were being marketed and labelled as foodstuffs suitable for particular nutritional uses, due to the broad definition laid down in the directive. It was pointed out that food regulated under the PARNUTS Framework Directive differed significantly between the Member States; similar food could at the same time be marketed in different Member States as food for particular nutritional uses and/or as food for normal consumption, including food supplements, addressed to the population in general or to certain subgroups thereof, such as pregnant women, postmenopausal women, older adults, growing children, adolescents, variably active individuals and others.[860] That state of affairs was considered to undermine the functioning of the internal market, to create

[856] See the website of DG Health and Consumers of the European Commission: http://tinyurl.com/2rehsg, and http://tinyurl.com/kr4zacf.

[857] See for a critical assessment of the argumentation of the Commission: A. Meisterernst, Foods for Particular Nutritional Uses – Death Sentence Passed for Sound Reasons? EFFL 2011, pp. 315-327.

[858] Recital 1 FSG Regulation.

[859] Report from the Commission of 27 June 2008 to the European Parliament and to the Council on the implementation of the notification procedure required by Directive 2009/39/EC; see Recital 9 FSG Regulation.

[860] Study report of 29 April 2009 by Agra CEAS Consulting concerning the revision of Directive 2009/39/EC; see also Recital 10 FSG Regulation.

legal uncertainty for competent authorities, food business operators – in particular small and medium-sized enterprises – and consumers, while the risk of marketing abuse and distortion of competition could not be ruled out. There appeared to be a need to eliminate differences in interpretation by simplifying the regulatory environment.[861] Moreover, experience showed that certain rules included in, or adopted under, the Framework Directive were no longer effective in ensuring the functioning of the internal market.[862]

17.4.2 Evolution

Furthermore, other recently adopted Union legal acts were considered more adapted to an evolving and innovative food market than the PARNUTS Framework Directive. More in particular reference was made to Directive 2002/46 (on food supplements), the NHCR and Regulation 1925/2006 (on the addition of vitamins and minerals and of certain other substances to foods – also known as 'fortified foods'). Besides, those Union legal acts were considered to adequately regulate a number of categories of food covered by the PARNUTS Framework Directive with less of an administrative burden and more clarity as to scope and objectives.[863]

17.4.3 Limitation of categories

It was established that a limited number of categories of food constitute a partial or the sole source of nourishment for certain population groups, and that such categories of food are vital for the management of certain conditions and/or are essential to satisfy the nutritional requirements of certain clearly identified vulnerable population groups. Those categories of food include infant formula and follow-on formula, processed cereal-based food and baby food, and food for special medical purposes. It was concluded that the provisions laid down in the existing legislative texts regarding these specific categories of foods were satisfactory and should therefore be retained.[864]

In addition, there are a number of infant and follow-on formula products which are sold as medical foods. There were concerns that there are a number of formula milks on the EU market that are inappropriately marketed as food for specific medical purposes, which do not respect the full provisions of the infant formula rules. Discussions under the FSG Regulation sought to address this by introducing a procedure to harmonise the categorisation of products and to review which of the labelling and advertising provisions in Directive 2006/141 should also apply to food for specific medical purposes for infants.

---

[861] Recital 10 FSG Regulation.
[862] Recital 12 FSG Regulation.
[863] Recital 11 FSG Regulation.
[864] Recital 15 FSG Regulation.

With regard to the category of food for specific medical purposes, the Commission indicated that the three types of medical foods currently defined under Article 1(3) of Directive 1999/21 should be retained. However, there continues to be widespread problems with categorisation of food for specific medical purposes and overlap with other foodstuffs such as food supplements. The European Commission has indicated that it would like to see the labelling of medical foods brought more into line with the general labelling rules as set out in the FIC Regulation, whilst maintaining a number of specific mandatory provisions where appropriate.

17.4.4 Reducing obesity and improving diet

Last but not least, one of the most pressing issues on the agenda for the years to come is overweight and obesity. Obesity is one of the greatest public health challenges of the 21st century. Its prevalence has tripled in many countries of the WHO European Region since the 1980s, and the numbers of those affected continue to rise at an alarming rate. In addition to causing various physical disabilities and psychological problems, excess weight drastically increases a person's risk of developing a number of non-communicable diseases, including cardiovascular disease, cancer and diabetes.[865] The WHO stated that both society and governments need to act to curb the epidemic; that national policies should encourage and provide opportunities for greater physical activity, and improve the affordability, availability and accessibility of healthy foods; and that they should also encourage the involvement of different government sectors, civil society, the private sector and other stakeholders.[866]

Thus far, the EU legislator has not found suitable instruments to deal with this problem.[867] Measures are currently limited to providing consumers with information directly and on food product labels. In view of the growing rates of people with problems related to being overweight or obese, an increasing number of foods are placed on the market as total diet replacement for weight control. Currently, for such foods present in the market a distinction can be made between products intended for low calorie diets and products intended for very low calorie diets. Given the nature of the foods in question the Commission considered it appropriate to lay down certain specific provisions for them. Experience has shown that the relevant provisions laid down in Directive 96/8 ensure the free movement of foods presented as total diet replacement for weight control in a satisfactory manner while ensuring a high level of protection of public health. It is therefore considered appropriate that the FSG Regulation focuses on the general

[865] See also Chapter 20.

[866] See: http://tinyurl.com/plhhezq.

[867] White Paper on A Strategy for Europe on Nutrition, Overweight and Obesity related health issues, 2007, available at: http://tinyurl.com/6crk2gy. Strategy for Europe on nutrition, overweight and obesity related health issues, implementation progress report, 2010, available at: http://tinyurl.com/3xs3cwt; http://tinyurl.com/kh85ev9.

compositional and information requirements for foods intended to replace the whole of the daily diet including foods of which the energy content is very low, taking into account the relevant provisions of Directive 96/8.[868]

17.5 The FSG Regulation in detail

More specifically, the FSG Regulation establishes the following rules regarding product, process and presentation, for business operators in order to protect consumers and guide the authorities on how to safeguard consumer protection.

17.5.1 General provisions

The food that is subject to the Regulation encompasses, according to Article 1:
- infant formula and follow-on formula;
- processed cereal-based food and baby food;
- food for special medical purposes;
- total diet replacement for weight control.

Furthermore, the Regulation establishes a Union list of substances that may be added to one or more of these categories of food.

Article 2 of the FSG Regulation provides for the definitions related to the specific categories of food that are the subject of the FSG Regulation: 'infant formula' and follow-on formula', 'processed cereal-based food' and 'baby food', as well as 'food for special medical purposes' and 'total diet replacement for weight control'.[869] For the definitions of 'food', 'food business operator', 'retail' and 'placing on the market' reference is made to the General Food Law.[870] Furthermore, reference is made to the FIC Regulation with regard to the definitions of 'prepacked food', 'labelling' and 'engineered nanomaterial', and to the NHC Regulation with regard to the definitions of 'nutrition claim' and 'health claim'.

In order to ensure the uniform implementation of the FSG Regulation, the Commission may adopt implementing acts regarding the decision whether a given food falls within the scope of the Regulation and to which specific category of food a given food belongs (Article 3).

Article 4 indicates that food that is subject to the FSG Regulation may only be placed on the market if it complies with the FSG Regulation and is only allowed on the retail market in the form of prepacked food. Member States may not restrict or forbid the placing on the market of food which complies with the Regulation, for reasons related to its composition, manufacture, presentation or labelling.

[868] Recital 16 FSG Regulation.
[869] Recital 17 FSG Regulation.
[870] Recital 14 FSG Regulation.

Reference is made in Article 5 to the precautionary principle. This principle is set out in Article 7 of the General Food Law (Regulation 178/2002); it provides the authority for Member States to take provisional risk management measures, even when it's not – scientifically – certain that harmful effects on health are present in given circumstances. The FSG Regulation specifically applies the precautionary principle, in order to ensure a high level of health protection in relation to the person for whom the food that is subject to the FSG Regulation is intended.

17.5.2 Compositional and information requirements

Food that is subject to the FSG Regulation must, according to Article 6, comply with any requirement of Union law applicable to food. However, in case of conflicting requirements, the requirements laid down in the FSG Regulation prevail.[871] For the purpose of the FSG Regulation, the EFSA should be consulted on all matters likely to affect the public. It shall therefore provide scientific opinions, which will serve as the scientific basis for any Union measure adopted pursuant to this Regulation which is likely to have an effect on public health (Article 7).[872]

In Article 9, the Regulation sets out general compositional and information requirements that apply to all the categories of food subject to the Regulation:

- The composition of food shall be such that it is appropriate for satisfying the nutritional requirements of, and is suitable for, the persons for whom it is intended, in accordance with generally accepted scientific data.[873]
- Food shall not contain any substance in such quantity as to endanger the health of the persons for whom it is intended. For substances which are engineered nanomaterials, compliance with the requirements referred to in the first subparagraph shall be demonstrated on the basis of adequate test methods, where appropriate.[874]
- On the basis of generally accepted scientific data, substances added to food shall be bio-available for use by the human body, have a nutritional or physiological effect and be suitable for the persons for whom the food is intended.[875]
- Food may contain substances covered by the Novel Foods Regulation, provided that those substances fulfil the conditions under that regulation for being placed on the market.[876]
- The labelling, presentation and advertising of food referred to shall provide information for the appropriate use of such food, and shall not mislead, or attribute to such food the property of preventing, treating or curing a human disease, or imply such properties; this will however not prevent the dissemination

[871] This conforms to the maxim: *lex specialis derogat legi generali*, i.e. more specific legislation takes priority over more general legislation.
[872] Recital 18 FSG Regulation.
[873] Recital 19 FSG Regulation.
[874] Recital 19 FSG Regulation
[875] Recital 19 FSG Regulation.
[876] Recital 23 FSG Regulation.

of any useful information or recommendations exclusively intended for persons having qualifications in medicine, nutrition, pharmacy, or for other healthcare professionals responsible for maternal care and childcare.[877]

The FSG Regulation furthermore provides, in Article 10, for additional requirements regarding the labelling, presentation and advertising of infant formula and follow-on formula:[878] this should not discourage breast-feeding and shall not include pictures of infants, or other pictures or text which may idealise the use of such formula (whereas graphic representation for easy identification of infant formula and follow-on formula and for illustrating methods of preparation are permitted).[879]

Furthermore, the Regulation (Article 11 in conjunction with Articles 18 and 19)[880] empowers the Commission to adopt – by 20 July 2015 – and update delegated acts[881] with respect to:

- the specific compositional requirements applicable to the food that is subject of the Regulation;[882]
- the specific requirements on the use of pesticides in products intended for the production of the food that is the subject of the Regulation and on pesticide residues in such food;[883]
- the specific requirements on labelling, presentation and advertising of food that is the subject of the Regulation, including the authorisation of nutrition and health claims in relation thereto;[884]
- any eventual notification requirements for the placing on the market of food that is the subject of the Regulation;
- the requirements concerning promotional and commercial practices relating to infant formula;
- the requirements concerning information to be provided in relation to infant and young child feeding in order to ensure adequate information on appropriate feeding practices;
- the specific requirements for food for special medical purposes developed to satisfy the nutritional requirements of infants, including compositional requirements and requirements on the use of pesticides in products intended for the production of such food, pesticide residues, labelling, presentation, advertising and promotional and commercial practices, as appropriate.[885]

[877] Recital 25 FSG Regulation.
[878] Recital 24 FSG Regulation.
[879] Recital 26 FSG Regulation.
[880] Recital 45 FSG Regulation.
[881] On the concept of delegated acts, see Chapter 5 Section 5.5.2.
[882] Recital 27 FSG Regulation.
[883] Recitals 20-22 FSG Regulation.
[884] Recital 28 FSG Regulation.
[885] Recital 30 FSG Regulation.

By 20 July 2015, the Commission is obliged, based on Articles 12 and 13, to present to the European Parliament and the Council two reports (if necessary, accompanied by a legislative proposal):

1. on the necessity, if any, of special provisions for milk-based drinks and similar products intended for young children regarding compositional and labelling requirements and, if appropriate, other types of requirements (taking into consideration, for instance, the nutritional requirements of young children, the role of those products in the diet of young children and whether those products have any nutritional benefits when compared to a normal diet for a child who is being weaned);[886]
2. on the necessity, if any, of provisions for food intended for sportspeople.[887]

Article 14 provides for the Commission to adopt technical guidelines to facilitate compliance by food business operators, in particular small and medium-sized enterprises.[888]

17.5.3 Union list

The FSG Regulation provides for one single Union list of substances that may be added to one or more of the categories of food that is subject to the FSG Regulation (Article 15). The list refers to substances belonging to the following categories of substances:

- vitamins;
- minerals;
- amino acids;
- carnitine and taurine;
- nucleotides;
- choline and inositol.

The list, included in the Annex to the FSG Regulation, will contain the specific category of food to which the substances may be added, the name and the description of the substance and the specification of its form, as well as the conditions of use of and the purity criteria applicable to the substance (whereas Member States may maintain national rules setting stricter purity criteria). The Commission is empowered to remove from or add to the Union list a category of substances, or to add, remove or amend the categories, the name, the conditions of use and the purity criteria of the substances on the list (Articles 15 and 16).[889]

[886] Recitals 29-31 FSG Regulation.
[887] Recitals 32-33 FSG Regulation.
[888] Recital 34 FSG Regulation.
[889] Recitals 35-39 FSG Regulation.

17.5.4 Procedural provisions

The Commission is assisted by the Standing Committee on the Food Chain and Animal Health (Article 17).[890]

The power to adopt delegated acts is conferred on the Commission – for a period of five years from 19 July 2013 – subject to the conditions laid down in Article 18. A delegated act only enters into force if no objection has been expressed either by the European Parliament or the Council within a period of two months of the notification of that act or if the European Parliament and the Council have both informed the Commission within that time period that they will not object. In case of urgency, Article 19 provides for an urgency procedure. In that case delegated acts enter into force and apply without delay as long as no objection is expressed by either the European Parliament or the Council. If and when an objection is made, the Commission will repeal the delegated act without delay following the notification of the objection.

17.5.5 Final provisions; transitional measures

The Regulation applies from 20 July 2016, unless otherwise indicated (Article 22).[891] Article 20 sets out that from this date Directives 2009/39, Directive 92/52 and Regulation 41/2009 are repealed. Directive 96/8 shall not apply from this date on to foods presented as a replacement for one or more meals of the daily diet. Regulation 953/2009 and Directives 96/8, 1999/21, 2006/125 and 2006/141 are repealed from the date of application of the delegated acts referred to.[892]

Article 21 provides for adequate transitional measures.[893] Food that is subject to the Regulation, which does not comply with the Regulation, but complies with the repealed regulations and which is placed on the market or labelled before 20 July 2016, may continue to be marketed after that date until stocks of such food are exhausted. Where the date of application of the delegated acts referred to is after 20 July 2016, food that is subject to the FSG Regulation and complies with the FSG Regulation as well as with the regulations the delegated acts will replace, but does not comply with those delegated acts, and which is placed on the market or labelled before the date of application of those delegated acts, may continue to be marketed after that date until stocks of such food are exhausted. Food that is not subject to the FSG Regulation, but which is placed on the market or labelled in accordance with the – to be – repealed regulations before 20 July 2016, may continue to be marketed after that date until stocks of such food are exhausted.

[890] Recital 40 FSG Regulation.
[891] Articles 11, 16, 18 and 19 apply from 19 July 2013; Article 15 and the Annex shall apply from the date of application of the delegated acts.
[892] Recitals 13, 46 and 47 FSG Regulation.
[893] Recital 48 FSG Regulation.

17.6 Future developments

We welcome the effort of the Commission to try and simplify the PARNUTS legislation, by way of other existing, more generally applied legislation, such as the FIC Regulation and the NHCR, as well as by creating the FSG Regulation. Although the FSG Regulation seems to be functional in this more simplified approach to foods for specific groups, much of its effectiveness will in our opinion depend on how the Commission uses its power to adopt delegated acts. Much of the simplification will become void if and when the Commission adopts complex, diversified and detailed acts. We will therefore have to wait and see whether or not the objective of the FSG Regulation will indeed be met.

18. Importing food into the EU

Cecilia Kühn and Francesco Montanari

18.1 Introduction

18.1.1 Food regulatory compliance in the European Union

With more than 500 million consumers of relatively high average income per capita, the European Union (EU) is one of the most important markets in the world. Currently comprising 28 Member States, over time the EU has developed a considerable amount of food-related legislation, mainly in the form of directives and regulations, which applies to its whole territory. However, in certain instances or areas, each Member State may still have its own national rules in place. This is the case, for example, for food labelling, in spite of the recently adopted Regulation 1169/2011 on food information to consumers (FIC). Thus, food business operators (FBOs) must comply with rules at EU level as well as at Member State level. For example, for certain aspects cheese must comply with FIC provisions (e.g. ingredients list, allergens, nutritional information, 'best before' date), while for others it must fulfil requirements laid down in national legislation of each Member State (e.g. legal name, permitted usage of quality descriptors like 'full fat', of production processes, sometimes recipe, etc.).

Furthermore, with more than 15,000 intertwined pieces of legislation and other published official documents,[894] food law is one of the most regulated areas in the EU. This makes regulatory compliance an uneasy task for FBOs. EU food law applies at various levels and regulates raw materials, additives and ingredients, intermediates and finished products; product composition, formulation, production processes, packaging and traceability; labelling, claims, advertising and communication. EU food law requirements may be of horizontal or vertical nature; while horizontal requirements apply to all foods, vertical ones apply to specific food product groups or categories. Furthermore, for certain products (e.g. novel foods, genetically modified organisms) or certain product-related information (e.g. health claims) scientific evidence must be gathered and submitted in order to obtain pre-market authorisations.

EU food law requirements apply to both food produced in the EU or in non-EU countries. However, there is an extra set of requirements applying to food originating from non-EU countries destined for import into the EU.[895] These

[894] For the figures, see Chapter 8.

[895] By 'food destined for import into the EU' is meant any food originating from countries outside the EU and intended for placing on the EU market. When referring to 'placing on the market', Article 3(8) GFL means '*the holding of food or feed for the purpose of sale, including offering for sale or any other form of transfer, whether free of charge or not, and the sale, distribution, and other forms of transfer themselves*'.

requirements may be foreseen for sanitary (to protect human health), veterinary (for animal health) or phytosanitary reasons (for plant health). In conclusion, food regulatory compliance in the EU is an intricate puzzle: a combination of food regulatory affairs with science and technology, industrial experience and know-how. In the EU with 28 Member States, knowledge of differences with regards to laws, languages, approaches and functioning is therefore essential.

18.1.2 Scope and purpose of this chapter

Several different sets of EU rules apply to imports – e.g. food, agriculture and environment requirements as well as customs and trade regulations. This chapter addresses the EU food regulatory regime that applies to import of food, its regulation, procedures and compliance. As EU food law aims at guaranteeing food safety but also the protection of consumers, animals, plants and the environment, this results in an extensive body of laws to be complied with, in addition to the extra requirements that need to be fulfilled to ensure food is imported lawfully. Even bringing in food into the EU for personal use or consumption (e.g. in luggage) or receiving parcels from abroad (e.g. through mail or internet orders) may be subject to specific rules.[896]

The purpose of this chapter is to provide the reader with a general understanding of the overall structure and functioning of the current EU regime for imports. However, it does not and cannot constitute an exhaustive coverage of all requirements that may be applicable to a given product or product category. Indeed, the reader must be aware that EU food law requirements largely vary depending, e.g. on the type, nature and intended use of each product. How to comply must be thus evaluated on a case-by-case basis, taking into account EU requirements but also those that the Member State(s) where the products are to be marketed may impose.

EU food law is a dynamic area and, as such, subject to constant normative evolution. From this angle, the EU regime for imports makes no exception. In particular, emergency measures are often established when serious risks arise in this area and, then, subject to periodical review based on risk prevalence. Moreover, the EU framework governing official controls, regulated by Regulation 882/2004,[897] is currently under review with changes being considered in order to ensure more consistency between the existing import procedures.[898]

[896] For example, Regulation 206/2009 establishes rules governing official controls on imports of small consignments of animal origin intended for personal use or consumption (thus, non-commercial), including products that are carried as travellers' luggage or ordered remotely (e.g. via mail, telephone or internet) by private persons.

[897] Regulation 882/2004 of 29 April 2004 on official controls performed to ensure the verification of compliance with feed and food law, animal health and animal welfare rules.

[898] See also Chapter 15.

18.1.3 Chapter structure

The following Diagram 18.1 assists in explaining the structure of this chapter. This Diagram complements the one for domestic products (Diagram 8.1 in Chapter 8) which also applies to food imports.

All foods to be placed on the market in the EU, whether produced in the EU or imported, must comply with a set of general requirements. These latter are discussed at length in various chapters of this book, but, since they also apply to import foods, a general overview is provided in Section 18.2. Section 18.3 addresses the legal basis on which the EU import regime rests. Section 18.4 describes requirements and procedures applying specifically to imports, whereas Section 18.5 sets out how official controls are organised and carried out at EU level. Section 18.6 provides examples of emergency measures that the EU has currently in place to

Diagram 18.1. European food law for imports.

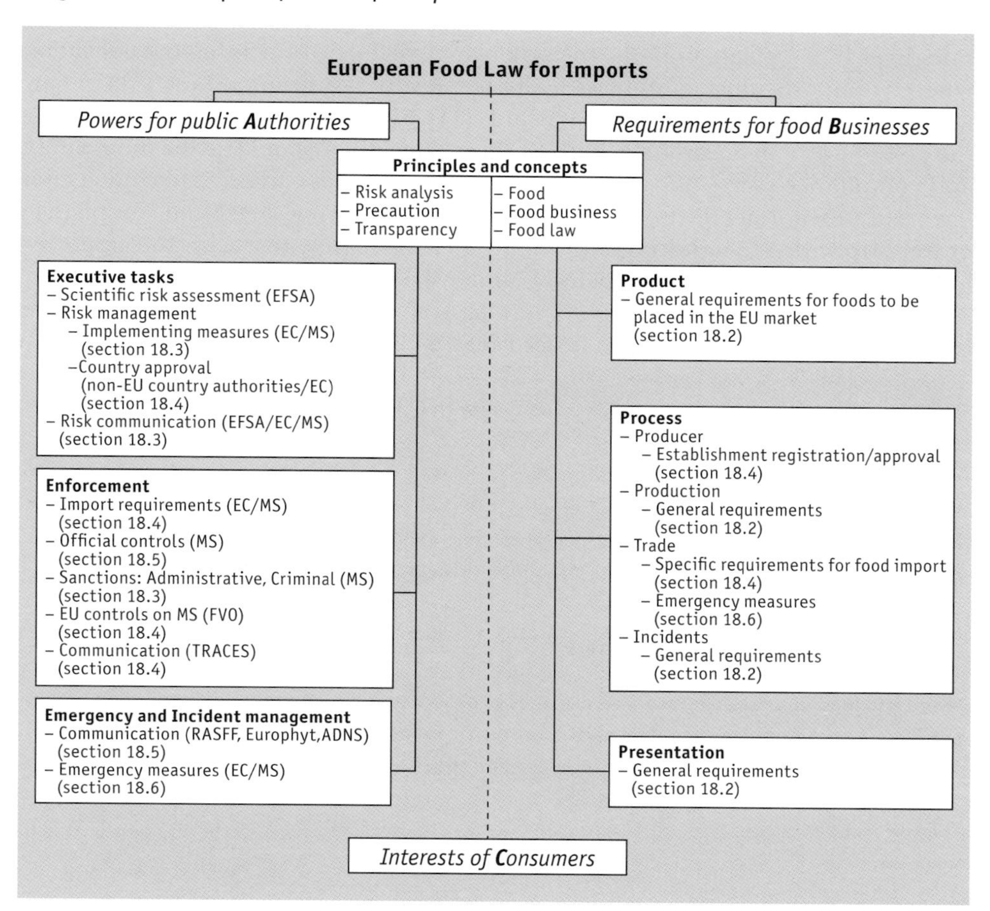

address serious risks associated with imports. Finally, Section 18.7 rounds off this chapter with an account of the ongoing legislative review of Regulation 882/2004 and, in particular, on the main changes affecting EU import policy.

18.2 General import requirements

18.2.1 GFL and imports: general principles

Overall, EU food law aims at the protection of consumers' interests (Article 8 GFL).Consumer protection can be achieved in a number of ways, including ensuring compliance with food safety requirements, providing accurate and truthful information so that the consumer can make informed choices, as well as discouraging misleading or fraudulent practices. Currently, several legal acts at EU level expressly pursue these objectives. Altogether, they constitute the 'general requirements' of EU food law since they apply to any food placed in the EU market, irrespective of the place where the food was produced.

However, in relation to imported foods, the GFL contains some overarching principles that regulate imports. In particular, Article 11 GFL sets out the obligation whereby imports must comply with all EU relevant requirements (see (1) in Law text box 18.1). On the other hand, Article 17(1) GFL provides the criteria to identify the person who, in accordance with EU law, is legally responsible for their safety and compliance (see (2) in Law text box 18.1). Since, as a rule, EU law does not have extraterritorial effects, the responsibility for ensuring safety and compliance of imports with EU requirements would, in most cases, rest with the importer located in the EU within the business under his control.

Law text box 18.1. General principles for food imports.

The legal obligation of imported foods to comply with EU food law:

Article 11 GFL

'Food and feed imported into the Community for placing on the market within the Community shall comply with the relevant requirements of food law or conditions recognised at least equivalent thereto or, where a specific agreement exists between the Community and the exporting country, with requirements contained therein'.

Establishing the responsible FBO:

Article 17 (1) GFL

'Food and feed business operators at all stages of production, processing and distribution within the businesses under their control shall ensure that foods or feeds satisfy the requirements of food law which are relevant to their activities and shall verify that such requirements are met.'

18.2.2 Two sets of requirements

Two sets of requirements apply to food intended for import into the EU market. On the one hand, imports must comply with 'general food requirements' that apply also to domestic products. On the other, they must also meet a second and more specific set of requirements laid down for import purposes,[899] which, overall, guarantee their suitability for placing on the EU market. These two sets of requirements co-exist and are applied at any time throughout the entire chain from the EU border to the final consumer. General food requirements are discussed below, while the specific requirements for food imports are presented in Sections 18.3 and 18.4. In addition, specific rules applying to imports with a history of risk are discussed in Section 18.6.

18.2.3 General Food Requirements

'General food requirements' comprise a vast array of rules and provisions covered in several chapters of this book and beyond. They may be horizontal (i.e. applying to all foods) or vertical (i.e. applying to certain foods or product categories), and cover various aspects of product, process and presentation (as set out in Chapter 8). Since they are of relevance also for food imports, some selected aspects are highlighted thereinafter:

- *Product*: composition and food improvement agents,[900] hygiene,[901] chemical and biological safety,[902] fortification, novel foods,[903] GMOs,[904] foods for special groups,[905] and food supplements.
- *Process*: hygiene,[906] food contact materials (FCM),[907] and traceability.[908]

[899] 'Lawful import' means not only ensuring compliance at the time the product enters the EU territory, but also that it remains compliant throughout further stages of processing and distribution. In accordance with that, official controls on imports are carried out at entry as well as at any other stage of the food chain. Control throughout the product life in the food chain is necessary because risks or non-compliances may also appear at a later stage of the product life-cycle.
[900] Chapter 10 for additives, enzymes and for flavourings.
[901] Chapter 13.
[902] Chapters 11 and 12.
[903] Chapter 10.
[904] Chapter 10.
[905] Chapter 17 on FSG.
[906] Chapter 13.
[907] FCM are materials and articles intended to come into contact with foods such as: packaging materials; cutlery and dishes; processing machines; containers. FCM must not transfer their components into the foods in unacceptable quantities (migration). Migration limits are set for plastic materials. For authorised substances, migration limits are established based on toxicological evaluations and in accordance with daily intake levels established by the Scientific Committee on Food. The general EU law on FCM is Regulation 1935/2004. Wherever EU legislation is not yet in place for all aspects of FCM, then national legislation applies. On FCM, see Chapter 19.
[908] See further responsibilities of the FBO in Section 18.2.3.

- *Presentation*: labelling, health and nutrition claims and food information to the consumer,[909] other claims on foods such as organic,[910] geographical indications.[911]

The following aspects of key importance to food imports are discussed here below: food safety, responsibility of FBOs, food information, and country of origin indication.

Food safety

Article 14(1) and (2) GFL stipulates: *'Food shall not be placed on the market if it is unsafe. Food shall be deemed to be unsafe if it is considered to be: (a) injurious to health; (b) unfit for human consumption'.* This provision implies an obligation for the FBO to monitor and keep under control any risk associated with the products he markets in the EU.

The provision means that food products can be placed in the market as long as there is no risk for human health. For certain products or product categories, however, EU food law foresees that safety must be proven before placing on the market. For this purpose, a pre-market authorisation or notification is required (e.g. novel foods, GMOs). If authorisation is lacking, import into the EU is not allowed. Lastly, even when all the elements point to full conformity with food law requirements, competent authorities may still take actions deemed appropriate towards a certain food product (e.g. by restricting its placing on the market or ordering its withdrawal from the market), whenever there are suspicions of unsafety despite conformity (Article 14(8) GFL).

Responsibilities of the FBO

From EU food law it follows that for every food product there will always be a responsible FBO.[912] In the case of imports, the FBO for the EU entry and importing business activities is the one under whose responsibility the food is imported (see also Section 18.2.1). As the product moves down through the food chain, subsequent activities related to the food may be carried out under the responsibility of other FBOs (thus, 'the retailer' is the FBO responsible for retailing activities whilst the importer remains the responsible FBO for the importing activities).

The FBO has wide-reaching legal responsibilities with respect to monitoring, traceability, food information, recalls and withdrawal and all other relevant aspects regulated under EU food law:

[909] Chapter 14.
[910] Chapter 14.
[911] Chapter 14.
[912] The most important exception is food that has been grown and prepared within a private household.

- The FBO must ensure food products fully comply with EU food law within her/his business activity (Articles 17(1)).
- The FBO placing on the market food products intended for the final consumer is responsible for food information to consumers (Article 8 FIC) and for the traceability of each of his products 'one step back and one step forward' (Article 18 GFL). Thus, s/he must be able to identify her/his suppliers as well as the FBOs he has supplied with her/his product. In this respect, it must be noted that the SCFCAH interprets this provision as not applying to FBOs located outside the EU'.[913]
- If the FBO considers that a food s/he has imported, manufactured or distributed is not in compliance with food safety requirements, s/he must initiate appropriate action which may include withdrawal and recall from the market, public information to consumers on the reasons for withdrawal, and information to and collaboration with the competent authorities. S/he must also pass information to trace the food to others in the food chain that must take action (importer, manufacturer, distributor, retail or other) (Article 19 GFL).
- The FBO is liable towards the consumer for defective products.[914]

Food information

As regards food information, Article 16 GFL provides that '*the labelling, advertising and presentation of food or feed, including their shape, appearance or packaging, the packaging materials used, the manner in which they are arranged and the setting in which they are displayed, and the information which is made available about them through whatever medium, shall not mislead consumers*.' By stipulating that '*food information shall not be misleading*', Article 6 FIC further reinforces this principle.

The EU legal definition of 'food information' is very broad and covers not only information provided through labelling, but also by means of e.g. advertising, internet or simply conveyed orally at the point of sale (Article 2(2)(a) FIC).

Any food intended for sale to consumers or supplied to mass caterers must display food information (Article 6 FIC). If the imported food is intended for consumers or mass caterers directly this obligation rests on the importer. If the customers are other FBOs the importer must ensure that these other FBOs are provided with sufficient information to enable them to meet their obligations.[915] The responsibility for the food information rests with the FBO under whose name or business name the food is marketed. When the FBO is not established in the EU, the responsibility for food information lies with the importer (Article 8(1) FIC).

[913] Guidance on the Implementation of Articles 11, 12, 14, 17, 18, 19 and 20 of GFL. Conclusions of the Standing Committee on the Food Chain and Animal Health. The document is available on DG Sanco's website at http://ec.europa.eu/food/food/foodlaw/guidance/index_en.htm.
[914] Directive 85/374 and Chapter 16.
[915] Article 8(8) FIC.

Country of origin and place of provenance

According to EU food law, FBOs are free to provide consumer information relating to origin or provenance of food products (Article 26 FIC). There are, however, a number of products for which EU legislation requires mandatory indication of country of origin, such as beef and beef products, sheep, goat, swine and poultry meat, fish, fruits and vegetables, honey, olive oil, wine and spirits.[916]

Currently, an extension of mandatory country of origin information is being considered at EU level for milk and dairy products, game meat, unprocessed foods (e.g. rice, wheat flour) or foods containing a single ingredient (e.g. sunflower oil, sugar) or a primary ingredient (e.g. tomato puree, wheat flour in bread, orange juice) (Article 26(5) FIC). Harmonised criteria for provision of country of origin information when given on a voluntary basis are also under consideration (Article 26(8) FIC).

18.3 Specific import requirements: definitions and legal basis

This section provides the definitions of the type of requirements that apply specifically to food from non-EU countries destined for import into the EU, in addition to indicating the legal basis that EU most commonly uses for their establishment.

18.3.1 Import requirements

As referred to above, Article 11 GFL requires food and feed imported into the EU to comply with general EU safety requirements. However, depending on the type of product (e.g. animal or non-animal origin), the risk that may be associated with them and the country of origin, the EU requires compliance with specific import requirements at EU level.

Import requirements may take the form of:

a. import conditions, i.e. requirements that food and feed destined for EU import must fulfil prior to their dispatch from the exporting country. The most recurrent conditions foreseen in EU legislation for import purposes are (1)

[916] See, for instance, Regulation 1760/2000 of 17 July 2000 establishing a system for the identification and registration of bovine animals and regarding the labelling of beef and beef products and Regulation 1337/2013 of 13 December 2013 laying down rules for the application of Regulation 1169/2011 as regards the indication of the country of origin or place of provenance for fresh, chilled and frozen meat of swine, sheep, goats and poultry.

official certificates or declarations (e.g. health certificate[917]) generally verified and signed by the competent authorities of the exporting country; and (2) report of results of analytical tests performed prior to export by an accredited laboratory in the exporting country that demonstrate compliance with EU requirements; and/or

b. import controls, i.e. control activities performed on consignments of food and feed from non-EU countries by staff of competent authorities of EU Member States. Import controls may be carried out at EU borders or within the EU territory at any time and stage of the food chain. They usually consist of documentary, identity and physical checks. The modalities for the performance of such controls, including their frequency, may vary depending on the nature of the imports (e.g. animal or non-animal origin) and the risk that they present (see further Sections 18.4 to 18.6).

18.3.2 Risk identification

The establishment of import requirements by the EU is generally triggered by the presence of known risks or emergence of new ones. Risks may be identified through a number of information sources. Diagram 18.2 lists the most common sources of information the EU relies upon for identifying risks associated with imports.

Diagram 18.2. Information sources most commonly used by the EU for the identification of risks associated with imports.

- Food incidents, outbreak of animal diseases or plant pests
- Notifications of non-compliant imports made by competent authorities of EU Member States through the Rapid Alert System for Food and Feed (RASFF), EUROPHYT or the Animal Disease Notification System (ADNS)
- Audits performed in non-EU countries by the Food and Veterinary Office (FVO) of the European Commission to evaluate the effectiveness of official controls on food and feed destined to the EU market
- Trade relevance (i.e. import volumes)
- New scientific or technical evidence by the European Food Safety Authority (EFSA) or any other national or international scientific body or organisation
- Other risk-related information communicated by competent authorities of EU Member States in other ways

[917] As regards certificates/declarations, it must be noted that terminology used in EU legislation may vary. Nevertheless, three main types can be identified: sanitary or health certificates (certifying compliance with public heath requirements), veterinary (certifying compliance with animal health requirements) or phytosanitary (certifying compliance with plant health requirements).

18.3.3 Legal basis for import requirements

Today, the EU relies on a wide range of legal tools for the establishment of the most appropriate import requirements. These provisions are currently scattered through several EU legal acts applying across all food categories (i.e. GFL and to some extent Regulation 882/2004) or to specific areas (e.g. food and feed of animal origin or of non-animal origin). Diagram 18.3 provides an overview, per product category, of the provisions on which EU import requirements are usually based.

Diagram 18.3. Overview table of provisions for the legal basis of EU import requirements.

Type of imports/ Import requirements	Import conditions	Import controls	Emergency measures
Food and feed of animal origin	Art. 7 Dir. 97/78 (veterinary certificate)	Art. 3-21 Dir. 97/78; Art. 14 Reg. 882/2004	Art 53 GFL and Art. 22 Dir. 97/78
Food and feed of non-animal origin (sanitary aspects)		Art. 15-17 Reg. 882/2004	Art. 53 GFL
Food and feed of non-animal origin (phytosanitary aspects)	Art. 13 Dir. 2000/29 (phytosanitary certificate, when applicable)	Art.12 and Art. 13e Dir. 2000/29	Art. 16 Dir. 2000/29

18.3.4 Emergency measures

With regard to emergency measures, these are legislative acts that the EU typically adopts when facing serious risks that may stem from domestic or imported products. For this reason, they generally lay down stringent requirements to the extent that is necessary to prevent, reduce or eliminate said risks. In relation to imports, Article 53(1)(b) GFL foresees that:

> *Where it is evident that food or feed [...] imported from a third country is likely to constitute a serious risk to human health, animal health or the environment, and that such a risk cannot be contained satisfactorily by means of measures taken by Member State(s) concerned, the Commission, acting in accordance with the procedure provided for in Article 58(2) on its own initiative or at a request of a Member State, shall immediately adopt one or more of the following measures, depending on the gravity of the situation:*
>
> *[...]*

(b) in the case of food and feed imported from a third country:
(i) suspension of imports of the food or feed in question from all or part of the third country concerned and, where applicable, from the third country of transit;
(ii) laying down special conditions for the food or feed in question from all or part of the third country concerned;
(iii) any other appropriate interim measures.

The broad scope of this provision 53(1)(b) enables the EU to design a wide range of risk-management scenarios to deal with imports, including trade bans, conditions to be met prior to export to the EU and/or controls to be performed before import. For this reason, Article 53(1)(b) GFL is often used when food scares occur. Article 22 Directive 97/78 on the organisation of veterinary official controls from non-EU countries[918] and Article 16 Directive 2000/29 on preventive measures for plant health[919] provide the EU with similar powers in the respective policy areas.

18.3.5 Other legal basis

Specific categories of food and feed

Article 48 Regulation 882/2004 empowers the EU to establish import requirements for specific categories of food and feed (including prior approval of establishment, import certificates and other import conditions), whenever detailed rules and procedures have not been established at EU level. Used so far to a relatively limited extent, import requirements developed in accordance with this provision are intended to be permanent. The requirements for import of sprouts and seeds for sprouting that the EU introduced in the aftermath of the EHEC crisis are based on this provision (see Section 18.6.1).

Legal basis for easier EU market access

Article 23 Regulation 882/2004 provides a mechanism for ensuring easier access to the EU market. It consists of the approval by the EU of the pre-export checks performed by the competent authorities of an exporting country. In order to obtain this approval, non-EU countries with an interest must submit a request to the European Commission. The approval is generally granted following a favourable FVO audit. Approval of pre-export checks has, as a consequence the reduction of the frequency of import controls on the EU side. To date, only the USA (for peanuts

[918] Council Directive 97/78 of 18 December 1997 laying down the principles governing the organisation of veterinary checks on products entering the Community from third countries.
[919] Council Directive 2000/29 of 8 May 2000 on protective measures against the introduction into the Community of organisms harmful to plants or plant products and against their spread within the Community.

and derived products) and Canada (for wheat and wheat flour) have requested and obtained an EU approval pursuant to this Article 23.[920]

Precautionary principle

For the sake of completeness, it should be recalled that EU decisions relating to imports may also be influenced by the application of the precautionary principle. Enshrined in Article 7 GFL, this principle enables the EU to adopt risk management measures also when scientific uncertainty persists in relation to a given risk. EU measures based on the precautionary principle are, by definition, temporary and meant to be subject to a more comprehensive risk assessment as soon as further scientific information becomes available.

18.4 Specific import requirements according to food category

In addition to the EU general food requirements discussed in Section 18.2, imports must also comply with specific import requirements. These latter substantially vary depending on the food category to which the product belongs. For this purpose, EU food law identifies the three following import regimes:

- food and feed of animal origin;
- food and feed of non-animal origin;
- composite foods.

Diagram 18.4 provides an overview of the specific import requirements per food category.

18.4.1 Food and feed of animal origin

Food and feed of animal origin covers products that have been derived from animals or coming from animals, whether processed or not. In certain cases, this may include live animals (e.g. lobsters or live bivalve molluscs) that are placed on the market for consumption. Import of food and feed of animal origin (e.g. meat, fish, honey, eggs, milk, derived products and any other product containing them) in the EU is strictly regulated. As a rule, EU import is allowed only if products of animal origin do not pose any risk for public or animal health. To this end, EU veterinary legislation foresees an import system consisting of positive lists of:

- non-EU countries that are authorised to export to the EU; and
- approved establishments located in said countries authorised for export to the EU.

[920] Commission Decision 2008/47 of 20 December 2007 approving the pre-export checks carried out by the United States of America on peanuts and derived products thereof as regards the presence of aflatoxins; Commission Implementing Regulation 844/2011 of 23 August 2011 approving the pre-export checks carried out by Canada on wheat and wheat flour as regards the presence of ochratoxin A.

Diagram 18.4. Overview table of specific import requirements per food category.

Type of imports/ Import requirements	Food of animal origin	Food of non-animal origin	Composite foods
Country authorisation to export to EU	Required	Not required	Regimes 1 and 2: required Regime 3: not required
Establishment approval	Required	Not required	Regime 1: required Regimes 2 and 3: not required
Certificate required	Veterinary certificate	Phytosanitary certificate is required for products listed in Annex V, Part B, Dir. 2000/29	Veterinary certificate for Regimes 1 and 2 Regime 3: not required
Prior notification	Common Veterinary Entry Document (CVED)	**Sanitary** Common Entry Document (CED) needed for products listed in Annex I Reg. 669/2009	CVED for products listed in Art. 4 Decision 2007/275
Mandatory presentation of consignment (at the entry point in the EU)	Yes (at the Border Inspection Post (BIP))	**Sanitary** Yes – for products listed in Annex I Reg. 669/2009 (at the Designated Points of Entry (DPE)) **Phytosanitary** Yes – for products listed in in Annex V Part B Dir. 2000/29 (at the Point of Entry (PoE))	Yes for products listed in Art. 4 Decision 2007/275 (at the BIP)

Country authorisation

In order to be able to export products of animal origin to the EU, non-EU countries must submit an official request to the European Commission. Following such an application, the Commission usually requires the competent authorities of the concerned non-EU country to provide guarantees that may be relevant for this purpose. For instance, these may include evidence regarding:

- the national legislation currently in force or under development;
- the reliability of the national system for official controls;

- the existence of sound risk-assessment and risk-management procedures;
- the membership or affiliation to international organisations (e.g. World Organisation for Animal Health – OIE).

Country authorisation to export to the EU usually depends on the successful outcome of an FVO audit. The promptness shown in complying with the recommendations that the FVO may formulate following the audit is also taken into account when assessing the guarantees provided by the country requesting the authorisation.

Establishment approval

In addition to the country authorisation, EU legislation may require that FBOs located in non-EU countries that want to export to the EU have their establishments approved by their national authorities. When the approval is granted, the establishment is listed amongst those that are allowed to export. The establishment may be subject to inspection on a regular basis in order to verify that compliance is ensured consistently over time. Competent authorities of non-EU countries are required to maintain lists of approved establishments up-to-date and keep the European Commission duly informed of any relevant change. The Commission in turn makes such lists publicly available on the internet per country and per sector.[921]

EU entry points for products of animal origin

As regards import controls, Directive 97/78 requires FBOs to introduce all consignments consisting of products of animal origin into the EU territory via Border Inspection Posts (BIPs), i.e. seaports, airports or terrestrial borders, where the relevant veterinary controls must be carried out. Operation of BIPs is subject to prior approval by the European Commission. Currently there are over 300 BIPs approved by the EU for the performance of import veterinary controls.[922]

[921] Lists of non-EU countries and establishments authorised to export to the EU are available on DG Sanco's website at the following webpage: http://ec.europa.eu/food/international/trade/third_en.htm.

[922] Approval of BIPs is, in particular, subject to FVO audit. The list of approved BIPs is available on the website of DG Health and Consumers of the European Commission (DG Sanco) at the following webpage: http://ec.europa.eu/food/animal/bips/approved_bips_en.htm. The contact details of the national BIPs are listed per Member State at: http://ec.europa.eu/food/animal/bips/bips_contact_en.htm.

Presenting food and feed of animal origin for EU import

Decision 2012/31 lists all products of animal origin that must be presented for official controls at BIPs.[923] Failure by a business operator to present the consignment subject to veterinary border controls to a BIP results in an illegal import, with the consequence that the consignment must be destroyed or re-dispatched.[924]

Furthermore, a veterinary certificate, to be issued by the competent authorities of the exporting country must always accompany the consignment when presented at a BIP. This certificate attests compliance of the products of animal origin with the legislation of the exporting country and that of the importing Member State(s).[925]

Arrival of consignments at BIP premises must be pre-notified at least one working day in advance. For the purpose of prior notification, FBOs must submit a Common Veterinary Entry Document (CVED), as laid down in Annex III to Regulation 136/2004,[926] through TRACES (TRAde Control and Expert System), the EU database for the monitoring of import and intra-EU trade of live animals and other products of animal origin.[927] TRACES makes available models of all veterinary certificates required by EU legislation in all EU official languages.

18.4.2 Food and feed of non-animal origin

Designed by EU legislation as a residual category, food and feed of non-animal origin includes e.g. fruits, vegetables, spices, certain bakery products and drinks. When considering certain imports of non-animal origin (e.g. fruit, vegetables, spices), it should be borne in mind that they may be carriers of sanitary and phytosanitary risks (thus a threat to health of humans as well as plants and plant products). For example, a FBO willing to export oranges to the EU must ensure that, when dispatched, such fruits neither contain pesticide residues in excess of maximum permitted residue limits nor show symptoms of plant pests. Therefore, for certain imports of non-animal origin, sanitary (a) and phytosanitary (b) aspects must be considered.

---

[923] Commission Implementing Decision of 21 December 2011 amending Annex I to Decision 2007/275 concerning the lists of animals and products to be subject to controls at border inspection posts under Council Directives 91/496 and 97/78.

[924] Article 17(1) Directive 97/87. For a definition of re-dispatching refer to Section 18.5.3 'Risks and Consequences of non-compliance'.

[925] It must be noted that EU veterinary legislation foresees specific import conditions for several product categories, such as fresh meat, meat products, poultry meat, game meat, aquaculture products and other products of animal origin.

[926] Commission Regulation of 22 January 2004 laying down procedures for veterinary checks at Community border inspection posts on products imported from third countries.

[927] TRACES can be accessed via the following webpage: https://webgate.ec.europa.eu/sanco/traces/jsp/index.jsp.

Sanitary aspects

Whereas imports of animal origin are subject to strict border surveillance having to be presented mandatorily to BIPs, imports of non-animal origin can enter the EU with relative freedom. As a rule, these products do not have to meet specific import conditions or be subject to import controls. They must however comply with the general food requirements applicable to all food (Section 18.2) and remain subject to official controls that competent authorities of EU Member States perform routinely on the market.

However, whenever an emerging or a known risk for public health is associated with certain imports of non-animal origin, the EU may decide to subject them to reinforced border surveillance pursuant to Article 15(5) Regulation 882/2004 and its implementing act, which is Regulation 669/2009.[928]

Imports of non-animal origin listed in the 'negative' list laid down in Annex I[929] to Regulation 669/2009 require prior notification before arrival and presentation of the goods to a Designated Point of Entry (DPE).[930] DPEs are seaports, airports or terrestrial borders of EU Member States authorised to perform control activities on imports of non-animal origin. Failure by a business operator to present a consignment that is subject to reinforced border surveillance to a DPE results in an illegal import, with the consequence that it will have to be either destroyed or re-dispatched.[931]

FBOs are required to give prior-notification to the competent authorities of the DPE at least one working day before the physical arrival of the consignment. Prior notification must be ensured by filling in a Common Entry Document (CED), as laid down in Annex II to Regulation 669/2009. CED must be submitted in one of the official languages of the importing Member State, although the vast majority of EU Member States accept the English version of the CED. A CED can be submitted electronically or by fax. TRACES is also being adopted in some Member States for the management of data relevant to imports of non-animal origin.

[928] Regulation 669/2009 implementing Regulation 882/2004 of the European Parliament and of the Council as regards the increased level of official controls on imports of certain feed and food of non-animal origin.
[929] This list is updated every three months, mainly on the basis of the results of official controls performed by EU Member States and other information sources listed in Diagram 18.2.
[930] Currently there are over 220 DPEs across the EU. The list of national DPEs and their contact details are available on DG Sanco's website at the following page: http://ec.europa.eu/food/food/controls/increased_checks/list_DPE_en.htm.
[931] Article 19(2)(b) Regulation 882/2004.

Phytosanitary aspects

The EU phytosanitary regime for import of plant products is currently laid down in Directive 2000/29. This allows entry of all plant products originating from non-EU countries without specific import restrictions except those that are included in the 'negative' list laid down in Annex V, Part B of Directive 2000/29. Of course, general phytosanitary aspects (i.e. mainly the absence of harmful organisms and pests) will still have to be complied with.

Plant products listed in the negative list provided for by Directive 2000/29 must undergo official controls before EU import in order to ensure that they are free from quarantine organisms and practically free from any other harmful organisms. To this end, they must be presented to Points of Entry (PoEs), i.e. seaports, airports and terrestrial borders in EU Member States for plant health controls. Consignments of listed plant products must be accompanied by a phytosanitary certificate (required format in Annex VII to Directive 2000/29). This certificate must be issued by the competent authorities of the exporting country in at least one of the official languages of the EU, it must have a limited legal validity (14 days from the date of issuing), and must attest that plant products to be exported to the EU:

- underwent official controls prior to export with a favourable outcome;
- are free from harmful organisms prohibited in the EU;
- comply with the national legislation of the importing EU Member State.

Once imported, plant products may be subject to official controls that competent authorities of EU Member States perform on the market. These controls are aimed at verifying compliance with the requirements of Directive 2000/29 and performed at random at any place in the product chain.

18.4.3 Composite products

Under EU law, composite products are defined as food products that contain both processed products of animal origin and products of non-animal origin.[932] Regulation 28/2012[933] lays down three different import regimes for certain composite foods:

Regime 1. The first regime covers composite products that contain processed meat or milk products as well as composite products where processed fish or

[932] Article 2(a) Commission Decision 2007/275 of 17 April 2007 concerning lists of animals and products to be subject to controls at border inspection posts under Council Directives 91/496/EEC and 97/78/EC.

[933] Commission Regulation 28/2012 of 11 January 2012 laying down requirements for the certification for imports into and transit through the Union of certain composite products and amending Decision 2007/275 and Regulation 1162/2009.

eggs products constitute the primary ingredient (i.e. represent 50% or more of the whole food); such composite products must:

a. originate from a non-EU country that is authorised for EU export and from an establishment which, located in that country, is approved to that effect; and
b. be accompanied by a veterinary certificate as laid down in Annex I to Regulation 28/2012 (Article 3(1) and (2)).

Regime 2. The second import regime applies to all other composite products in which the primary ingredient is a processed product of animal origin, which must:

a. originate from a non-EU country that is authorised for EU export; and
b. be accompanied by the veterinary certificate that EU legislation requires for the import of the products of animal origin in question (Article 3(3)).

Regime 3. For other composite products not falling under the categories above, there are no specific import requirements other than the general ones applying to all foods (see Section 18.2).

Presenting composite products for EU import

In terms of import controls, as a rule, composite products containing processed meat products and those where a product of animal origin represents the primary ingredient must be presented to BIPs for veterinary controls and require prior-notification of the consignment.[934] Composite products such as biscuits, confectionery, chocolate, pasta and noodles without meat or food supplements containing small amounts of meat products are not subject to mandatory veterinary controls at BIPs.[935]

18.5 Official controls

Once the consignment and accompanying documents have been presented at the appropriate EU entry point, official controls take place. This section illustrates how official controls on food imports are organised and carried out.

18.5.1 Types, frequency, place, costs

Types of official controls

Regulation 882/2004 provides a general framework for official controls, including definitions of the various control activities that may be conducted to verify compliance with EU legislation. These activities may include documentary, identity and physical checks. Diagram 18.5 portrays the relevant definitions as laid down in Article 2 Regulation 882/2004.

[934] Article 4 Decision 2007/275.
[935] Article 6 and Annex II Decision 2007/275.

Diagram 18.5. Definitions of official control activities.

Documentary check Article 2(17)	'means the examination of commercial documents and, where appropriate, of documents required under feed or food law that are accompanying the consignment'
Identity check Article 2(18)	'means a visual inspection to ensure that certificate or other documents accompanying the consignment tally with the labelling and the content of the consignment'
Physical check Article 2(19)	'means a check on the feed or food itself which may include checks on the mean of transports, on the packaging, labelling and temperature, the sampling for analysis and laboratory testing and any other check necessary to verify compliance with feed or food law'

Similar definitions may be found in EU veterinary or phytosanitary legislation.[936] EU Member States must ensure that staff performing official controls are suitably qualified (e.g. veterinarians in case of products of animal origin) and provided on a regular basis with adequate training in their area of competence (Article 6 Regulation 882/2004).

Although rules for official controls are largely harmonised at EU level by Regulation 882/2004, differences in their implementation may still exist in practice at the national level. Overall, EU Member States are free to organise the official controls and the staff performing them as they deem appropriate. Control tasks may be therefore carried out by different government branches or departments (e.g. Ministry of Agriculture, Health, Trade and Industry, etc.).

Frequency of official controls

Frequency of official controls may vary substantially depending on the type of products destined for EU import.

With regard to imports of animal origin, as a rule, official controls are systematic, meaning that documentary, identity and physical checks are performed on all incoming consignments.[937] However, reduced frequencies for physical checks may be foreseen for certain imports from certain non-EU countries, usually based on assurances provided by the competent authorities of the exporting country and results of previous official controls.

On the other hand, should a serious or repeated violation of EU requirements be ascertained during controls performed at one BIP, all the other BIPs will be

[936] Article 4(3) Directive 97/78 or Article 13a(2) Directive 2000/29.
[937] Article 4(3) and 7(2) Directive 97/78.

informed through TRACES. As a result, all concerned BIPs may perform reinforced sampling and analysis on consignments with the same origin.[938]

For imports of non-animal origin included in the negative list laid down in Annex I to Regulation 669/2009, documentary checks must be performed on all consignments, while the specific frequency of identity and physical checks is set in that Annex for each import.[939] As regards phytosanitary checks, official controls in this area are generally systematic. However, EU phytosanitary legislation may foresee reduced control frequencies for physical checks, under certain circumstances.[940]

When there is suspicion of non-compliance in a consignment intended for EU import or doubts about its actual destination, Article 18 Regulation 882/2004 sets out a general obligation for authorities to perform certain controls. EU veterinary and phytosanitary legislation also foresee similar obligations in such cases for their respective sectors.[941]

Place of official controls

EU legislation foresees different solutions with respect to the place where official controls are performed, depending on the products to be imported. As a rule, official controls pursuant to EU veterinary legislation are conducted at the premises of BIPs. In relation to imports of non-animal origin listed included in the negative list in Annex I to Regulation 669/2009, the relevant controls are generally performed at the premises of DPEs, unless transitional arrangements allow otherwise (e.g. inland control points). Compliance with phytosanitary requirements is normally verified at the premises of PoEs, although in certain cases, performance of physical checks may be authorised to be carried out at an inland location.

Cost of official controls

Fees for official controls may be charged to FBOs based on Regulation 882/2004. However, in certain cases (e.g. when an increased level of official controls needs to be performed due to specific EU import requirements), EU legislation requires the mandatory recovery of the costs due to the official controls.[942] Since EU legislation does not fully harmonise fees, the actual costs charged by different Member States may differ.

---

[938] Article 24(1) Directive 97/78.

[939] Article 8(1) Regulation 669/2009.

[940] Commission Regulation 1756/2004 of 11 October 2004 specifying the detailed conditions for the evidence required and the criteria for the type and level of the reduction of the plant health checks of certain plants, plant products or other objects listed in Part B of Annex V to Directive 2000/29.

[941] E.g. Article 22 Directive 97/78.

[942] See, for instance, Article 14 Regulation 669/2009, but also Article 10 Regulation 1152/2009 and Article 6 Regulation 1135/2009. These two measures are presented under Section 18.6.1 and 18.6.2.

18.5.2 Procedural guarantees

Regulation 882/2004 provides FBOs that are subject to official controls with some procedural rights of general nature. When laboratory tests are performed on a consignment, with a view to ensuring transparency and reliability of official controls, competent authorities of EU Member States must guarantee the right for the FBO to carry out counter-analysis (so-called 'supplementary expert opinion') and, to this end, to perform sampling (Article 11 Regulation 882/2004).

Whenever controls are carried out, competent authorities may provide a report detailing the purpose of the control, the control methods applied, the results and, where appropriate, the actions to be taken. However, in cases when non-compliance is established, the report must obligatorily be provided to the FBO (Article 9 Regulation 882/2004). This must contain the official notification of the decision, provide appropriate justification as well as information on rights to appeal, applicable procedures and time limits (Article 54(3) Regulation 882/2004).

18.5.3 Risks and consequences of non-compliance

If non-compliance is detected following an official control at the import stage, the competent authorities that have performed said control must place the consignment under official detention. Subsequently, they are required to take one of the following enforcement actions:
- where possible, change the intended use of the consignment, e.g. importing them as feed instead of as food;
- where possible, subject the consignment to special treatment (e.g. decontamination) in order to ensure it is in line with EU legislation;
- re-dispatch of the consignment to the country of origin or to another non-EU country provided that the relevant authorities of the country of destination are informed of the reasons for non-compliance in the EU;
- destruction of the consignment.

There are cases when EU law allows onward movement of the products to their final destination while the results of official controls are still pending; in such cases, if non-compliance is established, the products would have to be first withdrawn/recalled from the market before any of the measures referred to above is taken (Article 19(1)(b) Regulation 882/2004).[943]

[943] Article 17 Directive 97/78 and Article 13(c)(7) Directive 2000/29.

When the competent authorities of an EU Member State do not allow the introduction of a consignment into the EU territory, they must alert:

- the European Commission and the other Member States through the RASFF in case of food presenting a serious risk for human health, or, when listed animal diseases or plant pests are detected through ADNS or EUROPHYT, respectively;
- customs authorities.

18.5.4 Sanctions

Whenever non-compliance with one or more of requirements applicable to imports is established, the competent authorities of EU Member States are required to take the necessary enforcement actions. Enforcement measures may also include the application of sanctions. In this regard, Article 55 Regulation 882/2004 requires EU Member States to have in place:

- systems for sanctioning violations of relevant food and feed law requirements, and
- sanctions that are effective, proportionate and dissuasive.

In essence, depending on the gravity of the violation, sanctions foreseen at national level may range from fines to the suspension of the relevant commercial activity or withdrawal of licences. For the most serious infringements (i.e. criminal offences), business operators may be also sanctioned with imprisonment.

18.5.5 Release for free circulation and role of customs

Whenever results of official controls are favourable, the consignment can be presented for import clearance to the competent customs office. This may be the customs office of the point of entry or of the final destination of the consignment, depending on the import regime that applies.

Import clearance or 'free circulation in the EU' is granted by the competent customs office upon presentation of:

- a customs declaration (i.e. Single Administrative Document), which is common for the whole EU, either electronically, in writing or by lodging it directly at the competent customs office; and,
- where applicable, relevant documents required by EU legislation (e.g. sanitary, veterinary or phytosanitary certificates) and others that attest the successful completion of official controls (e.g. CVED, CED).

To grant entry, EU legislation may foresee the collection of applicable import duties and the verification of compliance with applicable commercial measures (e.g. quota restrictions).

When import clearance is granted, a customs clearance document is issued to the applicant and the products are released for free circulation, meaning that they enjoy free movement within the EU territory to the same extent as products of EU origin.

18.6 Emergency measures

Emergency measures are risk management tools that the EU resorts to when facing serious risks for human, animal or plant health. As mentioned earlier, they may concern virtually any type of food and feed (domestic or imported; of animal, non-animal origin as well as composite foods). Intended to be of limited duration and to be maintained until the risk persists, in certain cases they may well apply for several years and become *de facto* permanent.

The legal basis for adopting emergency measures is analysed in Section 18.3.4, whilst the information sources listed in Diagram 18.2 generally provide the evidence that justifies their introduction.

Emergency measures on import may apply to:
- all non-EU countries;
- a group of non-EU countries;
- a specific non-EU country.

Examples for each of these categories are given in the following sections.

18.6.1 Import requirements applicable to all non-EU countries

Sprouts and seeds for sprouts

In 2011 the EU faced the 'EHEC crisis' which affected thousands of consumers in Germany and France causing the death of almost 50 individuals.[944] Investigations led by the European Commission traced back the origin of the outbreak to a consignment of fenugreek seeds from Egypt. Following the adoption of a temporary ban on certain imports from that country,[945] early in 2013 the European Commission

[944] On the EHEC crisis and EU measures that followed, see M. Rodriguez Font, *The 'Cucumber Crisis': legal gaps and lack of precision in the risk analysis system in food safety*, Rivista di diritto alimentare, VI, 2, 2012, pp. 1-15.

[945] Commission Implementing Decision 402/2011 of 6 July 2011 on emergency measures applicable to imports of fenugreek seeds and certain seeds and beans from Egypt as modified by subsequent Commission Implementing Decision 718/2011 of 28 October 2011.

introduced permanent import conditions[946] for sprouts and seeds for sprouting from all non-EU countries through Regulation 211/2013.[947]

According to Regulation 211/2013, as of 1 July 2013, consignments of concerned products must be accompanied by a certificate, annexed to the Regulation, attesting production in compliance with:

- general hygiene standards for primary production as per Part I Annex I to Regulation 852/2004;[948]
- traceability requirements set out in Regulation 208/2013;[949]
- requirements for the approval of establishments laid down in Regulation 210/2013;[950]
- microbiological criteria set in Annex I to Regulation 2073/2005[951] as modified by Regulation 209/2013.[952]

18.6.2 Import requirements applicable to a group of non-EU countries

Aflatoxins

Following the adoption of emergency measures applicable to individual non-EU countries for possible aflatoxin contamination in the late 1980s, the EU eventually regrouped all the products under the same import regime. Regulation 1152/2009[953] currently applies to a wide range of dried fruits (e.g. pistachios, nuts, figs, peanuts) and products that contain them intended for human consumption originating

[946] While the initial ban (Decision 402/2011) was adopted a temporary measure, Regulation 211/2013, based on Article 48 Regulation 882/2004, sets import requirements that are meant to be permanent. For this reason, legally speaking, it cannot be considered a fully-fledged emergency measure.
[947] Commission Regulation 211/2013 of 11 March 2013 on certification requirements for imports into the Union of sprouts and seeds intended for the production of sprouts.
[948] Regulation 852/2004 of the European Parliament and of the Council of 29 April 2004 on the hygiene of foodstuffs.
[949] Commission Implementing Regulation 208/2013 of 11 March 2013 on traceability requirements for sprouts and seeds intended for the production of sprouts.
[950] Commission Regulation 210/2013 of 11 March 2013 on the approval of establishments producing sprouts pursuant to Regulation 852/2004 of the European Parliament and of the Council.
[951] Commission Regulation 2073/2005 of 15 November 2005 on microbiological criteria for foodstuffs.
[952] Commission Regulation 209/2013 of 11 March 2013 amending Regulation 2073/2005 as regards microbiological criteria for sprouts and the sampling rules for poultry carcasses and fresh poultry meat.
[953] Commission Regulation 1152/2009 of 27 November 2009 imposing special conditions governing the import of certain foodstuffs from third countries due to contamination risk by aflatoxins and repealing Decision 2006/504.

from various countries (e.g. Brazil, China, Egypt, Iran, Turkey). For the purpose of EU import, the following documents are required:

- health certificate signed and stamped by an authorised representative of the competent authorities of the exporting country guaranteeing, inter alia, compliance with EU legislation;
- results of analytical tests performed prior to export that detail sampling and analytical method followed while demonstrating compliance with relevant EU requirements.

FBOs are required to submit a CED as prior notification of the physical arrival of the consignment to a DPE, where control activities to be carried out under Regulation 1152/2009 take place. At the DPE, the consignment is subject to systematic documentary checks, and to identity and physical checks depending on the control frequency foreseen for each import listed.[954] Traceability of the consignments must be ensured by means of a code that must appear on the consignment itself and on all documents accompanying it.

18.6.3 Import requirements applicable to one specific non-EU country

China and products of animal origin

Products of animal origin from China are at the time of writing (2014) subject to emergency measures under Decision 2002/994. [955] This latter contains positive lists of products that are authorised for EU export. Whilst a few products can be exported without meeting specific requirements (e.g. wild fish and gelatine), others (e.g. meat, honey, shrimps and aquaculture products) must be accompanied by a veterinary certificate and report of results of analytical tests performed prior to export that demonstrate compliance. Products covered by this measure must be tested for possible presence of chloramphenicol and nitrofuran, two veterinary drugs whose residues can be harmful to humans, and, in case of aquaculture products, for malachite green and crystal violet.

[954] Identity and physical checks take place at the premises of a Designated Point of Import (DPI). This may be a DPE or a control point located inland.

[955] Commission Decision of 20 December 2002 concerning certain protective measures with regard to the products of animal origin imported from China, OJ L 348, 22.12.2002, p. 154. The decision was amended several times and lastly by Commission Implementing Decision 2012/482.

Melamine contamination

Following the adulteration of certain dairy-based products with melamine in China in 2008,[956] the EU adopted Regulation 1135/2009.[957] This Regulation prohibits import into the EU of milk, milk products, soy and soy products from China when destined for the nourishment of infants and young children. Also, it lays down specific import requirements for the following products of Chinese origin:

- ammonium bicarbonate (for human and animal consumption);
- feed and food containing milk, milk products, soy and soy products other than those that are subject to the import prohibition.

FBOs must give prior notification of the arrival of relevant consignments to the appropriate Control Points (CPs) designed for this purpose by EU Member States. Consignments must be accompanied by a report of analytical tests attesting that the products do not exceed the maximum levels set by EU legislation for melamine. At CPs, consignments must undergo systematic documentary checks, whilst identity and physical checks must be performed at 20% control frequency on all arriving consignments. Whenever laboratory tests performed at import indicate that the maximum levels of melamine have been exceeded, the consignment is to be refused entry into the EU and safely disposed of.

Fukushima

Following the incident that occurred in Japan at Fukushima nuclear plant on 11 March 2011, the EU promptly introduced emergency measures to prevent all potentially contaminated food and feed originating from Japan to enter its territory.[958] Three years after the incident, the implementation of strict surveillance both prior to export and at EU borders ensured effective containment of the risk involved (i.e. contamination with Cs-134 and Cs-137). For this reason, with effect from 1 April 2014, Regulation 322/2014[959] has foreseen an overall relaxation of

---

[956] For an account of the melamine case see: M.A. Pagnattaro, E.R. Peirce, *From China to your plate – An analysis of new regulatory efforts and stakeholder responsibility to ensure food safety*, The George Washington International Law Review, 42, 2010, pp. 1-55; See also: A. Alemanno, *L'approche européenne de la sécurité des importations – Concilier protection des consommateurs et accès au marché après l'affaire du lait chinois frelaté*, Revue du Droit de l'Union Européenne, 3, 2010, pp. 527-548.

[957] Commission Regulation 1135/2009 of 25 November 2009 imposing special conditions governing the import of certain products originating in or consigned from China, and repealing Commission Decision 2008/798.

[958] For an analysis of the first measures adopted by the EU following the Fukushima nuclear accident, see V. Paganizza, *Fukushima, RASFF and ECURIE – Condizioni speciali per l'importazione di alimenti e mangimi provenienti dal Giappone dopo l'11 marzo 2011*, in Rivista di diritto alimentare, V, 2011, pp. 1-13; and R. O'Rourke, *EU Measures on the safety of food imports from Japan following the nuclear accident at Fukushima*, in European Journal of Risk Regulation, 2012, pp. 82-86. See also Chapter 11.

[959] Commission Implementing Regulation 322/2014 of 28 March 2014 imposing special conditions governing the import of feed and food originating in or consigned from Japan following the accident at the Fukushima nuclear power station.

import requirements. This emergency measure applies to all products originating from or consigned by Japan, with a few exceptions (e.g. products harvested/processed before 11 March 2011, certain alcoholic beverages, food and feed intended for personal consumption). The emergency measure is revised periodically at each growing season.

FBOs must give prior notification of the arrival of the consignment to the relevant BIP (for imports of animal origin) or DPE (for imports of non-animal origin), by using the CED or the CVED as appropriate. All consignments must be accompanied by a declaration signed by the Japanese competent authorities attesting the prefecture of provenance of the products as well as their compliance with Japanese legislation. In addition, certain products – e.g. those listed in Annex IV to the emergency measure – must also be accompanied by a report of results of analytical tests performed prior to export.

At their arrival at the DPE/BIP, every consignment undergoes documentary checks, while identity and physical checks are performed only at random. Non-compliant products must be disposed of or returned to Japan.

18.7 Future developments

The European Commission has been evaluating the present policy on controls on imports of food, feed, animals and plants since October 2010.[960] As a result of this evaluation, two major areas of improvement have been identified:
1. better prioritisation of the risks associated with imports; and
2. greater consistency across the policy areas where import controls take place.

A package of legislative proposals, including a reviewed framework for official controls, is currently under scrutiny by the European Parliament and the Council.[961] Proposed changes to the present system include the following:
- Border Control Posts (BCPs) would replace the existing BIPs in the veterinary areas, PoEs in the plant health area and DPEs responsible for import surveillance of products of non-animal origin;
- one single document, the Common Health Entry Document (CHED), would be used for pre-notification of imports requiring surveillance at EU borders and to be submitted through TRACES;

[960] Report from the Commission to the European Parliament and the Council on the effectiveness and consistency of sanitary and phytosanitary controls on imports of food, feed, animals and plants, 21.10.2010, COM(2010) 785.
[961] Under the name of 'Smarter rules for safer food', the package of the four proposals and relevant impact assessments is available on DG Sanco's website at http://ec.europa.eu/dgs/health_consumer/pressroom/animal-plant-health_en.htm.

- for imports requiring surveillance at EU borders, documentary and identity checks would be carried out on all incoming consignments, while the frequency of physical checks would be more risk-based;
- in case of intentional infringements of food law requirements (i.e. food frauds), sanctions should at least offset the economic gain sought through the fraudulent behaviour.

18.8 Overview

The decision tree in Diagram 18.6 provides an overview of the most important rules discussed in this chapter. It may assist in identifying requirements that must be met in case of imports of certain food products into the EU.

Diagram 18.6. Decision tree for food import requirements.

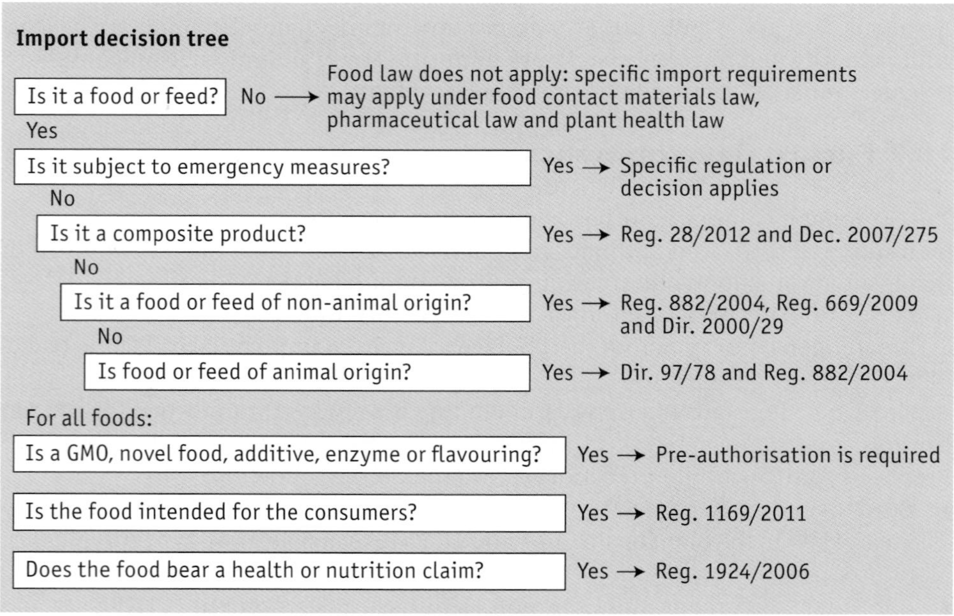

Import decision tree

Question	Answer	Consequence
Is it a food or feed?	No →	Food law does not apply: specific import requirements may apply under food contact materials law, pharmaceutical law and plant health law
Yes ↓		
Is it subject to emergency measures?	Yes →	Specific regulation or decision applies
No ↓		
Is it a composite product?	Yes →	Reg. 28/2012 and Dec. 2007/275
No ↓		
Is it a food or feed of non-animal origin?	Yes →	Reg. 882/2004, Reg. 669/2009 and Dir. 2000/29
No ↓		
Is food or feed of animal origin?	Yes →	Dir. 97/78 and Reg. 882/2004
For all foods:		
Is a GMO, novel food, additive, enzyme or flavouring?	Yes →	Pre-authorisation is required
Is the food intended for the consumers?	Yes →	Reg. 1169/2011
Does the food bear a health or nutrition claim?	Yes →	Reg. 1924/2006

19. Food contact materials

Karola Krell Zbinden

19.1 Introduction

In order to take a sufficiently comprehensive and integrated approach to food safety, there should be a broad definition of food law covering a wide range of provisions with a direct or indirect effect on the safety of food and feed, including provisions on materials and articles in contact with food ... (Recital 11 Regulation 178/2002).[962]

Ensuring food safety does not stop at testing the food itself. Everything that comes in contact with food as it is produced, packaged, transported, stored, prepared and consumed also needs to be safe. Food contact materials (FCMs) are materials and articles intended to come into contact with foods such as: packaging materials and containers, cutlery and dishes and domestic appliances; processing machines, and adhesives and inks for printing labels. The safety of such materials relies on ensuring that during contact there is no migration of unsafe levels of chemical substances from the material to the food.

Food contact material law builds the bridge between the packaging manufacturer and the food manufacturer. While the packaging manufacturer shall be responsible for the assessment of packaging for FCMs to ensure safety and full compliance of the food to be packed in the FCMs, the food manufacturer is responsible for the safety and compliance of the finished pre-packed food product. The final packaging materials involve many elements that need to be managed taking into account the food to be packaged or to come into contact with the materials. The risk of FCMs is that they can influence food safety throughout the whole value chain.

19.2 Legislation concerning food contact materials

The adopted legislative measures regarding FCMs can be divided into the following categories:

1. General measures relating to all materials and articles, in particular the Framework Regulation 1935/2004 of the European Parliament and of the Council of 27 October 2004 on materials and articles intended to come into contact with food (here-after Framework-Regulation 1935/2004) and the Commission Regulation 2023/2006 of 22 December 2006 on good manufacturing practice for materials and articles intended to come into contact with food (here-after GMP Regulation).

[962] EU Commission DG Health and Consumers. See: http://tinyurl.com/7sygkka.

2. Specific material measures regarding groups of materials and articles, e.g. plastics, recycled plastics, regenerated cellulose films and ceramics, as listed in the Framework Regulation 1935/2004.
3. Specific substance measures regarding individual or groups of substances, e.g. plastics and certain epoxy-derivatives (BADGE, BFDGE and NOGE[963]) in coatings.
4. National legislation covering groups of materials and articles for which EU legislation is not yet in place.

19.3 EU legislation

Harmonising legislation on FCMs at EU level aims to protect consumers' health and to remove technical barriers to trade. Food contact materials must not transfer their components into the foods in unacceptable quantities (migration).

19.3.1 Framework Regulation 1935/2004

The Framework Regulation 1935/2004 entered into force on the 3rd of December 2004 and provides for general principles and rules for certain groups of food contact materials and articles. Furthermore, it sets out procedures for authorisation and the establishment of a European Reference Laboratory[964] for food contact materials (Recital 23, Article 24, paragraph 3).

FCMs in the sense of the Framework Regulation are materials and articles, including active and intelligent food contact materials and articles, which in their finished state (a) are intended to be brought into contact with food; or (b) are already in contact with food and were intended for that purpose, or (c) can reasonably be expected to be brought into contact with food or to transfer their constituents to food under normal or foreseeable conditions of use (Article 1(2) Framework Regulation 1935/2004). The Regulation does not apply to materials or articles which are supplied as antiques, covering or coating materials, such as materials covering cheese rinds, prepared meat products or fruits, which form part of the food and may be consumed together with this food, as well as fixed public or private water supply equipment (Article 1(3) Framework Regulation 1935/2004).

[963] Bisphenol A diglycidyl ether (BADGE), bisphenol F diglycidyl ether (BFDGE), and novolac glycidyl ether (NOGE) are endocrine-disrupting compounds. Through the ingestion of food and beverage products tainted with these compounds the consumer is exposed to these substances. The compounds are used to make polycarbonate plastics and epoxy-based lacquers. Therefore, its ingestion is related to two food packaging types and their enclosed products: re-useable rigid containers made of polycarbonate plastic, commonly used for water bottles, baby bottles, plastic mugs, carboys, and storage containers, and metal cans with an internal epoxy-based lacquer coating; used to keep the foods or beverages from directly contacting the metal.

[964] Http://ihcp.jrc.ec.europa.eu/our_labs/eurl_food_c_m. European Commission Joint Research Centre Institute for Health and Consumer Protection, Food contact materials group, TP 260, Via E. Fermi 2749, I-21027 Ispra (VA), Italy.

The Regulation establishes 17 groups of materials and articles which may be covered by specific measures, in particular lists of substances authorised for use, purity standards for substances used, specific limits on migration (Article 5 and Annex I Framework Regulation 1935/2004). Up to now specific measures exist for active and intelligent materials, ceramics, regenerated cellulose, plastics, and recycled plastics. In the absence of such specific measures, the Member States may adopt national specific measures.

The Regulation lays down the procedure to be followed for authorisation of substances that are not included in the lists of substances to be used in specific food contact materials and articles. The applications are evaluated by the European Food Safety Authority (EFSA) (Article 7-14 Framework-Regulation 1935/2004). EFSA provides scientific advice on the safety evaluation of substances used in food contact materials including active and intelligent materials and of the recycling processes for recycled plastics used in FCM.[965] The procedures for submission of applications and the required information vary widely for each area depending on the specific legislation and applicable guidance. The Regulation establishes rules on the obligation to have a declaration of compliance for materials and articles covered by specific legislation.

The Framework Regulation establishes the general principles and provisions applicable to all FCMs:

a. Safety (Article 3(1) Framework-Regulation (EC) No 1935/2004)
   FCMs shall be safe. This requires that they shall be manufactured in compliance with 'good manufacturing practice' (GMP) so that, under normal or foreseeable conditions of use, they do not transfer their constituents to food in quantities which:
   - could endanger human health ('unsafe' migration – safety clause);
   - change the composition of the food in an unacceptable way (inertness clause); or
   - deteriorate the organoleptic characteristics of the food, e.g. taste and odour.

   These requirements do not exclude migration of substances in foods as such, but only as long as human health is not endangered. Furthermore, the Framework Regulation does not require a general effort on minimisation of migration, which means that safe migration is allowed even if it would be technically avoidable. It also points to the fact that migration as such does not necessarily lead to an unacceptable composition of the food. If, for example, FCMs unintentionally colour foods they are in contact with, these coloured foods are principally marketable as long as they are safe, the composition is still acceptable and the organoleptic characteristics are not deteriorated.

[965] Http://www.efsa.europa.eu/en/applicationshelpdesk/foodcontactmaterials.htm.

b. Labelling (Article 15 Framework-Regulation 1935/2004)
   Materials and articles, which are not yet in contact with food when placed on the market, shall be accompanied by:
   - appropriate labelling ('for food contact') or bear the symbol with a glass and fork (Diagram 19.1);[966]
   - if necessary, special instructions to be observed for safe or appropriate use;
   - the name and address of the responsible person in the EU; and
   - adequate labelling or identification to ensure traceability of the material or article.

   This information for the consumer or for food industry shall indicate the suitability of materials and articles to come in contact with food. In cases where the intention for food contact is obvious by the nature of the article e.g. knife, fork, wine glass, this labelling is not obligatory. Labelling, advertising and presentation of FCMs shall not mislead the consumer.

Diagram 19.1. The FCM symbol.

c. Declaration of compliance (Article 16 Framework Regulation 1935/2004)
   The declaration of compliance is a tool of compliance work. Throughout the manufacturing of FCMs a plurality of substances is used. Migration is unavoidable. The consumption of migrated substances in food is a risk for the consumer. The examination of conformity of a finished product without information on the substances used is almost impossible. The declaration of compliance ensures the transferral of safety-related information in the manufacturing chain.
   Thus Article 16 Framework Regulation stipulates that for FCMs covered by specific measures a written declaration stating that they comply with the rules applicable to them is required and must accompany the FCM. This is required for example for plastic materials in Article 16 Regulation 10/2011, for regenerated cellulose films in Article 6 Directive 2007/42, for the epoxy derivative BADGE and its derivatives in Article 5 Regulation 1895/2005, and for recycled plastics in Article 12 Regulation 282/2008, etc.

[966] Annex II Framework Regulation 1935/2004.

Appropriate documentation shall be available to demonstrate such compliance. That documentation shall be made available to the competent authorities on demand. This means, that the so-called 'supporting documents' are not part of the declaration of compliance. They may be kept as professional secrets and must only be disclosed to the competent authorities. However, it is not defined what these documents must comprise. It should certainly include information from compliance work of the preliminary stages (completed conformity, open questions), GMP documentation, worst case calculations, modelling of migration scenarios, analysis results of migration testing, and risk assessment
The declaration of compliance is not an independent warranty for ensured characteristics. However, it is the crucial element in the value chain for information exchange between each step. The declaration of compliance is a legal document. It should 'certify' that the packaging is suitable for food contact and it must be linked to a specification of the user. This certification should explicitly take into account the packed product. More concrete requirements on form and content of declarations of compliance are set in the specific measure Regulations and Directives, such as Regulation 10/2011 (see Section 19.3.3). For all other FCMs there are no formal requirements for a declaration of conformity. Here only the GMP documentation according to Article 7 Regulation 2023/2006 applies.
To strengthen the coordination and responsibility of the suppliers at each stage of manufacture, including that of the starting substances, the responsible persons should document the compliance with the relevant rules in a declaration of compliance which is made available to their customers. This is certainly applicable and demanded if the FCMs are sold as such. Since the traceability of materials and articles shall be ensured at all stages, this duty applies for example also to the retailer who imports FCMs and sells them to further purchasers who are not end users.
The information is mostly needed by the producer, user or converter of FCMs. To this end and subject to the requirement of confidentiality, food business operators should also be given access to the relevant information to enable them to ensure that the migration from the materials and articles to food complies with the specifications and restrictions laid down in food legislation. If the foods pre-packed in FCMs are then sold as such by a retailer to the final consumer, it does not seem appropriate that the retailer is obliged to provide a declaration of compliance. A declaration does not have to be attached to FCMs, if they are sold to the final consumer.

d. Traceability (Article 17 Framework Regulation 1935/2004)
   Business operators shall ensure the traceability of FCMs in order to facilitate control, the recall and the attribution of responsibility. This obligation stands 'with due regard to technological feasibility'.

19.3.2 GMP Regulation 2023/2006

The requirements of the GMP Regulation, applicable as from 1 August 2008, have to be applied at all stages of production of FCMs and in all sectors. The stages of production of starting substances and raw materials are excluded even if the business operators shall describe their products in the declaration of compliance.

As an example, for the plastics production chain the GMP requirements start with the plastic manufacturer, followed by the converter including the printing process of the packaging up to the production of the final article. All aspects of the GMP need to be adequately documented and the documentation should be available to control authorities. Imports from third countries should also apply adequate GMP systems in their production.

Conformity with GMP means the establishment of a quality assurance system (Article 5 GMP Regulation 2023/2006), a quality control system (Article 6), and the documentation with respect to specifications, manufacturing formulae and processing, which are relevant to compliance and safety of the finished material or article (Article 7). For two materials GMP requirements have been further detailed, i.e. for printing inks (Annex of the GMP-Regulation) and for recycled plastics (Regulation 282/2008).

19.3.3 Plastics – Regulation 10/2011

Commission Regulation 10/2011 of 14 January 2011 on plastic materials and articles intended to come into contact with food is applicable to printed or coated plastic materials and articles, printed or coated plastic multi-layer materials and articles held together by adhesives or other, plastic layers or coatings forming gaskets in caps and closures that make a set of 2 or more layers of different types of materials, and plastic layers in multi-material, multi-layer materials and articles. 'Plastic' means polymer to which additives or other substances may have been added, which is capable of functioning as a main structural component of final materials and articles (Article 3(2) Plastics Regulation).

The Regulation consists of:
- a Union list of authorised monomers and additives for use in plastics manufacture (Article 5) ;
- restrictions and specification for authorised substances including specific migration limits;
- rules on non-intentionally added substances, i.e. impurities and reaction products;
- overall migration limit;
- rules on compliance especially migration testing, listing simulants, testing conditions;

- the concept of functional barrier;
- authorisation for nano-materials before use;
- declaration of compliance and supporting documentation.

Regulation 10/2011 sets up a general overall migration limit of 10 mg of substances/ dm^2 of the food contact surface for all substances that can migrate from food contact materials to foods as well as specific migration limits (SML) for individual authorised substances fixed on the basis of toxicological evaluations. A SML is set according to the Acceptable Daily Intake or the Tolerable Daily Intake established by the Panel on Food Contact Materials, Enzymes, Flavourings and Processing Aids (CEF) of EFSA[967]. The limit is set on the assumption that every day throughout his lifetime, a person weighing 60 kg eats 1 kg of food packed in plastics containing the substance in the maximum permitted quantity.

The following Regulations amend the rules in Regulation 10/2011:
- Regulation 321/2011 bans the use of bisphenol A in plastic infant feeding bottles;[968]
- Regulation 1282/2011[969] and Regulation 202/2014[970] add new substances and amend restrictions and specifications of already authorised substances in the Union list.

Article 15 Regulation 10/2011 requires a declaration of compliance for plastic materials and articles, for products from intermediate stages of their manufacturing as well as for the substances intended for the manufacturing of those materials and articles. The declaration shall accompany the FCMs at the marketing stages other than the retail stage. The written declaration shall be issued by the business operator, which is in practice the producer or the importer. If a food business operator can be considered as the packaging importer, he needs to issue the declaration of compliance. The declaration shall permit an easy identification of the materials and shall be renewed when substantial changes in the composition or production occur that bring about changes in the migration from the materials or articles or when new scientific data becomes available. To identify the respective FSMs a reference in the delivery documents can be used. In practice it seems to be useful to limit the scope and time of the declaration. A declaration of compliance does not have to be enclosed with each delivery; it may also be deposited as long as amendments are not necessary. A declaration of compliance can be supplied in paper form or be made available on the internet, if a reference is included in the delivery documents.

[967] See http://www.efsa.europa.eu/en/panels/fip.htm.
[968] Commission Implementing Regulation 321/2011 of 1 April 2011 amending Regulation (EU) No 10/2011 as regards the restriction of use of Bisphenol A in plastic infant feeding bottles.
[969] Commission Regulation 1282/2011 of 28 November 2011 amending and correcting Commission Regulation (EU) No 10/2011 on plastic materials and articles intended to come into contact with food.
[970] Commission Regulation 202/2014 of 3 March 2014 amending Regulation (EU) No 10/2011 on plastic materials and articles intended to come into contact with food.

The declaration of compliance shall contain the information laid down in Annex IV Regulation 10/2011, in particular (1)-(4) formal information, (5) confirmation of conformity with Framework Regulation, (6) adequate information relative to substances with restrictions – in order to inform further converters in the value chain, (7) adequate information relative to dual-use substances (substances which are approved for use as plastic additives and as food additives (for example, antioxidants and emulsifiers)) – in order to be in compliance with the respective legal requirements, (8) specifications for the use of the materials (type or types of food with which it is intended to be put in contact, time and temperature of treatment and storage in contact with the food, ratio of food contact surface area to volume used to establish the compliance of the material or article) – in order to demarcate responsibility, (9) information relative to functional barriers – in the case of use of functional barriers the use of non-authorised substances is allowed, as long as a maximum level of 10 μg/kg is kept, the substances are neither mutagenic, carcinogenic or toxic to reproduction or are substances in nano size (see referral 27 of the Regulation 10/2011).

19.3.4 Active and intelligent materials

On active and intelligent packaging the Framework Regulation includes definitions and it specifies that these materials and articles may induce changes in the foodstuff, only if the food then still complies with the Union provisions applicable to food such as those on food additives. These materials and articles shall especially not be used to mask spoilage of the food and shall not mislead the consumer (Article 4 Framework-Regulation). Examples of active and intelligent materials are antimicrobial materials, bio-active materials, selective and adjusting barriers, indicating and sensing materials, flavour maintenance and enhancing materials. Only substances that are included in an EU list of authorised substances may be used in active and intelligent components.

According to Regulation 450/2009 active and intelligent materials and articles or parts thereof must be labelled (whenever they give the impression that they are edible) with the words 'DO NOT EAT' (font size of at least 3 mm), and with the active and intelligent material' symbol, where technically possible.

The EU importer must issue a declaration in which the compliance of the specific requirements of the Regulation on active and intelligent materials and the Framework Regulation are confirmed. This declaration must accompany active and intelligent materials and articles at each stage of the manufacturing process. Annex II of the Regulation stipulates further detailed requirements for the declaration of compliance, which exceed the requirements for plastic materials in Regulation 10/2011.

19.3.5 Ceramics

FCMs of ceramics are regulated by Directive 84/500 and 2005/31. Ceramic articles can transfer lead and cadmium released from decoration and/or glazing. The Directive imposes a threshold on the quantities of lead and cadmium allowed to pass into food, which depend on use and form of the articles, and gives the analytical method determining the migration of these substances.

The declaration of compliance must confirm that the specific requirements of the Directive on ceramics and the Framework Regulation are met, in particular concerning the transferral of lead and cadmium. The Directive also specifies the requirement of the supporting documentation to be made available to the national competent authorities on request. This documentation must contain the results of the analysis carried out, the test conditions and the name and the address of the laboratory that performed the testing. For other heavy metals the safety clause of the Framework Regulation applies.

19.3.6 Regenerated cellulose films

Directive 2007/42 on regenerated cellulose films (RCF) establishes a list of authorised substances (positive list) and restrictions on the composition of the material. It includes provisions for plastic-coated regenerated cellulose film. The restrictions in the positive list are usually expressed as residual content in the film because migration testing with pure cellophane film into liquid simulant is in general not feasible due to the absorption of water by the film.

19.3.7 Recycled plastics

Commission Regulation 282/2008 of 27 March 2008 on recycled plastic materials and articles intended to come into contact with foods and amending Regulation (EC) No 2023/2006 which applies to mechanical recycling of plastics, foresees the authorisation of the recycling process at Union level based on the safety evaluation of the recycling process performed by EFSA. Critical points in the recycling process are the sourcing of the material that is being recycled as well as the capacity of the process to reduce contamination. Only those plastics that respect the compositional requirements of the plastics Directive can be used as source material for mechanical recycling. As the recycling processes are unique based on the technology applied, individual authorisation dedicated to the applicant will be issued.

All recycling processes shall be accompanied by an adequate quality assurance system which should be audited by the Member States. Recycled plastic as well as the materials and articles containing recycled plastics need to be accompanied by a declaration of compliance. The Regulation also covers recycled plastics from third countries. These can only be used if the recycling process is authorised. Requests for authorisation have to be addressed to a Member State's contact point. Premises in third countries that use the authorised recycling processes have to be notified to the Commission.

19.3.8 Specific substances

For nitrosamines Commission Directive 93/11 of 15 March 1993 concerning the release of the N-nitrosamines and N-nitrosatable substances from elastomer or rubber teats and soothers sets out limits in teats and soothers made of rubber.

For certain epoxy derivatives Commission Regulation 1895/2005 of 18 November 2005 on the restriction of use of certain epoxy derivatives in materials and articles intended to come into contact with food restricts their use in coatings (NOGE, BFDGE) and sets limits in coatings (BADGE).

From 1 July 2011 kitchenware made of melamine or polyamide originating or consigned from China or Hong Kong must comply with the import rules of Commission Regulation 284/2011 of 22 March 2011 laying down specific conditions and detailed procedures for the import of polyamide and melamine plastic kitchenware originating in or consigned from the People's Republic of China and Hong Kong Special Administrative Region, China. Consignments must be notified to the competent authorities at the entry points at least 2 working days before arrival. Consignments must have a declaration and a laboratory report on the analysis of primary aromatic amines (for polyamide) and formaldehyde (for melamine).[971]

19.4 National legislation

The EU legislation on food contact materials is quite extensive. However, there are still many materials that are not yet covered by specific EU legislation, such as adhesives, inks, paper and steel (for cans). In such cases Member States' national legislation or specific industry codes of practice can apply (e.g. the European CEPE Code of Practice for food contact (metal) coatings).[972]

[971] For further guidance see: http://tinyurl.com/5vl4g3.

[972] National legislation covering groups of materials and articles for which EU legislation is not yet in place may be found at: http://tinyurl.com/ps6z29q. In such cases businesses should contact the Member State (MS) Competent Authority to ask for information on the specific requirements for preparing and submitting an application.

19.5 Further international private standards concerning FCMs

Since FCMs are part of the food chain, they need to be included in the HACCP Concepts of food businesses. ISO/EN 22000:2005 include packaging materials and ISO/EN 22005:2007 foresee the traceability of packaging materials just like for food 1 step up and 1 step down. EN 15593:2007 is applicable for the hygienic management during the production of food packaging material and sets up requirements on transport, storage, raw materials, packaging materials and products.

The International Featured Standard (IFS) Food, Version 6, includes requirements concerning product packaging. The IFS Food Packaging Guideline[973] addresses the problem that some of the IFS Food issues have been directly passed on to the packaging industry, without considering whether those issues fall within their area of responsibility. Considering the legal requirements apparently requires responsibilities within the supply chain to be differentiated; the purpose of this guideline is to increase the safety of product packaging in companies certified according to IFS Food. In addition, the professional knowledge associated with IFS Food shall be improved so that product packaging is practical and meaningfully ensured.

19.6 Study case: 'mineral oils' from cartons

In 2012 the European Food Safety Authority published an Opinion on Mineral Oil Hydrocarbons in Food[974] with among others the following conclusions: '*Mineral oil hydrocarbons occur in food both as a result of contamination and from various intentional uses in food production ... Because of their complexity it is not possible to resolve mineral oil hydrocarbons (MOH) mixtures into individual components for quantification. However, it is possible to quantify the concentration of total Mineral oil saturated hydrocarbons (MOSH) and mineral oil aromatic hydrocarbons (MOAH) fractions, as well as certain sub-classes, using methods based on gas chromatography ... The Panel identified numerous sources for the occurrence of MOH in food. Among food contact materials, sources are food packaging materials made from recycled paper and board, printing inks applied to paper and board, MOH used as additives in the manufacture of plastics, e.g. internal lubricants in polystyrene, polyolefins, adhesives used in food packaging, wax paper and board, jute or sisal bags with mineral batching oil, lubricants for can manufacture and wax coating directly applied to food. Food additives, processing aids and other uses contribute to MOSH levels, together with release agents for bakery ware and sugar products, and oils for surface treatment of foods, such as rice and confectionery ... All MOH are mutagenic unless they are treated specifically to remove MOAH ... Because of its potential carcinogenic risk, the CONTAM*

[973] See http://www.ifs-certification.com/index.php/en/ifs-certified-companies-en/ifs-standards/ifs-packaging-guideline.
[974] EFSA Journal 2012; 10(6): 2704.

Panel considers the exposure to MOAH through food to be of potential concern ... A significant source of dietary exposure to MOH may be contamination of food by the use of recycled paperboard as packaging material'.

The industry is confronted with the problem that FCMs from paperboard and cartons are not regulated: there is no list of admitted materials, there is no possibility for authorisation of substances, there are no maximum levels and there is no obligation to establish a declaration of compliance. Last but not least the packaging industry has learned to use recycled paper and carton for environmental reasons. This strategy seems to be endangered here, as substances from recycled paper bear a high risk of migration.

Since December 2011 one can find only one notification of the 7th of June 2013 in the European Rapid Alert System for Food and Feed (RASFF) from Germany concerning migration of mineral oil (MOAH: 77 mg/kg and MOSH: 411 mg/kg) from a carton box for pizza. The notification was classified as 'information for follow-up', due to the fact that no maximum levels for MOSH and MOAH exist, but yet such migration needs to be proven as safe.

Germany drafted a 'Mineral Oil Decree' on 2 May 2011, which would regulate specific migration limits for hydrocarbons, based on migration from recycled carton packaging. In November 2012 a German Consumer Magazine published the results of tests conducted on 24 chocolate products packaged in the seasonal advent calendars and found that in all chocolates MOH could be shown. However, 23 of the 24 carton calendars were not from recycled paper. It was found that the mineral oil contamination had taken place either during harvesting or transport of the cocoa beans in the country of origin, during the manufacturing procedures of the chocolate or during long-term storage of the final product. The German authorities realised that the planned Mineral Oil Decree would not have prevented the contamination in all these cases. Apparently one needs to differentiate between contamination of the food before and after the packaging stage, which in the state of a finished product cannot be practically controlled. Even secondary and tertiary packaging can influence the safety of the packaged food. And it is impossible to foresee the effects of migration until the end of the durability date of the food product.

19.7 Control of FCMs

FCMs are part of the EU food safety scheme. They are controlled by the national food control authorities. It needs to be noted that Food Control Authorities can easily control violations of the law. This is the case, if FCMs do not fulfil legal requirements regarding the allowed materials, exceed migration limits, contain non-authorised substances, or are not accompanied by an adequate declaration of compliance.

It is important that Food Business Operators and EU importers of food products require documentation on toxicology and risk assessment of chemical migration from food contact materials and/or declarations of compliance. EU competent authorities often have to reject consignments of FCMs purely on the basis of the absence of a declaration of compliance of the EU buyer. With this declaration the EU buyer is supposed to state that the consignment complies with the relevant legislation on the respective FCMs.

From the 1st of January 2014 until the 30th of June 2014 there have been 109 notifications in the European Rapid Alert System for Food and Feed (RASFF) concerning FCMs, mostly because of migration risks and primarily concerning FCMs made in countries outside the European Union. In 2013 food contact materials from China were among the top 10 notifications by country of origin: the migration of chromium 59 times and the migration of manganese 38 times.[975]

19.8 The challenge of FCMs: how to tackle safety without laws

The EU legislation concerning FCMs is not comprehensive. Given the wide variety of substances it will probably never be comprehensive. In all cases not covered by specific provisions, Article 3 Framework Regulation is applicable: materials and articles shall not transfer their constituents into food in quantities which could: (a) endanger human health; or (b) bring about an unacceptable change in the composition of the food; or (c) bring about a deterioration in the organoleptic characteristics thereof. Next to this provision Article 14 of the General Food Law applies: food shall not be placed on the market if it is unsafe. Food business operators are thus confronted with the dilemma how to evaluate the safety of food packaging in the case of missing SMLs or other reference data.

In order to solve the problems caused by the variety of FCMs materials and articles, the complexity of the value chain and the different stakeholders involved, the exchange of information between each step must be well defined. Each user of FCMs needs to specify and define requirements for the packaging material with respect to the product to be packed. The user needs to inform the packaging manufacturer of these facts regarding the packed product. The manufacturer needs to agree to these specifications taking into account the packed product and the type of contact (direct and/or indirect). Furthermore s/he provides a declaration of compliance for food contact applications, including a validation of all packaging components, like inks, adhesives, lacquer, support (film, cardboard, and tinplate), etc.

[975] See RASFF Annual Report 2013, available at: http://tinyurl.com/ozbutqx.

The following common mistakes in declarations of compliance should be avoided: unclear identification of the FCMs material and substance, missing information on substances with restrictions in use in foods (dual-use substances), missing information on substances with restrictions (SML substances), incomplete information on use of products (time and temperature of contact). The declaration of compliance can serve as an argument for demarcation of responsibility. However, the packaging manufacturer may not transfer her or his responsibility to the user of the packaging without violating his or her own due diligence for conformity.

20. Nutrition policy in the European Union

Martin Holle

20.1 Introduction

Nutrition is a relatively recent field of European policy making. The Treaty of Rome that established the European Economic Community in 1957 used the terms 'health' and 'public health' only in the context of grounds that justify national restrictions to the freedom of the Internal Market. Just a little more than a decade after the end of the Second World War nutrition policy was by and large synonymous with agricultural policy. For a large part of the population a constant and sufficient supply of food had still been a luxury in the years of post-war austerity and overweight was largely unknown.[976] It is therefore not surprising that the agricultural policy of the European Economic Area focused on increasing agricultural productivity and assuring availability of supplies at reasonable prices.[977] Consumer protection and public health aspects were only indirectly considered when harmonisation of food legislation started in the mid-seventies.[978] Still, the aim of this legislation was primarily to facilitate the free movement of goods in the Internal Market. It was only in the late eighties of the 20th century that public health, and in particular nutritional aspects of public health, made its way into EU legislation.

20.2 The establishment of public health and consumer protection as objectives in primary and secondary EU law

A first firm acknowledgement that a certain level of coordination on public health issues was required in a Common Market came with the Single European Act in 1987 which in Article 100a obliged the European Commission to observe a high level of protection with regard to health and consumer protection for its proposed harmonising measures. However, it was only with the Maastricht Treaty in 1993 that public health and consumer protection were for the first time established

[976] Rationing of meat in the United Kingdom only ended on 4 July 1954, see: http://tinyurl.com/dl4vk.

[977] Article 39 of the Treaty of Rome, Article 33 of the consolidated version.

[978] E.g. Council Directive 79/112 of 18 December 1978 on the approximation of the laws of the Member States relating to the labelling, presentation and advertising of foodstuffs for sale to the ultimate consumer and Council Directive 90/496 of 24 September 1990 on nutrition labelling for foodstuffs.

as policy areas for the European Union in its Articles 129[979] and 129a.[980] These provisions evolved further in the Amsterdam Treaty of 1997.[981] Article 152 gave the European Union the competence to take action on Community level in order to complement national policies in the area of public health. And in the area of consumer protection its mandate moved from a mere contribution to a high level of consumer protection to ensuring this horizontally in the definition and implementation of other Community actions and policies, as well as actively promoting the interest of consumers (Article 153). Today, Articles 168 and 169 of the Treaty on the Functioning of the European Union (TFEU)) still reflect this and at first glance it appears that apart from the recognition of an increasing need for co-ordination between Member States and the Commission in the relevant texts of the Treaty not a lot has changed in the last fifteen years. A closer look however shows that despite the only relatively minor changes in the competencies after the Amsterdam Treaty there was still significant momentum to strengthen the role of the European institutions in the area of health and consumer protection.

One key factor was the introduction of delegated and implementing acts in Articles 290 and 291 of the Lisbon Treaty[982] which entrusts a large number of day-to-day decisions on the implementation of EU legislation to the European Commission, thus extending the Commission's room to manoeuvre in Consumer and Health Policy. Secondly, from the 1990s onwards the European Commission significantly increased the number of proposals for directly applicable secondary legislation in the area of food law. This new legislative approach was first laid out in the European Commission's Green Paper on the General Principles of Food Law in the European Union,[983] followed by a White Paper setting out a 'Farm to Table' legislative action programme.[984] A cornerstone of the White Paper was the introduction of a European Food Safety Authority that would act as a 'first port of call when scientific information on food safety and nutritional issues is sought'[985]

[979] Article 129(1) of the Treaty on European Union: 'The Community shall contribute towards ensuring a high level of human health protection by encouraging co-operation between the Member States and, if necessary, lending support to their action. Community action shall be directed towards the prevention of diseases, in particular the major health scourges, including drug dependence, by promoting research into their causes and their transmission, as well as health information and education. Health protection requirements shall form a constituent part of the Community's other policies'. OJ C 191, 29 July 1992.

[980] Article 129a(1) of the Treaty on European Union (now Article 169 TFEU): 'The Community shall contribute to the attainment of a high level of consumer protection through:
(a) measures adopted pursuant to Article 100a in the context of the completion of the internal market;
(b) specific action which supports and supplements the policy pursued by the Member States to protect the health, safety and economic interests of consumers and to provide adequate information to consumers.', OJ C 191, 29 July 1992.

[981] Treaty of Amsterdam amending the Treaty on European Union, the Treaties Establishing the European Communities and related acts.

[982] Consolidated version of the Treaty on the Functioning of the European Union.

[983] COM(1997) 176 final, 30.04.1997.

[984] COM(1999) 719 final, 12.01.2000.

[985] COM(1999) 719 final, p. 19.

and which was eventually established by Regulation 178/2002.[986] At the same time the Commission announced it would look into the EU-wide regulation of nutrition and health claims as well as undertake a revision of the Nutrition Labelling Directive from 1990 to bring it into line with consumer needs and expectations.[987] Due to the highly controversial positions of the various stakeholders affected by the proposals it took much longer than initially expected to complete the relevant legislative processes but eventually Regulation 1924/2006 on nutrition and health claims[988] and Regulation 1169/2011 on food information to consumers[989] were adopted. These legislative acts were the culmination of a long development process from solely national policy initiatives on public health to a European nutrition policy agenda.

20.3 The birth of European Nutrition Policy

Nutritional science in the first half of the 20th century had focused on diseases caused by deficiency and undernutrition. Public nutrition programmes were aiming at increasing the consumption of milk, meat eggs and 'practically everything in the usual diet'.[990] Only in the 1950s did the link between affluent diets and a higher prevalence of non-communicable diseases (in particular heart disease) in the typical Western diet rich in animal fats become a major subject of nutritional research.[991] And already in 1951 the Joint FAO/WHO Expert Committee on Nutrition concluded: 'Malnutrition may, however, also result from excessive food intake. This is particularly evident in the case of excessive consumption of calories, but there is also reason to suppose that excessive consumption of carbohydrates and fats, quite apart from calories, may produce serious forms of malnutrition. While these are rare in many parts of the world, they may be of outstanding importance in regions in which food supplies are abundant and economic levels high. The association of obesity with a high incidence of 'degenerative' diseases, e.g. certain cardiovascular and metabolic disorders, suggests that in these regions malnutrition from the overconsumption of food is a problem of major significance.'[992] However, after the post-war years of austerity, at a time when people had just got back their freedom to choose whatever food they liked and were for the first time

[986] Regulation 178/2002 of the European Parliament and of the Council of 28 January 2002 laying down the general principles and requirements of food law, establishing the European Food Safety Authority and laying down procedures in matters of food safety.
[987] COM(1999) 719 final, p. 32.
[988] Regulation 1924/2006 of the European Parliament and of the Council of 20 December 2006 on nutrition and health claims made on foods, corrected version.
[989] Regulation 1169/2011 of the European Parliament and of the Council of 25 October 2011 on the provision of food information to consumers.
[990] O.W. Portman and D.M. Hegsted, Nutrition, Annual Review of Biochemistry, 1957, pp. 307-326.
[991] See for details K. Carpenter, A Short History of Nutritional Science: Part 4 (1945-1985), Journal of Nutrition, 2003, pp. 3331-3342.
[992] Joint FAO/WHO Expert Committee on Nutrition (1951), Report on the second session (WHO Technical Report Series, No. 44), p. 43.

experiencing 'the pleasures of a rich and varied diet,'[993] a statement like this could not fall on fertile ground. Perhaps the insight that food could not only nourish but also kill you seemed too inconceivable. In 1962, a WHO expert committee of cardiologists concluded that 'at present time there are no effective means by which the occurrence of ischaemic heart disease can be prevented. ... much further research is needed before public health authorities can recommend major alterations in the diet (...).'[994] This remained the final verdict on preventive nutrition for almost two decades and nutritionists went back to looking at deficiencies. Likewise, nutrition policy was understood as a vehicle for fighting undernutrition and therefore regarded as redundant for developed countries.[995]

20.3.1 First national initiatives

It was only in the mid-1970s and 1980s that, starting from the Nordic countries, nutrition policies were adopted in some European countries.[996] The intensification of agricultural production at that time had led to an abundance of nutrient-rich food at ever lower prices. However, despite the tremendous progress in nourishment of the population as well as in battling infectious diseases, around 25% of males and about 20% of females would still die before retirement.[997] The 'diseases of affluence' like heart disease and cancers thus became a major public health problem.[998] This led to a change of mind in the World Health Organization, too. The 1978 WHO Expert Committee Report on Arterial Hypertension recommended the further investigation of possible links between weight control, dietary factors like salt intake and physical activity with hypertension in order to develop preventive and general therapeutic measures.[999] Subsequently, the 1982 Report on Prevention of Coronary Heart Disease acknowledged the urgent need to develop policies and techniques of prevention as cardiovascular disease had become a mass disease and the most important cardiovascular cause of premature disability and mortality.[1000]

---

[993] W.P.T. James, Food and nutrition policy in this century, in WHO, Regional Office for Europe, European food and nutrition policies in action (WHO Regional Publications, European Series, No. 73), 1998, p. 21.

[994] WHO, Arterial hypertension and ischaemic heart disease: preventive aspects, Report of an Expert Committee 1962 (WHO Technical Report Series, No. 231).

[995] WHO, Regional Office for Europe, European food and nutrition policies in action (WHO Regional Publications, European Series,1998 No. 73), p. 3.

[996] E.g. in Norway, Sweden, Denmark, Finland, Iceland, the Netherlands, Romania and Malta.

[997] W.P.T. James, Food and nutrition policy in this century, in WHO, Regional Office for Europe, European food and nutrition policies in action (WHO Regional Publications, European Series, No. 73), 1998, p. 22.

[998] W.P.T. James, Food and nutrition policy in this century, in WHO, Regional Office for Europe, European food and nutrition policies in action (WHO Regional Publications, European Series, No. 73), 1998, p. 22.

[999] WHO, Arterial hypertension, Report of a WHO Expert Committee (WHO Technical Report Series, 1978 No. 628), p. 40 and 46.

[1000] WHO, Prevention of coronary heart disease, Report of a WHO Expert Committee (WHO Technical Report Series, 1982 No. 678), pp. 5, 12 *et seq.*

20.3.2 From national health plans to a multi-sectoral and multidisciplinary approach

Further work by the WHO led to comprehensive reports on a European as well as global level that described the major historical changes that led to the 'affluent diet', summarised the knowledge on the relationship between diet and chronic diseases, proposed population nutrient goals, analysed the various existing national nutrition policies and gave recommendations on how to successfully develop such policies.[1001] In particular the 1990 report of the WHO Study Group marks a clear shift in perspective from seeing work on the prevention of chronic diseases mainly as a task of national health authorities to a multi-sectoral and multidisciplinary approach. It acknowledges that a nutrition and food policy can only succeed if all relevant sectors of government like health, agriculture, economics, education, social welfare, planning and development are involved and the highest levels of decision-making are committed to providing their support. In addition, the report sees technical and operational assistance by the nutrition community, universities, non-governmental organisations but also the food industry, farmers' associations and caterers as key.[1002] It was also the first time a critical look was taken at the role of food producers and manufacturers in claiming that public campaigns on changing eating patterns could be foiled by the much greater advertising campaigns for individual foods by food companies.[1003] However, the report stopped short of recommending controls on the marketing of food similar to the ones in the tobacco sector, mainly because this would require agreement on the levels of saturated fatty acids that would warrant such restrictions first and such controls would likely meet opposition. Instead, nutrition labelling was advocated as the weapon of choice to improve the knowledge of the population on the content of certain nutrients in foods.[1004]

From all this it is apparent that by the end of the 1980s preventive nutrition had left its infancy and taken its first steps from being a small, slightly exotic field of public health policy towards a complex, horizontal task in all relevant policy areas. It is therefore not at all surprising that this change is also reflected in the introduction of public health and a high level of consumer protection as important considerations for all EU harmonising measures by the Single European Act of 1987.

[1001] W.P.T. James *et al.* Healthy Nutrition, preventing nutrition-related diseases in Europe (WHO Regional Publications, European Series, 1988 No. 73); WHO, Diet, nutrition and the prevention of chronic diseases, Report of a WHO Study Group (WHO Technical Report Series, No. 797), 1990.
[1002] WHO Technical Report Series, No. 797, p. 144.
[1003] WHO Technical Report Series, No. 797, p. 129.
[1004] WHO Technical Report Series, No. 797, p. 130.

20.4 Early childhood: the 1990s

20.4.1 1990: The first European Action Programme on Nutrition and Health

The first major document that the Council of the European Community adopted on nutrition policy starts with a confession that the Community up to that point had 'not given overall consideration to aspects of nutritional education and consumer information taken as a whole with the aim of promoting eating habits in keeping with individual needs'.[1005] This is not surprising, as the Council resolution was made almost three years prior to the ratification of the Maastricht Treaty, thus public health policy was still by and large a domain of the Member States. The resolution echoed the findings of the WHO Study Group from earlier that year by stressing the role of proper, balanced eating habits for the prevention of certain chronic diseases and the holistic approach required to successfully promote such habits. It recommended taking the activities of the FAO and WHO into account and cooperating with these organisations as far as possible. Furthermore, it proposed that an action programme on nutrition and health be initiated and coordinated by the European Commission. The objectives and measures to be taken, however, were rather modest, a fact that could not be disguised by including a nice-sounding call for a European Nutrition Year in 1994 in the text. The programme focused mainly on dissemination of information to highlight awareness of healthy eating habits in the population (a measure that had already shown to have only a very limited effect by the WHO experts) and some basic research projects. And even though the proposed actions were not very ambitious, the European Commission was apparently not very active in implementing them, causing the Council to reaffirm the importance of such action on a Community level in its conclusions on nutrition and health of 15 May 1992.[1006]

20.4.2 1992: The World Declaration and Action Plan on Nutrition

On a global level, a further milestone towards the alignment of national nutrition policies was the World Declaration and Action Plan on Nutrition which was the result of a meeting of representatives from 159 countries at the International Conference on Nutrition in December 1992. It stressed that access to a safe and healthy variety of food was a fundamental human right. Despite the fact that a major focus was still on the fight against undernutrition and deficiencies it also contained for the first time a pledge by the participating states to reduce diet-

[1005] Resolution of the Council and the representatives of the governments of the Member States, meeting within the Council, concerning an action programme on nutrition and health, OJ C 329, 3 December 1990, p. 1.

[1006] Conclusions of the Council and the Ministers for Health of the Member States, meeting within the Council, OJ C 148, 12.06.1992, p. 2.

related and non-communicable disease substantially within one decade.[1007] The conference stressed once more that 'full participation of all multilateral, bilateral, and non-governmental organizations to support activities at the country level will be essential for success in alleviating and eventually eliminating nutritional problems and promoting universal health and nutritional well-being.'[1008] Significant improvements could be gained by incorporating 'nutritional considerations into the broader policies of economic growth and development, structural adjustment, food and agricultural production, processing, storage and marketing of food, health care, education and social development.'[1009] With great clarity the conference summarised the situation faced by most developed countries, including the EU Member States:

'Non-communicable diseases related to unhealthy lifestyles and inappropriate diets are becoming increasingly prevalent in many countries. With greater affluence and urbanization, diets tend to become richer on average in energy and fat, especially saturated fat, have less fibre and complex carbohydrates and more alcohol, refined carbohydrates and salt. In urban settings exercise and energy expenditure frequently decrease, while levels of smoking and stress tend to increase. These and other risk factors, as well as increased life expectancy, are associated with the increased prevalence of obesity, hypertension, cardiovascular diseases, diabetes mellitus, osteoporosis and some cancers with immense social and health care costs.'[1010]

To improve the health status of the population and to promote appropriate diets and lifestyles the conference recommended that governments develop comprehensive nutrition policies and dietary guidelines, support healthy food consumption patterns, promote the knowledge of food and nutrition across all age groups, encourage the promotion of healthy diets in the food service and catering sector as well as advancing exercise programmes including the provision of recreation and sporting facilities. It called for nutrition education programmes, to include nutrition labelling, which would enable individuals and families to choose a healthy diet.[1011]

20.4.3 1993: The public health mandate of the Maastricht Treaty

In 1993, Article 129 of the Maastricht Treaty provided the European Community for the first time with a clear mandate to develop a coherent public health strategy on European level. Already at the end of the same year, the European Commission presented its framework for action in the field of public health which contained

[1007] FAO/WHO, International Conference on Nutrition, Final Report of the Conference (1992), p. 13.
[1008] FAO/WHO, International Conference on Nutrition (1992), p. 7.
[1009] FAO/WHO, International Conference on Nutrition (1992), p. 24.
[1010] FAO/WHO, International Conference on Nutrition (1992), p. 45.
[1011] FAO/WHO, International Conference on Nutrition (1992), p. 46 *et seq.*

the Commission's proposal on how to meet the objectives of this article, while highlighting that its communication was only a first step in the development of policies, programmes and actions designed to give full effect to Article 129.[1012] The Commission's communication defined four key objectives for Community action:

- to prevent premature death;
- to increase life expectancy without disability or sickness;
- to promote the quality of life by improving general health status and the avoidance of chronic and disabling conditions;
- to promote the general well-being of the population particularly by minimising the economic and social consequences of ill health.[1013]

Together with the Member States and the Council it identified a comprehensive list of possible fields for activities, which included amongst many others cardiovascular diseases, nutrition, fundamental health choices, pilot projects in prevention and early diagnosis, exchanges of information on Member States' policies and the establishment of centres of excellence.[1014] The European Community in this context was allocated a largely facilitating and coordinating role. It was probably not possible to achieve more than this at a time when Member States were very reluctant to hand over responsibilities in this policy domain to the European institutions even with the example of the 'Europe against cancer' programme launched in 1986 which according to the European Commission demonstrated the value of identifying common objectives and goals.[1015] Eventually, the first Community action programme lasted from 1993 to 2002 and comprised eight programmes. It included funding of the some nutrition-related research projects like Eurodiet,[1016] the European prospective investigation into cancer and nutrition (EPIC), a project to promote a Master's training in public health nutrition and a number of projects promoting physical activity.[1017] While the programme enhanced the knowledge base of nutritionists it had little impact on the eating habits of the European population. And so in June 2000 the Eurodiet report once more states: 'Community action in the field of public health has, to date, taken insufficient account of the importance of nutrition, diet and physical activity as a health determinant.'[1018]

[1012] COM(1993) 559 final, 24.11.1993, p. 2.
[1013] COM(1993) 559 final, 24.11.1993, p. 16.
[1014] COM(1993) 559 final, 24.11.1993, p. 15.
[1015] COM(1993) 559 final, 24.11.1993, p. 10.
[1016] The Eurodiet project was initiated in October 1998 with the aim of contributing towards a coordinated European Union and Member State health promotion programme on nutrition, diet, and healthy lifestyles, by establishing a network, strategy and action plan for the development of European dietary guidelines. The project presented its core report in June 2000: Eurodiet, Nutrition & Diet for Healthy Lifestyles in Europe, Science & Policy Implications, available at: http://tinyurl.com/mk74hjh.
[1017] European Commission, DG Sanco, Status report on the European Commission's work in the field of nutrition in Europe, October 2002, p. 5, available at: http://tinyurl.com/mk2nl6c.
[1018] Eurodiet, Nutrition & Diet for Healthy Lifestyles in Europe, Science & Policy Implications, p. 11, available at: http://tinyurl.com/mk74hjh.

20.4.4 1994: The Council of Europe Recommendation

Whilst the European Union institutions were still struggling with the development of a comprehensive and coherent nutrition policy, the Parliamentary Assembly of the Council of Europe in 1994 made a far more advanced proposal on how to tackle the serious health problems caused by unhealthy eating habits. In its report on food and health the Committee on Agriculture described the existence of a high level of confusion amongst consumers about nutrition and emphasised the consumers' right to receive better information on the foodstuffs they buy and what they do or do not contribute to a healthy diet.[1019] Even though the Report contains a number of simplifications, it fully embraces the concept of preventive nutrition and calls for a nutrition policy that aligns food production, distribution and consumption to consider nutritional priorities for the promotion of consumer health.[1020] In its Recommendation 1244 (1994) on food and health the Parliamentary Assembly asked the Committee of Ministers to:

- initiate activities which would promote public health by improving consumers' eating habits;
- to adopt nutritional policies, and to integrate such measures into other fields, such as food and agriculture, health, consumer policy, research and education;
- to give more emphasis to research into all aspects of the relationship between diet and health and the safety of food, and to work for more international co-operation and consensus between experts at national and international level in this field;
- to improve school education on the importance of a balanced and healthy diet for human health;
- to promote the production of healthy food through co-operation between producers, consumers and the food industry;
- to organise information campaigns on issues of importance for consumers' choice of a healthy diet; and
- to implement nutrition labelling.[1021]

The impact of these recommendations, however, was mainly limited to the national nutrition policies in the individual countries and it was only the new millennium that saw significant progress in the development of a European nutrition policy.

[1019] Council of Europe, Parliamentary Assembly, Report on food and health, adopted on 12 April 1994 (Doc. 7083), available at: http://tinyurl.com/l9xte7q.

[1020] Council of Europe, Parliamentary Assembly, Report on food and health, adopted on 12 April 1994 (Doc. 7083).

[1021] Council of Europe, Parliamentary Assembly, Recommendation 1244 (1994) on food and health, Text adopted by the Assembly on 28 June 1994 (18th Sitting), available at: http://tinyurl.com/m6lvm5v.

20.5 The school years: 2000-2003

20.5.1 2000: The First WHO Action Plan for Food and Nutrition Policy in Europe

It was the World Health Organization's Regional Office for Europe that gave new momentum to the development of a European nutrition policy by adopting the First Action Plan for Food and Nutrition Policy in September 2000.[1022] The action plan was based on the principles set out in the declaration of the World Health Assembly on a global strategy for the prevention and control of non-communicable diseases in May 2000.[1023] It highlighted that there was still a lot of work to do until a supply of safe and nutritious food was fully achieved in Europe and asked the WHO Member States, intergovernmental and non-governmental organisations to collaborate with the WHO's Regional Office to maximise region-wide efforts to promote public health through food and nutrition policy.[1024] The framework within which the public health issues should be addressed consisted of a food safety strategy, a nutrition strategy and a sustainable food supply strategy, which were all interrelated. The need for a new food safety policy was demonstrated by the emergence of new pathogens like bovine spongiform encephalopathy but also by the fact that each year an estimated 130 million Europeans were affected by episodes of foodborne diseases.[1025] Another concern was the potential transfer of antibiotic resistance caused by the use of veterinary drugs to human pathogens.[1026] To tackle this, the WHO recommended a strategy that prevented chemical and biological contamination at all stages of the food chain which could only be achieved with new food safety systems that took a 'farm to fork' approach.[1027] This integrated approach to food safety that embraces the whole food chain was the result of the lessons learned from the BSE crisis and can already be found in the European Commission's Green Paper on the General Principles of Food Law in the European Union.[1028] It was labelled the 'farm to table' approach in the European Commission's White Paper on Food Safety of January 2000.[1029] The second area of concern for the WHO was a prevalence of obesity in adults of up to 20-30% (and a significant increase also being observed in children), which was estimated to cost the relevant health services about 6-7% of their health care budget. Approximately one third of cardiovascular disease cases in Europe at the time were related to unbalanced nutrition and 30-40% of cancers according to

[1022] WHO, Regional Office for Europe (2001), The First Action Plan for Food and Nutrition Policy, WHO European Region, 2000-2005, available at: http://www.euro.who.int/document/e72199.pdf.
[1023] 53rd World Health Assembly, WHA53.17, Agenda item 12.11, 20 May 2000.
[1024] WHO, Regional Office for Europe, First Action Plan for Food and Nutrition Policy (2001), p. 41.
[1025] WHO, Regional Office for Europe, First Action Plan for Food and Nutrition Policy (2001), p. 1.
[1026] WHO, Regional Office for Europe, First Action Plan for Food and Nutrition Policy (2001), p. 11.
[1027] WHO, Regional Office for Europe, First Action Plan for Food and Nutrition Policy (2001), *p. 2*.
[1028] COM(1997) 176 final, 30.04.1997, p. 10.
[1029] COM(1999) 719 final, 12.01.2000, p. 3.

WHO could be prevented through better diet.[1030] The WHO therefore called for 'a nutrition strategy geared to ensure optimal health, especially in low-income groups ...'[1031] The two elements of food safety and nutrition should be complemented by a sustainable strategy of food supply in order to secure a sufficient supply of good quality food.[1032]

In the context of the action plan the WHO Regional Office worked on:

- collating existing knowledge and scientific evidence;
- stimulating research in areas where evidence is lacking;
- developing innovative ways to communicate scientific knowledge and information;
- collaborating with countries in translating knowledge into action;
- providing information, experience and expertise as required to national counterparts;
- developing cost-effective indicators for surveillance;
- producing updated lists of new information, documents and training materials;
- facilitating information sharing.[1033]

This included a comparative analysis of nutrition policies and food-based dietary guidelines in WHO European Member States.[1034] As for the proposed European Food and Nutrition Task Force, representatives from the WHO, the European Commission, the Council of Europe, UNICEF and the FAO decided that such a task force should be of a technical rather than a political nature and that it should primarily be created with the Member States.[1035] The decision shows that the institutions participating in the meeting still felt the time was not ripe to move to a European Nutrition Policy and that the Member States should remain the drivers of the agenda, a position that the Council of the European Community had confirmed in its Resolution on health and nutrition in 2000.[1036] Nevertheless, there was clearly a movement towards greater alignment of national nutrition policies and one of the main reasons why progress was relatively slow was the rather diverse consumption habits, in particular between Northern and Southern European countries and the lack of pan-European intake data – an issue that has still not been fully resolved today.

[1030] WHO, Regional Office for Europe, First Action Plan for Food and Nutrition Policy (2001), p. 1.

[1031] WHO, Regional Office for Europe, First Action Plan for Food and Nutrition Policy (2001), p. 2.

[1032] WHO, Regional Office for Europe, First Action Plan for Food and Nutrition Policy (2001), p. 2.

[1033] WHO, Regional Office for Europe (2001), Progress Report on The First Action Plan for Food and Nutrition Policy, WHO European Region, 2000-2005, available at: http://www.euro.who.int/document/e79034.pdf, p. 7.

[1034] WHO, Regional Office for Europe (2003), Comparative analysis of food and nutrition policies in WHO European Member States, available at: http://www.euro.who.int/document/e81507.pdf.

[1035] WHO, Regional Office for Europe, Progress Report (2001), p. 6.

[1036] Council Resolution of 14 December 2000 on health and nutrition, OJ C 20, 23.01.2001, p. 1. Recital 11 states: 'that the diversity of food cultures throughout the European Union constitutes a valuable asset that ought to be respected, and that it is necessary to take this into account when drawing up and implementing nutritional health policies, which must therefore be defined first of all at national level'.

20.5.2 The EU Health Framework for the New Millennium

In 1998 the European Commission had already outlined three main pillars of its health framework for the new millennium: better information exchange between Member States and European institutions, a rapid reaction to emerging health risks and tackling health determinants through health promotion and disease prevention, building on existing disease-specific programmes and including new issues like nutrition, in particular calling for more activities to address obesity.[1037] 'Food, nutrition and health' was a key area for Community funded research in the Fifth Framework Programme (FP 5) from 1998-2002.[1038] And the Commission's White Paper on Food Safety included in its action plan the objective to develop a comprehensive and coherent nutrition policy on Community level.[1039] 1999 also saw the creation of the Directorate General for Health and Consumers (DG SANCO) which was to become a major driver of European nutrition policy. By May 2001 it had presented its first discussion paper on nutrition claims and functional claims.[1040] However, the focus of the paper was still very much on the harmonisation of the rules for the internal market and the protection of the consumer from being misled rather than on making the claims legislation a tool of nutrition policy. The original intention of the Commission was also to facilitate innovation in the food sector by allowing claims on health benefits under specific conditions, a point reiterated later in the first legislative proposal.[1041]

20.5.3 2002: The Programme of Community Action in the Field of Public Health

In 2002 the European Parliament and the Council adopted a programme of Community action in the field of public health.[1042] They acknowledged the three pillars identified by the Commission and supported the Commission's view that 'actions at Community level should be set out in one overall programme to run for a period of at least five years and comprising [the] three general objectives.'[1043] Recital 2 of this Decision also marked a change of tack towards a more European health policy:

[1037] COM(1998) 230 final, 15.04.1998, reinforced by the Communication from the Commission to the Council, the European Parliament, the Economic and Social Committee and the Committee of the Regions on the health strategy of the European Community, COM(2000) 285 final, 15.06.2000.
[1038] As part of the 'Quality of Life' programme. Examples of research projects can be found at: http://tinyurl.com/pnub3sz and in Annex IV of the European Commission's October 2002 Status Report on its work in the field of nutrition in Europe, available at: http://tinyurl.com/qz86s5f.
[1039] COM(1999) 719 final, 12.01.2000, p. 41.
[1040] SANCO/1341/2001, available at: http://ec.europa.eu/food/fs/fl/fl03_en.pdf.
[1041] COM(2003) 424 final, 16.07.2003.
[1042] Decision 1786/2002 of the European Parliament and of the Council of 23 September 2002 adopting a programme of Community action in the field of public health (2003-2008).
[1043] Recital 6 of Decision 1786/2002.

'Health is a priority and a high level of health protection should be ensured in the definition and implementation of all Community policies and activities. Under Article 152 of the Treaty, the Community is required to play an active role in this sector by taking measures which cannot be taken by individual States, in accordance with the principle of subsidiarity.'[1044] Despite the clear reference to the principle of subsidiarity the Member States acknowledged a distinct and active role of the European Community in shaping public health policy that went beyond mere data collection and exchange of information. The Council Decision mandated the European Commission to implement the action plan and committed overall funding of 312 million euros. Part of the action plan was the promotion of health and prevention of disease through action on health determinants across all Community policies and activities, including life-style parameters such as nutrition and physical activity.[1045] In its Conclusions on obesity of 2 December 2002 the Council put particular emphasis on the health risks associated with obesity and asked the Commission to ensure the prevention of obesity was taken into account in all relevant Community policies and specifically in the framework of the programme of Community action on public health.[1046] The huge dimension of the action programme with its multiple projects in 2005 led to the establishment of a dedicated Executive Agency tasked with the implementation of technical projects that did not involve political decision making.[1047]

20.6 Into adulthood: 2003 and beyond

20.6.1 2002: The European Commission's Status Report

The European Commission started its work on a Community nutrition policy with the publication of a Status Report that summarised the challenges faced in Europe with regard to nutrition and health of the population, described the ongoing nutrition-related activities on Community level and defined operational goals to be achieved in the implementation of the Community's new public health programme from 2003 to 2008 as adopted in the Council Decision.[1048] It pointed out the fundamental changes to eating habits in the European Union that took place in the decades after the war and which were essentially caused by an improvement in production methods and yields, higher available incomes of consumers and changes in the way food was marketed. A positive effect of the easy access to low-cost food created by the EU Common Agriculture Policy in the post-war years was a significant decrease in nutritional deficiencies. On the other

[1044] Recital 2 of Decision 1786/2002.
[1045] Decision 1786/2002/EC of the European Parliament and of the Council of 23 September 2002 adopting a programme of Community action in the field of public health (2003-2008), Annex 3.1.
[1046] Council Conclusions of 2 December 2002 on obesity, OJ C 11, 17.01.2003, p. 3.
[1047] Commission Decision 2004/858 setting up an executive agency, the 'Executive Agency for the Public Health Programme', for the management of Community action in the field of public health – pursuant to Council Regulation 58/2003.
[1048] European Commission, Status Report on its work in the field of nutrition in Europe, October 2002.

hand, the change in consumption patterns together with a reduction in physical activity generated the phenomenon of the 'diseases of affluence,'[1049] with diet-related risk factors being responsible for almost 10% of the overall disease burden in Europe, an effect worse than the overall impact of smoking, which accounts for around 9%.[1050] Above all, obesity by 2002 had become a 'pan-European epidemic' with 135 million people affected.[1051] In many countries more than half of the population was overweight, with up to 30% being clinically obese. This affected citizens on lower income more than wealthier consumers as the former, even though they spent a higher percentage of their income on food, often followed a diet of poorer nutritional quality.[1052] In its more detailed analysis, the European Commission provided an overall inventory on diet and lifestyle related risk factors affecting the health of the European population, which was based on the earlier work of Eurodiet[1053], FAO and WHO[1054] (Diagram 20.1).

Diagram 20.1. Examples of probable correlations between certain pathologies and dietary risk factors.[1055]

Pathology	Dietary risk factor
Arterial hypertension	Inadequate fruit and vegetable consumption Excessive alcohol consumption Excessive salt consumption
Cerebro- and cardiovascular diseases	Inadequate fruit and vegetable consumption Excessive consumption of saturated fatty acids Inadequate consumption of food rich in fibre
Cancers (especially colon, breast, prostate and stomach cancer)	Excessive alcohol consumption Excessive salt consumption Inadequate fruit and vegetable consumption Inadequate consumption of food rich in fibre Inadequate physical activity, overweight
Obesity	Excessive energy intake Inadequate physical activity

►►

[1049] W.P.T. James, Food and nutrition policy in this century, in WHO, Regional Office for Europe (1998), European food and nutrition policies in action (WHO Regional Publications, European Series, No. 73), p. 22.
[1050] European Commission, Status Report (2002), p. 7.
[1051] International Association for the Study of Obesity, International Obesity Task Force and European Association for the Study of Obesity, Obesity in Europe – The Case for Action, 2002, p. 4 *et seq.*, available at: http://tinyurl.com/m2rkrog.
[1052] European Commission, Status Report (2002), p. 7.
[1053] Eurodiet, Nutrition & Diet for Healthy Lifestyles in Europe, Science & Policy Implications, available at: http://tinyurl.com/mk74hjh.
[1054] FAO/WHO (2003), Joint WHO/FAO Expert Consultation on Diet, Nutrition and the Prevention of Chronic Diseases, WHO Technical Report Series 916, Available at: http://tinyurl.com/m8klmtg.
[1055] European Commission, Status Report (2002), p. 23 *et seq.*

Diagram 20.1. Continued.

Non-insulin dependent diabetes (type 2)	Obesity Inadequate physical activity
Osteoporosis	Inadequate calcium consumption Inadequate vitamin D consumption Inadequate physical activity
Dental decay	Frequent consumption of fermentable carbohydrates/sugary foods or beverages
Dental erosion	Consumption of acidic foods, fruits or beverages
Iodine deficiency disorders	Inadequate consumption of fish or of iodine-enriched food
Prematurity and low birth weight	Inadequate intake of food nutrients
Iron deficiency anaemia	Inadequate or unavailable iron intake Inadequate vegetable, fruit and meat consumption
Neural tube defects (spina bifida)	Inadequate folate and folic acid intake Inadequate vegetable and fruit consumption
Lowered resistance to infections	Inadequate fruit and vegetable consumption Inadequate consumption of micronutrients Inadequate breastfeeding
Anorexia, bulimia, binge eating disorder	Self-starvation and excessive weight loss or obesity
Food allergies	Allergens contained in food
Infectious food poisoning	Pathogenic micro-organisms contained in food
Non-infectious food poisoning	Pathogenic substances contained in food: e.g. dioxin, mercury, lead and other heavy metals, agrochemical residues and other contaminants

In summary, the Report identified the following major nutritional challenges for the EU:[1056]

- low consumption of fruit and vegetables, in particular in the northern parts of Europe and in socioeconomically disadvantaged groups;
- a too high intake of fat, in particular saturated fat, caused by an increased consumption of meat and full-fat dairy products;
- a drop in the consumption of cereals; and
- a general imbalance of energy intake and energy expenditure, aggravated by a reduction in physical activity.

[1056] European Commission, Status Report (2002), p. 24.

20.6.2 The 2003-2008 Public Health Action Programme

One of the first activities initiated by the European Commission in the context of the 2003-2008 Public Health Action Programme was the foundation of a Network on Nutrition and Physical Activity in 2003. Its members from Member States, the WHO as well as non-governmental organisations from the health and consumer protection sector acted as advisers to the Commission on activities for better nutrition, promotion of physical activity and initiatives to reduce obesity.[1057] Around the same time measures were introduced in the internal processes within the Commission that helped to strengthen the coherence in the area of nutrition policy and to ensure health aspects were taken into account in other Community policy areas. These included a systematic consultation of the Health and Consumer Protection Directorate General DG Sanco on major policy proposals, the inclusion of health impacts in the Commission's impact assessment procedure for legislative proposals and the discussion of health-related issues by an inter-service group of the various Commission services.[1058] Furthermore, a large number of pan-European projects promoting healthy eating habits and physical activity were supported in the context of the Public Health Action Programme.[1059] Following its earlier Conclusions on obesity,[1060] the Council of the European Union adopted three more Conclusions (on healthy lifestyles: education, information and communication,[1061] on promoting heart health[1062] and on obesity, nutrition and physical activity[1063]) in which it asked the European Commission to develop further activities to promote healthier lifestyles and to continue the work towards a comprehensive and integrated European food and nutrition policy.[1064] This included a stronger link with other Community policies, initiatives ensuring consumers are not misled by product marketing, using the experience gained in the tobacco area but also to encourage food manufacturers, processors, retailers and caterers to contribute to the efforts to promote healthy lifestyles through their production, marketing and other related activities.[1065]

[1057] Green Paper, p. 6.

[1058] Green paper, p. 6.

[1059] An overview of the initiatives can be found in Annexes III and IV of the European Commission's Status Report on its work in the field of nutrition in Europe, October 2002.

[1060] OJ C 11, 17.01.2003, p. 3.

[1061] OJ C 22, 27.01.2004, p. 1.

[1062] Council conclusions of 14 May 2004 on promoting heart health, Council document 9627/04 from 18 May 2004.

[1063] Council conclusions of 3 June 2005 on obesity, nutrition and physical activity, Council document 9803/05 from 06.06.2005.

[1064] Council conclusions of 14 May 2004 on promoting heart health, Council document 9627/04 from 18 May 2004, p. 9.

[1065] Council conclusions of 2 December 2003 on healthy lifestyles: education, information and communication, OJ C 22, 27.01.2004, p. 2; Council conclusions of 3 June 2005 on obesity, nutrition and physical activity, Council document 9803/05 from 06.06.2005.

20.6.3 2005: The Commission's Green Paper on promoting healthy diets and physical activity

In its proposal for a Programme of Community action in the field of Health and Consumer protection for 2007-2013 of April 2005 the European Commission asked for a combination of the policies on public health and consumer protection under one framework.[1066] Even though this call was a thinly veiled attempt to enhance the political weight of DG SANCO, it should also be said that it echoed earlier demands uttered by the WHO and was the first substantial political attempt to create a genuinely European nutrition policy. It was spelled out in more detail in the Green Paper 'Promoting healthy diets and physical activity: a European dimension for the prevention of overweight, obesity and chronic diseases'.[1067] Here the Commission identified as areas for action:[1068]

- consumer information (including advertising and marketing);
- consumer education, with a particular focus on children;
- activities at the work place;
- better diet and lifestyle advice by the health services;
- changes to the environment to encourage physical activity;
- addressing socio-economic inequalities;
- an integrated and comprehensive approach for the promotion of a healthy diet and lifestyle;
- development of European dietary guidelines and intake recommendations.

These points resonated the WHO's Global Strategy on Diet, Physical Activity and Health, that had been adopted by the 57th World Health Assembly in May 2004,[1069] even though they did not yet adopt the WHO's recommendation to look into fiscal measures. Also, the alignment of agricultural policy with public health requirements proved a touchy subject that the Commission preferred to avoid, given that the Common Agricultural Policy subsidised the production of e.g. alcoholic beverages and tobacco, a contradiction that remains unresolved today.[1070]

[1066] COM(2005) 115 final, 06.04.2005.
[1067] COM(2005) 637 final, 08.12.2005.
[1068] COM(2005) 637 final, 08.12.2005, p. 7 *et seq.*
[1069] WHO, Global Strategy on Diet, Physical Activity and Health (2004), available at: http://tinyurl.com/5cyxzh.
[1070] Yet in March 2013 the European Parliament still voted to maintain the subsidies for tobacco farming in the EU, available at: http://tinyurl.com/necozon.

20.6.4 Carrot and stick: legislative proposals vs. European Platform on Diet, Physical Activity and Health

One of the most remarkable changes in the Commission's approach towards nutrition policy in these years was certainly the turn towards food legislation and its use as a vehicle to approach diet-related health issues. In the 2003 proposal for a regulation on nutrition and health claims made on foods[1071] the concept of nutrient profiles was introduced with the intention to prevent foods that do not meet certain benchmarks for nutrients like fat, saturated fat, sugar or salt from making health claims. The restriction was justified with the association of high consumption of these nutrients with non-communicable diseases like cardiovascular, disease, diabetes, cancer, obesity, osteoporosis and dental disease,[1072] thus precisely with the list provided in the Status Report less than a year earlier. At the same time in the same paragraph the Commission still stresses that: 'Although based on understandable concerns and important arguments, a number of scientific and policy arguments could challenge such restrictions. The concept of prohibiting the use of claims on certain foods on the basis of their 'nutritional profile' is contrary to the basic principle in nutrition that there are no 'good' and 'bad' foods but rather 'good' and 'bad' diets.'[1073] This document therefore seems to be the best available illustration of the fact that the Commission had reached a watershed in 2003/4 with a major policy shift to follow. The 2008 proposal for the Food Information Regulation[1074] should set the paradigm of this new approach. Rather than providing the technical rules for food labelling as in the food labelling directive 2000/13 and the nutrition labelling directive 90/496, the new proposal made explicit reference to the White Paper on a Strategy for Europe on Nutrition, Overweight and Obesity related health issues and stipulated that nutrition labelling should be used 'as a means to support consumers' ability to choose a balanced diet'.[1075]

A second major difference to earlier initiatives was the foundation of the European Platform on Diet, Physical Activity and Health in March 2005. Led by the Commission, it not only brought together key stakeholders like businesses, non-governmental organisations and members of the public sector in a common forum, but it also asked its members to make binding and verifiable commitments that could help to stop and reverse the trend of an increasing prevalence of overweight and obesity.[1076] The Commission saw this initiative as 'the most promising means of non-legislative action'[1077] even though the commitments made by the members of the Platform did not stop the Commission from proceeding with the comprehensive regulation of marketing and labelling of foods. The approach of the

[1071] COM(2003) 424 final, 16.07.2003.
[1072] COM(2003) 424 final, 16.07.2003, p. 5.
[1073] COM(2003) 424 final, 16.07.2003, p. 4.
[1074] COM(2008) 40 final, 30.01.2008.
[1075] COM(2008) 40 final, 30.01.2008, p. 2.
[1076] COM(2005) 637 final, 08.12.2005, p. 5.
[1077] COM(2005) 637 final, 08.12.2005, p. 5.

platform is action-oriented and cooperative and since 2005 has led to more than 300 commitments by the participating stakeholders, primarily from industry.[1078] An annual report ensures transparency and that the fulfilment of the commitments is constantly monitored. One of the first commitments in the context of the Platform was the introduction of a voluntary nutrition labelling scheme based on Guideline Daily Amounts (GDA) by the members of the European Food Industry Association CIAA[1079] and the European Modern Restaurants Association. The scheme was eventually transformed into legislation with the Food Information Regulation in 2011, albeit with some technical complications that make it less attractive for food business operators to use it. Another milestone was the announcement of the EU Pledge by Burger King, Coca-Cola, Danone, Ferrero, General Mills, Kellogg, Kraft Foods, Mars, Nestlé, PepsiCo and Unilever in 2007. In the Pledge the participating companies commit to two major principles:

- no advertising for food and beverage products to children under the age of twelve on TV, print and internet, except for products which fulfil specific nutritional criteria based on accepted scientific evidence and/or applicable national and international dietary guidelines; and
- no communication related to products in primary schools, except where specifically requested by, or agreed with, the school administration for educational purposes.

More companies joined in the subsequent years and today the EU Pledge members account for more than 80% of the total advertising expenditure for foods and beverages in the EU.

Since its inception the Platform has made quite a lot of progress towards achieving its objective of becoming a catalyst for voluntary action to tackle the obesity problem. While the first annual report still criticised too little effort on reformulation and physical activity and called for more specific, more measurable commitments that were better to monitor,[1080] the 2012 report confirms a 'sustained focus on reformulation' and indicates twenty initiatives on physical activity with the overall reach of the active commitments being 6.5 million children and adolescents, 7.5 million individuals in the general population and 80 million online users.[1081] While these numbers in absolutes terms look rather impressive it needs to be pointed out that they basically represent no more than the population of one major European capital like London or Paris against the background of a total EU population of around 500 million. At the same time the EU Pledge monitoring report for 2012 showed that in the five EU Member States surveyed (Germany, Hungary, Italy,

---

[1078] The commitments can be viewed in the Platform's database at: http://ec.europa.eu/health/nutrition_physical_activity/platform/platform_db_en.htm.

[1079] Now called: FoodDrink Europe.

[1080] EU Platform on Diet, Physical Activity and Health, First Monitoring Progress Report (2006), p. 3, available at: http://tinyurl.com/nz2wesl.

[1081] EU Platform on Diet, Physical Activity and Health, 2012 Annual Report (2012), p. vii, available at: http://tinyurl.com/p26dkua.

Poland and Portugal) in the first quarter 2012 alone 774,207 television commercials for foods and beverages were aired.[1082] Of these the (admittedly few) adverts that did not comply with the EU Pledge's nutritional criteria for marketing to children created 1,267 million impacts.[1083] Given that the five countries surveyed have a joint population of approximately 200 million, this means every citizen of these countries within a three month period was exposed to at least six advertising messages about a food or beverage that according even to the companies' view was not considered to be healthy.

20.6.5 2006: The WHO European Charter on Counteracting Obesity and the Second Action Plan

A next big step in forming European nutrition policy was the European Charter on Counteracting Obesity that was adopted by the WHO European Ministerial Conference in November 2006.[1084] It recalled the urgency of measures tackling the obesity epidemic and demanded a move to more interventionist measures than in the past. It saw the root cause of the obesity problem in the fast-changing social, economic and environmental factors affecting people's lifestyles, specifically the increased consumption of energy-dense, nutrient-poor foods and beverages combined with a lack of physical activity. If healthy diet and lifestyle choices were made more easily accessible to individuals, the obesity epidemic could be reversed as improvements in this area could make a substantial and rapid impact.[1085] Therefore, individuals should not be left solely accountable for their obesity; government and society should accept their part of the responsibility as well.[1086] Complementary to the partnership approach, as exemplified in the Platform on Diet, Physical Activity and Health, WHO called for more intervention by governments in a number of fields like the reduction of marketing pressure, economic measures to facilitate healthier food choices or the introduction of adequate nutrition labelling.[1087] Translated into plain English, this meant an appeal for advertising bans, targeted taxation of nutrients like fat, sugar and salt as well as the provision of mandatory nutrition information. Regardless of the fact that WHO also recommended actions like offers of affordable exercise facilities, the provision of free fruit in schools or the promotion of cycling and walking by better urban design[1088] it became apparent quite soon that the focus of government intervention would be on interventions targeted at manufacturers and distributors of food. The main reason for this development was the fact that

[1082] EU Pledge 2012 Monitoring Report, p. 22 *et seq.*, available at: http://tinyurl.com/prlzc5u.
[1083] EU Pledge 2012 Monitoring Report, p. 22 *et seq.*
[1084] WHO Ministerial Conference on Counteracting Obesity, European Charter on Counteracting Obesity (2006), p. 1, available at: http://tinyurl.com/qht4rt6.
[1085] European Charter on Counteracting Obesity (2006), p. 2.
[1086] European Charter on Counteracting Obesity (2006), p. 2.
[1087] European Charter on Counteracting Obesity (2006), p. 4.
[1088] WHO Ministerial Conference on Counteracting Obesity, European Charter on Counteracting Obesity (2006), p. 4 *et seq.*, available at: http://tinyurl.com/qht4rt6.

such measures came at basically zero cost to the national budgets, an argument that gained even more weight when Europe plunged into recession in 2008. In September 2007 the WHO Regional Committee for Europe endorsed a Second European Action Plan for Food and Nutrition Policy for 2007-2012.[1089] It advised setting population nutrition goals according to FAO/WHO recommendations, meaning a limitation of the daily energy intake from saturated fatty acids and free sugars to not more than 10%, of the daily energy intake from trans-fatty acids to less than 1% and of salt intake to a maximum of 5 g a day. Furthermore, at least 400g of fruit and vegetable should be consumed daily.[1090] The other core elements of the plan were:[1091]

- public campaigns informing consumers about food, nutrition, food safety, consumer rights and the opportunities of being physically active;
- regulations or other effective measures to warrant that marketing practices for foods and beverages are in line with internationally agreed guidelines and dietary recommendations;
- a food labelling that improves consumers' understanding of product characteristics, supports healthy choices (like front of pack signposting) and includes a set of nutrient profiles to assess the nutritional quality of food products;
- education of children in nutrition, sensory properties of foods, food safety and physical activities as part of the school curriculum, guidelines for healthy school meals, the provision of healthy options in school catering as well as fruit and vegetable distribution schemes;
- a revision of agricultural policies and a removal of trade barriers to make fruit and vegetable more available and affordable;
- the reduction of salt, added sugar, saturated fats and trans-fats in foods by reformulation of mainstream food products;
- a review of the locations where certain foods are put on offer, in particular the position of vending machines and catering establishments;
- the use of fiscal measures like taxes or subsidies to make products that are in line with food-based dietary guidelines more affordable;
- the creation of an intersectoral food safety system based on a 'farm to fork' approach and a reduction of the environmental contamination of the food chain;
- more opportunities for physical activity in the workplace and integration of physical activity in daily life, across all settings.[1092]

[1089] WHO Regional Office for Europe, WHO European Action Plan for Food and Nutrition Policy 2007-2012, available at: http://tinyurl.com/c6vzbwv.

[1090] WHO Regional Office for Europe, Action Plan for Food and Nutrition Policy 2007-2012, p. 4, referencing FAO/WHO (2003), Joint WHO/FAO Expert Consultation on Diet, Nutrition and the Prevention of Chronic Diseases, WHO Technical Report Series 916, available at: http://tinyurl.com/m8klmtg.

[1091] WHO Regional Office for Europe, Action Plan for Food and Nutrition Policy 2007-2012, p. 6 *et seq.*

[1092] WHO Regional Office for Europe, Action Plan for Food and Nutrition Policy 2007-2012, p. 17.

20.6.6 The Commission's White Paper on nutrition, overweight and obesity-related health issues

For the implementation of the Action Plan in the European Union the WHO explicitly referenced the European Commission's White Paper on nutrition, overweight and obesity-related health issues that had been published only a few months earlier in May 2007.[1093] In its White Paper, the Commission once again pointed out the necessity to integrate policies from all relevant fields, i.e. from food and consumer to sport, education and transport and to conduct actions on all levels of decision-making.[1094] It based its strategy on four guiding principles which require that actions should:

- target the root-causes of obesity and poor nutrition;
- work across policy areas and different levels of government;
- include private actors like food industry and civil society;
- be monitored and their effectiveness assessed.[1095]

The European Commission saw its focus in actions that were either within its competencies or where it perceived added value like facilitating the dialogue with international food industry stakeholders or the development of comparative indicators for monitoring.[1096] While it stressed that the development of effective partnerships must be the cornerstone of Europe's response to tackling nutrition, overweight and obesity and their related health problems and that it preferred in principle self-regulatory approaches in marketing and advertising,[1097] the Commission opted for a higher degree of intervention in the area of on-pack information about foods.[1098] For the first time the White Paper also addressed the role of the Common Agricultural Policy (CAP) in the promotion of public health goals by the provision of healthy options to consumers. However, action was by and large limited to the distribution of surpluses of fruit and vegetables to schools and other education centres.[1099] The question whether the existing allocation of subsidies within the framework of the CAP with its promotion of the production of alcoholic beverages, tobacco and foods high in saturated fats is appropriate to serve the cause of public health, however, remained unaddressed. Finally the Commission planned to encourage more physical activity, from organised sports to 'active commuting', which makes use of active ways of travel like walking or cycling to get to work and proposed European health survey systems to harmonise and improve the database on important indicators and nutrition measures related to diet and physical activity, like height and weight, Body-Mass-Index, performance of physical activity, consumption of fruit and vegetables, cholesterol levels and

[1093] COM(2007) 279 final, 30.05.2007.
[1094] COM(2007) 279 final, 30.05.2007, p. 3.
[1095] COM(2007) 279 final, 30.05.2007, p. 3 *et seq.*
[1096] COM(2007) 279 final, 30.05.2007, p. 4.
[1097] COM(2007) 279 final, 30.05.2007, p. 4 and 6.
[1098] COM(2007) 279 final, 30.05.2007, p. 5 *et seq.*
[1099] COM(2007) 279 final, 30.05.2007, p. 6.

hypertension.[1100] Many of these points were included in the Commission's White Paper on the EU Health Strategy for 2008-2013 in autumn 2007.[1101]

20.6.7 The stick is out: The Regulations on Nutrition and Health claims and on Food Information

The first example of a major regulatory intervention for the purpose of the implementation of a European Nutrition Policy was Regulation 1924/2006. The Regulation was adopted in December 2006 and applied from 1 July 2007. However, as some key elements of the regulation required implementing acts, like the adoption of a list of authorised health claims and the creation of nutrient profiles, the legislation remained merely a fragment for almost six years. It was only with the adoption of the list of authorised health claims in December 2012 that the general prohibition of unauthorised health claims became effective in practice. And even this was still not more than a partial success as a vast number of botanicals with health claims can continue to be placed on the market without an authorisation because they benefit from a transition provision that preserves the status quo until the scientific assessment of these claims by the European Food Safety Authority has been completed. Another core element of the regulation, the nutrient profiles, is still missing. As these profiles were supposed to set limits to the amounts of fat, saturated fat, trans fat, sugar and salt in foods or beverages that are allowed to bear a nutrition or health claim, without the profiles the contribution of the regulation to the promotion of healthier diets became rather limited if not a failure. Based on the current status quo, even products with e.g. a high sugar level could be advertised as having health benefits, e.g. because of their calcium content unless such advertising would be regarded as misleading under the general provisions of the legislation. Whether the latter would be the case is entirely left to the judgment of national courts and enforcement authorities, thus creating major inconsistencies in how the regulation is applied.

Another main area of regulatory attention was nutrition information and labelling. Here the Commission, after a consultation in 2006, presented first drafts for a Regulation on the Provision of Food Information to Consumers in late 2007 with the final proposal published in January 2008.[1102] The proposal contains an explicit reference to the White Paper, highlighting the consumers' need for 'clear, consistent and evidence-based information' and the role nutrition labelling should play in enabling consumers to make 'health conscious food choices' and in supporting balanced diets,[1103] points that had also been raised by consumer groups and public health NGOs during the stakeholder consultation. It also took on board the suggestion to have selected elements of nutrition labelling on the front-of-pack.

[1100] COM(2007) 279 final, 30.05.2007, p. 7 and 9.
[1101] COM(2007) 630 final, 23.10.2007.
[1102] COM(2008) 40 final, 30.01.2008.
[1103] COM(2008) 40 final, 30.01.2008, p. 2.

In the original proposal a mandatory declaration for energy, fat, saturates and carbohydrates with specific reference to sugars and salt, expressed as amounts per 100 g or per 100 ml was foreseen in the principal field of vision, which the Commission understood to be front-of-pack. The Commission justified this selection with the fact that these nutrients were the ones which had a relationship to risk factors for obesity and non-communicable diseases.[1104] During the legislative process a major controversy over the format of the nutrition labelling arose. While a majority of Member States and the food industry supported the system proposed by the Commission, that closely resembled the format used by the food industry in its voluntary commitment to front-of-pack labelling based on Guideline Daily Amounts (GDA) made in the context of the Platform on Diet, Physical Activity and Health, the United Kingdom and consumers associations advocated a colour coded traffic light system, which linked a low, medium and high content of certain nutrients to a respective green, amber or red colour. While the supporters of the GDA system argued that it was more factual, the followers of the traffic light approach highlighted the simplicity of the colour coding that would allow consumers to identify at a glance whether a product was high or low in calories, fat, sugar or salt.

Diagram 20.2. GDA system (source: Food Drink Europe).

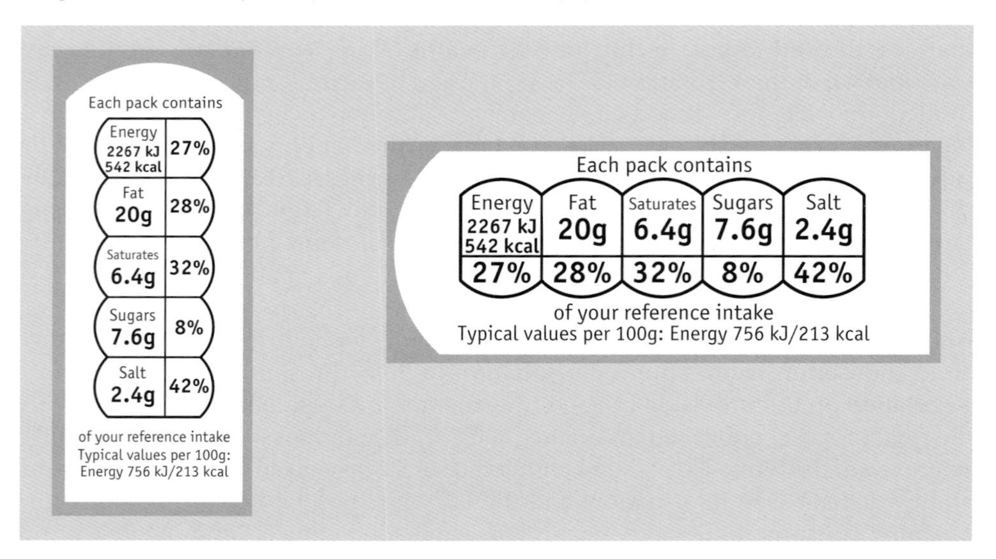

[1104] COM(2008) 40 final, 30.01.2008, p. 8.

Diagram 20.3. UK traffic light system (source: Food Standards Agency).

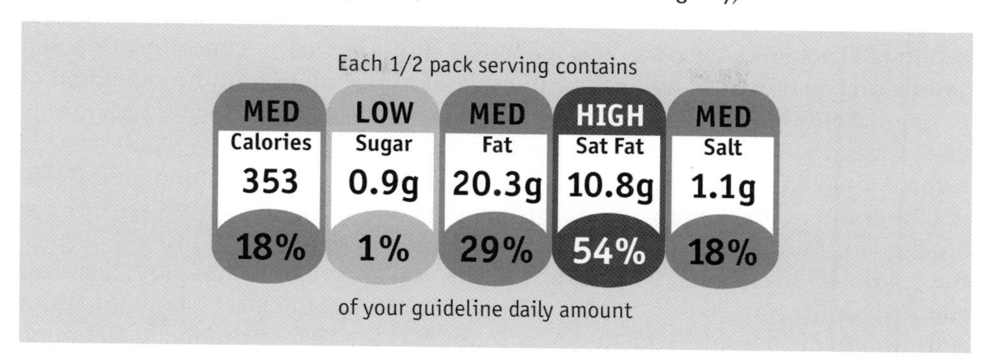

After long negotiations between the Commission, the Council and the European Parliament a compromise was reached and the labelling of seven nutrients became mandatory from December 2014 onwards, albeit not front-of-pack. The Food Information Regulation 1169/2011 however allows front-of-pack nutrition labelling using a GDA-based system.

20.6.8 Product reformulation and self-regulatory approaches in audiovisual media – the silence before the storm?

The Round Table had presented its report in July 2006 in which it had identified some best practices for effective self-regulation of advertising.[1105] As a result of the report, the Commission decided to maintain its voluntary approach for the time being but to review the situation in 2010. It called however for the identified best practices to be applied to marketing of food and beverages to children and encouraged Member States to ask broadcasters and other media service providers to introduce codes of conduct for food and beverage advertising targeted at children that limit the advertising in particular for products that contain a high amount of fat, trans-fatty acids, salt/sodium and sugars for which excessive intakes in the overall diet are not recommended.[1106] In 2010 this was turned into law in Article 9(2) of the Audiovisual Media Services Directive.[1107] However, in some countries, like the United Kingdom or Spain, such advertising restrictions had already been introduced by national self-regulatory bodies as early as 2007.[1108]

[1105] Available at: http://tinyurl.com/q3u52fp.

[1106] COM(2007) 279 final, 30.05.2007, p. 6.

[1107] Directive 2010/13 of 10 March 2010.

[1108] The Ofcom Code in the United Kingdom introduced broadcasting restrictions on foods and beverages high in fat sugar or salt (HFSS) based on nutrient profiles developed by the Food Standards Authority in April 2007 and a total ban on advertising for HFSS products in all children's programmes came into force in January 2009, available at: http://tinyurl.com/n2mwkeb; in Spain the PAOS Code was established in 2005, available at: http://tinyurl.com/o6c2ohl.

Another important theme in the European Commission's White Paper of 2007 was to encourage initiatives for the reformulation of processed foods to make diets healthier. It acknowledged the food industry's efforts but at the same time proposed conducting a study to fully explore the role of reformulation in the reduction of those nutrients that at high intakes could contribute to the development of chronic diseases and to assess all possible measures, including the option of regulation.[1109] It further called for the food industry to make healthier options more affordable and announced its support for campaigns on salt reduction.[1110] As part of the Strategy for Europe on Nutrition, Overweight and Obesity Related Health Issues, the High-Level Group on Nutrition and Physical Activity back in July 2008 had adopted an EU framework for national salt initiatives which set a benchmark of a minimum of 16% salt reduction over four years for all food products, including salt consumed in restaurants and catering.[1111] It followed a five-step process:[1112]

1. Review existing data on salt intakes.
2. Define benchmarks and food categories to focus on for salt reduction efforts.
3. Get broad endorsement of the common vision on salt reduction through product reformulation with food producers and their local federations; in order to maximise the impact of salt reformulation, food manufacturers were invited to prioritise products with the largest market share. Furthermore, efforts were to be designed for salt reformulation to take place across the full range of food products from premium to economy items, so that all population groups could benefit.[1113]
4. Raise public awareness, preferably in partnerships with NGOs, industry, media, the health sector and national platforms.
5. Monitor progress and evaluate effectiveness.

Even though the interim report on the framework noted that salt intake reduction is a slow process with technological barriers to be addressed, it also said that the framework proved to be a catalyst for action.[1114] Based on these learnings, the High Level Group widened the scope and in February 2011 established an EU Framework for National Initiatives on Selected Nutrients, which introduced in particular saturated fat as another target nutrient for reduction initiatives. The agreed benchmarks are:[1115]

- no increase in absolute amounts of trans-fat, sugars, salt and caloric content;

[1109] COM(2007) 279 final, 30.05.2007, p. 7.

[1110] COM(2007) 279 final, 30.05.2007, p. 7.

[1111] European Commission, Survey on Members States' Implementation of the EU Salt Reduction Framework, 2012, p. 5, available at: http://tinyurl.com/bxj59qn.

[1112] EU framework for national salt initiatives, available at: http://tinyurl.com/kt2fw7y.

[1113] European Commission, Survey on Members States' Implementation of the EU Salt Reduction Framework, 2012, p. 5.

[1114] Available at: http://tinyurl.com/mevuvn9, p. 14.

[1115] Reformulation Frameworks Progress in the Member States, Presentation of the European Commission to the Platform on Diet, Physical Activity and Health, 28 February 2013, available at: http://tinyurl.com/kdxyqlp.

- a minimum reduction in saturated fat intake of 5% in four years, an additional 5% until 2020;
- main product categories in focus:
  - school meals;
  - ready meals;
  - dairy products (including cheese);
  - meat products;
  - fats, oils margarines;
  - meals served in modern restaurants;
  - breakfast cereals.

The Council of the European Union and the European Parliament welcomed the Commission's initiative. They supported the inter-sectoral approach and the challenge put to food manufacturers, retailers and advertisers to be more active in the field of product reformulation and in the development of codes of conduct for the marketing of foods high in calories, fat, sugar and salt to children.[1116] The two institutions also welcomed the establishment of a High Level Group on Nutrition and Physical Activity composed of representatives of the Member States to link up national initiatives with the activities of the EU Platform for Action on Diet, Physical Activity and Health.[1117] It agreed as well with the other priorities set by the Commission, namely salt reduction, promotion of healthy nutrition in schools and workplaces and increasing the level of physical activity in the European population, and asked the Commission to report back on the progress made in 2010.[1118] The European Parliament made it clear though that if the voluntary approaches didn't show effects by that time, regulation should be introduced.[1119] Moreover, it called for restrictions on the sale of foods and beverages high in fat, sugar and salt in schools.[1120]

20.6.9 2010: The year of truth for the effectiveness of the EU nutrition policy measures

With two major reports on the implementation of the European nutrition policy being due, the year 2010 became a moment of truth for the new EU nutrition policy as it was laid out in the 2007 White Paper. In that year, the Commission published their report on the implementation of the activities proposed in the White Paper and another one on the effectiveness of the European Platform for Action on Diet, Physical Activity and Health.

[1116] Council conclusions on the Commission White Paper on a strategy for Europe on nutrition, overweight and obesity-related issues, 5 December 2007, Council document 9803/05, p. 2 et seq; European Parliament resolution of 25 September 2008, OJ C 8E, 14.01.2010, p. 97.
[1117] Council conclusions (2007), p. 3; European Parliament resolution (2008), p. 100.
[1118] Council conclusions (2007), p. 2 *et seq.*
[1119] European Parliament resolution (2008), p. 104.
[1120] European Parliament resolution (2008), p. 104.

20.6.10 The implementation report on the Commission's White Paper

The Commission's implementation report was the result of a collaboration with the WHO Regional Office for Europe where improvements in nutrition, physical activity and eventually in preventing obesity as a result was monitored over a three-year period.[1121] Based on the WHO European Charter on Counteracting Obesity the overall benchmark for progress was defined as 'visible progress' in reducing the prevalence of obesity, in particular in children and adolescents with the aim to reverse the obesity trend by 2015.[1122] A more detailed assessment was made based on sixteen indicators that were allocated to the seven action areas defined in the White Paper.[1123] The report firstly acknowledged that a three-year period was too short to see major changes in the general trends on obesity, in particular as survey data was only available for a few countries until 2009, but that the overall situation remained worrying with 30-70% of adults being overweight and 10-30% being obese.[1124] Measures taken in the area of consumer information included the proposal for the Food Information Regulation and the work on the implementation of the Regulation on Nutrition and Health Claims. However, the Commission's expectation to adopt nutrient profiles early on for the use of nutrition and health claims appears, in hindsight, to have been overly optimistic. Overall it still remains to be seen whether these legislative acts will have any major impact on consumer behaviour. Progress was certainly made in the area of marketing to children where most Member States had agreed a code of conduct with food business operators.[1125] The availability of healthier options was strengthened by the EU School Fruit Scheme[1126] launched in 2009 that was subsidised by the EU at 50% of the cost with an annual budget of 90 million euros. 25 countries participated in 2010-2011 reaching over 8 million children in more than 54,000 schools.[1127] A short-term positive impact on fruit and vegetable consumption was identified in the final report of the scheme. However it was also stressed that continuous free availability and parental behaviour were key factors to success while the limited availability of funds for the scheme on national levels was seen as a major threat.[1128] To ensure the full effectiveness the report recommended a high frequency of distribution over a long period for as many children as possible. For this, however, the funding of the scheme had to be improved,[1129] a call that as

[1121] European Commission, Strategy for Europe on nutrition, overweight and obesity related health issues, Implementation Progress Report, December 2010, available at: http://tinyurl.com/k9vn2qr.
[1122] Progress Report, p. 7; WHO Ministerial Conference on Counteracting Obesity, European Charter on Counteracting Obesity (2006), p. 2, available at: http://tinyurl.com/qht4rt6.
[1123] Annex 6 of the Progress Report, available at: http://tinyurl.com/pmh5x7c.
[1124] Progress Report, p. 8.
[1125] Progress Report, p. 17.
[1126] EU School Fruit Scheme, available at: http://tinyurl.com/qgohsde.
[1127] European Commission, Evaluation of the School Fruit Scheme, Final Report, 2012, p. 104, available at: http://tinyurl.com/m62xq8a.
[1128] School Fruit Scheme Evaluation Report, p. 105 *et seq.*
[1129] School Fruit Scheme Evaluation Report, p. 112.

yet has remained unanswered as the funding of the scheme for the year 2013/14 stayed at a constant level.[1130] On the support of physical activity the Commission referred to the actions laid out in its White Paper on Sport.[1131] These included the adoption of EU Physical Activity Guidelines[1132] and measures taken on safe walking and cycling in the context of the CIVITAS initiative.[1133] Based on the insight that the prevalence of obesity and overweight is above average in low socio-economic groups and that over 20% of the obesity in men and 40% of the obesity in women in Europe could be attributed to social inequalities[1134] the Commission in 2009 presented a Communication with suggestions for action on the Community level.[1135] The proposals concentrated on the improvement of the database and the exchange of best practices, e.g. the continuation of the school fruit and school milk schemes as well as the distribution of food products to the most deprived groups in the European Union with an annual budget of 480 million euros.[1136]

20.6.11 The progress report on the European Platform for Action on Diet, Physical Activity and Health

A major part of the progress report dealt with the collaborations and initiatives that were established with food business operators on healthy diet and lifestyle promotion, reformulation and marketing to children, mainly within the EU Platform on Diet, Physical Activity and Health. As far as reformulation was concerned, the Commission concluded that Platform commitments could have a significant impact, provided that the reformulation efforts were taken beyond the national level and monitored for compliance. It also identified the need for public authorities to take the lead in setting national standards for recommended levels of sugar, salt, saturated fats and trans-fats and to define common thresholds.[1137] The latter is somewhat surprising, given that the European Commission had presented its working document on setting the nutrient profiles in the context of the EU regulation on Nutrition and Health Claims in February 2009[1138] but might have been a reaction to the political deadlock that followed the proposal and prevented the profiles from being adopted on an EU level. For the self-regulatory commitments on advertising to children, mainly represented by the EU Pledge, a case study initiated by the Commission indicated that the Pledge had led to a reduction in TV

[1130] Commission allocates € 88 million for 2013/2014 School Fruit Scheme, available at: http://tinyurl.com/cgudj8l.
[1131] COM(2007) 391 final, 11.07.2007.
[1132] EU physical activity guidelines 2008, available at: http://tinyurl.com/l498zjr.
[1133] See: http://tinyurl.com/n8fu3n4.
[1134] A. Robertson, T. Lobstein and C. Knai, Obesity and socio-economic groups in Europe: Evidence review and implications for action, 2007, available at: http://tinyurl.com/mqok3sf.
[1135] Solidarity in health: Reducing Health Inequalities in the EU, COM(2009) 567 final, 20.10.2009.
[1136] Regulation 945/2010 of 21 October 2010.
[1137] Progress Report, p. 30.
[1138] Working document of 13 February 2009, available at: http://tinyurl.com/mr6l4qw.

advertising by member companies of one third and a discontinuation of advertising for ice-cream, confectionery chocolate, soft drinks or non-reformulated snacks to children under 12.[1139] Despite these developments, the Case Study Report stressed that the overall exposure of children to marketing for foods and beverages high in fat, sugar and salt had gone down by a significantly lesser degree.[1140] To increase the impact the report suggested including young people up to the age of 16 in the definition of 'children' and applying stricter thresholds for what are considered children audiences, e.g. by applying a time watershed rather than a certain percentage of viewership.[1141] Moreover, it was recommended that the members of the Pledge applied stricter nutritional criteria that meet national and international standards.[1142] In summary, the overall Evaluation Report on the effectiveness of the Platform, of which the Case Study formed a part, took the view that for both the voluntary efforts of food business operators in the areas of reformulation and marketing to children it was too early to judge whether a significant impact on public health could be made.[1143] As a result of these findings, the Commission clearly warned that if its review of the self-regulatory rules created under the framework of the Audiovisual Media Services Directive (AVMSD) in 2011 found them to be insufficient, it would consider regulation.[1144] When the AVMSD review came out, it concluded that substantial progress had been made in the establishment of self-regulatory codes of conduct in general, but that further work was required on age and audience thresholds, more consistent nutritional benchmarks and in general in the area of commercial communications for sweet, fatty or salty foods or drinks in children's programmes.[1145]

In its overall conclusions of the Progress Report on the implementation of the European Strategy on Nutrition, Overweight and Obesity Related Issues, the Commission found that in defiance of numerous activities, 'there is little sign of decrease in the recently identified negative trends on overweight and obesity' and that the situation could get even worse as a result of the economic crisis that had haunted Europe by that time.[1146] It stressed that it would stay committed to the Platform's voluntary and non-prescriptive approach but expected the delivery of proven results by the end of 2013.[1147]

[1139] European Commission, Evaluation of the European Platform for Action on Diet, Physical Activity and Health, Case Study Report: Advertising and marketing to children, July 2010, http://tinyurl.com/3abvqvz.
[1140] Case Study Report, p. 49.
[1141] Case Study Report, p. 50.
[1142] Case Study Report, p. 50.
[1143] European Commission, Evaluation of the European Platform for Action on Diet, Physical Activity and Health, July 2010, p. 4, available at: http://tinyurl.com/osfcvaz.
[1144] Progress Report, p. 28 *et seq.*
[1145] First Report from the Commission to the European Parliament, the Council, the European Economic and Social Committee and the Committee of the Regions on the application of Directive 2010/13/EU 'Audiovisual Media Service Directive', COM(2012) 203 final, 04.05.2012, p. 9.
[1146] Progress Report, p. 34.
[1147] Progress Report, p. 35.

20.6.12 The WHO Global Action Plan 2008-2013

The developments in Europe can only be fully understood in the overall global context. Already back in 2008, the World Health Organization had adopted a 2008-2013 Action Plan for the implementation of its Global Strategy for the Prevention and Control of Non-communicable Diseases, which once more underlined the need to tackle the high prevalence of cardiovascular diseases, diabetes, cancers and chronic respiratory diseases by targeting those risk factors the diseases had in common, in particular physical activity, unhealthy diets and the use of alcohol and tobacco. [1148] It also stressed that non-communicable diseases were a subject not only for the developed world but also for low- and middle-income countries,[1149] with overweight and obesity in 2010 ranking as the fifth leading risk of death globally.[1150] The global dimension of the problem eventually led to a Resolution by the United Nations General Assembly in 2011 where governments made multiple commitments. These included:

- multi-sectoral, cost-effective and population wide interventions to reduce the impact of the common risk factors for non-communicable diseases, including education, legislative, regulatory and fiscal measures;
- work on public policies for health-promoting environments which empower individuals, families and communities to make healthy choices and lead healthy lives;
- further implementation of the Global Strategy on Diet, Physical Activity and Health and the WHO Recommendations on the Marketing of Foods and Non-alcoholic Beverages to Children;
- interventions to reduce salt, sugar and saturated fats, to remove industrial trans-fats from foods and to discourage the production and marketing of foods that contribute to an unhealthy diet;
- promotion of the production of foods that support a healthy diet, including a greater use of local products.[1151]

Given that the WHO recommendations called for a reduction in the exposure of children to marketing of foods high in saturated fats, trans-fats, sugar or salt, a general prohibition of such marketing in places where children gather like nurseries, school or playgrounds and a limitation of the impact of cross-border marketing,[1152] it was not surprising that the World Federation of Advertisers

[1148] WHO, 2008-2013 Action Plan, 2008, Available at: http://tinyurl.com/cfztxns.
[1149] 2008-2013 WHO Action Plan, p. 2.
[1150] WHO, Set of recommendations on the marketing of foods and non-alcoholic beverages to children, 2010, p. 4, available at: *http://tinyurl.com/cewsq3k*.
[1151] United Nations, General Assembly, Resolution on the Political Declaration of the High-level Meeting of the General Assembly on the Prevention and Control of Non-communicable Diseases, 19 September 2011, A/RES/66/2, available at: http://tinyurl.com/q5w6ljt.
[1152] WHO, Set of recommendations on the marketing of foods and non-alcoholic beverages to children, 2010.

announced a strengthening of the EU Pledge as of 2012.[1153] As a result, the members of the Pledge committed not to broadcast advertising for products that don't meet their nutritional criteria in programmes where more than 35% of the audience are children under 12 (instead of the previous threshold of 50%) and not to make websites for such products appealing for children under the age of 12.[1154] Despite these efforts it appears that the time for voluntary, self-regulatory commitments is running out.

20.6.13 Nutrition Policy in Europe – the interim balance and outlook

Since the modest beginnings of the first European Action Programme on Nutrition and Health in 1990 there has been an increasingly dynamic move towards a common European nutrition policy. While in the early 1990s the calls to tackle the issue of unhealthy diets were mainly political in nature and could at best be characterised as 'soft law', since the start of the new millennium the field of human nutrition has become increasingly juridified by the adoption of legal acts and self-regulatory initiatives. What used to be Recommendations, Health Programmes, Resolutions and Action Plans evolved to Green Papers, White Papers and eventually legislation, e.g. on nutrition and health claims or on consumer information.

Diagram 20.4. Key initiatives on EU nutrition policy and their legal classification.

Year	Measure	Subject matter	Classification
1987	Single European Act	Introduction of public health and high level of consumer protection as basic principle	Legal Act
1990	Council Resolution on Nutrition and Health	First Action Programme on Nutrition and Health	Political declaration
1993	Maastricht Treaty	Commission mandate for EU public health strategy	Legal Act
1994	Council of Europe Report on Food and Health	Call for a comprehensive European nutrition policy based on preventive nutrition	Political declaration
1998	Commission Communication COM(1998) 230 final	Development of a public health policy in the EU	Policy paper
1999	Commission White Paper on Food Safety	Development of a comprehensive and coherent nutrition policy	Policy paper

▶▶

[1153] High Level Group on Nutrition and Physical Activity, Flash Report, 28 November 2011, available at: http://tinyurl.com/pxkndxp.
[1154] High Level Group on Nutrition and Physical Activity, Flash Report, 28 November 2011.

Diagram 20.4. Continued.

2000	WHO Europe First Action Plan for Food and Nutrition Policy	Three pillar strategy of food safety, healthy nutrition and sustainable food supply	Political declaration / Recommendation
2000	Council Resolution on Health and Nutrition	Call for more alignment of national nutrition policies	Political declaration
2001	Commission Discussion Paper on Nutrition and Functional Claims	Stakeholder consultation on regulation of nutrition and health claims	Preparatory document for legislative act
2002	European Parliament and Council Community Action on Public Health	5 year action programme, acknowledging an active role of the EU in shaping public health policy; foundation of Network on Nutrition and Physical Activity	Political declaration and 'soft' measures
2002	Council Conclusions on Obesity	Prevention of obesity acknowledged as an issue to be considered in all relevant EU policies	Political declaration
2002	Commission Status Report	Analysis of nutritional status of EU population; collection of correlations between certain pathologies and dietary risk factors	Policy paper
2003	Commission Proposal for a Regulation on Nutrition and Health Claims	Restriction of nutrition and health claims, linked to nutritional benchmarks (profiles)	Legislative proposal
2005	Commission Green Paper on Promoting Healthy Diets and Physical Activity	Outline of areas for action in EU nutrition policy	Policy paper
2005	European Platform on Diet, Physical Activity and Health	Round table with stakeholders to achieve binding and verifiable commitments that help to stop the trend of overweight and obesity	Self-regulation
2006	Regulation (EC) No. 1924/2006 on Nutrition and Health Claims	Restriction of nutrition and health claims, liked to nutritional benchmarks (profiles)	Legal Act

►►

Diagram 20.4. Continued.

2006/7	WHO European Charter on Counteracting Obesity and Second Action Plan	Call for more interventionist measures in tackling obesity	Political declaration
2007	Commission White Paper on Nutrition, Overweight and Obesity-related Health Issues	Proposal of integrative approach touching all policy fields relevant to nutrition; review of food labelling legislation	Policy paper and preparatory document for legislative act
2007	EU-Pledge by major food companies	No advertising for foods and beverages high in sugar, fat or salt to children under the age of twelve; no advertising in primary schools	Self-regulation
2008	WHO Global Action Plan 2008-2013	Implementation of the global strategy for the prevention and control of non-communicable diseases	Political declaration
2008	Proposal for a Regulation on Food Information to Consumers	Introduction of mandatory nutrition labelling; labelling as a tool to facilitate health conscious food choices	Legislative proposal
2008	EU Framework for National Salt Initiatives	Five step process to achieve a salt reduction of at least 16% over four years	Political declaration combined with self-regulation
2011	EU Framework for National Initiatives on Selected Nutrients	Reduction of saturated fat intake by 5% in four years and an additional 5% by 2020; no increase in trans-fat, sugar, salt and calories	Political declaration combined with self-regulation
2011	Regulation (EU) No. 1169/2011 on Food Information to Consumers	Introduction of mandatory nutrition labelling; labelling as a tool to facilitate health conscious food choices	Legal Act
2012	Strengthened EU-Pledge by major food companies	Extension of TV programmes that are considered to have an audience <12 years; inclusion of websites	Self-regulation

The reason for the increasing juridification in European nutrition policy is simple. The 2012 OECD report 'Health at a glance'[1155] showed that regardless of the efforts taken by the EU institutions, Member States and private stakeholders like NGOs and food industry, the prevalence of overweight and obesity in the population as a major indicator for the effectiveness of nutrition policy didn't show any improvement. By the time of the report, levels of overweight had increased to 52% of the adult population, with 17% being obese, despite an overall increase in life-expectancy, albeit with significant variations between different socio-economic groups. [1156] On a positive note the OECD reported that alcohol and tobacco consumption had significantly fallen in many EU Member States compared to the 1980s. This effect was mainly attributed to advertising bans or restrictions, sales restrictions and tax increases.[1157] Whereas admittedly alcohol and tobacco are very specific products with often dedicated sales channels and therefore easier to tackle than a sector as large as the total food consumption of a population, the OECD report suggests that regulatory intervention seems to produce more tangible results than self-regulation.

It must also be noted that the discussions in the High Level Group on Nutrition and Physical Activity increasingly circle around questions of regulatory intervention. The meeting in February 2012 looked at experiences with fat and sugar taxes in Member States like Denmark, France and Hungary and the Group reconfirmed in its February 2013 meeting that the exchange of experience in this field should continue.[1158] In November 2012 the Group looked into guidelines for policy interventions encouraging a healthy diet that were developed by the EU-supported EATWELL project.[1159] Within the EATWELL project two major policy options were identified.[1160] One strand is to create an environment supportive of informed choice by providing or limiting certain information, e.g. through nutrition labelling or by limiting the advertising for certain product groups. Such measures are considered to have positive but relatively small effects on diets, because knowing about the healthier choice does not necessarily mean a change in consumer behaviour and purchasing decisions.[1161] The other, more promising strand for achieving a substantial effect on consumer's diets is interventions that change the market environment by enhancing or limiting the availability of certain foods to consumers. Measures could also include changing relative prices through taxes or subsidies

[1155] OECD (2012), Health at a Glance: Europe 2012, OECD Publishing, DOI: http://dx.doi.org/10.1787/9789264183896-en.

[1156] OECD (2012), p. 8 *et seq.*

[1157] OECD (2012), p. 8 *et seq.*

[1158] High Level Group on Nutrition and Physical Activity, Flash Report, 2 February 2012, available at: http://tinyurl.com/ort74sj and 7 February 2013, available at: http://tinyurl.com/b6anjmx.

[1159] High Level Group on Nutrition and Physical Activity, Flash Report, 15 November 2012, available at: http://tinyurl.com/muzt43w.

[1160] Effectiveness of Policy Interventions to Promote Healthy Eating and Recommendations for Future Action: Evidence from the EATWELL Project, p. 7, available at: http://tinyurl.com/myflkp2.

[1161] EATWELL report, p. 7.

and aims to set-off the social costs of unhealthy diets.[1162] Even though the High Level Group still in November 2012 committed to achieving reformulation targets primarily in collaboration with industry,[1163] the mid-term perspective appears to be a move towards more regulation affecting product composition. This trend is confirmed by the recently adopted WHO Global Action Plan for 2013-2020, which proposes the use of economic tools like taxes and subsidies to encourage the consumption of healthier food products and discourage the consumption of less healthy options.[1164] It is flanked by an ever increasing push for the provision of information to consumers, exemplified by the EU Regulation on the provision of food information to the consumers.[1165] Rather than providing the technical rules on the labelling of pre-packaged food the legislative act became an instrument for consumer education far beyond the product label. Article 1(2) of the Regulation describes its subject matter as laying down 'the means to guarantee the right of consumers to information and procedures for the provision of food information, taking into account the need to provide sufficient flexibility to respond to future developments and new information requirements'. This leaves a lot of room for future initiatives on nutritional education.

20.7 Conclusions

In summary it must be concluded that despite a massive increase in the knowledge about the root causes and numerous action plans over the last decades, the prevalence of obesity in the European population is still on the rise. It appears that neither the carrots of educational campaigns or free fruit and vegetables nor the stick of regulatory intervention in the field of food information had any enduring impact on consumer behaviour. One of the main reasons for this could be the fact that most activities in the context of the European Nutrition Policy so far have aimed at educating the consumer and at providing the consumer with an informed choice. It seems, however, that even if consumers are provided with the necessary information to make a healthier choice, many of them still seem to prefer the unhealthy choice. As a consequence, the next likely step by European policy makers will be to increasingly target the supply side of the food market with measures that improve the nutritional composition of the food products that are on offer. The first legislative acts introducing certain thresholds or taxes on nutrients like (saturated or trans-)fat, sugar or salt in some Member States[1166] are

[1162] EATWELL report, p. 9; this finding is supported by the WHO Regional Committee for Europe's Health 2020 policy framework and strategy, p. 22 *et seq.*, available at: http://tinyurl.com/kzop4xf.
[1163] Flash Report, 15 November 2012.
[1164] WHO, Global Action Plan for the Prevention and Control of Noncommunicable Diseases 2013-2020, p. 25, available at: http://tinyurl.com/mr8mkfo.
[1165] Regulation (EU) 1169/2011, OJ L 304, 22.11.2011, p. 18.
[1166] E.g. Danish Executive Order No. 160 of 11 March 2003 on the Content of Trans Fatty Acids in Oils and Fats and Austrian Regulation of 20 August 2009 on the Content of Trans Fatty Acids in Foods, both setting an upper limit of 2% for industry-produced trans-fats and oils in food which could be sold to consumers; Hungarian tax on foods with a high sugar or salt content.

an indicator for that. The challenge for such measures will be to strike the right balance between effectiveness and the restriction of freedom of choice. Only if consumers don´t feel patronised and food business operators are allowed to meet the demands of their customers, can regulatory intervention on the supply side lead to a healthier portfolio of food products available in the shops. If the level of regulation becomes excessive, the case of alcohol prohibition in the United States in the 1920s is testament to the fact that human ingenuity will always find a way around such measures. And a black market for chocolate bars or ice cream would be a rather disturbing thought.

21. EU Feed Law

Dionne Chan

21.1 Introduction

The holistic approach to food law introduced by the General Food Law brings feed for food producing animals fully within the scope of food law. The objective of this chapter is to provide an overview of the structure of feed law.

Within the 'farm to fork' policy feed is important both for the health of animals and for the health of humans. For food to be safe, the animals it comes from have to be safe, as contaminated or unhealthy animals may subsequently affect the human health. An outline of EU feed law will be given in this chapter. This chapter closely follows the framework set out in Chapter 8 for the analysis of food law. Adapted to feed this framework is shown in Diagram 21.1

Diagram 21.1. Framework for the analysis of EU feed law.

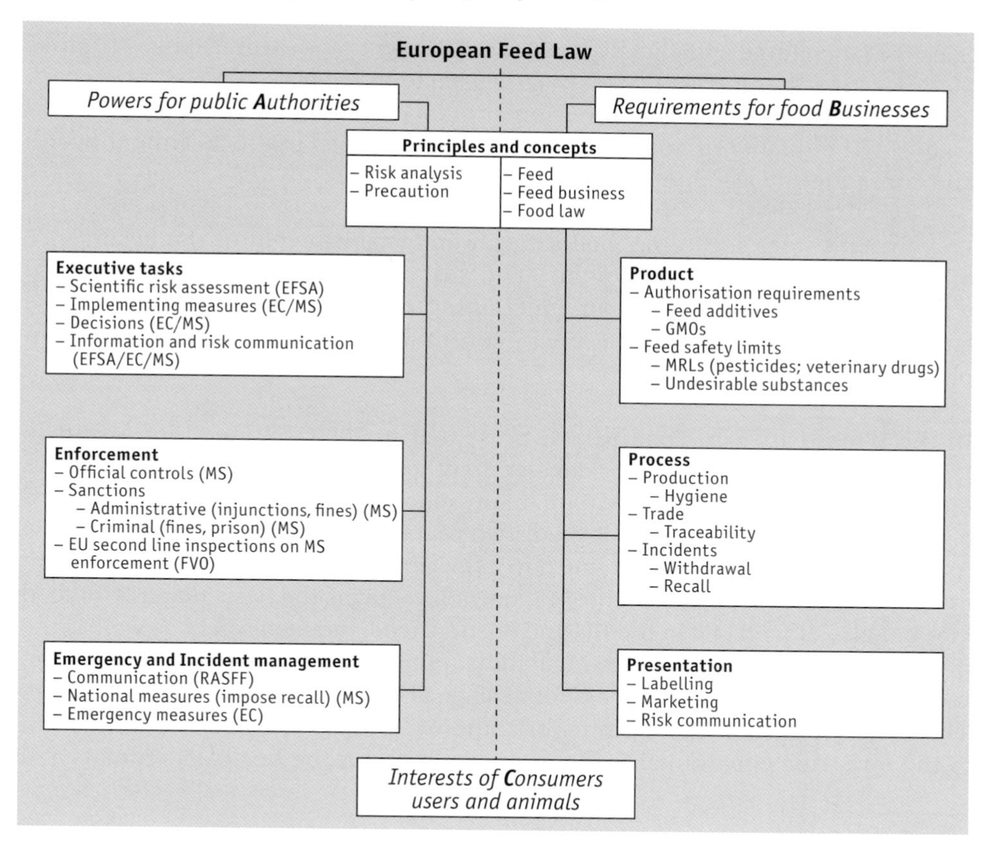

This chapter is organised as follows. Section 21.2 provides a short historical overview of the development of feed law. Section 21.3 addresses the general concepts and principles. Section 21.4 focuses on product specific requirements regarding authorisation and purity. Section 21.5 discusses process requirements: hygiene, traceability and recall. Section 21.6 addresses presentation requirements: labelling and marketing. Section 21.7 focuses on the role of public authorities in enforcement and Section 21.8 presents the conclusions.

21.2 Development of EU Feed law

From the early 1960s until the eruption of the BSE crisis in the mid-1990s, European food law was principally directed at the creation of an internal market for food products in the EU. However, food issues influenced not only the development of EU food law, but also the development of EU feed law as many food safety problems were caused by the feed given to livestock animals.

The spread of mad cow's disease, for example, was caused by cattle feed containing BSE-infected meat. The contamination of chicken and chicken eggs with dioxin in the Belgian dioxin crisis resulted from chickens that were fed feed containing maize or fat polluted with dioxin. Residues of the synthetic contraceptive hormone Medroxy Progesterone Acetate (MPA) has been found in pork and was traced back to the feed. These are just a few examples of food safety issues caused by animal feed.[1167] The quality of animal nutrition is essential as it affects animal health and consequently also human health.

Preventative rules regarding food safety were established after the occurrence of large-scale food safety problems. The aim of EU legislation regarding food was mainly to support a proper functioning of the EU market through legal harmonisation. During the BSE crisis, consumer confidence in the food chain plummeted.

In the White Paper on Food Safety, presented in 2000, the need for a shift in legislative priority from market development towards food safety assurance was emphasised.[1168] Since its publication, many pieces of legislation considering food and feed safety have been established. Two years hereafter, the General Food Law (GFL), Regulation 178/2002, laying down the general principles and requirement of food law, was established. The GFL nowadays forms the basis not only of food law, but also of feed law in the EU and ensures food and feed safety as well as the protection of public health. The regulation applies to all stages of the production, recognising animal feed as an essential stage at the beginning of the food chain. This is also known as the 'farm to fork' approach. As food comes from the farm to the fork, rules on animal feed are crucial both for the health of animals and

[1167] See also Chapter 7.
[1168] COM(1999) 719 final. White Paper on Food Safety.

for the health of humans. Legislation regarding feed for food-producing animals takes both into account. Legislation regarding feed for other animals (such as pets) regards only the former.

21.3 General concepts and principles

21.3.1 Concepts

The General Food Law defines the scope of food and feed law. Feed is included in the scope of food law. According to Article 3(1) GFL 'food law' means 'the laws, regulations and administrative provisions governing food in general, and food safety in particular, whether at Community or national level; it covers any stage of production, processing and distribution of food, *and also of feed produced for, or fed to, foodproducing animals*' (emphasis added). This does not mean that it is also included in the concept of food.[1169] Article 2 GFL makes a clear distinction between the definitions 'food' and 'feed'. The GFL explicitly excludes feed in the definition of food with the words: 'food shall not include feed'. Subsequently, Article 3(4) of the GFL provides the definition of 'feed'. Pursuant to this article, 'feed' or 'feedingstuff' means any substance or product, including additives, whether processed, partially processed or unprocessed, intended to be used for oral feeding to animals. Oral feeding of animals means the introduction of feed into an animal's gastrointestinal tract through the mouth with the aim of meeting the animal's nutritional needs.[1170]

Feed legislation addresses four types of feed: feed materials, compound feed including pet food, feed additives and medicated feedingstuffs. The definitions of these concepts determine which legislation is applicable.

The first type of feed is 'feed materials' and is defined in Article 3(1)(g) of Regulation 767/2009. Feed materials means products of vegetable or animal origin, whose principal purpose is to meet animals' nutritional needs, in their natural state, fresh or preserved, and products derived from the industrial processing thereof, and organic or inorganic substances, whether or not containing feed additives, which are intended for use in oral animal-feeding either directly as such, or after processing, or in the preparation of compound feed, or as carrier of premixtures.[1171]

The second type of feed is 'compound feed'. As stated by Article 3(1)(h) of Regulation 767/2009, compound feed means a mixture of at least two feed materials, whether

[1169] This is different in the USA. The definition of food in the Federal Food, Drug and Cosmetic Act covers 'food for humans and other animals'.
[1170] Article 3(2)(b) of Regulation 767/2009.
[1171] Article 2(1)(e) Regulation 1831/2003: 'Premixtures' means mixtures of feed additives or mixtures of one or more feed additives with feed materials or water used as carriers, not intended for direct feeding to animals.

or not containing feed additives, for oral animal-feeding in the form of complete or complementary feed. Whereby complete feed means compound feed which, by reason of its composition, is sufficient for a daily ration and complementary feed means compound feed which has a high content of certain substances but which, by reason of its composition, is sufficient for a daily ration only if used in combination with other feed.[1172]

The third type of feed, 'feed additive', is defined in Article 2(2)(a) of Regulation 1831/2003. Based on this provision and Article 5(3) of this Regulation, feed additives means substances, microorganisms or preparations, other than feed material and premixtures, which are intentionally added to feed or water in order to perform, in particular, one or more of the following functions: (a) favourably affect the characteristics of feed, (b) favourably affect the characteristics of animal products, (c) favourably affect the colour of ornamental fish and birds, (d) satisfy the nutritional needs of animals, (e) favourably affect the environmental consequences of animal production, (f) favourably affect animal production, performance or welfare, particularly by affecting the gastrointestinal flora or digestibility of feedingstuffs, or (g) have a coccidiostatic or histomonostatic effect. There are five categories of feed additives: technological additives, sensory additives, nutrition additives, zootechnical additives, and coccidiostats and histomonostats.[1173]

The definition of feed additives is more explicit and more extensive than the definition of food additives. Food additives are defined as substances that are intentionally added to foods, whether or not it has nutritive value for a technical purpose. The legal definition of feed additives spells out certain technological purposes and also includes substances that do not have any technological purpose, but instead a nutritional or zootechnical purpose.[1174]

The last type of feed is 'medicated feedingstuffs'. This type of feed is described as any mixture of a veterinary medicinal product or products and feed or feeds which is ready prepared for marketing and intended to be fed to animals without further processing, because of its curative or preventive properties or other properties as a medicinal product covered by the concept of a veterinary medicinal product. Veterinary medicinal products are substances presented for treating or preventing disease in animals. Any substance or combination of substances which may be administered to animals with a view to making a medical diagnosis in animals is likewise considered a veterinary medicinal product.[1175]

[1172] Article 3(2)(i) and 3(2)(j) of Regulation 767/2009.
[1173] Coccidiostats and histomonostats are substances intended to kill or inhibit certain parasites.
[1174] See Article 3(2) of Regulation 1333/2008; and B. Klaus, 'Distinction between Feed Materials and Feed Additives in Consideration of the General Principles of Law', European Food and Feed Law Review 1, 2011, pp. 6-7.
[1175] Article 1(6) and Article 1(2) of Directive 2001/82.

Feed law by no means applies the rigorous separation from medicine that we find in food law. A product can be a feed and a veterinary medicine at the same time.

21.3.2 Objectives

The aim of the GFL is to ensure a high level of protection of human health and consumers' interest, whilst ensuring effective functioning of the internal market.[1176] The general objectives of EU food law are described in Article 5 of the GFL:

(1) Food law shall pursue one or more of the general objectives of a high level of protection of human life and health and the protection of consumers' interests, including fair practices in food trade, taking account of, where appropriate, the protection of animal health and welfare, plant health and the environment.

(2) Food law shall aim to achieve the free movement in the Community of food and feed manufactured or marketed according to the general principles and requirements in this Chapter.

The rules established by the GFL apply to all stages of the production, processing and distribution of food, and also of feed produced for, or fed to, food-producing animals, the farm to fork approach.[1177] In the farm to fork approach, feed is an essential phase as it is the start of the food chain. Consequently, feed law shares the similar objectives as food law: a high level of food and feed safety, thus the protection of public health.

The objective of Regulation 767/2009, in accordance with the general principles laid down in the GFL, is to harmonise the conditions for the placing on the market and the use of feed, in order to ensure a high level of feed safety and thus a high level of protection of public health, as well as to provide adequate information for users and consumers and to strengthen the effective functioning of the internal market.[1178]

21.3.3 Principles

According to Article 6 of the GFL, food and feed law protecting human health should be based on risk analysis, and therefore on science. For circumstances where the possibility of harmful effects on health is identified but scientific uncertainty persists, the precautionary principle of Article 7 of the GFL applies. According to this article, provisional risk management measures may be adopted necessary to ensure high level of health protection, pending further scientific information for a more comprehensive risk assessment. These measures should be proportionate, no more restrictive to trade than required and reviewed within a reasonable period of time.

[1176] Article 1(1) of the GFL.
[1177] See also Article 4(1) of the GFL.
[1178] Article 1 of Regulation 767/2009.

21.4 Product-focused provisions

21.4.1 Market access requirements

Unsafe food products may not be placed on the market.[1179] The same condition applies to feed products. To this effect Article 15(1) GFL states:[1180] 'Feed shall not be placed on the market or fed to any food-producing animal if it is unsafe'. Feed products that have an adverse effect on human or animal health or make the food derived from food-producing animals unsafe for human consumption are considered unsafe feed products.[1181] It is the responsibility of food and feed business operators to ensure that food or feed products satisfy the requirements of food law and to verify that such requirements are met.[1182] Feed that is safe, does not have a direct adverse effect on the environment or animal welfare, is sound, genuine, unadulterated, fit for its purpose and of merchantable quality and also correctly labelled, packaged and presented, may be freely marketed without pre-market approval.[1183]

There is no general authorisation requirement for feeds that are 'novel'. However, for reasons of transparency materials used for feed are registered in a catalogue, the European Feed Materials Register. A person who, for the first time, places on the market a feed material that is not listed in the Catalogue should notify its use.[1184] Only specifically designated categories of feed – feed additives and genetically modified feeds – need prior authorisation. The legislator considers that these feedingstuffs may involve potential hazards.[1185] Feed additives are only allowed to be used or placed in the market with an authorisation.[1186] The authorisation granted in accordance with Regulation 1831/2003 is valid for ten years and can be renewed.

The placing on the market of genetically modified feed is regulated by a specific authorisation procedure. Rules for genetically modified organisms (GMOs) and food or feed containing GMOs, whether intended for human or animal consumption, are laid down in the Regulation 1829/2003 on genetically modified food and feed. GM feed can be only placed on the market if it is covered by an authorisation.

[1179] Article 14(1) of the GFL.
[1180] See also B. Klaus, 'Distinction between Feed Materials and Feed Additives in Consideration of the General Principles of Law', European Food and Feed Law Review, 2011, p. 3.
[1181] Article 15(1)(2) of the GFL.
[1182] Article 17(1) of the GFL.
[1183] Article 4 of Regulation 767/2009 and B. Klaus, 'Distinction between Feed Materials and Feed Additives in Consideration of the General Principles of Law', European Food and Feed Law Review, 2011, p. 3.
[1184] Article 24(6) of Regulation 767/2009. See also Regulation 68/2013. The register is available at: http://www.feedmaterialsregister.eu.
[1185] B. Klaus, 'Distinction between Feed Materials and Feed Additives in Consideration of the General Principles of Law', European Food and Feed Law Review, 2011, p. 3.
[1186] Article 3(1) of Regulation 1831/2003.

The approval of the application authorises only the applicant to bring the product on the market.[1187] Another business operator who wants to place the same GM feed on the market would have to obtain authorisation as well. It is a so-called specific authorisation.

21.4.2 Purity requirements

All food products for human consumption and animal feedingstuffs are subject to a maximum residue level (MRL) of pesticides. Regulation 396/2005 lays down rules on maximum residue levels of pesticides in or on food and feed products of plant and animal origin. The objective of this regulation is to ensure that the residue level of pesticides in food and feed products does not constitute an unacceptable risk to human and animal health.

What is more, undesirable substances in feed should be prevented or limited in order to constrain negative effects on animal or human health, the environment or the production of livestock. Directive 2002/32 sets maximum limits on undesirable substances in feed materials, feed additives and feedingstuffs. Any substance or product, with the exception of pathogenic agents, which is present in and/or on the product intended for animal feed and which presents a potential danger to animal or human health or to the environment or could adversely affect livestock production is specified as an undesirable substance.[1188] Directive 2002/32 includes maximum limits for heavy metals and materials such as dioxins, aflatoxin, certain pesticides, and botanical impurities.[1189]

Additionally, the administering of hormones to stock farming is limited by Directive 96/22. While substances having a hormonal or thyrostatic action and beta-agonist substances may not be administered to farm animals at all, the use of certain substances for therapeutic and zootechnical purposes may be authorised but must be strictly controlled.

21.5 Process-focused provisions

Feed safety needs: that the involved business processes are under control such that wherever possible safety issues are prevented through proper hygiene; that businesses are prepared to respond to safety issues by having traceability in order; and finally that businesses quickly and effectively respond in case a safety issue occurs.

[1187] Article 15 and 16 of Regulation 1829/2003.
[1188] Article 3(l) of Directive 2002/32.
[1189] See also Annex I of this Directive.

21.5.1 Prevention

In order to realise a high level of protection of human and animal health, food and feed safety should be ensured. Feed hygiene contributes to this objective as it aims at the prevention of food and feed safety risks.[1190] Regulation 183/2005 lays down requirements for feed hygiene to ensure a high level of consumer protection with regard to food and feed safety. Feed safety throughout the food chain starts with primary production of feed, up and including, the feeding of food-producing animals.[1191] Regulation 183/2005 requires feed business operators (FeBOs) to comply with obligations relating to hygiene requirements. Feed business operators carrying out operations other than those at the level of primary production of feed must apply the principles of the Hazard Analysis and Critical Control Points (HACCP) system.[1192] The intention of this system is to prevent hazards in production by among other things requiring food and feed business to make an analysis of their processes.

21.5.2 Preparedness

The traceability requirement of Article 18 GFL includes feed and food-producing animals. Regulation 183/2005 also lays down (besides general rules on feed hygiene) conditions and arrangements ensuring traceability of feed. The traceability of feed and feed ingredients is an essential element in ensuring feed safety.[1193] Traceability means the ability to trace and follow food, feed, food-producing animals or substances that will be used for consumption, through all stages of production, processing and distribution.[1194] Feed business operators must be able to identify where their products come from and where their products end up.[1195] The traceability of the manufacturer to final user must be ensured.

21.5.3 Response

If a FeBO has reason to believe that a feed which it has imported, produced, processed, manufactured or distributed does not satisfy the feed safety requirements, it is under the obligation to immediately initiate procedures to withdraw the feed in question from the market and inform the competent authorities thereof. The

[1190] Article 3(a) of Regulation 183/2005 gives a definition of feed hygiene: 'feed hygiene means the measures and conditions necessary to control hazards and to ensure fitness for animal consumption of a feed, taking in to account its intended use.

[1191] See Recital 6(b) of Regulation 183/2005. Furthermore, see Article 3(f) of Regulation 183/2005 for the definition of primary production of feed: '"primary production of feed" means the production of agricultural products, including in particular growing, harvesting, milking, rearing of animals (prior to their slaughter) or fishing resulting exclusively in products which do not undergo any other operation following their harvest, collection or capture, apart from simple physical treatment'.

[1192] See Article 6 and 5(1) of Regulation 183/2005.

[1193] See Recital 21 of Regulation 183/2005.

[1194] Article 3(15) of the GFL.

[1195] Article 18 of the GFL.

operator has to inform users of the feed of the reason for its withdrawal, and if necessary, recall the product from the consumers.[1196]

21.6 Presentation

21.6.1 Labelling

Feed labelling requirements for EU Member States are established in different kinds of legislation. Diagram 21.2 gives an overview of legislation regarding feed labelling requirements and to which type of feed they apply.

Diagram 21.2. Feed labelling.

Regulation	Labelling requirements for
Regulation 178/2002	Feed in general
Regulation 767/2009	Feed materials and compound feed
Regulation 1831/2003	Feed additives and premixtures
Directive 90/167	Medicated feed
Regulation 1829/2003	Genetically modified food and feed
Regulation 1830/2003	Products consisting of/or containing GMO

The general basis for labelling food or feed is laid down in Articles 8 and 16 of Regulation 178/2002, the GFL. Article 8 aims at the protection of the interests of consumers. The law shall provide a basis for consumers to make informed choices in relation to foods they consume. According to Article 16 feed labelling and other information may not mislead consumers.

21.6.2 Marketing of feed

The rules on the marketing of feed materials and compound feed are established in Regulation 767/2009 on the placing on the market and use of feed. Following the GFL, Article 11(1) of Regulation 767/2009 prohibits misleading of the user:

> *The labelling and the presentation of feed shall not mislead the user, in particular:*
>
> *(a) as to the intended use or characteristics of the feed, in particular, the nature, method of manufacture or production, properties, composition, quantity, durability, species or categories of animals for which it is intended;*
>
> *(b) by attributing to the feed effects or characteristics that it does not possess or by suggesting that it possesses special characteristics when in fact all similar feeds possess such characteristics; or*

[1196] Article 20(1) of the GFL.

> (c) *as to the compliance of the labelling with the Community Catalogue and the Community Codes referred to in Articles 24 and 25.*[1197]

The FeBO that first places feed on the market or the FeBO under whose name the feed is marketed is responsible for the labelling.[1198] According to Article 14 of Regulation 767/2009, the mandatory labelling particulars must be given in their entirety in a prominent place on the packaging, the container, on a label attached thereto or on the accompanying document. It must be conspicuous, clearly legible and indelible. The product must be labelled in the official language or at least one of the official languages of the Member State in which the product is placed on the market. This might be a problem for pre-packed products that are internationally delivered. It is usually not clear what the destination of the product is during the production and packaging, while relevant information is required. The mandatory use of the national language can lead to extra costs for re-labelling.[1199] Another requirement of Article 14 is that the mandatory labelling particulars are easily identifiable and not obscured by any other information. They must be displayed in a colour, font and size that does not obscure or emphasise any part of the information, unless such variation is to draw attention to precautionary statements. General mandatory labelling requirements are set forth in Article 15 of Regulation 767/2009. A feed material or compound feed can only be placed on the market if the following particulars are indicated by labelling: the type of feed, the name or business name and the address of the feed business operator responsible for the labelling, the batch or lot reference number, the net quantity expressed in units of mass in the case of solid products and in units of mass or volume in the case of liquid products, the list of feed additives preceded by the heading 'additives', and the moisture content. If the establishment approval number of the person responsible for the labelling is available, it should also be included. [1200] In addition to the mandatory labelling requirements, the labelling of feed materials and compound feed may include voluntary labelling particulars, on condition that the general principles laid down in Regulation 767/2009 are complied with.[1201]

Additionally, in accordance with Article 25 and 26 of Regulation 767/2009 a Code for good labelling practice of pet food was elaborated by the European Pet Food

---

[1197] As indicated above, a person who, for the first time, places on the market a feed material that is not listed in the Catalogue should notify its use as stated by Article 24(6) of Regulation 767/2009.

[1198] Article 12 of Regulation 767/2009.

[1199] I. Scholten-Verheijen and H. Tychon, 'The Marketing and Use of Feed – a new Regulation', European Food and Feed Law Review, 2010, p. 160.

[1200] Article 15(c) of Regulation 767/2009: 'if available, the establishment approval number of the person responsible for the labelling granted in accordance with Article 13 of Regulation (EC) No 1774/2002 for establishments authorised in accordance with Article 23(2)(a), (b) and (c) of Regulation (EC) No 1774/2002 or Article 17 of Regulation (EC) No 1774/2002 or with Article 10 of Regulation (EC) No 183/2005. If a person responsible for the labelling has several approval numbers he shall use the one granted in accordance with Regulation (EC) No 183/2005'.

[1201] Article 22 of Regulation 767/2009.

Industry Federation (FEDIAF), a federation that represents the pet food industry in the EU.[1202] The FEDIAF Code was validated as the first such European Code in December 2011. The Code is intended to provide practical guidance for the labelling and marketing of pet food and should be read in conjunction with the relevant EU and national legislation. The Code addresses three basic functions of product communication: consumer information on product use, control and enforcement, and marketing and retail. The Code will regularly be reviewed and updated.

21.7 Enforcement

The Member States are responsible for the enforcement of food law, and shall monitor and verify that the relevant requirements of food law are fulfilled by food and feed business operators at all stages of production, processing and distribution. For that purpose, they shall maintain a system of official controls and other activities as appropriate to the circumstances, including public communication on food and feed safety and risks,[1203] food and feed safety surveillance and other monitoring activities covering all stages of production, processing and distribution. The responsibility of the Member States is set out in Article 17(2) of the GFL. If a Member State identifies the non-compliance of a FeBO, it shall take action to ensure that the operator remedies the situation. When deciding which action to take, the competent authority shall take account of the nature of the non-compliance and that operator's past record with regard to non-compliance.[1204] The Member State entitles the competent authorities responsible for performing the official controls.

21.8 Conclusion

To a large extent feed law addresses different business than food law. The interests at stake, however, are fully integrated. The same is true for the law. Food law and feed law are intimately connected. They follow the same principles and apply the same legal models. As a consequence they can generally be approached on the basis of the same framework of analysis.

[1202] The code is available at: http://www.fediaf.org.

[1203] See also Article 10 GFL.

[1204] Article 54(1) of Regulation 882/2004.

22. Intellectual property rights in the agro-food chain

Bram De Jonge and Bernard Maister

22.1 Introduction

Intellectual property rights (IPRs) are everywhere. Almost all objects surrounding you are, or were once, protected by IPRs. The clothes you wear, the book you are reading, the cellphone in your pocket, they are all protected by IPRs including design protection (which protects the visual appearance of a product), patents (protect the functional aspects) and trademarks (identify the unique manufacturer). This book and even its individual articles and references, are copyright protected. Your cellphone alone contains many thousands of different IPRs including patents (hardware and software[1205]), industrial design rights (e.g. the shape, the appearance of the screens), trademarks (e.g. the manufacturer's logo) and copyrights (e.g. instruction manual, screen instructions).

How important are IPRs to the economy of the EU? A study conducted by the European Patent Office (EPO) and the Office for Harmonization in the Internal Market (OHIM) published in September 2013 reported that 39% of total economic activity in the EU (around € 4.7 trillion annually) is generated by industries in which IPRs are particularly important. These same industries provide for 56 million jobs in the EU.[1206]

This chapter reviews IPRs relevant to the food industry, with a focus on European legislation. It will describe the different IPRs that play a role in the production and marketing of food products, from the seed that is sown to the beer that is produced from it. But before analysing this production chain, the chapter will explain what IPRs are and what they are good for (or not). It will provide a historic overview of the key international treaties and agreements on which national IP laws are based, and more specifically discuss the key European patent treaties. The chapter also briefly discusses the administration and enforcement of IPRs.

[1205] Computer programs (i.e. software) are considered patentable subject matter if such a program, when running on a computer, produces a further technical effect such as, for example, controlling a technical process. Of course, the usual patentability criteria (discussed below) must be satisfied.
[1206] EPO & OHIM, 2013. Intellectual Property Rights intensive industries: Contribution to economic performance and employment in Europe. Available at: http://tinyurl.com/ktwno64.

22.2 What are intellectual property rights, and what are they good for?

The term 'intellectual property' (IP) has no official definition. IP may be considered, broadly speaking, as the creative intangible product of the human mind. Examples of these creative products include various technical and artistic creations such as inventions, production methods, processes and artistic works encompassing expressions, music, performances or fine art.

IPRs are the legal mechanism whereby IP is protected. Individual IPRs relate to a specific form of IP, each IPR having different characteristics related to the nature of the IP it protects. Modern IPRs fall into two broad categories, those related to manufacturing (industrial technology) and those related to the arts. The former group includes patents, trademarks, industrial designs, geographical indications, plant breeder's rights, and trade secrets. Copyrights are the IPR related to various forms of artistic creation such as writings, music, fine art, movies and performances.

IPRs are part of a bargain between inventors or artists and the public benefiting from their work. The traditional rationale (and justification!) for IPRs is that by balancing the conflicting interests of control versus access through the means of a limited monopoly, the creativity and hard work of inventors and artists is rewarded and encouraged. Society at large becomes the beneficiary. Considered from another perspective, IPRs represent the legal means whereby the economic and moral rights of innovators and artists are protected yet still making their creations accessible to the greater society.

This is not too different from the way that society deals with private tangible property such as land or objects. Owning a piece of land, for example, gives a private person certain rights such as being able to exclude others or to use the land for their own purposes. However, such ownership is not unlimited – society imposes certain restrictions or limitations on an individual's powers regarding his own property. Such restrictions may include building codes, noise abatement regulations and compulsory easements for public use. So too, the monopoly provided by IPRs is limited – for patents, there is compulsory licensing, for copyright, fair use and for PBRs, the breeders' exemption.

The theory is that the short-term cost of IP (the monopoly given to the rights' holder) is offset by the long-term benefits to society. This occurs when the IP-protected innovation enters the public domain and can be freely used by all. Even before this happens, society has already benefited from the fact that, for example, to be granted a patent, complete disclosure of the invention is required of the inventor applicant. By publishing the patent application and, once granted, the patent itself, details of the invention become available to the public. It is intended that further development of the invention by others will occur as a consequence of publication.

But not everyone agrees with the above justifications for IPRs. One counter-argument is that monopoly rights do not make artists and inventors more creative or inventive. On the contrary, it may be argued that the temporary monopoly rights established by IPRs may block creativity and innovation because the possibility to freely access and build upon all existing knowledge, artefacts and technologies is restrained. Others argue that IPRs have mainly become a strategic tool for companies to acquire market power, creating monopolies and a few multinationals that control the market in many fields of production.[1207]

Scientific evidence on the impact of IP protection on innovation and economic growth provides contrary findings and conclusions. Certain industries in developed countries, such as pharmaceutical, chemical, petroleum and biotechnology, consider the patent system to be essential for innovation, but there remains controversy regarding the ideal duration and extent of IP protection. Studies suggest that the impact of IPRs on innovation in developed and developing countries is different.[1208]

Since its inception in Venice in the 15th Century, IP law has continued to develop and evolve. The emergence of new technologies, particularly in the past three decades, has necessitated modifications in existing IPRs and even, in some cases, new forms of IPRs. Biotechnology is a good example of a new technology stirring changes in IP law. Not so long ago, life-forms were not eligible for patent protection because they were not considered human inventions. Discoveries are not patentable and life was considered sacrosanct and not to be the subject of human ownership. But one of the first man-made bacteria, which was invented by Chakrabarty in the late seventies, changed all of that. Chakrabarty filed an application for patent protection in the USA, which was initially rejected by the US patent office on the grounds portrayed above. The case was finally brought before the Supreme Court which ruled, with 5 votes against 4, in favour of Chakrabarty, holding that a live, man-made microorganism is a patentable subject matter and that 'anything under the sun that is made by man' can be patented in the USA.[1209]

[1207] For more information on the various arguments for and against IP protection, see e.g. M. Gollin, Driving Innovation. Intellectual Property Strategies for a dynamic world, Cambridge University Press, New York, NY, USA, 2008.

[1208] See e.g. W. van Genugten *et al.*, Harnessing Intellectual Property Rights for Development Objectives, Wolf Legal Publishers, Nijmegen, the Netherlands, 2011.

[1209] Diamond v. Chakrabarty 447 US 303 (1980).

Since then, the US patent office has granted numerous patents for newly created microorganisms, genes and living animals. Many other countries in the world, including the EU, have since tweaked their patent laws in order to facilitate the patentability of biotechnological inventions, including plant varieties and human genes.[1210] It might not come as a surprise that these developments have been the subject of much social and political debate between proponents and opponents of IPRs in general, and regarding the patentability of living matter in particular.

22.3 Historic overview of the key international treaties on IP

22.3.1 The Paris Convention for the Protection of Industrial Property

As countries became industrialised in the nineteenth century and the products of this industrialisation entered international trade, it was recognised that international rules to protect these inventions and creations, whether industrial or artistic, were necessary.

While there existed some form of national IP law in most industrialised countries, it was inconsistent from country to country. A further problem was that an inventor seeking worldwide protection had to submit applications in several countries almost simultaneously. This was necessary in order to prevent copying in a country where the invention was not protected and to ensure the eligibility of patent protection by maintaining the novelty requirement.

Following a series of negotiations regarding the protection of international IP, the Paris Convention was signed in 1883, coming into effect in July 1884. It has been revised on several occasions, most recently in 1967 in Stockholm (amended in 1979).[1211] The Convention covers patents, trademarks, industrial designs, and geographical indications. Provisions found in the Paris Convention have been incorporated into TRIPs (discussed below) making aspects of it still significant today. One of these provisions is the 'National Treatment' provision, which is a means of protecting foreigners from discrimination in terms of the application of national IP laws. This provision requires that each country provides to nationals of other members the same treatment afforded to its own citizens.[1212]

Another important provision is the 'right of priority'. This provision is particularly important to an inventor seeking protection for the same invention in different countries. When an application for a patent is first made in one country, the date of that application applies to applications for the same invention in other

[1210] This will be discussed further below.
[1211] You can find the Paris Convention at: http://tinyurl.com/owy2rkw.
[1212] Articles 2 and 3 Paris Convention.

countries provided they are filed within a specified period (usually 6 or 12 months). Thus later applications are considered to have been filed on the same day as the original.[1213] This is of particular significance due to the importance of the filing date in terms of prior art claims and establishing priority particularly in rapidly changing fields such as electronics.

The Paris Convention also acknowledges that national laws and how they are administered differs from country to country. Merely because one country grants, or rejects, a patent application for a specific invention does not mean that all other members are required to do the same. The converse is also true.[1214]

22.3.2 The Berne Convention for the Protection of Literary and Artistic Works

With the invention of the printing press around 1450 and the consequent greater availability of books which no longer had to be copied by hand, a system protecting the rights of authors was deemed necessary. With the passage of the Statute of Anne in 1710, the British Parliament recognised the importance of the authors' rights by giving to them the right to print and reprint a book within 14 years of its initial publication.

Authors faced similar issues to their creative cousins, inventors, having limited copyright protection across borders. In the same way that the Paris Convention was necessary, the Berne Convention was enacted in 1866 in order to provide for the international recognition of copyright between different states. Under the Berne Convention, literary and artistic works are protected without any formalities (i.e. automatically on their creation) although some countries have added protection in the form of registration of artistic works. Like the Paris Convention, various provisions of the Berne Convention have been incorporated into TRIPs, such as 'national treatment' (a state must grant to works originating in another contracting state the same protection it provides its own nationals), 'automatic' protection (no formalities are necessary to protect a work), and 'independence' of protection (protection granted by a state is independent of the presence of protection in the country of origin).

[1213] Article 4 Paris Convention.

[1214] Article 4*bis* Paris Convention.

Copyrights ©

Copyrights protect the expression of creative ideas, not the ideas themselves. In order to be protected by copyright, expression must be recorded in tangible form in a fixed medium. Examples include an author's hand-written notes, published novels, movies, sheet music, drawings, and computer software. Copyright protection is automatic – as soon as expression has been recorded in a fixed medium, copyright protection applies unlike, for example, patent protection which has to be applied for and granted by an examining office. Copyright registration is available in most jurisdictions and it is recommended that authors should seek such registration which enhances their ability to litigate alleged infringement. Note that copyright protection does not protect ideas, procedures, methods or information contained in a writing or movie but only their fixed expression. Some written materials, government legislation being a good example, are by law specifically not copyright-protected. As a result of the EU Duration Directive,[1] the standard for copyright duration is generally the life of the author plus 70 years.

[1] Council Directive 93/98/EEC of 29 October 1993 harmonizing the term of protection of copyright and certain related rights. Available at: http://tinyurl.com/qjeswsn.

22.3.3 The World Intellectual Property Organization (WIPO)

In 1893, the two Secretariats, or international *bureaux* (one for industrial property, the other for copyright) created by and managing the Paris and Berne Conventions united to create the United International Bureaux for Protection of Intellectual Property (BIRPI). Following a series of transitions, the 'Convention Establishing the World Intellectual Property Organization' entered into force in 1970 and WIPO, now one of the specialised agencies of the United Nations, was created.[1215]

The core mission of WIPO is to 'promote the protection of intellectual property throughout the world through cooperation among States, and, where appropriate, in collaboration with any other international organisation.'[1216] Part of WIPO's function includes administering various international treaties including IP protection treaties (e.g. Paris and Berne Conventions), global protection treaties (e.g. Patent Cooperation Treaty[1217]) and classification treaties (e.g. Strasbourg Agreement Concerning the International Patent Classification[1218]).

WIPO also plays an important role in assisting developing and least developed countries in the implementation of international and national IP law through its assistance in passing legislation and helping establish the necessary administrative infrastructure. In this role WIPO has developed an important cooperative role with the WTO particularly in the implementation of the TRIPs agreement (discussed

[1215] The WIPO Convention is available at: http://tinyurl.com/obdsn7w.
[1216] Article 3 WIPO Convention.
[1217] Available at: http://tinyurl.com/kvwu45.
[1218] Available at: http://tinyurl.com/nkchjey.

below). WIPO input has also been a factor in those fields impacted by IP such as traditional knowledge, folklore, biological diversity and climate change.

22.3.4 WTO and TRIPs

In the period following the Second World War, the nature of industry and international trade changed. Former colonies were now independent, technology was rapidly changing, new products were emerging and the Paris- and Berne-based international IP system was found inadequate, at least by the major industrial nations of North America and Europe. A major concern was the growing technological sophistication of upcoming countries in Asia and Latin America. Countries that did not shy away from pirating IP from abroad and which made 'reverse engineering' a national development policy.[1219] A related problem was that existing IP bodies, notably WIPO, had very limited power when it came to such matters as the enforcement of IPRs and settling IP-related disputes between different countries. Heavily IP-dependent industries located in industrially-developed countries began to seek alternative means of protecting their IP.

Their solution was to make the protection of IP a trade-related issue under the ongoing GATT negotiations. Such an approach was appealing for a few reasons: the various agreements in the Uruguay round of multilateral trade negotiations would have to be accepted as a package and a dispute resolution system was part of GATT negotiations. Further, 'linked-bargaining' was already part of the Uruguay round meaning that the major proponents of IPRs (USA, Japan and Europe) could link their IPR proposals to various concessions in agriculture and textiles. By the time the parties met at the Punte del Este trade round a coordinated strategy had emerged that effectively tied the protection of IP to international trade.

Out of these negotiations emerged the Agreement Establishing the World Trade Organization (WTO Agreement) and the Agreement on Trade-Related Aspects of Intellectual Property Rights (TRIPs) both of which came into force on 1 January 1995. In addition to its own provisions, TRIPs incorporates the substantive provisions of the Paris and Berne Conventions.[1220] Current membership of the WTO accounts for over 97% of world trade.[1221]

TRIPs' stated objective is the protection of IP and the enforcement of IPRs in such a manner as to promote 'technological innovation and the transfer and dissemination

[1219] See e.g.: Commission on Intellectual Property Rights, Integrating Intellectual Property Rights and Development Policy, 2000. Available at: http://tinyurl.com/ohejgfy.

[1220] For example, with regard to the Paris Convention, TRIPs requires that members comply with Articles 1 through 12 and 19, (Article 2(1) TRIPs).

[1221] J. Watal and R. Kampf, The TRIPs Agreement and Intellectual Property in Health and Agriculture. In: A Krattiger *et al.* (eds.). Intellectual Property Management in Health and Agricultural Innovation: A Handbook of Best Practices. MIHR, Oxford, UK, 2007, p. 253. Available at: www.iphandbook.org.

of technology.'[1222] This objective is intended to benefit both the producers and users of technology by enhancing 'social and economic welfare' while balancing 'rights and obligations.'[1223] TRIPs is binding on all WTO members. However, as not all members were considered to be at the same level of economic development, concessions regarding the timing of obligations ('transitional arrangements') were made for developing and least developed countries).[1224]

The significance of TRIPs is that it introduced the principle of minimum standards for intellectual property laws and regulations for all WTO members who are nevertheless free to implement its provisions as they see fit within their own legal systems. Members may implement 'more extensive protection than is required' provided these do not contravene other provisions within the agreement.[1225] Next to the principle of 'national treatment',[1226] which is discussed above, TRIPs includes the principle of 'most favoured nation'. This requires that any concession made by a country to another member must be provided to all others 'immediately and unconditionally.'[1227] Another important aspect of TRIPs is enforcement of IPRs. Countries are expected to introduce legal and administrative procedures, including criminal procedures and penalties, to deal with episodes of infringement.

Next to these procedural provisions, TRIPs sets minimum standards of protection for most IPRs, including patents, trademarks, geographical indications and trade secrets. With respect to patents, TRIPs requires members to provide patents for 'products and processes in all fields of technology, provided they are new, involve an inventive step and are capable of industrial application.'[1228] In granting patents, national offices may not discriminate as to 'the place of invention, the field of technology and whether products are imported or locally produced.'[1229]

Certain inventions may be excluded from patentability if the member believes that their commercial exploitation would be a threat to 'human, animal or plant life or health or to avoid serious prejudice to the environment.'[1230] Members may also exclude from patentability 'plants and animals other than micro-organisms, and essentially biological processes for the production of plants or animals other than non-biological and microbiological processes.'[1231] While plants may be excluded

[1222] Article 7 TRIPs.
[1223] Article 7 TRIPs.
[1224] Articles 65 and 66 TRIPs.
[1225] Article 1 TRIPs.
[1226] Article 3 TRIPs.
[1227] Article 4 TRIPs.
[1228] Article 27(1) TRIPs.
[1229] Article 27(1) TRIPs.
[1230] Article 27(1) TRIPs.
[1231] Article 27(3)(b) TRIPs.

from patentability, IP protection for plant varieties is required by TRIPs either 'by patents or by an effective sui generis system' or by a combination of both.[1232]

22.4 Administration and enforcement of IPRs

In addition to legislating their own IPRs, countries must provide the means whereby such laws are administered and enforced. Most countries have set up a government agency, usually called a 'Patent Office' to do this. Despite the name, Patent Offices often also handle matters related to other IPRs such as trademarks and industrial designs. In addition to national Patent Offices, there are also regional offices whose mandate covers a number of countries. A patent application can be made in such an office with the applicant requesting patent rights in a number of countries within the relevant region. The European Patent Office (EPO) is one of these.

Patent Offices are usually staffed by civil servants within a specific government department or ministry. Virtually all industrialised and developed countries with patent examination systems will have specialised staff whose job is to examine patent applications prior to granting them. Particularly in busy offices with many thousands of applications in many scientific fields requiring examination, these examiners are highly specialised in their respective fields often qualified at the Master's or PhD level.

The primary tasks of a Patent Office include receiving and processing applications, examining an application to ensure it satisfies the necessary formal requirements in addition to the core conditions of patentability (novelty, inventive step, industrial applicability, disclosure) and communicating with the applicant (usually represented by an attorney or patent agent). Depending on the individual national statutory requirements, the Patent Office may also be responsible for many other IP-related tasks such as publishing applications and granted patents, collecting fees, dealing with challenges to the application by outsiders (so-called 'oppositions') and patent renewals.

An important aspect of the 'national' nature of IPRs is that the protection it provides is limited to the country or countries in which it was granted. For example, a patent granted in the Netherlands, can only be enforced within the territorial limits of the Netherlands. If the inventor wishes to protect the invention in other countries, specific patent protection must be sought in those countries through one of the available pathways (as discussed below in the patents section).

While IPRs are granted and administered by national governments, their enforcement depends on individual rights holders. In order for them to do this, countries must provide the necessary infrastructure including, for example, a court

[1232] Article 27(3)(b) TRIPs.

system staffed by IP-experienced judges and an administration that can enforce judgments. Such judgments may include action against infringers to both recover losses caused by their actions and prevent further infringement. As well as the court system, there have to be attorneys familiar with filing formalities and the criteria for the granting of an IPR, who can file patents and if necessary litigate IP issues. These attorneys, particularly if in the patent field, usually have strong backgrounds in other areas such as electronics, pharmaceuticals or biotechnology. States must also have in place police mechanisms to deal with counterfeits and pirated goods both internally and across borders.

According to the TRIPs agreement, the goal of enforcement procedures is to 'permit effective action against any act of infringement of intellectual property rights covered by this Agreement, including expeditious remedies to prevent infringements and remedies which constitute a deterrent to further infringements'.[1233] Application of these procedures should be in a manner that avoids creating barriers to 'legitimate trade'.[1234] Recognising that not all rights holders have deep pockets, TRIPs requires that enforcement procedures should be 'fair and equitable' and not 'unnecessarily complicated or costly'.[1235]

22.5 Patent protection in Europe

In Europe, the inventor who wishes to protect a new invention by means of a patent can choose one of four options. Patents can be applied for, examined and granted under the national law of individual national patent offices, through the European Patent Office (EPO) under the European Patent Convention (EPC), and in the future under the new European Unitary Patent system. The EPO is an independent, intergovernmental organisation and not a body of the European Union. The contracting states to the EPC, which is the legal source of the EPO, are different from those forming the European Union.[1236] The Unitary Patent system, on the other hand, has been established by the EU Member States and will provide patent protection for inventions across the European Union. The fourth option is to make an 'international application' through the Patent Cooperation Treaty (PCT) system, which is administered by WIPO.

The choice of approach depends on the economic and strategic needs of the inventor. For example, if the inventor is seeking protection in a single country then a direct application through its national patent office might be sufficient and the cheapest approach. Once protection is sought in a number of European countries, it is better to apply either through the EPO, which can provide protection

[1233] Article 41(1) TRIPs.
[1234] Article 41(1) TRIPs.
[1235] Article 41(2) TRIPs.
[1236] A list of the Member States of the EPO can be found at: http://www.epo.org/about-us/organisation/member-states.html.

in all designated states, or the Unitary Patent system. Patent protection outside of Europe can be achieved either by direct application to national offices or through the international and national phases of the PCT system. In this section each of these is briefly discussed.

22.5.1 European Patent Convention and the European Patent Office

The nations of Europe share close legal, economic and political relationships covering a number of areas. IP law is no exception. The European Patent Convention (EPC) was established in 1973 with the goals of strengthening 'co-operation between the States of Europe in respect of the protection of inventions' and providing specific 'standard rules' for the patents granted through this system.[1237] The EPC acts as the legal framework for granting patents including aspects such as what can be patented.

Under the current system the European Patent Office (EPO) grants patents. The EPO accepts patent applications under both the European Patent Convention and the Patent Cooperation Treaty. The official languages of the EPO are English, French and German. Applications filed in other languages must be translated into one of these.

22.5.2 Patent Cooperation Treaty

The Patent Cooperation Treaty (PCT), which came into force in 1978, is an international treaty, administered by WIPO, which makes it possible to seek patent protection for an invention in 148 countries throughout the world. It is important to note, however, that there is no such thing as an 'international patent'. The PCT system merely facilitates the application process while the actual granting of patents remains under the control of the national or regional patent offices.

During the first stage of the PCT process the application is examined to ensure that all formalities are completed. Additionally an international search report is completed by the examiner following a search of the prior art. The applicant can also request a preliminary examination report. With this information, the applicant is better able to evaluate the patentability of the invention and decide whether to proceed to the National Phase of the process during which the application is examined by the national patent offices of those countries designated by the applicant.

With the opinions provided by the PCT Search and Examination reports, the candidate can decide whether to pursue the application as is or modify it, for example by rewriting the claims. The PCT system is also of benefit to those

[1237] Preamble EPC.

countries whose patent offices are unable to complete their own patentability examination process but can be guided by the international search report and preliminary examination.

22.5.3 The European Unified Patent System

Despite the general 'European' nature of its mandate, the legal rights provided by patents granted through the EPO are limited to individual countries in which the patent was granted, meaning that the holder of the patent rights must be prepared to enforce the patent in each country individually. Having to enforce patent rights in different countries can become very expensive. Not only is expense an issue but, because different countries have their own national patent laws and courts act independently, legal action could produce conflicting decisions. For example, a court in one country may decide that a rival invention is infringing in that country whereas a court in another country may reach the opposite conclusion in the same case. In response to these concerns, the Agreement on a Unified Patent Court was signed on 19 February 2013 with the goal of creating an efficient unified patent right and court within the European Union.[1238]

European 'Unitary' patents will be legally enforceable in all participating European states. It will be granted by the EPO based on the requirements of the EPC.[1239] Not only will there be a single 'European patent' but in addition a new court system is to be established that will have jurisdiction throughout the participating States. This means that decisions by this Court on questions such as patent infringement will be binding on all, thereby ensuring greater predictability and consistency for example for inventors who are working in the same field and do not want to infringe active patents. The arrangement of the new court system includes a Court of First Instance separated into regional divisions. Different branches will focus on specific areas. For example, the London branch will deal with pharmaceutical and life science patents, the Munich branch will cover patents in the fields of mechanical engineering and weapons and the branch in Paris will handle electricity, transportation and physics-related cases. A single Appeals Court will be based in Luxembourg.

22.5.4 Biotechnology Directive

Directive 98/44/EC of the European Parliament and the Council deals with the patentability of biotechnological inventions. EU Member States are required to

[1238] Agreement on a Unified Patent Court, available at: http://tinyurl.com/kjcj58p. At the time of writing this chapter it was anticipated that the system will be functional by the end of 2014.

[1239] As noted above, the 'unitary patent' will be another option for users besides already-existing national patents and the EPO patents. A unitary patent will be granted by the EPO under the provisions of the EPC and will be legally enforceable in the territory of the EU Member States. See: http://tinyurl.com/pgk895n.

modify their national law in accordance with the requirements of the Directive (Article 1).

Of relevance to the discussion of food and IP is Article 4 of the Directive which states:

1. The following shall not be patentable:
   (a) plant and animal varieties;
   (b) essentially biological processes for the production of plants or animals.
2. Inventions which concern plants or animals shall be patentable if the technical feasibility of the invention is not confined to a particular plant or animal variety.
3. Paragraph 1(b) shall be without prejudice to the patentability of inventions which concern a microbiological or other technical process or a product obtained by means of such a process.

It may seem strange that while Article 4(1) (see above) states that plants and animals are not patentable, Article 4(2) holds that plants and animals are patentable 'if the technical feasibility of the invention is not confined to a particular plant or animal variety'. The reason for this construction is that Article 53 of the EPC states that European patents shall not be granted in respect of 'plant or animal varieties'. The EPC came into force in 1977, just before the biotechnology revolution took off. One decade later, when the patenting of genes and plants had become common practice in the USA, the Biotechnology Directive was being developed in order to clarify and facilitate the patenting of life forms in the EU. The above construction (i.e. paragraphs 1 and 2) was then included to circumvent the EPC provision that plant and animal varieties *per se* are not patentable, thereby making it possible to patent, for example, transgenic plants carrying a newly developed trait as long as that invention can be incorporated into more than one variety.

The Biotechnology Directive is the subject of many social, political and legal debates, which have continued from the initial negotiations till today. These debates range from ethical concerns about the patenting of human genes to court cases where the scope of what should be considered patentable is being reviewed and decided upon. In relation to plant breeding, for example, what constitutes 'essentially biological processes for the production of plants' is still to be decided by the courts.[1240]

[1240] See for an overview of the main European court cases e.g.: J.A. Kemp, Exclusion from Patentability of Plant Varieties and essentially biological processes for the production of plants, 2013. Available at: http://tinyurl.com/nboo5g3. For more information on the social debate in this issue, see e.g. the website 'No patents on Seeds' at: http://www.alt.no-patents-on-seeds.org/index.php.

22.6 IPRs in the food chain

For purposes of this chapter, let us consider a simplified description of the making of beer beginning with the seed and ending with the glass of beer ready for the consumer. Once the chain of production is outlined, we can then evaluate how IPRs impact the different steps in this process.

The basic ingredients of beer are water (90% of beer), malt (usually barley but could be wheat or rye), hops and yeast. All beer is brewed using essentially the same process. Prior to planting the grain, the soil must be prepared. Once the soil is ready, the seed is sown. At various times during the growth period fertilizer must be added, often with some kind of mechanical device, and weeds removed usually by means of a commercial weedkiller. As the plants grow, a pesticide is often applied. Once the plants are grown, the barley, for example, is harvested and the grain is separated from the stalks.

Next comes the malting process during which barley kernels are allowed to germinate in a controlled environment. The malting process maximises the starch within the seed. Hops are a type of vine whose flowers are used as a preservative and also add flavour to the beer. Yeast, actually unicellular fungi, functions to convert (ferment) the sugars thereby producing alcohol. Each of these different components has a role in producing the different types and tastes of beer. The germinated grain is dried in a kiln in order to remove moisture and stop the natural enzyme activity. The malting process determines the colour and flavour of the malt. This process is completed outside of the breweries which then purchase the malt. The ground malt is crushed and combined with water to produce wort which is cooled before adding yeast to initiate the process of fermentation which is the essence of the brewing process by finally converting the mixture to beer.

IPRs are relevant at many stages of this process. For example, plant breeders' rights and patents could be implicated in the seed. Patents are also likely to play a role in the production of fungicides and insecticides, and in the various machines that are used on the farms and breweries. Some of the procedures involved in the brewing of beer might be the subject of process patents, while trade secrets are likely to be relevant as well. Trademarks and, possibly, Geographical Indications (GIs) are important to the various products along the way as well as marketing the final product.

22.6.1 Patents

In order for an invention to be considered 'patentable' it can belong to any field of technology and must be new (or novel), involve an inventive step and be susceptible of industrial application.[1241] Considering the making of beer one notes that the barley or wheat seeds, the mechanical devices used in the various stages of planting, reaping, processing and brewing, and even the procedures or processes involved at various stages of the manufacturing chain could be protected by patents, depending of course on whether they satisfy the criteria of patentability.

Patenting procedure

While procedures may differ slightly from national office to office, the essential process of examination by which a patent is granted remains consistent throughout the world. The first step is to ensure that the various 'formal' requirements such as the format and contents of the application have been completed correctly. Once this is done the examiner, who is usually an expert in the field of the invention, must determine whether the invention satisfies the essential requirements of patentability.

This determination begins with clearly understanding what is 'inventive' about the applicant's work. To do this the examiner must review and understand the 'claims', i.e. that parts of the patent application that clearly describe or delineate the actual invention to be patented. To determine whether the invention is 'new', the examiner must search the relevant prior art. This search might include granted patents, technical journals and other published materials. In this case the examiner is searching for a single source containing the invention as described in the claims. Regarding the question of an 'inventive step' the examiner uses prior art to see whether the invention contained in the application would be 'obvious' to an individual with 'ordinary skill' in the relevant art. The EPO uses the 'problem-solution' method, i.e. was there a technical problem in the prior art that is 'solved' by the new invention? Different pieces of prior art may be used to answer this question which is probably the most difficult aspect of the examining process.

Once the search is completed, relevant documents are reported in a 'Search Report' which is made available to the applicant who can then decide whether pursuing patenting is appropriate technically and financially, depending on the nature of the objections raised by the examiner.

[1241] Articles 54 (Novelty), 56 (Inventive Step) and 57 (Industrial Applicability), EPC.

Rights granted

The granting of a patent provides the owner with the legal right to exclude others from making, using, selling, offering for sale or importing the patent-protected invention except, of course, with the owner's consent. This patented invention may be either a product or a process that is 'new' for example, a new method of doing something, a new technical solution to an existing problem, a new chemical or molecule. The rights granted to the patent owner are not absolute. Patent laws may include certain exemptions to the rights described above. Most national patent laws include, for example, a research exemption, which allows a patented invention to be used for academic purposes without prior approval of the patent holder.

The protection provided by the granting of a patent extends for 20 years from the date of the original application although in certain circumstances this can be adjusted. All the rights enjoyed by the patent owner end after the term of the patent is over and the invention then enters the public domain to be available, and exploited, by anyone without requiring the owner's permission. In return for the limited monopoly provided by the patent grant, the owner is required to disclose the details of the invention to the public.

The patent[1242]

The granting of a patent begins with a formal application which must include certain specific items such as a request for the granting of a patent, a description of the invention, one or more claims, relevant drawings and an abstract. There should be a description of the background of the invention, the technical problem it solves and the best way known to the inventor of practicing the invention. If necessary for clarity, the application should include drawings, plans, tables or diagrams. The completeness of the application, or the 'disclosure', is judged by the standard of an average person working in the field of the invention. There should be enough explanation and detail so that this average person could read the application and practice the invention without additional experimentation or testing for example.

As an example, US Patent No. 6,637,447 is simply titled 'Beerbrella.'[1243] The invention is described as providing 'a small umbrella ('Beerbrella') which may be removably attached to a beverage container in order to shade the beverage container from the direct rays of the sun'.

[1242] In order to illustrate this section, the authors have selected a simple mechanical patent granted in the USA.

[1243] This patent can be found at the USPTO website at http://www.uspto.gov/. To find the patent, go to the 'Patent' Section, click on Search by Number and insert the patent number, 6637447. The Specification, Drawings and Claims as referenced can be found in the text available at this site.

The 'problem' that the invention solves relates to the fact that, while there are a number of bottle or glass insulators available to keep drinks cool, 'they do not shield the beverage from the direct rays of the sun. A beverage left out in the sun, even if insulated or cooled with ice, quickly warms due to the effect of the intense infrared radiation from the sun, particularly on hot, sunny summer days.'[1244] The invention, the placement of an umbrella over the drink container, solves the problem by shielding it from the sun, as illustrated in the patent drawings shown in Diagram 22.1.

Diagram 22.1. Beerbrella.[1245]

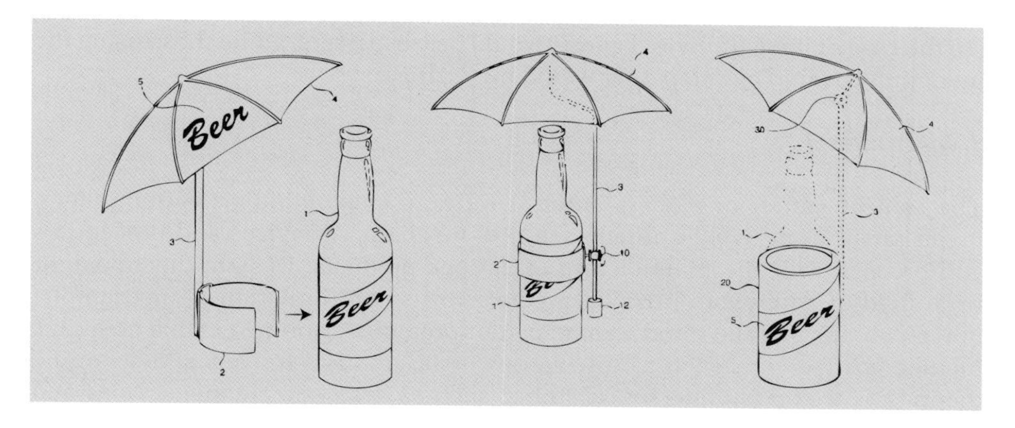

At the 'heart' of the application are the 'claims'. These are the precise description of the invention setting out its limits within the type of art. The claims identify exactly what the inventor owns, i.e. what the patent protects. Each word of the claims is significant and, should there be any litigation for example, could have a major impact on whether a related invention does or does not infringe a patent.

While a detailed discussion of 'claim writing' is beyond this chapter, some aspects of the first Claim of this patent (the first 'independent' claim) are worth noting. The first Claim of this patent reads as follows:

1. A combined beverage container and shading apparatus, comprising:
   - a beverage container, for containing a beverage;
   - a means for removably attaching the apparatus to the beverage container;
   - a shaft, coupled to the means for removably attaching the apparatus, and extending vertically with respect to the beverage container;
   - an umbrella, coupled to the shaft at a point above the means for removably attaching, so as to shade the beverage container,

[1244] Column 1, line 45, US Patent 6,637,447.

[1245] See Drawings, US Patent 6,637,447.

- wherein the means for removably attaching comprises a clip provided to attach to the beverage container by means of spring action and friction.[1246]

While this is written in patent 'legalese', if you look carefully at the figures from the patent you will see that it accurately describes the invention. Note how some terms are carefully chosen. For example, by 'claiming' the 'shading apparatus' comprises a 'beverage container' the inventor has widened the scope of the invention. Had it merely said 'a bottle' then other containers could be used without infringing the patent. The term 'beverage container' instead covers a range of options beyond just bottles such as glasses or mugs. The claim, through the use of a wider term, has set the limits of the patent, rather like a fence sets the limits of a piece of land. (In the case of land, uninvited guests would be trespassers, while those using the invention covered by the patent would be infringers.)

Subject matter

The subject matter of patents is the invention. Depending on national laws, it can range from living entities (seed) to mechanical devices and procedures and processes for the manufacture of various products. Despite this apparent wide field of patentable inventions, there are some limitations. For example, not all 'products of the mind' can be patent-protected. Non-patentable creations include abstractions, such as discoveries, mathematical models and scientific theories as well as non-technical items, such as aesthetic creations, rules and methods for performing mental acts, programs for computers and presentations of information.[1247]

Even though a product may qualify as an invention, there are certain 'exceptions' to patentability. For example, in the EPC an invention whose 'commercial exploitation' would be contrary to 'ordre public' or morality is denied patent protection based on the concern that such inventions could induce public disorder or criminal behaviour.[1248]

Patentability

For an invention to be patentable it must be new (or novel), have an inventive step (non-obvious) and of practical utility.[1249] Regarding the requirement that it be novel, the invention is compared to existing knowledge in the relevant field. This is called 'prior art'. Exactly what constitutes prior art is usually defined by an individual country's legislation and legal precedence. Generally, an invention

[1246] See, Claims, Column 4, US Patent 6,637,447.
[1247] Article 52(2) EPC. Note that, depending on the nature of the "invention" and national law, other Intellectual Property legal mechanisms may be available for protection.
[1248] Article 53(a) EPC.
[1249] Article 52(1) EPC.

is considered 'new' provided that 'it does not form part of the state of the art'.[1250] The state of the art comprises 'everything made available to the public by means of a written or oral description, by use, or in any other way, before the date of filing'.[1251] The standard by which non-obviousness is measured is based on considering whether the invention shows an 'inventive step' that would not have been obvious to the average person working in the relevant field, that is 'skilled in the art'.[1252] The invention must be disclosed in terms of a solution to a 'technical problem' with reference to 'any advantageous effects' in light of the prior art.[1253] Ultimately, whether an invention is considered non-obvious, will depend on how national legislation is interpreted by the patent office, in the person of the examiner, and how case law has been decided.

22.6.2 Trade secrets

Patenting is not the only way to protect an invention. 'Trade Secrets' is a specialised form of intellectual property protection.[1254] A 'trade secret' comprises essential or key information that companies use to maintain their competitive edge. Trade secrets could be either 'technical', involving, for example, the manufacturing processes and formulas for brewing a particular beer, or 'commercial', involving such things as customer lists or marketing studies. As can be appreciated, it is to a company's benefit to keep this kind of information confidential.

Individual country legislation is designed to protect this information by allowing a company to bring legal action against competitors who illegally acquire trade secret information. To qualify as a trade secret, the information must be shown by a company to be important to their product or manufacturing process, and the company must prove that it has taken all possible steps to protect that information.[1255] The advantage of trade secrets is that as long as they can be maintained, they remain under control of their holder unlike patented information which must be published upon application and which enters the public domain after the patent has expired. However, a disadvantage is that if the information is acquired by a competitor legally, e.g. by reverse-engineering, there is no legal redress and it is deemed to have become available to all. There is currently a proposal to align national legislation on trade secrets throughout the EU.[1256]

[1250] Article 54(1) EPC.
[1251] Article 54(2) EPC.
[1252] Article 83 EPC.
[1253] Rule 42(1)(c) EPC.
[1254] Known as 'undisclosed information' in TRIPs, protection of trade secrets is seen as 'ensuring effective protection against unfair competition'. Article 39 TRIPs.
[1255] Article 39(2) TRIPs.
[1256] European Commission 2013. Proposal for a 'Directive of the European Parliament and of the Council on the protection of undisclosed know-how and business information (trade secrets) against their unlawful acquisition, use and disclosure', Brussels, 28.11.2013, COM(2013) 813 final, 2013/0402 (COD).

Protection of 'regulatory data'

In the agro-food industry, there is a trend towards greater protection of 'regulatory data'. This is all the information that companies or research institutions are required to collect and submit in order to comply with various regulations. For example, before a new genetically modified organism (GMO) is allowed on the market a company needs to submit a range of scientific data and evaluations covering such matters as the (environmental) risk assessments of the GMO and the derived food or feed. The information documents that are submitted as part of, for example, the approval requirements for a GMO or a novel food,[1257] as well as some information related to health claims used in food labelling,[1258] are potentially valuable to competitors and therefore increasingly considered 'proprietary information', even though no formal IPR protection is applicable.[1259] This can have major consequences for, amongst others, competitors in the market. For example, once a patent on a GM trait expires the trait cannot be used if the patent holder decides not to maintain market approval for that trait. In such case, competitors would have to start the regulatory process for market approval from scratch, which is a very time-consuming and costly process.

22.6.3 Plant breeders' rights

The barley and wheat varieties planted for the production of the beer in our example, are likely to be protected by Plant Breeders' Rights (PBRs). In the EU, the Community Plant Variety Office (CPVO) is the agency responsible for implementing a system for plant variety protection. In operation since 1995, the CPVO exists in parallel with national systems as EU Member States may also have their own national PBR office. The main benefit of the CPVO system is that one application to the CPVO leads to one decision that is valid in all EU Member States. At the time of writing, 499 varieties of barley (*Hordeum*) were granted a Community Plant Variety Right with, on average, almost 60 new varieties being granted each year. For wheat (*Triticum*), 867 varieties were protected with 100 new PBRs granted each year.

New plant varieties, whether produced by traditional breeding or biotechnology methods, present a unique problem to IP. Unlike other 'products' protectable by IP, plant varieties are often capable of self-reproduction. Once even a small quantity

[1257] See for more information on the EU regulations on these matters the website of the European Food Safety Authority. Available at: http://tinyurl.com/ojxewc5.

[1258] See for an interesting discussion on proprietary information in the European Union's regulation of nutrition health claims: S. Carlson *et al.*, Publish and Perish: A Disturbing Trend in the European Union's Regulation of Nutrition Health Claims Made on Foods. FDLI Update Magazine, September/October 2010. Available at: http://tinyurl.com/ot2xq8h.

[1259] TRIPs obliges Member States to treat confidential data related to the market approval of pharmaceuticals and agricultural chemical products which utilize new chemical entities, see Article 39(3).

of plant reproductive material has been made available to the public, there exists the opportunity for self-replication. Nevertheless, provision has been made for breeders to protect their innovations on the principle that, in accordance with the core principles of IPRs, breeders should be rewarded for their efforts.

The International Union for the Protection of New Varieties of Plants (UPOV), which adopted its first convention in 1961, is the main international instrument providing intellectual property protection for plant varieties.[1260] UPOV provides legal protection based on requirements (e.g. distinctiveness, uniformity) and exceptions (e.g. breeders' exemption) that are different from patent law as it takes into account the biological nature of the protected subject matter and agricultural traditions. The UPOV convention was updated in 1972, 1978 and 1991 as a consequence of developments in agriculture, in particular the professionalisation of the breeding sector in the member countries.

For a long time, these member countries were all industrialised countries but since the adoption of TRIPs, which demands that members 'provide for the protection of plant varieties either by patents or by an effective *sui generis* system or by any combination thereof',[1261] several least developed to lower middle income countries have joined UPOV as well, amounting to 23 of the 71 member countries in 2013. New member countries can only subscribe to the latest UPOV Act of 1991. The CPVO is based on the principles of this 1991 Act.

Criteria for protection and scope

To be eligible for protection a variety must be new, distinct, uniform and stable. According to UPOV, the granting of a plant breeder's right (PBR) shall not be 'subject to any further or different conditions'.[1262] For a plant variety to be considered novel, it must not have been previously marketed for a specific time period. So, UPOV provides for a 'grace period' during which prior commercialisation will not destroy the 'novelty' needed to obtain a PBR for a specific variety. Distinctiveness means that the new plant variety must be clearly distinguishable from any other variety whose existence is a matter of common knowledge. Uniformity is important in order to ensure that a new plant variety can be defined for the purpose of protection, which means that individual plants of the same variety must be sufficiently uniform in their relevant characteristics. And the new variety needs to be stable so that it remains true to type after repeated cycles of propagation. In addition, applicants must comply with specific regulatory and administrative formalities and pay the prescribed fees.

[1260] International Convention for the Protection of New Varieties of Plants (UPOV). Available at: http://tinyurl.com/q29fwe8.

[1261] Article 27(3)(b) TRIPs.

[1262] Article 5 UPOV 1991.

UPOV 1991 requires members to provide protection for all genera and species.[1263] The period of protection is set at a minimum of 25 years for grapevines and trees and 20 years for all other plants. In Europe under the CPVO, this is 25 years and 30 years for vine, trees and potato varieties. With respect to the scope of plant breeders' rights, the authorisation of the plant breeder is required for, amongst others, the production or multiplication of the protected variety, for exporting and importing, and for its selling, marketing or offering for sale.

Exceptions to the breeder's right

Like patent law, UPOV also includes some limitations or exceptions to the rights of the right holder. The two most characteristic exceptions are the breeders' exemption and the farmers' privilege. The breeders' exemption allows protected varieties to be freely used for the purpose of breeding new varieties in order to allow any breeder to have access to the latest improvements and new variation. The farmers' privilege is an optional exception through which farmers are allowed to save and reuse seed of a protected variety on their own farm 'within reasonable limits and subject to the safeguarding of the legitimate interests of the breeder'.[1264]

A breeders' exemption in patent law?

For several years now, there has been a strenuous debate in the breeding sector as to whether or not the breeders' exemption should also be included in patent law. The problem is that in patent law, any use for which there is no specific exception provided, is prohibited to non-patent holders. Therefore, if a seed/plant is protected under both patent law and PVP law, the breeders' exemption is essentially nullified. Proponents of the breeders' exemption argue that the lack of such an exemption in patent law stimulates concentration in the global seed industry, reducing innovation and, eventually, may endanger global food security. Opponents emphasise instead that such an exemption in patent law would reduce innovation in the breeding sector because R&D in biotechnology is very expensive and many biotechnologies are easy to copy.[1265]

Plant variety protection and developing countries

More and more countries are considering becoming a member of UPOV as TRIPs requires them to establish a *sui generis* system for plant variety protection. Although least developed countries are exempt from this requirement till at least 2021,

---

[1263] There is a transition period of five (existing members) or 10 (new members) years for member countries to comply with this provision, Article 3 UPOV 1991.

[1264] Article 15(2) UPOV 1991.

[1265] For more information on this debate see e.g. N.P. Louwaars *et al.,* Breeding Business: The future of plant breeding in the light of developments in patent rights and plant breeder's rights. CGN, Wageningen, the Netherlands, 2009. Available at: http://tinyurl.com/ld2svbw.

many are anticipating joining UPOV in the hope that membership will stimulate the introduction of foreign varieties in the country, strengthening trade and food security. Yet, many civil society organisations are strongly criticising this trend for not being conducive to the agricultural realities in most developing countries, undermining smallholder farmers' agricultural practices and, ultimately, threatening food security. One key issue is that smallholder farmers rely heavily on farm-saved seed and seed exchange and trade amongst themselves in order to have access to seed for the next planting season. Through the farmers' privilege in UPOV, countries can permit the saving of seed but not seed exchange or trade. Whether another UPOV exemption, which exempts 'acts done privately and for non-commercial purposes' (Article 15(1)(i) UPOV), should be interpreted to include these farmer practices is currently under debate.[1266]

Diagram 22.2. A comparison of patents and plant breeders rights in Europe.[1267]

Patents	Plant variety rights
Plant defined by one or more inventive characteristics, not by the whole genome	A variety is defined by the whole genome or gene complex (i.e. the criteria of distinctiveness, uniformity and stability)
All plants with the inventive feature are protected, however obtained: foreign genes from related species or bacteria; gene mutations	Only a single variety and varieties essentially derived from it are protected (entire phenotype as defined by the genome)
Patentability criteria include novelty and inventiveness	Variety must be novel without an inventive step requirement
Plant protected for all users – no EU-wide breeders' exemption	Breeders' Exemption allows free use of the protected variety for further breeding; free commercialization of new varieties unless essentially derived.

22.6.4 Trademarks

Once brewed and ready for consumption the bottles or barrels of beer are sent to retailers. Like most products it is likely to be 'branded' with the particular brand identified by a trademark. Trademarks are intended to provide information to the consumer by identifying the commercial source of the product (or services). A consumer is encouraged to assume all products associated with a specific trademark share similar standards. Using the making of beer as an example, customers expect

[1266] B. De Jonge, 2014, Plant Variety Protection in Sub Saharan Africa. Balancing commercial and smallholder farmers' interests. Journal of Politics and Law 7(3). pp. 100-111.

[1267] S. Yeats, Latest Developments in Patenting Plant Inventions in Europe. EPO presentation, Brussels, Belgium, 11 October 2011. Available at: http://tinyurl.com/l6q32ks.

that every beer they drink bearing the recognized trademark will be of the same quality and taste.

Trademarks can be words, symbols, letters, numerals or any combination of these. Trademarks do not have to be only two-dimensional – the shape or packaging of a product could be the subject of a trademark. The most significant requirement for a trademark is that it must be distinctive so that it can clearly be distinguished by the public and is in no way misleading or deceptive.[1268]

Those intending to use a trademark must file an application with the relevant national or regional trademark office (often the same entity as the Patent Office of the country or region). In most jurisdictions, the application should include a clear reproduction of the proposed trademark and a list of the applicant's goods or services for which the mark will be used. In order to ensure that there is no likelihood of confusion with another mark, the trademark office will conduct a thorough search and examination of all marks in use or registered. In most countries trademark applications are published prior to grant in order for third parties who are concerned about confusion with their marks to oppose its grant.

For manufacturers it is important that their brand gain international recognition. To achieve this, and to avoid having to register trademarks separately in multiple trademark offices, there exists a system of international registration governed by the Madrid Agreement Concerning International Registration of Marks and the Madrid Protocol.[1269]

In our case here, it is likely that most of the members of the seed-to-beer chain, use trademarks to identify their products. The individual trademarks might not be of equal significance to everyone along the chain. For example, it probably doesn't mean much to consumers whether the farm vehicles are International Harvester® or John Deere®. A consumer might prefer a food product made by a particular company for its taste and quality and thereafter seek out other products by that company. These products would only be identified by the company's mark which, to maintain its distinctiveness, it would have to vigorously defend against use of the same or a confusingly similar mark. One example is Heineken's lawsuit against Olm beer for its similar label.

[1268] See, for example, Article 15 TRIPs 'Any sign, or any combination of signs, capable of distinguishing the goods or services of one undertaking from those of other undertakings, shall be capable of constituting a trademark'.

[1269] The Madrid System: International system for the registration and management of trademarks worldwide. Available at: http://tinyurl.com/49jq3xf.

Diagram 22.3. Trademarks of Heineken and Olm beer (source: www.biernet.nl).

Madrid Agreement Concerning the International Registration of Marks and the Protocol Relating to the Madrid Agreement

The two Madrid treaties handle the international registration of trademarks and service marks and are both administered by the International Bureau of WIPO. One benefit of the system is that an applicant seeking registration of a mark in a number of countries can make one application to the International Register and designate each of the specific countries desired. Once approved, the mark holder only has one administration to deal with and administrative tasks such as renewal, recording a change of ownership or changing the goods and services covered by the mark can be done with a single procedural step. In order to use the International Registration system, a mark first has to be registered in the applicant's Office of Origin, i.e. the Trademark Registration Office with which the applicant has a connection either by domicile or business location. A registered trademark is recognised by the ® logo. An unregistered trademark may carry the ™ symbol.[1270]

22.6.5 Industrial designs

Although sometimes confused with trademarks, a different form of protection is available known as 'industrial design' rights. The term 'industrial design' refers to the non-functional aspects of an article, specifically its ornamental or aesthetic aspects. Protection of industrial design could be obtained for a product's three-dimensional features (e.g. shape) or features such as its distinctive colour or

[1270] Trademark Symbols, International Trademark Association. Available at: http://tinyurl.com/nlzsep9.

patterns. In most countries industrial designs must be registered before protection is obtained. The Hague System for the international registration of industrial designs offers a means of obtaining protection in several countries through one filed application.[1271] Of course, the design of a product may function in a manner similar to a trademark, which identifies the manufacturer – consider the original Coca-Cola bottle, for example, a shape that readily identifies the product.

22.6.6 Geographical Indications[1272]

Agricultural products such as wines, cheeses, hams and beer can be identified by a special form of mark known as a Geographic Indication (GI) which could take the form of a name, sign or symbol. GIs serve a similar function to trademarks. Trademarks distinguish goods of one manufacturer from those of others while GIs inform consumers that a product originates in a certain geographical location and has characteristics, a reputation or qualities unique to that location. Because of its regional nature, the same GI could be used by a number of companies within the relevant area. Exactly what constitutes a GI, and how it may be used by local companies, is a matter of national law.

GIs usually include the name of the place they identify. This is particularly meaningful in the case of agricultural products, which are usually significantly affected by such factors as local climate or soil. But non-agricultural products can also be protected by GIs, such as 'Swiss watches' or 'Longquan Porcelain'. In Europe, several regulations provide a European protection system for agricultural products and foodstuffs,[1273] wines[1274] and spirits.[1275] So far, there is no unitary protection system available for non-agricultural products.

The EU, given its extensive international agricultural interests, is particularly concerned about a strong international GI system. WIPO administers key international treaties regarding GIs, specifically the Paris Convention (discussed above) and the 'Lisbon Agreement for the Protection of Appellations of Origin and Their International Registration'.[1276] Articles 22 to 24 of TRIPs, administered by WTO, also cover the international protection of geographical indications. GIs are also an important component of a number of bilateral treaties involving the EU.

'Appellations of Origin' (AO) is a unique kind of geographical indication that emphasises specific qualities or characteristics that are due to natural or human

[1271] The Hague System for the International Registration of Industrial Designs. Available at: http://tinyurl.com/4sf8haq.
[1272] A more detailed overview of the EU legislation on Geographical Indications is provided in Chapter 14.
[1273] Council Regulation 510/2006.
[1274] Council Regulation 479/2008.
[1275] Regulation 110/2008.
[1276] Available at: http://tinyurl.com/kcojn8x.

factors related to the geographical environment. AOs are covered internationally specifically under Article 2 of the Lisbon Agreement which allows international protection through a single registration procedure.[1277] In addition to the international system, bilateral agreements between countries also cover the protection of their AOs. As in other areas, national legislation in individual countries has evolved to keep up with the changes in national and international trade. For example, in France the AO for wines was created in 1919 and was extended to all agricultural food products in 1990.

Bavaria NV v Bayerischer Brauerbund eV

As the above discussion suggests, the possibility for overlap between various forms of 'trademark-like' protection exists and issues often have to be resolved through legal action. An important European case in this regard is *Bavaria NV v Bayerischer Brauerbund eV*, which dealt with the problem of a Dutch beer manufacturing company that produced 'Bavaria' beer even though the beer was not produced in Bavaria. The Bayerischer Brauerbund is a trade organisation which functioned as an association for Bavarian brewing companies and had registered the term 'Bayerisches Bier' (Bavarian Beer) as a GI and challenged the Dutch company's use of the name 'Bavaria Holland Beer' (apparently the beer was manufactured using a modified form of the Bavarian beer-making process). In its decision, the European Court of Justice noted that the issue was the relationship between the geographical name and other elements of the mark. For example, in this case, it was considered that the use of the name 'Holland', as well as various other features of the label meant that consumers would not be confused by the origin of the beer. As with all legal issues related to 'marks', whether trademarks, GIs or AOs, the core issue is always whether consumers would be confused about the origin of the product.[1278]

22.7 Concluding remark

The purpose of this chapter was to introduce and present an overview of IPRs as they relate to the agro-food industry with particular attention to Europe. As the reader has no doubt realised by now, this is an extremely complex area. Most of the topics touched upon deserve full chapters, if not whole books, in order to be fully explored.

It is also important to recognise that IPRs, like any legal system, are constantly evolving to keep up with developments in science and the changing demands of industry and society at large. In addition, IPRs are not a legal system existing in isolation. The manufacture and exploitation of products along the agro-food chain are also governed by a range of laws and regulations including those related to

[1277] Article 5, Lisbon Agreement. Available at: http://tinyurl.com/ozdz55n.
[1278] CJEU 22 December 2010, Case C-120/08, Bayerische Brauerbund v Bavaria.

contracts, bio-safety, competition and human rights. Most of these aspects are discussed in more detail in other chapters of this book, in order to provide an encompassing overview of the entire field of European Food Law.

Diagram 22.4. Summary of forms of industrial property that are most important for the food industry.

IPR	Description
Patents	Patents provide IP protection for inventions. Inventions eligible for protection may be found in all fields of technology and could be either products or processes. To qualify for patent protection, an invention must be new, involve an inventive step and be capable of industrial application.
Trademarks	Trademarks are words, phrases, symbols, designs, or any combination of these, that distinguish the goods or services of one business from those of another. The essential requirement of a trademark is that it be distinctive so that consumers can readily make an association between the company and the service or product it identifies.
Industrial designs	Industrial design protection is applicable to the ornamental and non-functional features designed as part of mass produced industrial articles.
Geographical indications	GIs are used to indicate that a particular good originates in a specified territory, region or locality which gives it certain qualities or characteristics. Often a GI will be the place name of origin of the goods e.g. Champagne, Roquefort cheese. A special type of GI is known as an 'Appellation of Origin' is usually reserved for those producers who are able to meet certain specifications such as a unique geographical production area and method of production.
Trade secrets	Trade secrets are items of information lawfully owned by an individual or a company that are not generally known or readily accessible to those in the particular field. It is their secrecy that makes them commercially valuable. Trade secrets can be legally protected from competitors provided the owner takes reasonable steps to maintain the secrecy of the information.
Plant Variety Protection (PVP)/ Plant Breeders' Rights (PBRs)	A *sui generis* system of IP protection geared towards plant breeders and designed to provide IP protection for new plant varieties. To be eligible for PBR protection, a variety must be new, distinct, uniform and stable.

23. Private food law

Rozita Spirovska Vaskoska and Bernd van der Meulen

23.1 Introduction

23.1.1 Overview

This chapter elaborates the legal structure of private food law, its anatomy so to speak. To this end the next section (Section 23.2) paints a picture of the way private food law has or might have developed. Sections 23.3 till 23.11 discuss the elements that together make up the structure of private food law: forms of chain orchestration (Section 23.3), ownership of private schemes (Section 23.4), enforcement of private food law (Section 23.5), adjudication and conflict resolution (Section 23.6), the role of audits (Section 23.7), certification (Section 23.8), accreditation (Section 23.9), the emergence of private alternatives to accreditation (Section 23.10) and standard setting (Section 23.11). Section 23.12 summarises the findings in graphic form. Diagram 23.2 may well be seen as the skeleton of private food law. Sections 23.13 and 23.14 discuss connections among private schemes (Section 23.13) and between private schemes and public law (Section 23.14). Section 23.15 goes into motives underlying private food law. Sections 23.16 till 23.22 describe the content of the currently most important examples of private regulation: the underlying concepts of good agricultural practices, good manufacturing practices and HACCP (Section 13.17), GlobalGAP (Section 23.18), BRC (Section 23.19), IFS (Section 23.20), SQF (Section 23.21) and ISO 22.000 (Section 23.22). Section 23.23 addresses the attempts at harmonisation of private food law through the Global Food Safety Initiative. Sections 23.24 and 23.25 analyse the relevance of EU and WTO law respectively for private food law. This chapter ends with some concluding remarks in Section 23.26.

23.1.2 Voluntary rules

In food law, one encounters several types of rules to which no legal obligation applies to comply with them; they are in other words not binding or at least not made binding by legislation. This is true for example for guides to good hygiene practices discussed in Chapter 13, for the codes of conduct elaborated by the Codex Alimentarius Commission and also for quite a lot of other texts on food.

A food business may choose to apply such rules thinking this is a good way to comply with the mandatory requirements of the law (e.g. apply hygiene guides to comply with HACCP), that this may improve the safety or other quality aspects of the product or for many other reasons.

If the owner of a business decides to apply non-binding rules, this may establish an internal obligation: the premises owned by the concern and the people working there may come under the obligation to apply these rules. The legal mechanism behind such obligations is not in the rules themselves, but may be found in property law (the owner has a say over his business), corporate law (the articles of association bind the partners in the company) or labour law (the labour contract gives the employer a certain power to impose duties upon the employee).

If a business communicates to its trading partners – its customers in particular – that it applies certain rules in the production of its products, this may become part of the offer it makes in the market: a guarantee that the product meets a certain level of quality as expressed in the rules applied to it(s production). If this feature is important to the customer, it may become part of the contract between the producer and his customer. A contract is binding upon the parties to it. Therefore, if a set of rules not binding by themselves is included in a contract, it thus becomes binding for the parties to the contract.

Customers may require their suppliers to apply certain rules/meet certain standards.[1279] If the suppliers agree, again this becomes part of the contract and thus a binding obligation. A contract requires mutual agreement ('a meeting of minds'). The specific content may be suggested by either party.

As long as there is no contract, there is no obligation to apply the non-binding rules. If most of the purchasers on the market (or a very powerful purchaser) make the application of certain rules a strict condition to enter into contractual relations, there still is no legal obligation for producers to apply these rules. They will know, however, that they can only acquire contracts if they are willing to accept the obligation. In a way the obligation is in the air. It is hovering over the contracts yet to be concluded. Legally there is no obligation to apply the rules, but if you do not apply them, you are out of business. In such a situation we may call the rules concerned 'de facto' binding. No obligation from the law (*de iure*) to apply, but such a necessity to apply from the facts in case (*de facto*) that in practice it feels like an obligation.

The combination of these elements – the power to create obligations by means of contracts and the power of certain players to dictate the terms of contract to such an extent that they have to be fulfilled even before the contract is concluded – forms the legal basis of a development where the private sector creates norms that apply to the food sector in addition to and even in competition with food law

[1279] Such as Fair Trade.

found in legislation, e.g. in public law. Contract is a part of private (civil) law.[1280] For this reason we have labelled the entirety of rules that in this way confront the food sector: 'private food law'.

23.2 The (hi)story of private standards

Most accounts on public European food law start with an account of the BSE crisis and the measures that have been taken since then. Many aspects of food law can best be understood if we take into account how it has developed historically. Also for private food law it helps to take its development into consideration. This development, however, is scattered and as yet little recorded. For these reasons and for simplicity's sake, in this section we will make a composition of developments that have taken place at divers times and places or that could have taken place and weld this into a 'story' of private food law that is only partly *hi*story, but that may help provide an understanding in a similar way to history.

One can imagine a development of contracts in the food chain from simple to complex. A simple contract states the identity, the quantity and price of the product to be sold by the vendor to the purchaser. The contract becomes more complex if it defines safety or other quality requirements like levels of contamination. Experience has shown that the safety of the product is influenced by the way it is handled (hygiene). The contract may therefore go beyond the product as such and include aspects of product handling, etc., etc.

Now one can imagine a second development. A company that has invested time and resources (maybe to pay for legal and technical advice) in elaborating a complex contract, and is happy with the results, will try to use the same provisions again. If this company has the bargaining power to impose the contractual provisions on its partners, it may start to use this contract as a model or even as general provisions to all the contracts it concludes.

[1280] In this connection the expression 'self-regulation' is often applied. The interinstitutional agreement on better law-making of the European Parliament, Council and Commission (OJ 31.12.2003, C 321/01), for example, defines self-regulation as: 'the possibility for economic operators, the social partners, non-governmental organisations or associations to adopt amongst themselves and for themselves common guidelines at European level (particularly codes of practice or sectoral agreements)'. While this definition applies to most of the private schemes discussed in this chapter, we avoid the expression 'self-regulation'. In our opinion the word, as well as the definition, convey too much of an impression of harmony and mutual understanding while – as we will see – in practice the creation of private schemes may well be based on market power of some businesses and dependence of others. In such situations legal theory still recognises equality, but in common speech this word is likely to create an incorrect image.

It takes a lot of expertise to develop a good model. Consultants may come to the market or businesses may cooperate to develop model contracts of ever increasing quality and complexity. And isn't it much easier instead of exchanging huge stacks of paper every time a contract is concluded to simply indicate the model that applies? Today it is sufficient to indicate in the contract the identity, quantity and price of the product and to stipulate that it conforms to ISO 22.000, BRC or SQF.[1281]

23.3 Chain orchestration

23.3.1 Contracts

The impact of the private scheme can go beyond the immediate contractual relation. A contract is a relationship between two parties. The safety and quality of a food product supplied by one business to the next, may largely depend, however, on the way it has been dealt with earlier up in the chain. So the customer depends on the agreements reached between his or her supplier and the previous supplier. S/he is, however, an outsider in this agreement. By demanding that in this relationship a certain standard applies, a purchaser can exercise considerable influence over contractual relationships upstream. A core issue, for example, in the Fair Trade scheme[1282] is that the workers at the very beginning of the chain (employees and smallholders in third world countries) receive reasonable remuneration for their efforts. The businesses expressing such a wish are often not the ones who have financial relations with these people. For this reason they demand that their suppliers provide proof (through certification) of applying fair conditions. In this way private standards can be used as an instrument for what is called 'chain orchestration'.

23.3.2 Vertical integration

Another approach to ensuring performance upstream in the chain – to which chain orchestration on the basis of contracts is the modern alternative – is through vertical integration. Businesses are vertically integrated if the different links of the chain are part of one concern. This can be achieved by setting up or acquiring businesses taking care of supply or to enter into a joint venture with a supplier. In such situations, the legal instrument of governance is found in property and/ or associations law.

[1281] The meaning of these abbreviations is explained hereafter.

[1282] See: http://fairtrade.nl/EN/MainContent/Home.aspx.

23.4 Owning a standard

The models described above (and below) represent a certain value. Sometimes the texts are available for free.[1283] It can be in the interest of the businesses that apply it if they are easily accessible. It is also possible that access to the standard is given only for payment.[1284] The legal instrument that makes it possible to claim ownership of the text of the standard is copyright. The copyright holder is entitled to decide on the conditions and price for circulation.

23.5 Enforcement

Generally speaking, legislation creates obligations towards the public authorities representing society as a whole. In case of non-compliance, sanctions may come as a consequence.[1285]

Contracts create obligations between parties and provide these parties with instruments to deal with non-compliance by the other parties. In case of non-compliance liability for damages arises, contractual relationships may be ended and all kinds of consequences may arise that have been agreed upon in the contract (like contractual fines). The difference in consequence of non-conformity with public law and non-conformity with private law can make it sensible to create obligations by contract that are similar or even identical to obligations that are already present in public law. We see for instance that all the major private food law standards include the obligation to apply the Hazard Analysis and Critical Control Point system (HACCP). This does not create an obligation for the supplier that s/he does not already have, but it turns it from an obligation towards society into an obligation towards the purchaser and provides the purchaser with civil law instruments to enforce this obligation.

23.6 Adjudication

Contractual rights and duties can be invoked in civil courts of law. However, the scheme at issue may also provide for other dispute settlements structures, such as arbitration. In case of requirements in the interest of third parties such as consumers, complaint mechanisms for these third parties may be included.

[1283] Like GlobalGAP. This conforms to Article 5.1(4) of the EU best practice guidelines for voluntary certification schemes for agricultural products and foodstuffs, OJ 2010 C 341/5.
[1284] Like ISO.
[1285] Indeed according to Article 6.2 of the EU best practice guidelines for voluntary certification schemes for agricultural products and foodstuffs, OJ 2010 C 341/5, unsatisfactory inspection results *should* lead to appropriate action.

23.7 Audits

In the contractual relationship it can be agreed that the customer has the right to visit the suppliers' premises to check if the agreed practices are being applied with the agreed results. Such inspection visits are generally known as 'audits'. It has been recorded that big production companies had special divisions where several people had full time jobs receiving auditors. In due course it was noted that all these auditors were checking for similar requirements. Along with the development described above where standard contracts were welded into harmonised and commonly used models, another development took place.

Instead of every customer of a certain producer auditing the producer, it became accepted that one player would audit on behalf of all.[1286] If the auditor considered the requirements of the applicable standard to be met, s/he would provide a certificate to the audited business to give it an instrument to communicate its compliance to its customers.

In case the audit shows non-compliance, no certification is provided and/or the right to use the mark representing the certification is withdrawn. As a consequence the company can no longer do business with customers that demand certification.

In this way the formulation of norms (standard), the inspection of their fulfilment and the proof of their fulfilment in the form of a certificate and the sanction on non-compliance in the form of withdrawal of certification, developed into structures known as certification schemes. If the customers whom the certification communicates are other businesses, the certification scheme is called B2B (business to business). If the customers are the final consumers, the certification scheme is called B2C (business to consumer).

[1286] This player would be an independent third party trusted by all parties concerned. Three types of audits can be distinguished: one (or first) party audits (self-controls or internal audits), second party audits (audits by the customer) and third party audits (audits by auditors independent from both other parties). Generally in legalese, the expression 'third' is used to indicate parties outside the relationship at issue. This applies regardless if the 'we' relationship consists of two parties or more. Talking about a contractual relationship, players outside this relationship are 'third'. Talking about the EU, states who are not members are 'third'. Talking about the Cold War, the world outside the opposing political blocs is 'third'.

Diagram 23.1. Steps in the certification process.[1287]

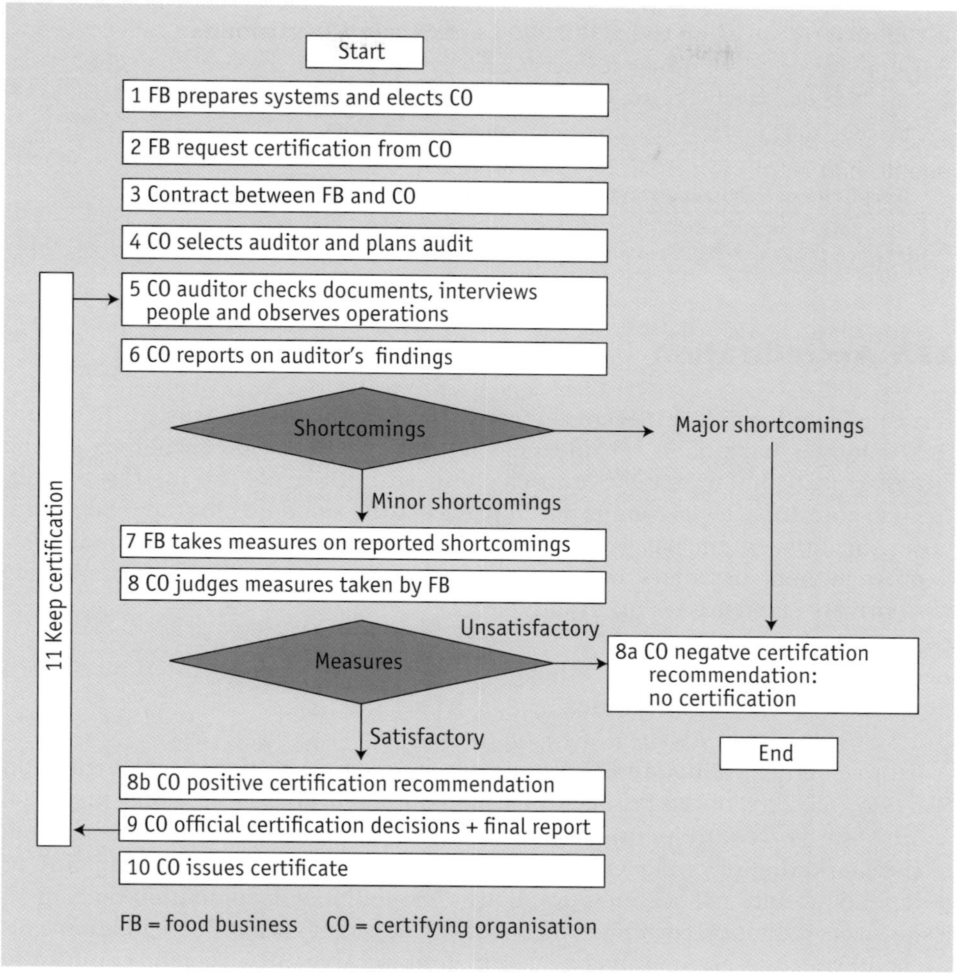

Start
1 FB prepares systems and elects CO
2 FB request certification from CO
3 Contract between FB and CO
4 CO selects auditor and plans audit
5 CO auditor checks documents, interviews people and observes operations
6 CO reports on auditor's findings
Shortcomings
Major shortcomings
Minor shortcomings
7 FB takes measures on reported shortcomings
8 CO judges measures taken by FB
Measures
Unsatisfactory
8a CO negatve certifcation recommendation: no certification
End
Satisfactory
8b CO positive certification recommendation
9 CO official certification decisions + final report
10 CO issues certificate
11 Keep certification

FB = food business CO = certifying organisation

23.8 Certification mark

Connected to a certification scheme is often a symbol that is owned by the owner(s) of the scheme. This ownership usually takes the form of a trademark. The owner of a trademark has the right to allow or forbid its use by others. On this basis the owner (both of the scheme and the related trademark) has the legal power to impose compliance with the intended use as proof of fulfilment or sanction for non-compliance.

[1287] This frame has been adapted from Nynke Bergsma, Voedselveiligheid: certificatie en overheidstoezicht, Praktijkgids Warenwet Sdu The Hague, 2010, p. 40.

Law text box 23.1. SQF on use of certification trade mark.

3 Conditions for using the SQF 2000 Certification Trade Mark

3.1 A producer shall, for the duration of its certification, prove to the satisfaction of SQFI[1] or the LCB[2] that its quality system satisfies the requirements set forth in the current edition of the SQF 2000 Code; and

3.2 A producer must only use the SQF 2000 Certification Trade Mark in accordance with its Certificate of Registration and these rules.

[1] SQF Institute (footnote added).
[2] Licensed certification Body (footnote added).

23.9 Accreditation

It is considered of vital importance that certifying bodies can be trusted. To ensure this, schemes have been set up to certify the certifiers. Such schemes can be strictly private, but in many countries public authorities take an interest as well. In such countries, legislation is put in place setting standards for the recognition of certification. On the basis of such standards, certifiers can ask for accreditation to prove to their customers that the product they deliver (e.g. proof of compliance with private standards) is up to standard.

Accreditation is therefore the official certification of certifiers often by or with the consent of public authorities.

As from 1 January 2010, an EU framework for accreditation is place. Regulation (EC) No 765/2008 of the European Parliament and of the Council of 9 July 2008 setting out the requirements for accreditation and market surveillance relating to the marketing of products,[1288] requires the EU Member States to appoint an independent national accreditation body. The national accreditation body shall issue an accreditation certificate when it evaluates that – what the regulation calls – a conformity assessment body is competent to carry out a specific conformity assessment activity (Article 5).

23.10 Beyond accreditation

Accreditation takes place independently of schemes and scheme owners. Not all scheme owners are fully satisfied that accreditation ensures that certifying bodies only certify businesses with a high level of compliance with their scheme.

[1288] OJ L 218, 13.8.2008, pp. 30-47.

For GlobalGAP questions arose regarding the efficacy of accreditation to ensure up to standards certification when GlobalGAP certified products appeared on the market not complying with the applicable maximum residue levels for pesticides.

To improve the output of audits and the quality of certification, GlobalGAP set up an Integrity programme. The programme is governed by the Integrity Surveillance Committee (ISC) which was made operational in 2009.[1289] The Integrity programme consists of a Brand Integrity Programme (BIPRO) and a Certification Integrity Programme (CIPRO). BIPRO provides an online database of certified businesses with a publicly available search site. This enables businesses to ensure themselves if a claim to GlobalGAP certification is valid.

The aim of CIPRO is to ensure that each certified producer meets the same level and to make sure that the control of these producers has been done consistently and each certification body applied the GlobalGAP rules the same way. In addition to accreditation, certification bodies need approval. Results of the CIPRO assessment of certification bodies are available to accreditation bodies. Ultimately, the license of certification bodies can be cancelled if they do not perform to CIPRO standards.

23.11 Standard setting

In the story described above the initiative to set up private schemes is placed with businesses holding a dominant position in the market. Initially these businesses were mainly the famous brand holders requiring B2B certification to be able to ensure a constant product quality in order to uphold the reputation of the brand. Such brand-related quality schemes still play an important role. Of late, however, retailers brands (the so-called private labels) have acquired a position on the market that has made them leading players in the creation of private food law.[1290] Most of the examples discussed below originate from retailer initiatives.

Private standards, however, are not always based on market power. They may also result from agreements reached with various stakeholders, for example in round tables.[1291] In particular schemes dealing with general interests like protection of the environment and corporate social responsibility take account of opinions of third parties (third from the point of view of the parties to the contract to which the scheme applies) including NGOs. Some schemes – such as GlobalGAP originated from a power base and are moving to more representative forms including other stakeholders in decision making.

[1289] See: http://www.globalgap.org/cms/front_content.php?idcat=129.

[1290] On the power of private labels, see F. Bunte *et al.*, The impact of private labels on the competitiveness of the European food supply chain, European Union 2011.

[1291] See O. Hospes, Regulating biofuels in the name of sustainability or the right to food? In: O. Hospes and B.M.J. van der Meulen (eds.) Fed up with the right to food? The Netherlands' policies and practices regarding the human right to adequate food, Wageningen Academic Publishers, Wageningen, the Netherlands, 2009, pp. 121-135.

23.12 Structure of private food law

The structure of private food law as set out in the above, is summarised in Diagram 23.2.

Diagram 23.2. The structure of private food law.

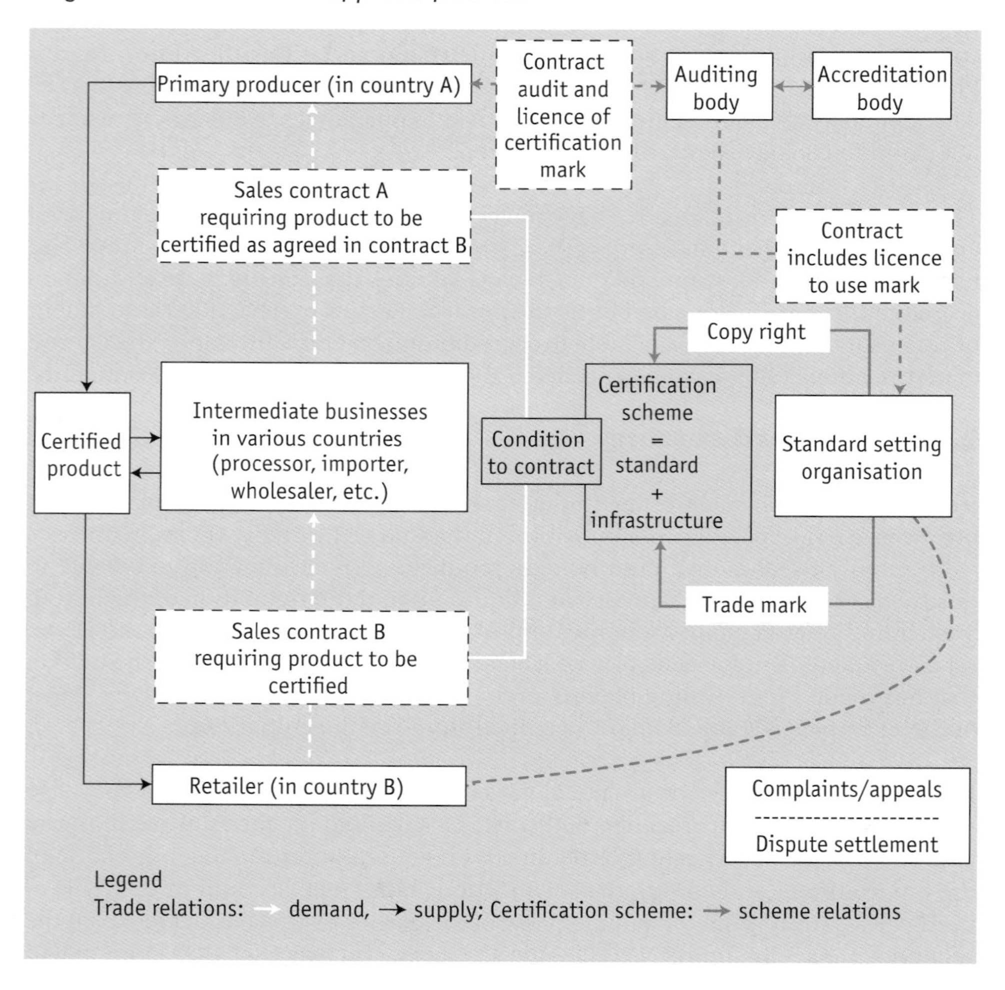

Rules have been agreed upon in the context of the standard-setting organisation. This organisation is the owner (copyright holder) of these rules. The standard-setting organisation can consist of businesses using (or imposing) the standard, but it can also be independent. Connected to the rules can be a certification symbol protected as a trademark. The standard setting organisation agrees (maybe through a licence contract or – as far as the rules are concerned – by putting them in the public domain) to the use of these rules and the trademark by auditors and food businesses. One of the actors in the chain decides on applicability of the scheme. It includes the scheme in its contractual requirements and requires it to be included in contracts earlier in the chain as well. The auditor checks for compliance with the scheme, certifies (or refuses to certify) the audited business and decides to license (or withhold) the right to use the certification mark. The quality of the auditor in turn is ensured through accreditation.[1292]

23.13 Interconnected private schemes

Diagram 23.2 shows two contractual relationships (representing a longer chain) where in each link the same private scheme applies. In practice also, examples are found where the scheme applied in one link requires another link to apply another scheme. In the Dutch dairy sector, for example, dairy-producing businesses require dairy farmers to be certified. The applicable certification scheme requires them to use as feed for milk-producing animals only products obtained from businesses certified for a certain feed quality scheme. In this way private food law obtains a web-like structure of interrelated schemes.

Law text box 23.2. Dutch HACCP on application of other schemes.

HACCP Certification Regulations 2006
Article 4
Where these regulations do not stipulate any other requirements with respect to the HACCP certification process, the certification bodies have to apply the procedures set in force for the certification of quality systems, that are accredited on the basis of EN 45012 or ISO/IEC Guide 62, excepting article 3.3. of ISO/IEC Guide 62. This article is replaced by article 5.3 of ISO/IEC Guide 66:1998.
With the information/application phase of the certification process, these regulations have to be submitted by the certification body to anyone requesting HACCP certification in accordance with clause 3.1.1.1 of EN 45012:1998 and ISO/IEC Guide 62:1996.

23.14 Public – private interconnections

Private schemes are not only interconnected among themselves, but also with public law. The vast majority of private certification schemes refers to public law requirements that have to be complied with. Less common is the inverse where public law provisions require compliance with private schemes. Legislation on community reference laboratories or on methods of sampling refer to private law technical standards.

[1292] Most private schemes require accreditation as laid down in the International Standard ISO/IEC Guide 65.

Law text box 23.3. The official controls regulation referring to private CEN standards.

Regulation 882/2004 *Article 11*
Methods of sampling and analysis

1. Sampling and analysis methods used in the context of official controls shall comply with relevant Community rules or,
   (a) if no such rules exist, with internationally recognised rules or protocols, for example those that the European Committee for Standardisation (CEN) has accepted or those agreed in national legislation;

Yet another public law approach to private regulation is where legislation imposes upon stakeholders the duty to regulate for themselves. The best-known example is the HACCP requirement in Regulation 852/2004. The very essence of HACCP is that a business sets up rules for its own processes. In the case of HACCP it is not a voluntary choice but a public law obligation that will be enforced by public authorities. Such situations are known as imposed self-regulation or enforced self-regulation. An alternative for the application of HACCP is the application of a hygiene code. The expression hygiene code is used to refer to the national or community guides of good practice.[1293] Member States approve the national guides. In this way this form of private regulation acquires status under public law. Compliance with the private standard is deemed to imply compliance with the legal HACCP requirement.

Some food safety inspection agencies acting on the basis of risk-based policies reduce the intensity of inspections for businesses operating under private schemes that are trusted to provide good results in ensuring food safety.[1294] They mainly limit themselves to an assessment of the quality of the private scheme. Such controls of the quality of private control schemes are known as meta-controls.[1295]

[1293] Articles 8 and 9 of Regulation 852/2004.

[1294] The official controls regulation (882/2004) takes such practices into account in Article 27(6): When, in view of own-check and tracing systems implemented by the feed or food business as well as of the level of compliance found during official controls, for a certain type of feed or food or activities, official controls are carried out with a reduced frequency or to take account of the criteria referred to in paragraph 5(b) to (d), Member States may set the official control fee below the minimum rates referred to in paragraph 4(b) (...).

[1295] The Dutch ministry of agriculture is strongly in favour of such policies. Publications on this topic with a summary in English language are for example: N. Bondt et al. Voedselveiligheid, ketens en toezicht op controle, Rapport 5.06.01, LEI The Hague 2006; V. Beekman *et al.*, Stimulering eigen verantwoordelijkheid. Zorgen dat producenten en consumenten zorgen voor voedselveiligheid, Rapport 5.06.05, LEI, The Hague, the Netherlands, 2006; E. de Bakker *et al.*, Nieuwe rollen, nieuwe kansen? Een programmeringsstudie voor toezicht op controle in het agro-foodcomplex, Rapport: 6.07.08, LEI, The Hague, the Netherlands, 2007.

Law text box 23.4. Regulation 2073/2005 on microbiological criteria, in Annex I referring to private EN/ISO standards.

Food category	Micro-organisms/ their toxins, metabolites	Sampling plan		Limits		Analytical reference method	Stage where the criterion applies
		n	c	m	M		
1.1 Ready-to-eat foods intended for infants and ready-to-eat foods for special medical purposes	*Listeria monocytogenes*	10	0	Absence in 25 g		EN/ISO 11290-1	Products placed on the market during their shelf-life
1.2 Ready-to-eat foods able to support the growth of *L. monocytogenes*, other than those intended for infants and for special medical purposes	*Listeria monocytogenes*	5	0	100 cfu/g		EN/ISO 11290-2	Products placed on the market during their shelf-life
		5	0	Absence in 25 g		EN/ISO 11290-1	Before the food has left the immediate control of the food business operator, who has produced it
1.3 Ready-to-eat foods unable to support the growth of *L. monocytogenes*, other than those intended for infants and for special medical purposes	*Listeria monocytogenes*	5	0	100 cfu/g		EN/ISO 11290-2	Products placed on the market during their shelf-life

Finally, we find examples where public authorities partake in private standard-setting to achieve objectives in foreign countries that could not be achieved by means of public law instruments. Hospes, for example, analyses principles formulated by Dutch authorities for the sustainable production of biomass (in countries such as Brazil).[1296] These principles are operationalised through private certification schemes. The Netherlands regard the need for certification as a condition for the import of biofuels.

23.15 Motives

What are the driving forces behind private food law? They are probably too numerous to list in full and motives may differ from stakeholder to stakeholder, but at least some points can be identified.

The motive mentioned in most private food schemes is food safety. Food safety is important for the protection of consumers, to comply with consumers wishes and also to comply with public law requirements.

Compliance with public law requirements can be a motive in itself.[1297] To comply with their own legal obligations, businesses depend on how the product has been dealt with upstream. Therefore they may want to ensure themselves with private law instruments that legal obligations are being complied with, or to impose these obligations on producers working in countries where different legal requirements apply, thus using private law to bridge the gap between different legal systems. Connected to compliance is liability. On the one hand businesses may try to pass on liability to other links in the chain (some require insurance and guarantees from producers); on the other hand explicit agreements are a way of showing that everything possible has been done to avoid non-compliance. In civil and criminal cases this may be used in what – in the UK – is called a due diligence defence.

Further private law is used to discourage the legislator from taking charge. If businesses solve problems there will be less urgency for the legislator to intervene. Businesses prefer private law to public law as it reflects their own wishes better and is easier to change if the need arises. Also, private law can be used to supplement or repair public law. For example, public law on traceability (Article 18 of Regulation 178/2002) is not certain about whether internal traceability (within a business) is

[1296] O. Hospes, Regulating biofuels in the name of sustainability or the right to food? In: O. Hospes and B.M.J. van der Meulen (eds.) Fed up with the right to food? The Netherlands' policies and practices regarding the human right to adequate food, Wageningen Academic Publishers, Wageningen, the Netherlands, 2009, pp. 121-135. See for another example of the use of private schemes by public authorities to influence behaviour abroad, his chapter in this book.

[1297] Empirical research shows that private standards are helpful in complying with public law requirements. The reason for this is probably the embeddedness of private standards in audit schemes that provide feedback on performance. See B.M.J. van der Meulen, Reconciling food law to competitiveness, Wageningen Academic Publishers, Wageningen, the Netherlands, 2009.

required.[1298] ISO 22.000 explicitly requires internal traceability. While EU legislation exempts the primary sector from HACCP, private schemes such as GlobalGAP impose it on this sector as well.

On the basis of Regulation 882/2004, official controls should be risk-based and control intensity can be related to compliance history. As discussed above, certification may be an instrument to convince inspection agencies that the level of compliance is high and therefore the urgency with regard to official controls low.[1299]

Private standards that go beyond compliance, that is to say apply higher safety and/or quality standards than public law,[1300] may help a business to distinguish itself on the market and acquire a share of the (top end) market. Or, to put it another way, raising standards may be used to protect markets from competitors.

Finally, moral considerations such as religion and corporate social responsibility are a driving factor of private regulation. With a view to showing their commitment to making a contribution to sustainable development, businesses bind themselves to private schemes that elaborate on these interests.

23.16 Examples

Below are some examples of the private schemes that currently seem to be leading on the market.[1301] The discussion of the examples focuses mainly on the content of the standards and less on the governance and certification structure of the schemes. The objective of these examples is to make, in addition to the structure set out above, the content of private food law more concrete. The examples have all been taken from the area of private food safety law. As we have seen above, many other areas of concern are covered by private food law as well.

[1298] The Standing Committee on the Food Chain and Animal Health has published a more or less official interpretation of the General Food Law, where they argue that 'the Regulation does not expressly compel operators to establish a link (so called internal traceability) between incoming and outgoing products. Nor is there any requirement for records to be kept identifying how batches are split and combined within a business to create particular products or new batches'. See Guidance on the Implementation of Articles 11, 12, 14, 17, 18, 19 and 20 of Regulation (EC) N° 178/2002 on General Food Law of 26 January 2010.

[1299] On the stacking of private and official controls see: B.M.J. van der Meulen and A.A. Freriks, Millefeuille. The emergence of a multi-layered controls system in the European food sector, Utrecht Law Review 2 (1), pp. 156-176, 2006.

[1300] According to Article 5.2(3) of the EU best practice guidelines for voluntary certification schemes for agricultural products and foodstuffs (OJ 2010 C 341/5), it should be clearly indicated where this is the case.

[1301] The overview is based on the information provided by the owners of these schemes. No critical comparison is intended at this stage of the research.

Diagram 23.3. Possible framework for the analysis of EU (public) food law.

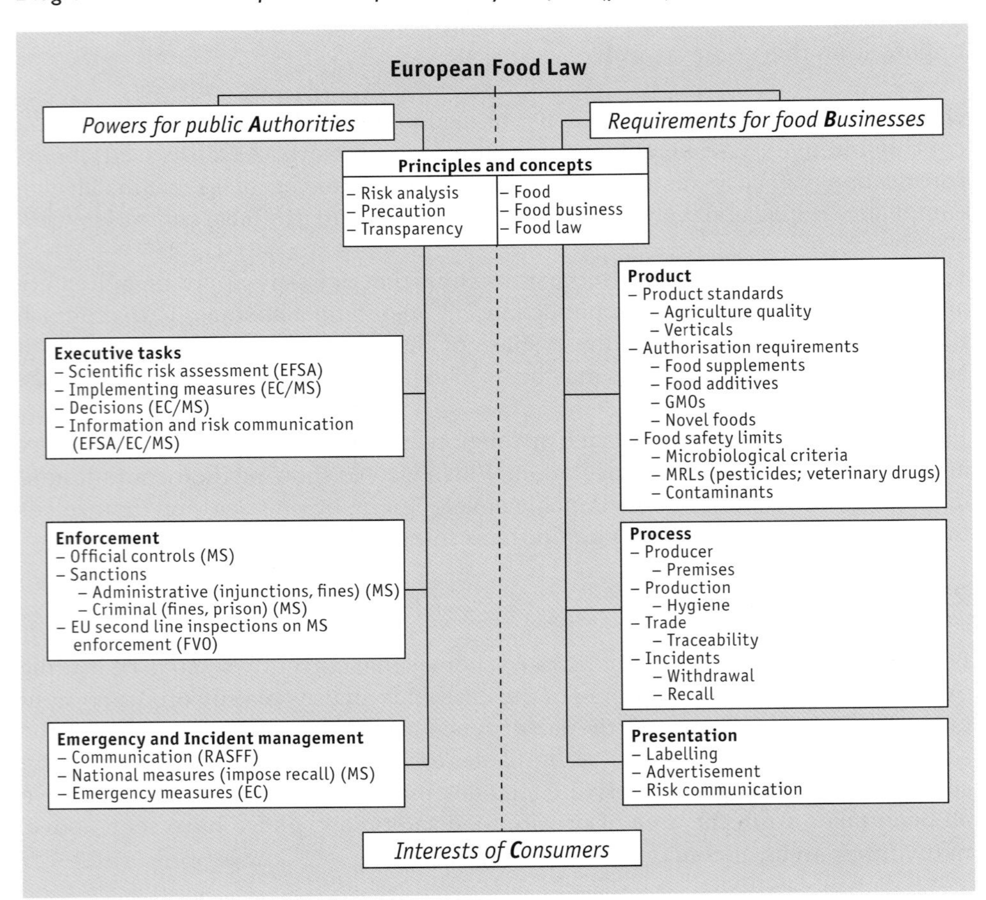

Public law requirements on food businesses can be divided[1302] into rules regarding the product (vertical standards, market approval requirements for certain ingredients and safety objectives in the form of maximum levels of contaminants and residues), rules regarding the process (hygiene, traceability and incident management), rules regarding the presentation (labelling and advertisement) and public powers (enforcement and incident management by authorities). In private food law we find similar ingredients, but in a different mix. Product-related rules mainly concern safety and quality objectives. Rules on the process (hygiene, traceability and risk management) are the core. Labelling provisions will usually be limited to the use of the certification mark. The place held by public powers of inspection and enforcement in public food law, is taken in private food law by the powers granted by the businesses themselves to the auditors and certifiers.

[1302] As set out in Chapter 8.

Diagram 23.4. Possible framework for the analysis of private food law.

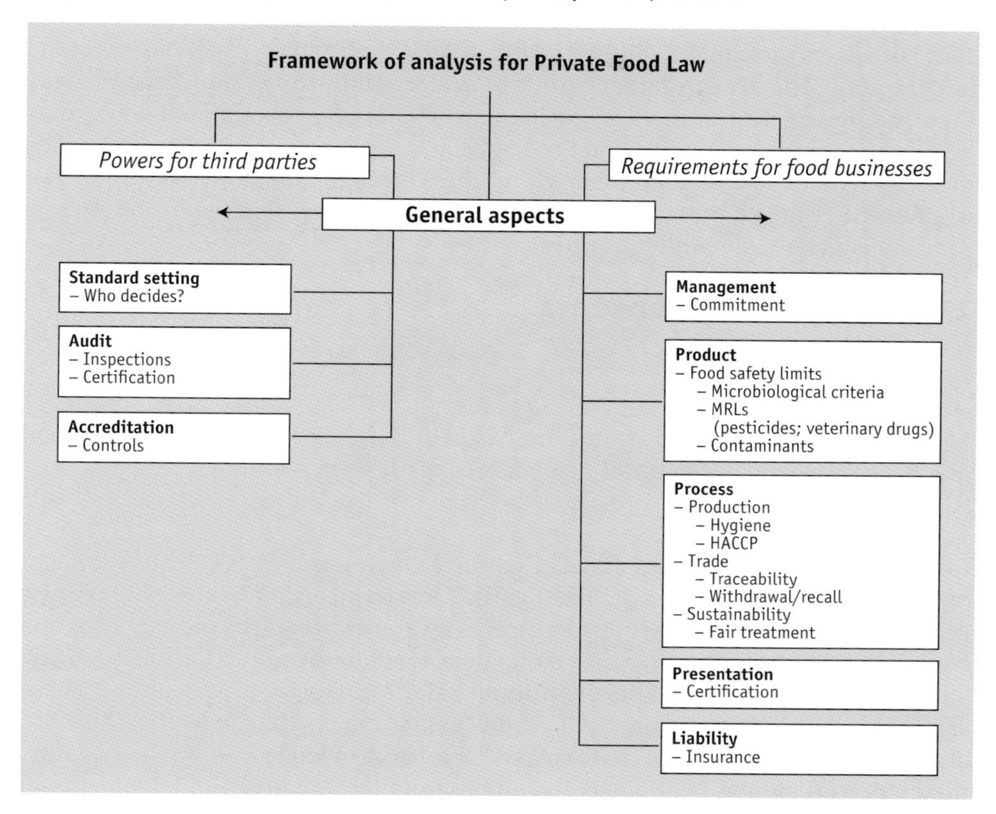

The most important additions to the types of food rules we have encountered in public law, are provisions governing business organisation and management systems including management commitment and provisions on information-sharing within the food chain (so-called chain transparency).[1303]

Our first impression of comparing private standards to legislation is that the drafting is sloppy. It seems that lawyers experienced in drafting legislation are not involved. Nevertheless, stakeholders seem to understand the meaning of

[1303] On private schemes, see: W. van Plaggenhoef, M. Batterink and J.H. Trienekens, International Trade and Food Safety. Overview of legislation and standards, Wageningen University, Wageningen, the Netherlands, 2003; G. Chia-Hui Lee, Private Food Standards and their Impacts on Developing Countries, European Commission, DG Trade Unit G2; M. Will and D. Guenther, Food Quality and Safety Standards, as required by EU Law and the Private Industry With special reference to the MEDA countries' exports of fresh and processed fruits & vegetables, herbs & spices A Practitioners' Reference Book, 2nd edition, GTZ, 2007; Working Party on Agricultural Policies and Markets, Final Report on Private Standards and the Shaping of the Agro-Food System, OECD, 2006.

private standards better than they do legislation.[1304] Maybe this can be explained by a different attitude towards private standards than towards legislation. Private standards have their place within a business relationship that stakeholders intend to continue. This is a strong motivator to understand private standards the way they are meant. In the case of legislation, by contrast, misunderstanding may justify non-compliance with what the legislator envisaged. Lawyers are trained to search for alternative meanings within the wording of the law, an attitude that would mean the end of the business relationship if it were applied to contractual provisions prior to a conflict.

23.17 Underlying concepts

23.17.1 GAP/GMP

Many private food safety systems are based on HACCP and some form of good practices, like good agricultural practices (GAP) or good manufacturing practices (GMP).

In the words of the UN Food and Agricultural Organisation FAO,[1305] the concept of Good Agricultural Practices has evolved in recent years in the context of a rapidly changing and globalising food economy and as a result of the concerns and commitments of a wide range of stakeholders about food production and security, food safety and quality, and the environmental sustainability of agriculture. These stakeholders include governments, food processing and retailing industries, farmers, and consumers, who seek to meet specific objectives of food security, food quality, production efficiency, livelihoods and environmental benefits in both the medium and long term. GAP offers a means to help reach those objectives. Broadly defined, GAP applies available knowledge to addressing environmental, economic and social sustainability for on-farm production and post-production processes resulting in safe and healthy food (and non-food agricultural products). Many farmers in developed and developing countries already apply GAP through sustainable agricultural methods such as integrated pest management, integrated nutrient management and conservation agriculture. These methods are applied in a range of farming systems and scales of production units, including as a contribution to food security, facilitated by supportive government policies and programmes.

GAP represents the state of the art in sustainable agriculture. Something similar is true for GMP. For this reason, its content in terms of dos and don'ts is continuously

[1304] On this topic see: B.M.J. van der Meulen, Reconciling food law to competitiveness, Wageningen Academic Publishers, Wageningen, the Netherlands, 2009.
[1305] FAO Committee on Agriculture, Development of a Framework for Good Agricultural Practices, Seventeenth Session, Rome, Italy, 31 March-4 April 2003.

developing and cannot be captured in one permanent description. It is not law in itself,[1306] but through private schemes it can acquire legal relevance.

23.17.2 HACCP

The version of HACCP mandatory for food business operators, is the one codified in Regulation 852/2004 of the European Union. The version referred to in private schemes is often the one laid down in the Codex Alimentarius.[1307] The former is based on the latter but it is not as elaborate. Through their inclusion in private schemes, the non-binding Codex acquires a measure of legal effect. Inclusion of HACCP in private schemes adds for European businesses applicability in sectors exempted from Regulation 852/2004 (the primary sector in particular), enforcement through private law instruments and visibility through certification.[1308] For non-European businesses it brings an obligation that may not or not in the same way follow on from their own public law system.

23.18 EurepGAP/GlobalGAP[1309]

EurepGAP started in 1997 as an initiative by retailers belonging to the Euro-Retailer Produce Working Group (EUREP). British retailers in conjunction with supermarkets in continental Europe were the driving forces. They reacted to the growing concerns of consumers regarding product safety, environmental and labour standards and decided to harmonise their own often very different standards for agricultural products.

The development of common certification schemes was also considered to be in the interest of producers. Those in contractual relationships with several retailers explained that they had to undergo multiple audits for different criteria every year. With this is in mind, EUREP started working on harmonised standards and procedures for the development of Good Agricultural Practices[1310] in conventional agriculture including highlighting the importance of Integrated Crop Management and a responsible approach to worker welfare. This resulted in what was initially called the European Retailers Protocol for Good Agricultural Practice (EurepGAP).

[1306] P.A. Luning, W.J. Marcelis and W.M.F. Jongen, Food Quality Management. A techno-managerial approach, Wageningen Academic Publishers, Wageningen, the Netherlands, 2002, p. 225 refer to Good Practices as 'self-discipline'.
[1307] The Codex Basic Text on hygiene, is a standard work comprising of: Recommended International Code of Practice General Principles of Food Hygiene CAC/RCP 1-1969, Rev. 4 (2003); Hazard Analysis and Critical Control Point (HACCP) System and Guidelines for its Application; Principles for the Establishment and Application of Microbiological Criteria for Foods CAC/GL 21-1997, and Principles and Guidelines for the Conduct of Microbiological Risk Assessment CAC/GL-30 (1999).
[1308] See for example the Dutch HACCP certification scheme (based in Apeldoorn, the Netherlands). See: www.foodsafetymanagement.info.
[1309] Information from the GlobalGAP website: www.globalgap.org.
[1310] Hence the GAP part in the name.

Over the next ten years a growing number of producers and retailers around the globe signed up to this idea because it matched the emerging pattern of globalised trading: EurepGAP began to gain global significance. To align EurepGAP's name with the proposition as the pre-eminent international GAP standard and to prevent confusion with its growing range of public sector and civil society stakeholders, the Eurep Board decided to undertake the step to re-brand. It was considered a natural path and evolution that led EurepGAP to become GlobalGAP. The decision was announced in September 2007 at the 8th global conference in Bangkok.

GlobalGAP has established itself as a key reference for Good Agricultural Practices in the global market-place, by translating consumer requirements into agricultural production in a rapidly growing list of countries – currently more than 80 across all continents.

Diagram 23.5. GlobalGAP structure Integrated Farm Assurance Standard.[1311]

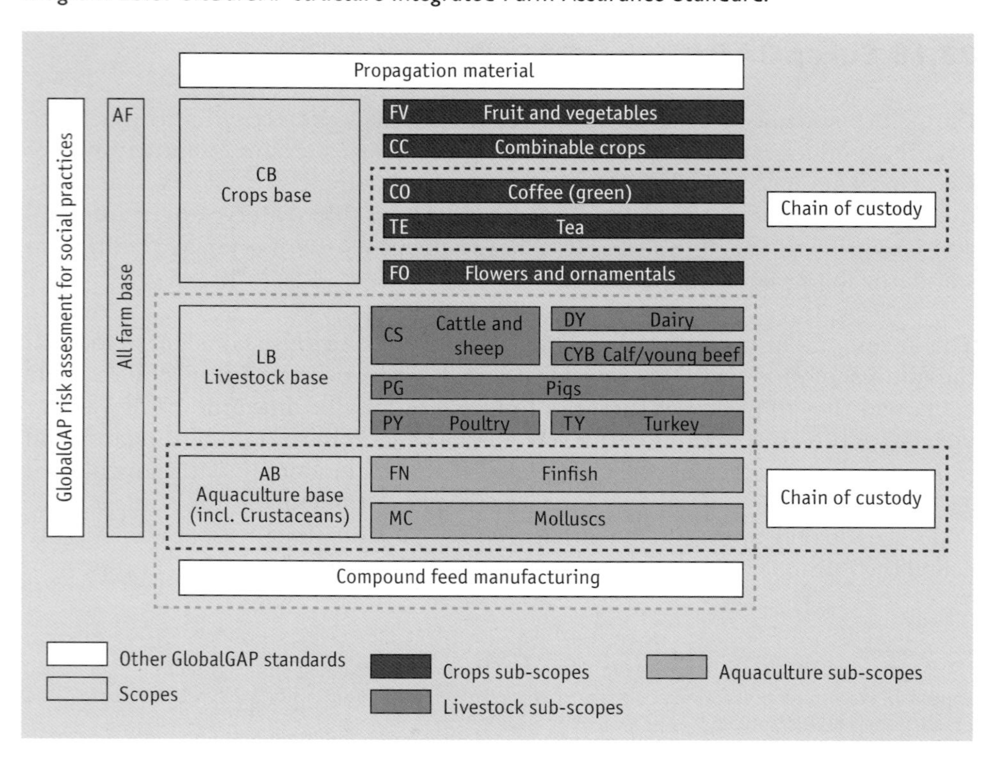

[1311] Source: http://www.globalgap.org/cms/front_content.php?idcat=176.

GlobalGAP is a private sector body that sets standards for the certification of agricultural products around the globe. The aim is to establish one single standard (the Integrated Farm Assurance (IFA) Standard) for Good Agricultural Practice with different product applications capable of fitting into the whole of global agriculture. Governance is by a Board whose decisions are based on a structured consultation process. Retailers and suppliers are represented on the board. Sector-specific interests and multi-stakeholder input are consolidated to ensure global acceptance. Sector Committees discuss and decide upon product- and sector-specific issues. All committees have 50% retailer and 50% producer/supplier representation.

GlobalGAP is a pre-farm-gate standard, which means that the certificate covers the process of the certified product from farm inputs like feed or seedlings and all the farming activities until the product leaves the farm. GlobalGAP is a business-to-business (B2B) label and is therefore not directly visible to consumers. Its certification is carried out by more than 140 independent and accredited certification bodies in more than 80 countries. It is open to all producers worldwide. GlobalGAP includes annual inspections of the producers and additional unannounced inspections.

GlobalGAP consists of a set of normative documents. These documents cover the GlobalGAP General Regulations, the GlobalGAP Control Points and Compliance Criteria and the GlobalGAP Checklist.

As many other on-farm assurance systems had been in place for some time prior to the existence of GlobalGAP, a way had to be found to encourage the development of regionally adjusted management systems so as to prevent farmers from having to undergo multiple audits. Existing national or regional farm assurance schemes can seek recognition as equivalent to GlobalGAP through independent benchmarking.

The GlobalGAP standard is subject to a three-year revision cycle of continuous improvement to take into account technological and market developments.

Law text box 23.5. GlobalGAP on traceability and record keeping.[1312]

No. b	Control point	Compliance criteria	Level
PM 1	Traceability		
PM 1.1	Is GLOBALGAP (EUREPGAP) registered product traceable back to and trackable from the registered nursery (and other relevant registered areas) where it has been grown?	There is a documented identification and traceability system that allows GLOBALGAP (EUREPGAP) registered plants to be traced back by individual batch numbers which relate to customer orders, on a per batch basis to inputs (such as Seed Lot/Growing Media Batch/Growing or Germination Temperature Regimes/Crop protection materials applied/plant movements within the nursery) to the registered nursery, and tracked forward to the immediate customer. No N/A.	Major must
PM 1.2	Do all propagators have a documented procedure to manage the withdrawal of registered products from the market?	All propagators must have access to documented procedures which identify the type of event that may result in a withdrawal, persons responsible for taking decisions on the possible withdrawal of product, the mechanism for notifying customers and the GLOBALGAP (EUREPGAP) CB (if a sanction was not issued by the CB and the propagator or group recalled the products out of free will) and methods of reconciling stock. The procedures must be tested annually to ensure that it is sufficient. Procedure must be demonstrated.	Major must
PM 2	Record keeping and internal self-assessment		
PM 2.1	Are all records requested during the external inspection accessible and kept for a minimum period of time of two years, unless a longer requirement is stated in specific control points?	Propagators keep up to date records for a minimum of two years from the date of first inspection, unless legally required to do so for a longer period. No N/A.	Minor must

▶▶

[1312] Source: GlobalGAP Control Points and Compliance Criteria Plant Propagation Material March 2008.

Law text box 23.5. Continued.

PM 2.2	Does the propagator take responsibility to undertake a minimum of one internal self-assessment per year against the GLOBALGAP (EUREPGAP) Standard?	There is documentary evidence that the GLOBALGAP (EUREPGAP) or benchmarked standard internal self-assessment under responsibility of the propagator has been carried out and are recorded annually. No N/A.	Major must
PM 2.3	Are effective corrective actions taken as a result of non-conformances detected during the internal self-assessment?	Effective corrective actions are documented and have been implemented. No N/A.	Major must

23.19 BRC[1313]

In 1998 the British Retail Consortium (based in London), responding to industry needs, developed and introduced the BRC Food Technical Standard to be used to evaluate manufacturers of retailers own-brand food products. In BRC the British supermarkets Tesco, Sainsbury, Safeway and Summerfield participate. In the early days each retailer inspected his own suppliers. These common efforts to inspect suppliers have huge cost advantages for retailers, because a supplier fulfils the requirements of all British retailers at once.

BRC is designed to be used as a pillar to help retailers and brand owners with their 'due diligence' defence, should they be subject to a prosecution by the enforcement authorities. Under EU food law, retailers and brand owners have a legal responsibility for their brands.[1314]

In a short space of time, the BRC Standard became invaluable to other organisations across the sector. It is regarded as a benchmark for best practice in the food industry. This and its use outside the UK has seen it evolve into a global standard used not just to assess retailer suppliers, but as a framework upon which many companies have based their supplier assessment programmes and the manufacture of some branded products.

The majority of UK, and many continental European and global retailers, and brand owners will only consider business with suppliers who have gained certification to the appropriate BRC Global Standard.

[1313] Information from BRC website: www.brc.org.uk.

[1314] See for example Article 17(1) of Regulation 178/2002 (the General Food Law) and Directive 85/374 on product liability.

Following the success and widespread acceptance of the Global Standard – Food, the BRC published the first issue of the Packaging Standard in 2002, followed by Consumer Products Standard in August 2003, and the BRC Global Standard – Storage and Distribution in August 2006. In 2009, the BRC partnered with the Retail Industry Leaders Association (RILA) to develop the Global Standard for Consumer Products North America edition. Each of these Standards is regularly reviewed and each standard is fully revised and updated at least every 3 years after extensive consultation with a wide range of stakeholders.

Law text box 23.6. BRC principle on management commitment.[1315]

2.1.1 Senior Management Commitment
Within a food business, food safety must be seen as a cross-functional responsibility, including activities that draw on many departments using different skills and levels of management expertise in the organisation. Effective food safety management extends beyond technical departments and must involve commitment from production operations, engineering, distribution management, procurement of raw materials, customer feedback and human resource activity such as training. The starting point for an effective food safety plan is the commitment of senior management to the development of an all-encompassing policy as a means to guide the activities that collectively assure food safety. The Global Standard for Food Safety places a high priority on clear evidence of senior management commitment.

23.20 IFS[1316]

In 2002, in order to create a common food safety standard, German food retailers from the HDE (Hauptverband des Deutschen Einzelhandels) developed a common audit standard called International Food Standard or IFS. In 2003, French food retailers (and wholesalers) from the FCD (Fédération des entreprises du Commerce et de la Distribution) joined the IFS Working Group.

The aim of the IFS (now based in Paris) is to create a consistent evaluation system for all companies supplying retailer-branded food products with uniform formulations, uniform audit procedures and mutual acceptance of audits, which will create a high level of transparency throughout the supply chain. Its scope now extends beyond the food sector alone: 'IFS' has become to mean: International Featured Standard. Among its standards IFS food still holds a prominent position.

The IFS food defines requirements in content, procedure and evaluation of audits and a requirement profile for the certification bodies and auditors.

[1315] From BRC Global Standard for Food Safety, issue 5 2008, Section 2.1 Principles of the Global Standard for Food Safety.
[1316] The information in this section has been taken from the IFS website: www.ifs-certification.com.

The IFS food standard (the so-called catalogue of requirements) consists of six parts called chapters:

- Senior Management Responsibility;
- Quality and Food Safety Management Systems;
- Resource Management;
- Planning and Production Process;
- Measurements, Analysis, Improvements;
- Food Defence and External Inspections.

The auditor will audit against the IFS food standard which is divided into two levels plus recommendation on higher level. The chapter 'Senior Management Responsibility' deals with the corporate policy and structure, the management review and the customer focus. In the chapter 'Quality and Food Safety Management System' requirements concerning the quality management (documentation requirements and records keeping) and the food safety management (HACCP system, the HACCP team and the HACCP analysis) are defined. The chapter 'Resource Management' addresses human resources management and issues (such as hygiene, clothing, and medical screening), training and instruction as well as sanitary, staff facilities and equipment for personal hygiene. The chapter 'Planning and Production Process' is the most extensive one. It considers topics about contract agreements as well as detailed rules on e.g. specifications for products, factory environment, pest control, maintenance, traceability, GMOs and allergens. The next chapter, 'Measurements, Analyses and Improvements', deals with e.g. internal audit, all kind of controls during production steps, product analysis, incident management, and corrective actions. The last chapter 'Food defence and external inspections', which recently became mandatory, includes requirements on defence assessment, security of site, personnel and visitors, and external inspections.

The requirements for auditors and the certification bodies are strictly regulated. All certification bodies shall have an accreditation against EN 45011 on IFS food.[1317] Only authorised auditors who have passed a written and oral examination can audit against the standard. The auditor shall have professional knowledge of the IFS food. The auditors can only audit against their competence in a certain sector (at least 2 years' professional experience in the specific sector or at least 10 audits in this sector). Finally, auditors who comply with these requirements shall only work for one IFS certification body accredited for auditing against the IFS food.[1318]

[1317] Again an example of interconnected private schemes as discussed above.

[1318] The question if such requirement is compatible with competition law (Article 101 TFEU) is outside the scope of this chapter.

23.21 SQF[1319]

Safe Quality Food (SQF; now based in Arlington, USA) is an Australian initiative. Taking over this system seems to have been the American answer to the mainly European initiatives described above.

Besides food safety, SQF focuses on product quality and stimulation of improvement strategies. The main goal of SQF is to control the whole chain. However, SQF believes that one standard does not work for all companies in the chain and that most other standards only work for big companies. Most procedures associated with the standards are considered too elaborate and laborious for small companies. So SQF developed two different norms, the SQF 1000 and the SQF 2000. The SQF 2000 Code was developed in consultation with food industry and quality professionals. HACCP guidelines, as developed by the Codex Alimentarius Commission, form the basis of the Code. Unlike other well-recognised quality systems like BRC, HACCP and ISO 9000,[1320] SQF combines a management quality system, like ISO 9000 and a food safety system (HACCP) with requirements for tracking and tracing. Besides the Critical Control Points (CCP) for food safety, Critical Quality Points are also identified, which makes SQF an integrated system.

The SQF codes (in particular the 1000 and 2000 Code) provide the food sector (primary producers, food manufacturers, retailers, agents and exporters) with a food safety and quality management certification program that is tailored to its requirements and enables suppliers to meet regulatory, food safety and commercial quality criteria in a cost effective manner. In 1994 the Code was developed and pilot programs implemented to ensure its applicability to the food sector. It was circulated in draft form for comment to experts in quality management, food safety, and food regulation, food processing, agriculture production systems, food retailing, food distribution and HACCP.

The Food Marketing Institute (FMI) acquired the rights to the SQF Program in August 2003 and has established the SQF Institute (SQFI) Division to manage the program. The SQF 2000 Code is recognised by the Global Food Safety Initiative[1321] as a standard that meets its benchmark requirements.

The SQF 2000 Code can be used by all sectors of the food industry. The Code is a HACCP-based quality management system that encapsulates NACMCF[1322] and CODEX HACCP Principles and Guidelines, proven methods used by the food

[1319] The information for this section has been taken from the SQF 2000 Code 6th edition issued in August 2008 (and updated in July 2010) and the SQF 2000 Certification Trade Mark rules for use 7th edition amended in November 2005. See also www.sqfi.com.
[1320] On quality management in general, see hereafter Section 23.22 for ISO 22.000 on food.
[1321] Discussed in Section 23.23.
[1322] National Advisory Committee on Microbiological Criteria for Foods.

Textbox 23.7. SQF 2000 on HACCP.

9. Principles and applications of HACCP

Table 1. A description of the 12 HACCP steps that comprise the HACCP method (Adapted from Codex Alimentarius Commisssion – recommended International Code of Practice Principles of Food Hygiene, CAC/RCP 1-1969, Rev. 4-2003)

Preliminary Steps	1. Assemble HACCP team with expertise in product and processes 2. Describe product 3. Identify intended use 4. Construct flow diagram 5. Confirm flow diagram against process in operation (or planned process)
HACCP Principle	**HACCP Application**
1. Conduct a hazard analysis.	6. List all potential hazards associated with each step and consider any measures to control identified hazards.
2. Determine Critical Control Points (CCPs).	7. Determine CCPs.
3. Establish critical limit(s).	8. Establish critical limits and tolerance levels. Determine at what point critical limit is exceeded based on known limits or risk assessment if unknown.
4. Establish system to monitor control of CCP(s).	9. Establish a monitoring system for CCP that is able to detect loss of control i.e. when critical limits are exceeded. Consider continuous monitoring and/or periodic audit.
5. Establish corrective action to be taken when monitoring indicates CCP(s) are not under control.	10. Establish corrective actions that are able to deal with loss of control when it occurs and is capable of determining when CCP has been brought under control.
6. Establish procedures for verification to confirm that the HACCP system is working effectively.	11. Establish procedures for verification or audit that include review of HACCP system and records, records of deviations and actions taken in order to confirm that CCPs are kept under control.
7. Establish documentation covering all procedures and records appropriate to these principles and their application.	12. Documentation and record keeping should be appropriate to the nature and scale of the operation.

industry to reduce the incidence of unsafe food reaching the marketplace. It is designed to support industry or company-branded products and to offer benefits to suppliers at all links in the food supply chain.

The SQF 2000 Code enables a supplier to demonstrate that they can supply food that is safe and that meets the quality specified by a customer. Certified SQF 2000 suppliers receiving raw materials from suppliers who have implemented the SQF 1000 Code can ensure that, through these complimentary systems, product is traceable from the producer to the consumer.[1323]

The SQF 2000 Code also provides a mechanism for the food sectors of developing countries seeking to effectively enter the global food market to implement a management system that addresses their needs and the needs of their customers.

23.22 FS22000[1324]

The youngest of the main private food safety schemes is ISO 22000. When GFSI evaluated ISO 22000 for approval, they requested more requirements identified for Prerequisite Programs. To fill this gap, the British Standards Institution wrote a document called PAS 220. The combination of PAS 220 and ISO 22000 has been approved by GFSI for a registration scheme, and is called FS22000 (previously FSSC 22000). This scheme is run by the Foundation for Food Safety Certification (FSSC).

The international organisation ISO (International Organization of Standardization) has decades of experience developing standards for many different types of applications. One of the most popular and most recognised is the Quality Management System standard ISO 9001. This standard was developed to provide a uniform standard worldwide for quality management. A buyer in one part of the world would have a degree of confidence in the quality practices of a registered company in another part of the world. This standard was used as the basis for other more specific standards for quality management in the automotive industry, the medical device industry, and the aerospace industry.

Now this approach has been taken for food safety management. ISO and its member countries used the Quality Management System approach, and tailored it to apply to food safety, incorporating the HACCP principles into the quality management system. The resulting standard is ISO 22000.

ISO 22000 requires that the business design and document a Food Safety Management System (FSMS). The standard contains the specific requirements to be addressed by the FSMS.

[1323] Here we find an example where a private scheme goes beyond compliance. Unlike EU food law, traceability is not mandatory in US food law.
[1324] Http://www.22000-tools.com.

Generally the standard addresses:

- Having an overall Food Safety Policy for the organisation, developed by top management.
- Setting objectives that will drive companies' efforts to comply with this policy.
- Planning and designing a management system and documenting the system.
- Maintaining records of the performance of the system.
- Establishing a group of qualified individuals to make up a Food Safety Team.
- Defining communication procedures to ensure effective communication with important contacts outside the company (regulatory, customers, suppliers and others) and for effective internal communication.
- Having an emergency plan.
- Holding management review meetings to evaluate the performance of the FSMS.
- Providing adequate resources for the effective operation of the FSMS including appropriately trained and qualified personnel, sufficient infrastructure and appropriate work environment to ensure food safety.
- Following HACCP principles.
- Establishing a traceability system for identification of product.
- Establishing a corrective action system and control of non-conforming products.
- Maintaining a documented procedure for handling withdrawal of products.
- Controlling monitoring and measuring devices.
- Establishing and maintaining an internal audit program.
- Continually updating and improving the FSFM.

ISO 22000 provides brand holders with a quality system equivalent to the other systems mentioned in this chapter that have primarily been developed with a view to retailers' brands.

23.23 GFSI[1325]

The Global Food Safety Initiative (GFSI) co-ordinated by CIES (Comité International d'Entreprise à Succursales; The Food Business Forum), was launched in May 2000 as a reaction to the proliferation of private standards. Much of the advantages of private food law will be lost, if every important player in the marketplace defines a separate set of private standards.

GFSI started to benchmark private standards. This gives businesses the opportunity to send out the message in the market that for them it does not matter which particular standard is applied, as long as it is GFSI-endorsed. In this way GFSI is developing into the standard of standards. The most powerful retailers in Europe, Asia and the USA have agreed to demand GFSI-compliant certification for their own private label products.

[1325] The information for this section has been taken from the GFSI pages on www.ciesnet.com and http://www.mygfsi.com/.

The GFSI mission is: 'To provide continuous improvement in food safety management systems to ensure confidence in the delivery of safe food to consumers worldwide'.

The GFSI objectives are:
1. Reduce food safety risks by delivering equivalence and convergence between effective food safety management systems.
2. Manage cost in the global food system by eliminating redundancy and improving operational efficiency.
3. Develop competencies and capacity building in food safety to create consistent and effective global food systems.
4. Provide a unique international stakeholder platform for collaboration, knowledge exchange and networking.

In the light of the plethora of food safety standards, the GFSI Task Force decided not to write a new standard. Instead, they compiled a set of 'Key Elements' to serve as the requirements against which existing food safety standards will be benchmarked. The 'Key Elements' as defined by the Task Force are:
1. Food Safety Management Systems.
2. Good Practices for Agriculture, Manufacturing and Distribution.
3. HACCP.

Under the umbrella of the Global Food Safety Initiative, seven major retailers have come to a common acceptance of initially four GFSI benchmarked food safety schemes.

Retailers accept certificates based on standards in order to be able to make an assessment of their suppliers of private-label products and fresh products and meat, to ensure that production is carried out in a safe manner. There are many of these standards and suppliers with many customers may be audited many times per year, at a high cost and with little added benefit.

The GFSI Guidance Document Sixth Edition contains commonly agreed criteria for food safety standards, against which any food or farm assurance standard can be benchmarked. In addition to the 'Key Elements' the Guidance Document holds 'requirements for the delivery of food safety management systems' regarding certification and accreditation. GFSI does not undertake any accreditation or certification activities.

The benchmarking work undertaken by the standard owners and other key stakeholders on four food safety schemes (initially BRC, IFS, Dutch HACCP and SQF) has reached a point of convergence. Each scheme aligns itself with common criteria defined by food safety experts from the food business, with the objective of making food manufacturing as safe as possible. As a result, this will also drive cost efficiency in the supply chain and reduce the duplication of audits.

The GFSI vision of 'once certified, accepted everywhere' has become a reality in the sense that Carrefour, Metro, Migros, Ahold, Wal-Mart and Delhaize have agreed to reduce duplication in the supply chain through the common acceptance of any of the GFSI benchmark schemes. Tesco, however, has withdrawn. It was unwilling to accept other schemes as equivalent to its own 'Nature's choice'. The number of schemes recognised by GFSI has continuously grown. See Law text box 23.7 for an overview.

Law text box 23.7. GFSI recognised schemes.[1326]

Manufacturing schemes:

- BRC Global Standard Version 6
- Dutch HACCP (Option B)
- FSSC 22000 – October 2011 issue
- Global Aquaculture Alliance BAP Issue 2 (GAA Seafood Processing Standard)
- Global Red Meat Standard Version 4
- International Food Standard Version 6
- SQF 2000 Level 2 Edition 7
- Synergy 22000
- China HACCP

Primary production schemes:

- CanadaGAP – Scheme Version 6 Options B and C and Program Management Manual Version 3
- GlobalG.A.P IFA Scheme V4
- SQF 1000 Level 2 Edition 7

Primary and manufacturing scheme:

- PrimusGFS

Packaging schemes:

- IFS PACKsecure
- BRC/IOP Global Standard for Packaging and Packaging Materials Issue 4

The GFSI Foundation Board, a retailer-driven group, with manufacturing advisory members, provides the strategic direction and oversees the daily management. Membership of the Board is by invitation only.

The GFSI Technical Committee was formed in September 2006 and is composed of retailers, manufacturers, standard owners, certification bodies, accreditation bodies, industry associations and other technical experts. It provides technical

[1326] As published on http://www.mygfsi.com/about-gfsi/gfsi-recognised-schemes.html, 1 March 2011.

expertise and advice for the GFSI Board and replaces a previous GFSI retailer-only Task Force. Membership of the Technical Committee is by invitation only.

A crucial moment in its development was the entry of Wal-Mart into the scheme (February 2008). Noteworthy is Wal-Mart's comment: 'GFSI Standards provide real time details on where suppliers fall short in food safety on a plant-by-plant basis and go beyond the current FDA[1327] or USDA[1328] required audit process'. Wal-Mart is the first US-based grocery chain to require GFSI, the company claims. The company published a schedule to suppliers requiring completion of initial certification between July and December 2008, with full certification required by July 2009.

Through GFSI benchmarking private food law seems to be achieving what public food law (through the Codex Alimentarius) never could: truly global harmonisation of food safety standards.

23.24 Public law on private food law

The European Commission has reacted to the emergence of private food law. A study conducted for DG Agri identified 441 different schemes.[1329] The European Commission decided that legislative action was not warranted to address the potential drawbacks in certification schemes at this stage. Instead, drawing on comments from stakeholders, the Commission undertook to develop guidelines for certification schemes for agricultural products and foodstuffs. So on 16 December 2010, it published '*Commission Communication – EU best practice guidelines for voluntary certification schemes for agricultural products and foodstuffs*' (2010/C 341/04).[1330]

These guidelines are designed to describe the existing legal framework and to help improve the transparency, credibility and effectiveness of voluntary certification schemes and ensure that they do not conflict with regulatory requirements.

The guidelines warn the Member States to respect the state aid rules when they give support to certain schemes. Businesses are reminded of the rules on competition. Certification schemes may not lead to anti-competitive behaviour.

From food law provisions are quoted that consumers may not be misled. We are mystified by the quote in full of Article 5(1) of Regulation 178/2002 'Food law

[1327] Food and Drug Administration. The food regulatory agency in the USA.
[1328] The US Department of Agriculture.
[1329] See: http://ec.europa.eu/agriculture/quality/index_en.htm.
[1330] OJ C 341, 2010, pp. 5-11. Also noteworthy is the Communication from the Commission Contributing to Sustainable Development: The role of Fair Trade and non-governmental trade-related sustainability schemes, Brussels 5.5.2009 COM(2009) 215 final, annex II to this book.

shall pursue one or more of the general objectives of a high level of protection of human life and health and the protection of consumers' interests, including fair practices in food trade, taking account of, where appropriate, the protection of animal health and welfare, plant health and the environment'. Is the Commission implying that private certification schemes are within the scope of the concept of 'food law' and have to comply with the objectives of food law? It does not seem a very likely interpretation, but no other interpretation readily presents itself. In fact we would be inclined to believe that private regulation is not limited to serving the objectives listed in Article 5, but can also legitimately serve other interests including those of the business sector itself.[1331]

The question about the public status of private food law also presents itself in the context of the WTO.

23.25 WTO

Private food law is rapidly replacing public law as the determining factor in international food chains. Within the WTO the question has been raised as to whether private standards constitute a new generation of trade barriers that Member States of the WTO should take on.[1332]

Two WTO agreements may be of relevance in this context; the TBT Agreement and the SPS Agreement.

23.25.1 TBT

In the Agreement on Technical Barriers to Trade, WTO members have set requirements for technical regulations and standards to ensure that they will not constitute unjustified barriers to international trade. WTO members are strongly encouraged to use international standards and to support international standardising bodies. In the agreement ISO/IEC is positioned as an overarching international standardising body. Other standardising bodies within the ambit of the TBT Agreement must be notified to ISO/IEC. This seems to imply that WTO members relying on ISO standards may believe they are complying with TBT requirements.

23.25.2 SPS

The Agreement on the application of Sanitary and Phytosanitary measures (the SPS Agreement) deals with measures aiming to protect the health of humans,

[1331] On Article 5 of Regulation 178/2002, see: B.M.J. van der Meulen, The function of Food Law. On the objectives of food law, legitimate factors and interests taken into account, European Food and Feed Law Review 2, 2010, pp. 83-90.
[1332] 29-30 meeting of the Committee on Sanitary and Phytosanitary (SPS) Measures.

animals and plants. Such measures are acceptable if they do not go beyond what is necessary and do not constitute disguised trade barriers. Measures to protect food safety are by definition SPS measures.

Some members of the WTO believe that private standards are within the ambit of the SPS Agreement and furthermore do not conform to the requirements.

St. Vincent and the Grenadines, supported by Jamaica, Peru, Ecuador, and Argentina, complained that 'EurepGAP' SPS standards imposed by the Euro-Retailer Produce Working Group, composed primarily of food retailers, were more strict than EU governments' requirements. Referring to Article 13 of the SPS Agreement, which says that member governments 'shall take such reasonable measures as may be available to them to ensure that non-governmental entities within their territories (...) comply with the relevant provisions of this agreement', these countries argue that only the public law EU rules should apply to the private sector.[1333]

Law text box 23.8. Article 13 SPS Agreement.

Article 13
Implementation
Members are fully responsible under this Agreement for the observance of all obligations set forth herein. Members shall formulate and implement positive measures and mechanisms in support of the observance of the provisions of this Agreement by other than central government bodies. Members shall take such reasonable measures as may be available to them to ensure that non-governmental entities within their territories, as well as regional bodies in which relevant entities within their territories are members, comply with the relevant provisions of this Agreement. In addition, Members shall not take measures which have the effect of, directly or indirectly, requiring or encouraging such regional or non-governmental entities, or local governmental bodies, to act in a manner inconsistent with the provisions of this Agreement. Members shall ensure that they rely on the services of non-governmental entities for implementing sanitary or phytosanitary measures only if these entities comply with the provisions of this Agreement.

So far it is believed that Article 13 of the SPS Agreement aims at entities that in reality are governmental in a private law guise. The text, however, leaves room for the interpretation that it also applies to 'real' private actors, in particular when they take on the role of regulator traditionally reserved for governments.[1334] From a legal point of view there is a fundamental difference between the situation where

[1333] See ICTSD Bridges weekly news digest Volume 9 Number 24 6 July 2005 and Volume 7 Number 13, 6 July 2007.

[1334] As we have seen above, governments may participate in the formulation of private standards aiming to influence behaviour abroad. In such situations it is less evident that private standards originate from 'real' private actors and applicability of the SPS Agreement becomes likely, even in its more limited interpretation.

a product may not be brought to the market (because it does not comply with public law requirements) and the situation where a product legally brought to the market is not bought by its intended customers (because it does not comply with these customers' private law requirements). From an economic point of view and for all practical purposes these two situations amount to the same thing where the customers concerned dominate the market.

This discussion pinpoints a weak aspect of private food law. It seems at present underdeveloped in the area of checks and balances.

23.26 Conclusions

The content of private food safety schemes, just like public food law, is based on HACCP. The private sector is in the process of achieving what the public sector never could: world-wide harmonisation of food safety standards.

The underlying legal structure is straightforward. Contractual requirements – flanked with instruments from (intellectual) property and business law – are used by dominant players in the food chain to impose 'voluntary' requirements on all players upstream, regardless of which country they are situated in. Contractual requirements, audits and certification can be applied across national borders. In this sense, private food law is more global than international food law (such as the SPS Agreement and the Codex Alimentarius). International (public) food law does not govern the behaviour of stakeholders, but sets a meta-framework for (national) food law that in turn applies to stakeholders' behaviour. Private food law does govern stakeholders' behaviour and in this sense private food law is more law than international food law.

24. Conclusions

Bernd van der Meulen

Food law cannot be understood in isolation of the general aspects of EU law. Nevertheless in several countries food law as a functional area of law has become an academic specialisation in its own right. From the 2000 White Paper on Food Safety onwards, an impressive body of food (safety) law has come to fruition in the EU. The (2002) General Food Law provides the general principles. The most important of these is a duty of care for food safety resting on food business operators. Food business operators have to address the legal limits to their liberty in the choice of raw materials, their obligations in organising their production processes and trade relations and the rules concerning their communication with consumers.

To study EU food law in further detail, it is helpful to distinguish substantive food law and procedural food law. The main part of substantive food law addressing food businesses consists of three categories of rules: rules on the product, which is what the food should be; rules on the process, which is what the food business should; and rules on the presentation of the food, which is what the food business says. Rules addressing public authorities grant these authorities responsibilities and powers of enforcement and incident management.

Substantive rules have to function within the legal system that produces them. For food law and food safety law the legal environment has become increasingly international, which increases the complexity of the law. The actions of the Codex Alimentarius Commission and other United Nations' organisations have been sketched to indicate the many rules and actions on food law that originate at the international level outside the EU. These inputs have to be channelled through the complex system of the Union and its Member States. Their internal relations in import, export, law making and policy development and execution have been described and were found to be perpetually changing. The shift in legislative instruments after the White Paper from directives to regulations signals a centralisation of food law at the EU level with less room for national adaptation.

Substantive rules need procedures for their implementation. In this book some procedures have been discussed, mainly in the context of pre-market approval and authorisation of food. Attention has been given to enforcement, both under public law and private liability law.

Scholars and practitioners in food law may want to keep their eyes wide open for developments in the private sector. It may well be that private food law, in particular at the global level, will make a bid for the dominant position currently held by public food law as a framework for food business relations.

Appendix A. Finding sources of EU law: legislation and case-law databases

Sofie van der Meulen[1335]

A.1 Introduction

Access to up-to-date documentation on EU law, legislation and case law is important for anyone involved in law. The Publications Office of the EU provides various databases that contain all available documents of the institutions, bodies and agencies of the EU, in all official languages of the EU. Several of the EU institutes also provide databases, with documentation within their own area.[1336] To make sure that you can find the necessary information within an acceptable timeframe, this annex provides some tips and tricks to find your way. Because all official documents can easily be found in the official databases, in this book we refrained from making reference to the Official Journal and the European Court Reports as was commonly done in the past.

A.2 EUR-Lex, CURIA and the Legislative Observatory

The two most important databases when searching for sources of EU law are EUR-Lex and CURIA. These databases respectively provide information on the legal documents of the European Union and the documents concerning the Court of Justice of the European Union (CJEU). Another useful database is the Legislative Observatory, offered by the European Parliament. This database provides access to all records for all procedures still on-going and all procedures that have completed their passage through the EU Parliament since July 1994.

Because of the large number of documents included in these databases, it is important to know how to perform an effective search. The following sections explain the major functions of each of these two databases and how to put the search results to use.[1337]

A.3 EUR-Lex

The EUR-Lex database can be accessed via http://eur-lex.europa.eu/. The language of the database and the documents can be changed to each of the EU's official

[1335] The author is very grateful to Lisanne van Kouterik, intern at Axon Lawyers, for her valuable contribution to this annex.
[1336] For food law and related policies, the most important website is the one of DG Sanco, available at: http://ec.europa.eu/food/index_en.htm.
[1337] This annex should be used with a little caution because the EU very often changes the 'look and feel' of its webpages.

languages, as far as the content is available in your preferred language.[1338] The database is updated daily and contains more than 3 million documents with texts dating back to 1951. These documents include:

- The Official Journal of the European Union
- EU law (EU treaties, directives, regulations, decisions, consolidated legislation, etc.)
- Preparatory acts (legislative proposals, reports, green and white papers, etc.)
- EU case law (judgments, orders, opinions of the Advocate-General, etc.)
- International agreements
- EFTA documents

A document can be displayed in up to three languages simultaneously. EUR-Lex also provides the option to follow legislative procedures.

A.3.1 Finding documents: search methods

The EUR-Lex website provides for different search methods to find documents and procedures. A search box is provided in the middle of the homepage, which offers three different search methods: 'simple' search, search 'By document reference' and search 'By ELI'. The homepage search box also provides the option for an 'Advanced search' and the default setting is to search 'By document reference'.

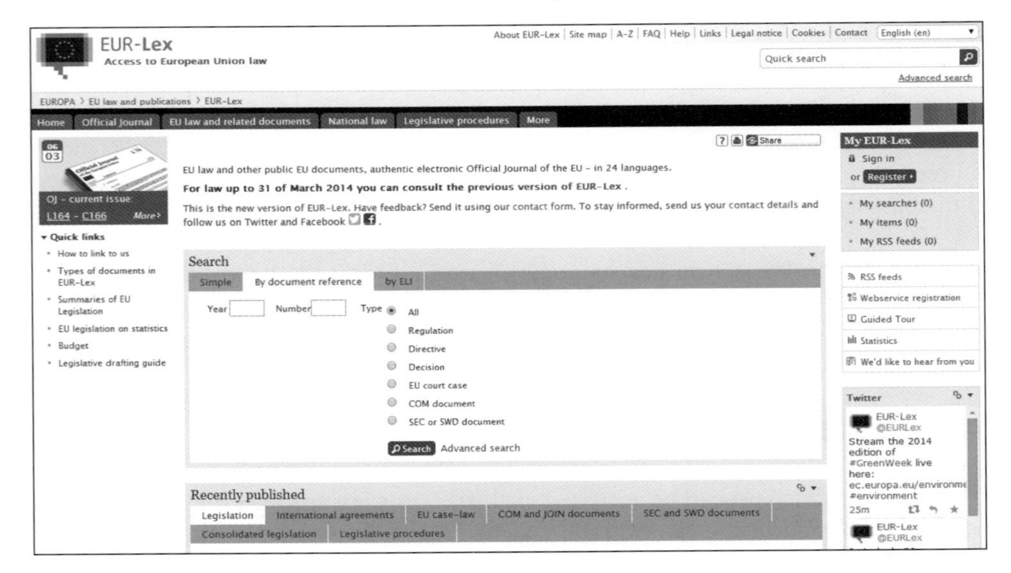

[1338] Older documents are often only available in the languages of the countries that were members at that time.

Simple search

The simple search, offered by the search box on the homepage, is the best method when one or more parameters of the required document or procedure are known. The simple search can be performed by entering a word, phrase or number in the entry field for 'text search' or by using the CELEX or ECLI[1339] number. The latter two parameters refer to the identifier systems providing the unique numbers for, respectively, the documents and legal documents available in EUR-Lex.

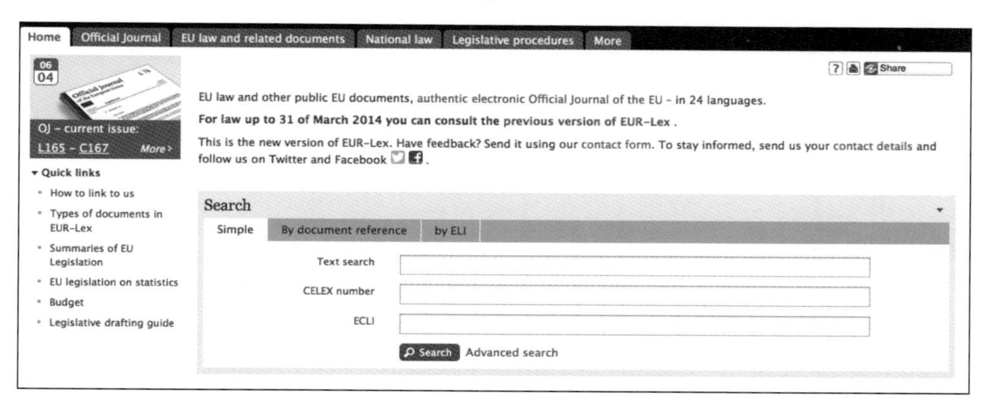

Search by document reference

The search by document reference provides for three entry fields: the year, the document number and the type of document. Although each field can be used independently, the search will be most effective when all fields are used. It is therefore recommended to use this search method only when the document reference is known.

In this book the document numbers are provided wherever possible for the sources quoted. This would therefore be the best approach when consulting a source of EU law mentioned in this book. Here below in Section A.3.3 the numbering system is explained in some more detail.

[1339] European Case Law Identifier. The ECLI system is also introduced at Member State level.

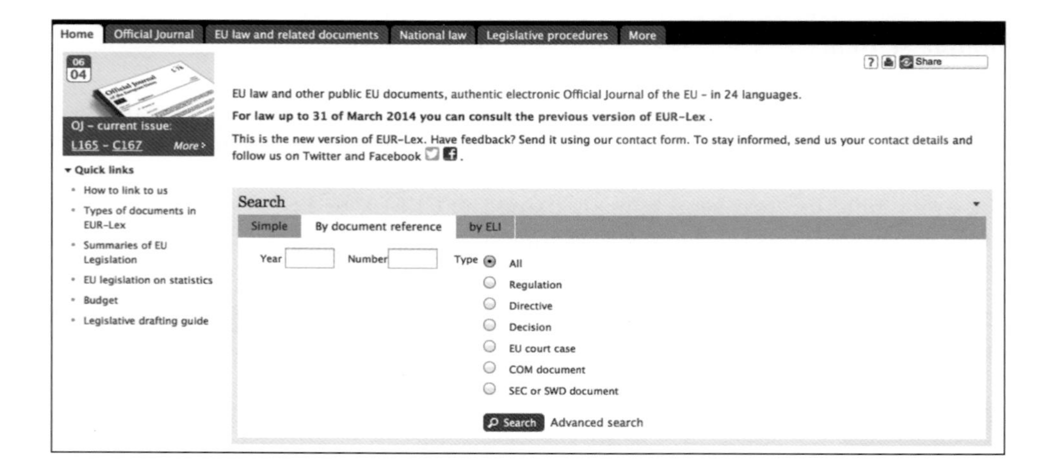

Search by ELI

ELI, or European Legislation Identifier, is a recently introduced semantic web solution that aims to create a flexible and unique way to reference legislation across different legal systems. ELI enables direct access to specific national and European legislation through a structured, flexible identifier. To perform the search by ELI:

1. Select a search language.
2. Enter a word, phrase or number. A list of suggested search terms will appear. Click any suggestion to insert it in the search box.
3. The type of legislation or author may be selected to narrow down the search. You do not have to fill out all fields.
4. Click on the 'Search' button to launch the search. If necessary, the search can be repeated with additional criteria by pressing the advanced search option.

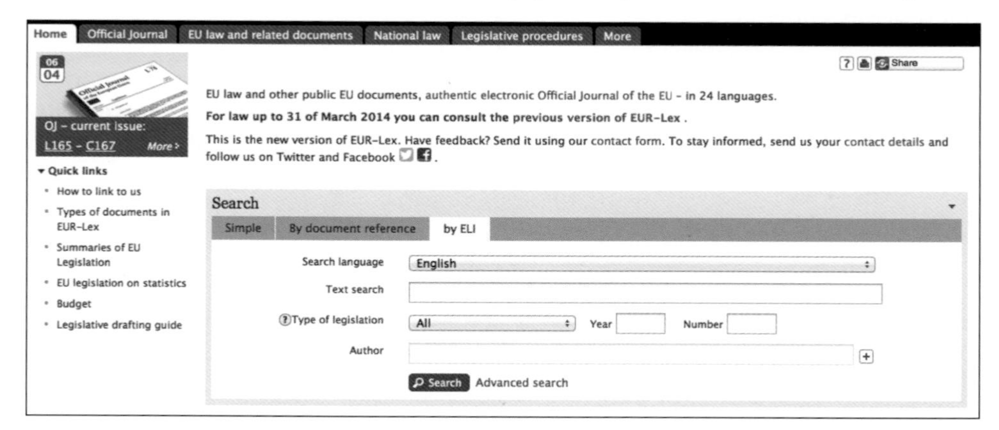

Further, every page has a 'Quick search' box in the upper right corner with the option to select an 'Advanced search' below it.

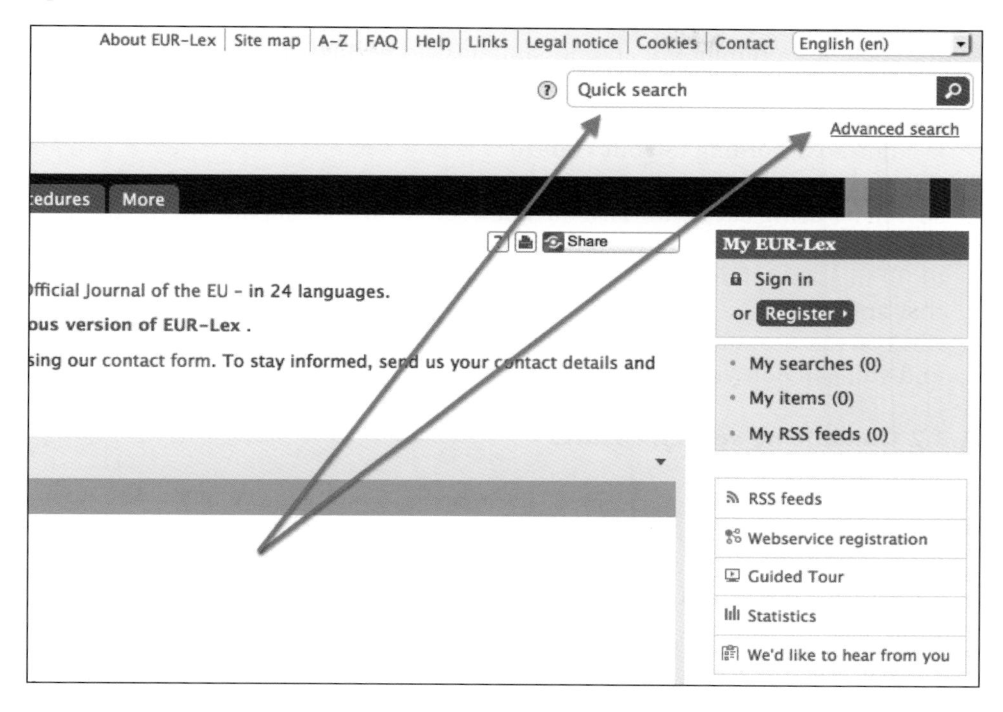

Quick search

The quick search offers a fast, full-text search in the EUR-Lex database. To perform the quick search:

1. Enter a word, phrase or number. A list of suggested search terms will appear. Click any suggestion to insert it in the search box.
2. To launch the search, click on the lens icon or press the Enter key on your computer. Once you type or insert a term in the quick search box, it is retained – allowing you to directly edit your search by adding or deleting terms.

Advanced search

To perform a more precise search, the advanced search option can be used. This search includes a number of additional fields. To perform the advanced search:
1. Select a domain. By selecting one or more domains, you can limit your advanced search to documents from one or several specific sectors. The search screen is adjusted according to your choice.
2. Fill in the search form according to your needs. You can narrow an advanced search by using date criteria. The date types available depend on the domain you select. You do not have to fill out all fields. For more information on how to fill in a field, hover over its name and click on the 'question mark' button that appears.
3. Click on the 'Search' button to launch the search. The search results are displayed and you can edit your search if necessary.

A.3.2 Consulting search results

When a search has been performed, the relevant results are displayed in a list. By selecting additional criteria on the left side of the results, the query can be further refined. To consult a document or legislative procedure, select the title from the list. There are up to four views available, presented as tabs:
- *About this document*: contains information about the document you are consulting.
- *Text*: access all available languages and formats of an item. The multilingual display function can be used to compare texts in different languages.
- *Linked documents*: contains information about other documents and search results related to the item you are currently viewing.
- *All*: provides an overview of the information presented in the other tabs.

A.3.3 Document structure: referencing

Documentation of the EU institutions, bodies and agencies is numbered using a referencing system that has been in place since the early 1960s. Each number consists of two elements: a reference to a year (indicated by two or four figures, such as '97' of '1999') and a reference number. The sequence of the elements depends on the kind of document.

The Official Journal of the European Union (OJ)

The Official Journal is published every day from Tuesday to Saturday and consists of several sub-series. The L-series contains Community legislation, the C-series information and notices. There is also an electronic section to the C-series that contains preparatory decisions in the legislative process. The supplement, the S series, contains notices of public tenders.

Legislation

The text of the legislation, which is available through EUR-Lex has been taken from the Official Journal and is supplemented with various data: the bibliographical information.

Directives and Decisions are numbered using first the year, and then the reference number. For example, the Product Liability Directive, Directive 85/374, was the 374th directive published in the OJ in 1985. Regulations are numbered by the year of publication in the Official Journal preceded by a classification number. This means that the General Food Law, Regulation (EC) 178/2002 is the 178th regulation that was published in the OJ in 2002.

Consolidated legislation

Legislation is often changed. The final product of a procedure changing an existing law is a text stating how the law is changed. The legislator does not usually publish a new version of the changed law where the changes have been included. As a consequence it is not always easy to assess the latest version of the law. EUR-Lex provides a service called 'consolidated legislation'. A consolidated version of the legislation is the text of the original law with all changes included. Consolidated texts are not official documents and EUR-Lex officially declines all liability for the content. This being said, for practical use the consolidated versions in EUR-Lex provide an accurate indication of the content of the law in one easy-to-read document. The consolidated legislation can be accessed via http://eur-lex.europa.eu/collection/eu-law/consleg.html.[1340]

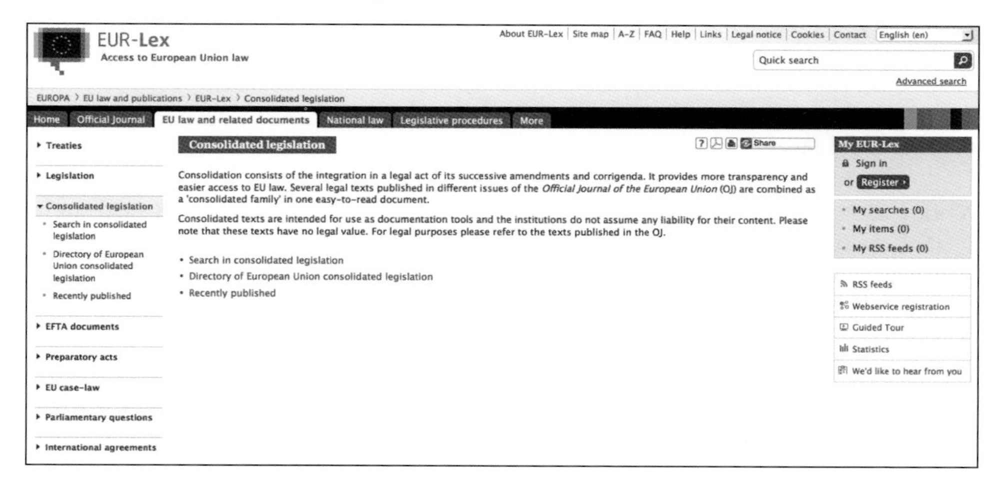

[1340] The search by document reference function discussed above yields a list of documents including the original piece of legislation and the existing consolidated versions of different age. That is to say the most recent version, but also older versions. It is important to make a conscious choice.

COM Documents

The European Commission publishes legislative proposals, Communications, Reports, White Papers and Green Papers as 'COM documents'. In EUR-Lex, COM documents can be found under 'preparatory acts'.

Policy documents from the European Commission are indicated: COM(year) number. 'COM' refers to the European Commission. The year represents the year in which the document was finalised and the number indicates the sequence. For example, the White Paper on Food Safety is numbered COM(1999) 719, indicating that it was finalised towards the end of 1999. It was published on the 12th of January 2000.

Case-law

The case law that is available through EUR-Lex has been taken from the European Court Reports (ECR) and is supplemented with recent case law that has not yet been published in ECR. One advantage of consulting case law through EUR-Lex is that hyperlinks allow access to related documents and the sources of annotations to the judgments are easily accessed. Case law becomes available through EUR-Lex almost immediately after a judgment has been rendered.

Court cases receive a number that consists of the reference number followed by the year when the case was submitted. The date of the ruling can be several years later. The number is preceded by a letter to indicate which court was addressed. The letters 'T', 'C' or 'F', respectively indicate the General Court,[1341] the Court of Justice and the Civil Service Tribunal.[1342] The case number applies similarly to the ruling, the opinion of the Advocate-General (A-G) and other documents pertaining to the case.

A.4 CURIA

While case law of the European Courts is made available through EUR-Lex, there is also an online database that provides access exclusively to all documents published in the European Court Reports since 1954, from the Court of Justice of the European Union.

The database, CURIA, can be accessed via http://curia.europa.eu/. Before entering the website, you can choose between one of the EU official languages and then proceed to the content of this database.

[1341] The General Court was previously known in English as the Court of first instance and in French as Tribunal, hence the 'T'.
[1342] In French called the 'Tribunal de la fonction publique'. This explains the 'F'.

It is important to mention that the European Court consists of three courts: the Court of Justice, the General Court (created in 1988) and the Civil Service Tribunal (created in 2004). Each of these courts has a different jurisdiction. Therefore, a case may be handled at one of the courts or may be presented to another court for appeal. In order to have a better understanding on the jurisdiction of each court, you may go to the 'Presentation' directory under each court's name (list on the left side) and read the information under 'Jurisdiction'.

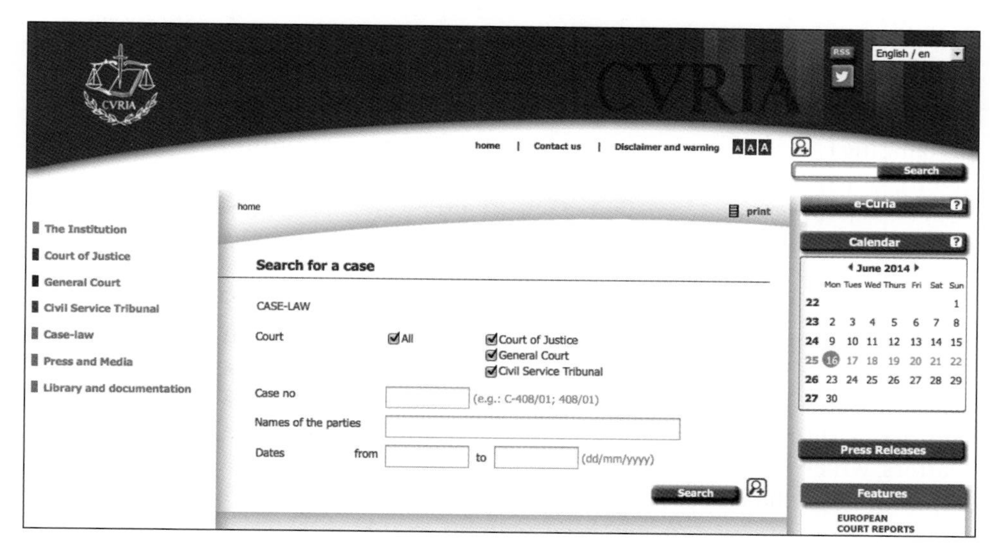

A.4.1 Finding documents: quick search

CURIA database provides a quick and an advanced search option. The quick search can be used when you either know the case number, the names of the parties involved or the specific date of the decision.

First of all you need to indicate which Court's decisions you wish to search. You can either select one or two courts, or select all of them if you are not sure which Court was involved in the case you are looking for. It is advisable to include in your selection the General Court and the Court of Justice as well, because they are higher instance courts (appeal or final) with regard to Civil Service Tribunal and/or General Court, respectively.

A.4.2 Finding documents: advanced search

The advanced search can be used when you do not know the case number or if you are looking for one or more cases on a specific subject. Clicking on the advanced search button, the 'lens' icon enables access the advanced search.

In the advanced search section you may continue to choose your search criteria as follows.

Initially, you may select the status of the case you are looking for. You can either opt for (1) closed cases (cases on which the Court(s) have given a final decision), (2) pending cases (cases on which a final decision by the Court(s) has not yet been given), or (3) 'all cases' if you wish to have the results of all the cases matching your search criteria, independent of whether they are closed or not. Consequently, you can continue choosing your search criteria by selecting either one or more of the courts or all of them.

If you do not have a specific case number or the involved parties' names, you may leave the respective boxes blank and directly jump to the 'period or date' criteria. Here you can select a time period or enter a specific date. Afterwards, you may continue by categorising your search by ticking one or more subject matters, which most closely match your case-law search. These are the four most important search criteria (case status, court, period or date, subject matter) that you can use to execute an advanced search.

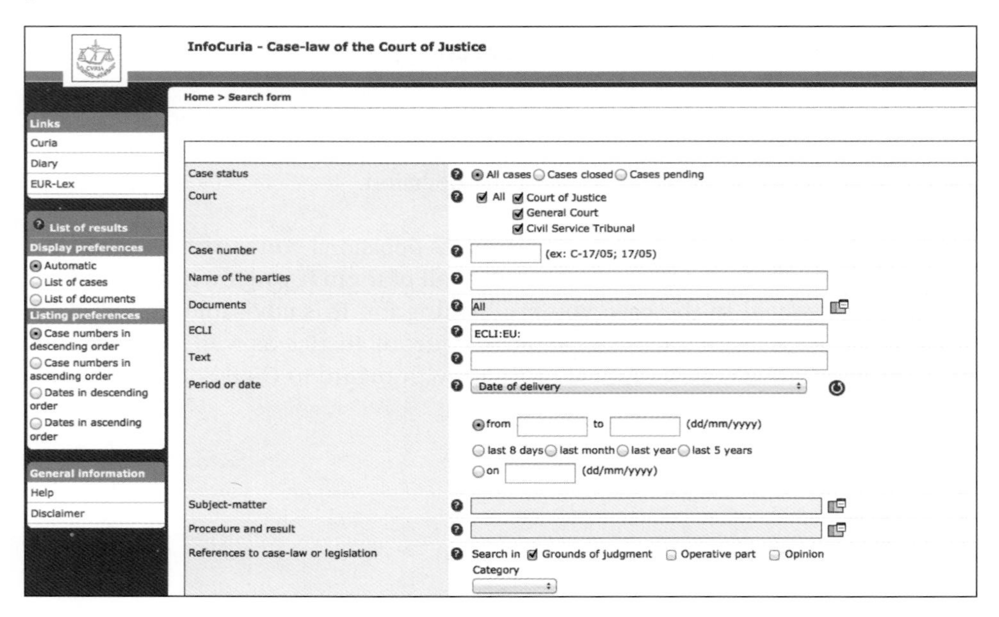

A.4.3 Consulting search results

The search results list provides two tabs: 'list of results by case' and 'list of documents'. The first tab provides all the documents that fulfil the selected criteria. This may provide one or more cases, and optionally the opinion of the A-G. Within this tab you can choose to access the document via EUR-Lex or CURIA. When clicking on the icon, you can select a language in which you prefer the document to be presented.

The tab 'list of documents' provides an overview of all available documents relating to the case. These documents may be accessed either via CURIA or EUR-Lex.

A.5 Legislative Observatory

The European Parliament had the Legislative Observatory database set up in 1994, as a tool to monitor the EU institutional decision-making process. The database provides access to the documents, records and additional information concerning all on-going and completed procedures that have passed through the EU Parliament. In the past few years, some extra features have been added to the database. Among the features available are:

- a search tool that enables users to combine different search options to explore the procedure records;
- a plenary calendar that provides information on the EU Parliament's past, present and future part-sessions;
- procedure files with more information on key players and the sequence of events, with additional links to various sites.

The documents can be accessed in English and French, but some documents provide for additional language versions. The database is updated daily to ensure that any new data is rapidly added. The website of the Legislative Observatory is accessible via http://www.europarl.europa.eu/oeil/home/home.do.

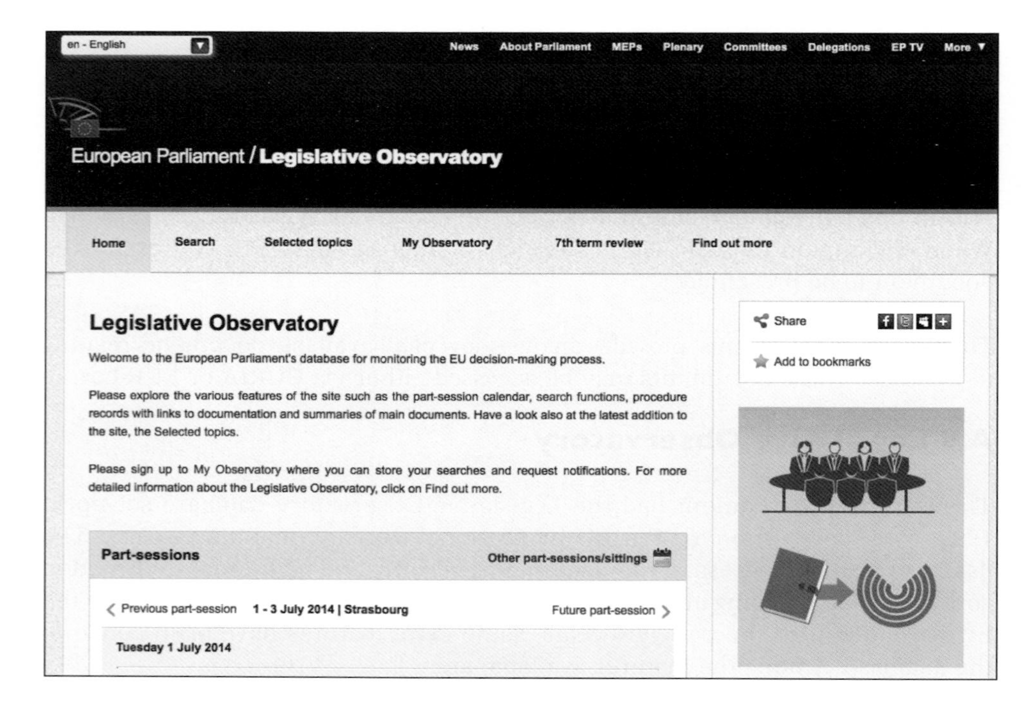

A.5.1 Documentation

The Legislative Observatory database uses procedure files to categorise its information. Each procedure file provides a centralised, frequently updated record of information on the different key players, events and documents in an individual procedure dossier. The procedure files also contain factual, politically neutral summaries of major documents and events in a procedure and, for on-going procedures, also forecasts for future stages.

A.5.2 Finding documents: homepage

Information provided by the Legislative Observatory database can be accessed through the homepage of the website or via a search tool. The homepage first shows a calendar with the past sessions of the EU Parliament. If a date is selected, the results show the procedure files that have been debated, or decided on in the session of that particular day. The homepage further provides a list of the procedure files that have been published most recently in the Legislative Observatory, as well as any procedure files containing the latest reports tabled for plenary by Parliament's committees. Also, there is a section linking to the most recent 'information document files'. These files contain documents with information sent to the Parliament by the EU Commission on an official basis and referred internally to the parliamentary committees responsible for the relevant subject areas. Lastly, the banners on the right-hand side of the home page provide direct links to a number of websites belonging to either the European Parliament or the other EU institutions.

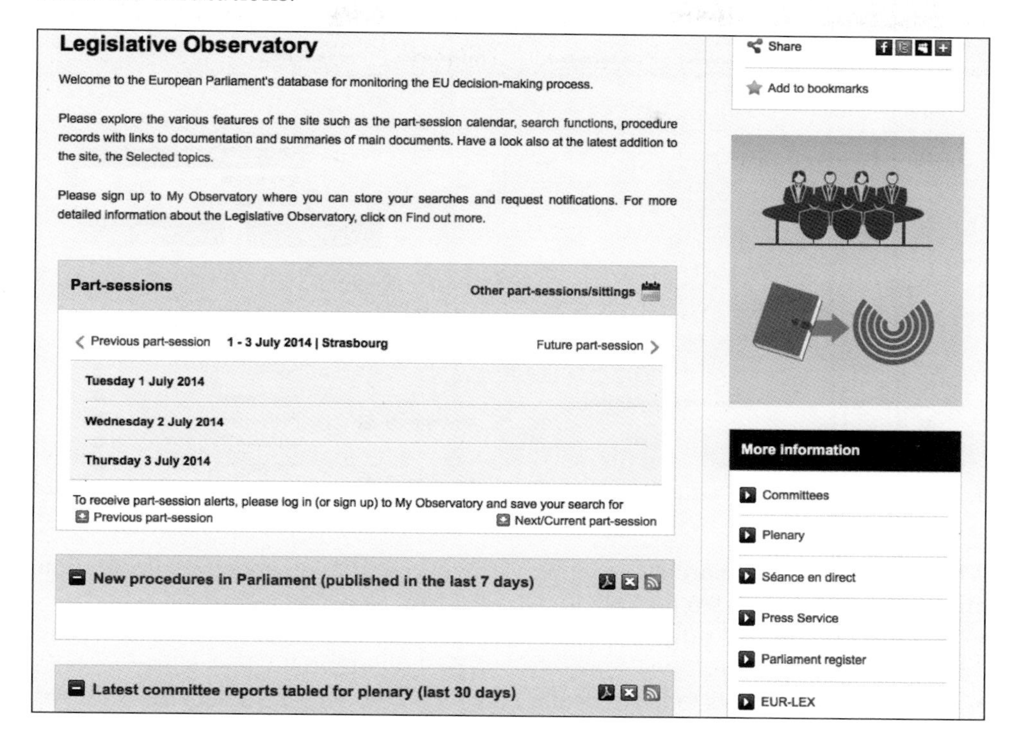

A.5.3 Finding documents: search tool

To access the Legislative Observatory search tool, you should select the search tab on the website. The search page will show the search box. The search box can be used for a general search or a specific reference search. The general search can be used to search for procedure files that relate to the search term. To perform a general search:

- Enter words, numbers or phrases in the search box.
- Leave the box empty and click on 'Search' to see the search options available.

The box on the right of the search box allows you to search for the 'exact word/ phrase', 'any of these words' or 'all of these words'. The list of results can be filtered using the search facets that appear on the right-hand side of the screen. Alternatively, you can click on the 'Search' button (or on Enter) to see all the search filters available for the database:

- Select a specific parliamentary term to narrow down the search results.
- Select search options to further refine the search.

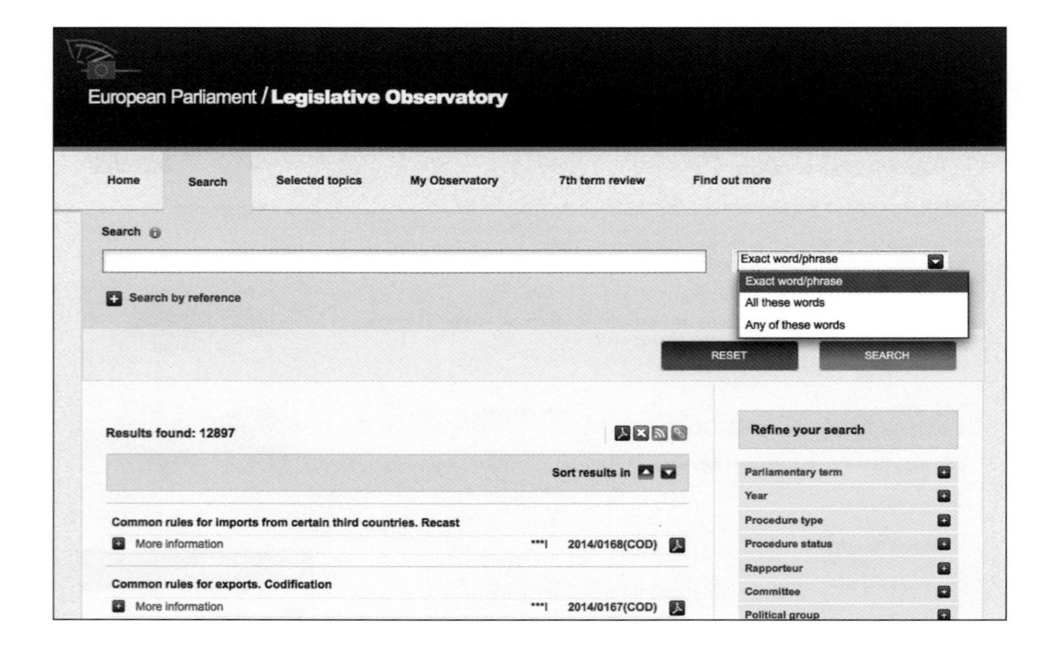

A 'specific reference search' can be performed to search on specific information on the document you are looking for. To perform a specific reference search:

- Click on 'Search by reference', select the relevant option(s) and fill in all the appropriate fields.
- To obtain results lists for a particular category of document, select the relevant option(s) and click 'Search', leaving the other fields empty.

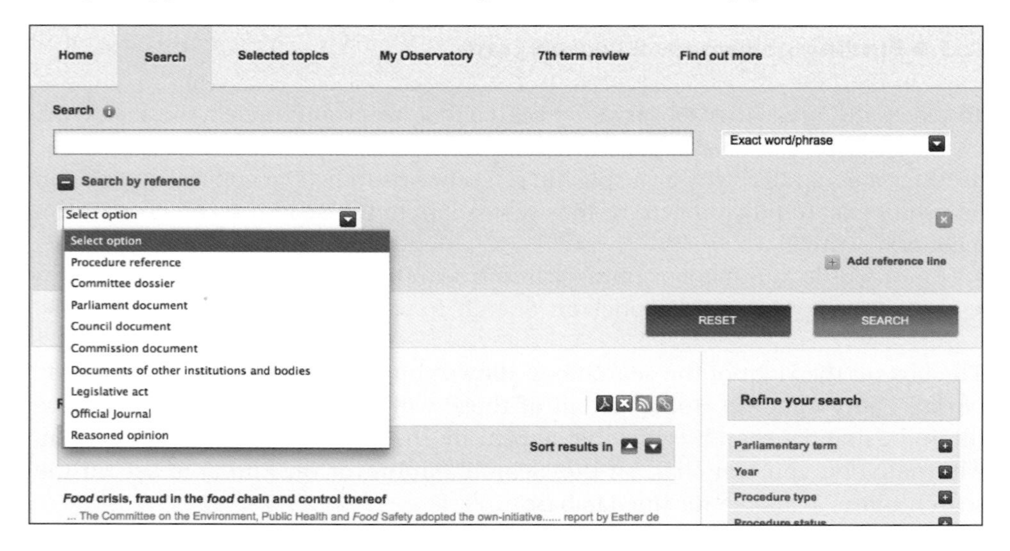

A.5.4 Consulting search results

The search tool does not search the content of the individual documents linked to the files. It selects the text that forms part of the Legislative Observatory procedure files (titles, references, summaries, etc.). Therefore, you have to check the reference to the procedure file to see what kind of document is displayed (such as a COD, RSP, INI).

When you select a procedure file, the following sections will be typically displayed:

- *Basic information*: title, subject classification, procedure reference, type of procedure and stage reached. This section may also contain links to related procedure files.
- *Key players* involved in the decision making concerning the procedure, including EU Parliament committees and rapporteurs, the Commissioner responsible, the relevant Commission Directorate-General, etc. Furthermore, links to various web pages of the institutions are provided wherever possible.
- *Key events*: listing the sequence of events in a procedure in chronological order, with links to relevant documents and summaries.
- *Technical information*: indicating the legal basis, legislative instrument, etc.
- Links to the *final act* of a procedure as published in the Official Journal of the European Union and the Commission's database of legislation; whenever a final act contains provisions for delegated acts, this is explicitly mentioned in the procedure file and the summaries.
- *Documentation gateway* showing all documents relating to a procedure, in chronological order, grouped by institution or in a full list, with references, links and summaries of the major documents.
- Links to two other sites: *IPEX*, a platform for information exchange between the European Parliament and the national parliaments, and *PreLex*, the European Commission's database for monitoring legislative decision-making at EU level.
- On-going procedure files may also contain a section showing forecasts and deadlines for forthcoming activities, such as the adoption of a report in the European Parliament committee, the debate or vote in plenary, etc.
- Some procedure files may contain links to the related delegated acts procedures (DEA), if any (a legislative act may delegate to the Commission the power to adopt delegated acts). The two co-legislators, Parliament and Council, can revoke a delegation of power or oppose to its tacit renewal and may initiate an examination procedure and eventually object to a delegated act. These possibilities are grouped under the broad category of delegated acts procedures.[1343]

[1343] Explanation based on the information in http://www.europarl.europa.eu/oeil/info/info2.do.

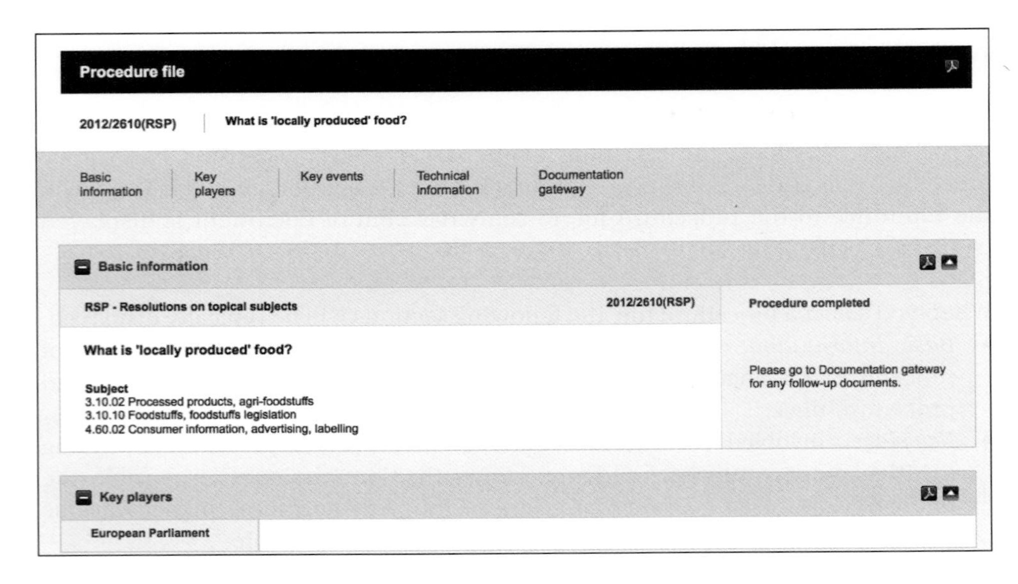

The best way to get acquainted with these useful sources of information is to simply start using them. For example, by checking some of the references in this book that have caught your eye.

Appendix B. Attribution of powers for the General Food Law, Regulation 178/2002

The first and last page of the General Food Law, Regulation 178/2002. Clauses indicating how the regulation was made, which powers were attributed to whom that made the regulation possible, are highlighted. (See next two pages)

(first page)

L (legislation) series of Offical Journal for binding law

Number of the Official Journal

Pagenumber

I

(Acts whose publication is obligatory)

Regulation is based on EC Treaty

Co-decision procedure

REGULATION (EC) No 178/2002 OF THE EUROPEAN PARLIAMENT AND OF THE COUNCIL

of 28 January 2002

laying down the general principles and requirements of food law, establishing the European Food Safety Authority and laying down procedures in matters of food safety

THE EUROPEAN PARLIAMENT AND THE COUNCIL OF THE EUROPEAN UNION,

Attribution: powers to make legislation are indicated

Having regard to the Treaty establishing the European Community, and in particular Articles 37, 95, 133 and Article 152(4)(b) thereof,

Exclusive right of initiative

Having regard to the proposal from the Commission [1],

Having regard to the opinion of the Economic and Social Committee [2],

Having regard to the opinion of the Committee of the Regions [3],

Co-decision procedure

Acting in accordance with the procedure laid down in Article 251 of the Treaty [4],

Whereas:

(1) The free movement of safe and wholesome food is an essential aspect of the internal market and contributes significantly to the health and well-being of citizens, and to their social and economic interests.

(2) A high level of protection of human life and health should be assured in the pursuit of Community policies.

(3) The free movement of food and feed within the Community can be achieved only if food and feed safety requirements do not differ significantly from Member State to Member State.

(4) There are important differences in relation to concepts, principles and procedures between the food laws of the Member States. When Member States adopt measures governing food, these differences may impede the free movement of food, create unequal conditions of competition, and may thereby directly affect the functioning of the internal market.

(5) Accordingly, it is necessary to approximate these concepts, principles and procedures so as to form a common basis for measures governing food and feed taken in the Member States and at Community level. It is however necessary to provide for sufficient time for the adaptation of any conflicting provisions in existing legislation, both at national and Community level, and to provide that, pending such adaptation, the relevant legislation be applied in the light of the principles set out in the present Regulation.

(6) Water is ingested directly or indirectly like other foods, thereby contributing to the overall exposure of a consumer to ingested substances, including chemical and microbiological contaminants. However, as the quality of water intended for human consumption is already controlled by Council Directives 80/778/EEC [5] and 98/83/EC [6], it suffices to consider water after the point of compliance referred to in Article 6 of Directive 98/83/EC.

(7) Within the context of food law it is appropriate to include requirements for feed, including its production and use where that feed is intended for food-producing animals. This is without prejudice to the similar requirements which have been applied so far and which will be applied in the future in feed legislation applicable to all animals, including pets.

(8) The Community has chosen a high level of health protection as appropriate in the development of food law, which it applies in a non-discriminatory manner whether food or feed is traded on the internal market or internationally.

[1] OJ C 96 E, 27.3.2001, p. 247.
[2] OJ C 155, 29.5.2001, p. 32.
[3] Opinion delivered on 14 June 2001 (not yet published in the Official Journal).
[4] Opinion of the European Parliament of 12 June 2001 (not yet published in the Official Journal), Council Common Position of 17 September 2001 (not yet published in the Official Journal) and Decision of the European Parliament of 11 December 2001 (not yet published in the Official Journal). Council Decision of 21 January 2002.

Necessary steps to create the regulation are registered in "Having regard to ..." and notes 1-4

[5] OJ L 229, 30.8.1980, p. 11. Directive repealed by Directive 98/83/EC.
[6] OJ L 330, 5.12.1998, p. 32.

(final page)

The Management Board of the Authority shall examine the conclusions of the evaluation and issue to the Commission such recommendations as may be necessary regarding changes in the Authority and its working practices. The evaluation and the recommendations shall be made public.

2. Before 1 January 2005, the Commission shall publish a report on the experience acquired from implementing Sections 1 and 2 of Chapter IV.

3. The reports and recommendations referred to in paragraphs 1 and 2 shall be forwarded to the Council and the European Parliament.

Article 62

References to the European Food Safety Authority and to the Standing Committee on the Food Chain and Animal Health

1. Every reference in Community legislation to the Scientific Committee on Food, the Scientific Committee on Animal Nutrition, the Scientific Veterinary Committee, the Scientific Committee on Pesticides, the Scientific Committee on Plants and the Scientific Steering Committee shall be replaced by a reference to the European Food Safety Authority.

2. Every reference in Community legislation to the Standing Committee on Foodstuffs, the Standing Committee for Feedingstuffs and the Standing Veterinary Committee shall be replaced by a reference to the Standing Committee on the Food Chain and Animal Health.

Every reference to the Standing Committee on Plant Health in Community legislation based upon and including Directives 76/895/EEC, 86/362/EEC, 86/363/EEC, 90/642/EEC and 91/414/EEC relating to plant protection products and the setting of maximum residue levels shall be replaced by a reference to the Standing Committee on the Food Chain and Animal Health.

3. For the purpose of paragraphs 1 and 2, 'Community legislation' shall mean all Community Regulations, Directives and Decisions.

4. Decisions 68/361/EEC, 69/414/EEC and 70/372/EEC are hereby repealed.

Article 63

Competence of the European Agency for the Evaluation of Medicinal Products

This Regulation shall be without prejudice to the competence conferred on the European Agency for the Evaluation of Medicinal Products by Regulation (EEC) No 2309/93, Regulation (EEC) No 2377/90, Council Directive 75/319/EEC [1] and Council Directive 81/851/EEC [2].

Article 64

Commencement of the Authority's operation

The Authority shall commence its operations on 1 January 2002.

Article 65

Entry into force

This Regulation shall enter into force on the 20th day following that of its publication in the *Official Journal of the European Communities*.

Articles 11 and 12 and Articles 14 to 20 shall apply from 1 January 2005.

Articles 29, 56, 57 and 60 and Article 62(1) shall apply as from the date of appointment of the members of the Scientific Committee and of the Scientific Panels which shall be announced by means of a notice in the 'C' series of the Official Journal.

This Regulation shall be binding in its entirety and directly applicable in all Member States.

Done at Brussels, 28 January 2002.

For the European Parliament
The President
P. COX

For the Council
The President
J. PIQUÉ I CAMPS

Co-decision procedure: consent of European Parliament and Council are necessary. So both presidents must sign.

3 dates are necessary to get a valid working regulation
1. date of formal decision
2. date of publication see also top page 1 "Acts whose publication is obligatory"
3. date of entry into force see Article 65

(1) OJ L 147, 9.6.1975, p. 13. Directive amended by Directive 2001/83/EC of the European Parliament and of the Council (OJ L 311, 28.11.2001, p. 67).
(2) OJ L 317, 6.11.1981, p. 1. Directive amended by Directive 2001/82/EC of the European Parliament and of the Council (OJ L 311, 28.11.2001, p. 1).

References

EU law

Legislation and soft law

Treaties and related texts

Charter of Fundamental Rights of the European Union. OJ C 364 of 18 December 2000.

Ioannina Compromise. EEC Bulletin no. 3 1994, page 79. http://tinyurl.com/ov2rqtv.

Luxembourg Compromise. EEC Bulletin no. 3 1966, page 5. http://tinyurl.com/oe25czv.

Protocol on the application of the principles of subsidiarity and proportionality. OJ C 83, 30 March 2010, pp. 206-210.

Single European Act. OJ L 169, 29 June 1987.

Schuman declaration, 9 May 1950. http://tinyurl.com/d6b4s3u.

Treaty of Paris (Treaty establishing the European Coal and Steel Community), 18 April 1951. http://tinyurl.com/k42uh3u.

Treaty of Rome (Treaty establishing the European Economic Community and Euratom Treaty). 25 March 1957.

Treaty of Maastricht (Treaty on European Union). OJ C 191, 29 July 1992.

Treaty of Amsterdam. OJ C 340 of 10 November 1997.

Treaty of Lisbon (Treaty on European Union and Treaty on the Functioning of the European Union). OJ C 306 of 17 December 2007.

Directives

Council Directive (EEC) on the approximation of the rules of the Member States concerning the colouring matters authorised for use in foodstuffs intended for human consumption. OJ 115, 11 November 1962, pp. 2645-2654.

Council Directive 65/65/EEC of 26 January 1965 on the approximation of provisions laid down by Law, Regulation or Administrative Action relating to proprietary medicinal products. OJ 22, 9 February 1965, pp. 369-373.

Council Directive 73/173/EEC of 4 June 1973 on the approximation of Member States' laws, regulations and administrative provisions relating to the classification, packaging and labelling of dangerous preparations (solvents). OJ L 189, 11 July 1973, pp. 7-29.

Council Directive 73/241/EEC on the approximation of the laws of the Member States relating to cocoa and chocolate products intended for human consumption. OJ L 228, 16 August 1973, pp. 23-35.

Council Directive 76/211/EEC of 20 January 1976 on the approximation of the laws of the Member States relating to the making-up by weight or by volume of certain prepackaged products. OJ L 46, 21 February 1976, pp. 1-11.

Council Directive 76/768/EEC of 27 July 1976 on the approximation of the laws of the Member States relating to cosmetic products. OJ L 262, 27 September 1976, pp. 169-200.

Council Directive 76/895/EEC of 23 November 1976 relating to the fixing of maximum levels for pesticide residues in and on fruit and vegetables. OJ L 340, 9 December 1976, pp. 26-31.

Council Directive 79/112 of 18 December 1978 on the approximation of the laws of the Member States relating to the labelling, presentation and advertising of foodstuffs for sale to the ultimate consumer. OJ L 33, 8 February 1979, pp. 1-14.

Council Directive 79/117/EEC of 21 December 1978 prohibiting the placing on the market and use of plant protection products containing certain active substances. OJ L 33, 8 February 1979, pp. 36-40.

Council Directive 80/778/EEC of 15 July 1980 relating to the quality of water intended for human consumption. OJ L 229, 30 August 1980, pp. 11-29.

Council Directive 81/602/EEC of 31 July 1981 concerning the prohibition of certain substances having a hormonal action and of any substances having a thyrostatic action. OJ L 222, 7 August 1981, pp. 32-33.

Council Directive 84/500/EEC of 15 October 1984 on the approximation of the laws of the Member States relating to ceramic articles intended to come into contact with foodstuffs. OJ L 277, 20 October 1984, pp. 12–16.

Council Directive 85/358/EEC of 16 July 1985 supplementing Directive 81/602/EEC concerning the prohibition of certain substances having a hormonal action and of any substances having a thyrostatic action. OJ L 191, 23 July 1985, pp. 46-49.

Council Directive 85/374/EEC of 25 July 1985 on the approximation of the laws, regulations and administrative provisions of the Member States concerning liability for defective products. OJ L 210, 7 August 1985, pp. 29-33.

Council Directive 86/362/EEC of 24 July 1986 on the fixing of maximum levels for pesticide residues in and on cereals. OJ L 221, 7 August 1986, pp. 37-42.

Council Directive 86/363/EEC of 24 July 1986 on the fixing of maximum levels for pesticide residues in and on foodstuffs of animal origin. OJ L 221, 7 August 1986, pp. 43-47.

Council Directive 88/146/EEC of 7 March 1988 prohibiting the use in livestock farming of certain substances having a hormonal action. OJ L 70, 16 March 1988, pp. 16-18.

Council Directive 88/299/EEC of 17 May 1988 on trade in animals treated with certain substances having a hormonal action and their meat, as referred to in Article 7 of Directive 88/146/EEC. OJ L 128, 21 May 1988, pp. 36-38.

Council Directive 88/344/EEC of 13 June 1988 on the approximation of the laws of the Member States on extraction solvents used in the production of foodstuffs and food ingredients. OJ L 157, 24 June 1988, pp. 28-33.

Council Directive 89/107/EEC of 21 December 1988 on the approximation of the laws of the Member States concerning food additives authorized for use in foodstuffs intended for human consumption. OJ L 40, 11 February 1989, pp. 27-33.

Council Directive 89/108/EEC of 21 December 1988 on the approximation of the laws of the Member States relating to quick-frozen foodstuffs for human consumption. OJ L 40, 11 February 1989, pp. 34-37.

Council Directive 89/398/EEC of 3 May 1989 on the approximation of the laws of the Member States relating to foodstuffs intended for particular nutritional uses. OJ L 186, 30 June 1989, pp. 27-32. corrigendum in OJ L 275, 5 October 1990, p. 42.

Council Directive 89/622/EEC of 13 November 1989 on the approximation of the laws, regulations and administrative provisions of the Member States concerning the labelling of tobacco products. OJ L 359, 8 December 1989, pp. 1-4.

Council Directive 89/662/EEC concerning veterinary checks in intra-Community trade with a view to the completion of the European market. OJ L 395, 30 December 1989, corrigendum in OJ L 151, 15 June 1990, pp. 13-22.

Council Directive 90/219/EEC of 23 April 1990 on the contained use of genetically modified micro-organisms. OJ L 117, 8 May 1990, pp. 1-14.

Council Directive 90/167/EEC of 26 March 1990 laying down the conditions governing the preparation, placing on the market and use of medicated feedingstuffs in the Community. OJ L 92, 7 April 1990, pp. 42-48.

Council Directive 90/220/EEC of 23 April 1990 on the deliberate release into the environment of genetically modified organisms. OJ L 117, 8 May 1990, pp. 15-27.

Council Directive 90/425/EEC of 26 June 1990 concerning veterinary and zootechnical checks applicable in intra-Community trade in certain live animals and products with a view to the completion of the internal market. OJ L 224, 18 August 1990, pp. 29-41.

Council Directive 90/496/EEC of 24 September 1990 on nutrition labelling for foodstuffs. OJ L 276, 6 October 1990, pp. 40-44.

Council Directive 90/642/EEC of 27 November 1990 on the fixing of maximum levels for pesticide residues in and on certain products of plant origin, including fruit and vegetables. OJ L 350, 14 December 1990, pp. 71-79.

Commission Directive 91/321/EEC of 14 May 1991 on infant formulae and follow-on formulae. OJ L 175, 4 July 1991, pp. 35-49.

Council Directive 91/496/EEC of 15 July 1991 laying down the principles governing the organization of veterinary checks on animals entering the Community from third countries and amending Directives 89/662/EEC, 90/425/EEC and 90/675/EEC. OJ L 268, 24 September 1991, pp. 56-68.

Council Directive 92/52/EEC of 18 June 1992 on infant formulae and follow-on formulae intended for export to third countries. OJ L 179, 1 July 1992, pp. 129-130.

Council Directive 92/73/EEC of 22 September 1992 widening the scope of Directives 65/65/EEC and 75/319/EEC on the approximation of provisions laid down by Law, Regulation or Administrative Action relating to medicinal products and laying down additional provisions on homeopathic medicinal products. OJ L 297, 13 October 1992, pp. 8-11.

Council Directive 92/102/EEC of 27 November 1992 on the identification and registration of animals. OJ L 355, 5 December 1992, pp. 32-36.

Commission Directive 93/11/EEC of 15 March 1993 concerning the release of the N-nitrosamines and N- nitrosatable substances from elastomer or rubber teats and soothers. OJ L 93, 17 April 1993, pp. 37-38.

Council Directive 93/43/EEC of 14 June 1993 on the hygiene of foodstuffs. OJ L 175, 19 July 1993, pp. 1-11.

Council Directive 93/98/EEC of 29 October 1993 harmonizing the term of protection of copyright and certain related rights. OJ L 290, 24 November 1993, pp. 9–13.

European Parliament and Council Directive 94/34/EC of 30 June 1994 amending Directive 89/107/EEC on the approximation of the laws of Member States concerning food additives authorized for use in foodstuffs intended for human consumption. OJ L 237, 10 September 1994, pp. 1-2.

Commission Directive 96/8/EC of 26 February 1996 on foods intended for use in energy-restricted diets for weight reduction. OJ L 55, 6 March 1996, pp. 22-26.

Council Directive 96/22/EC of 29 April 1996 concerning the prohibition on the use in stockfarming of certain substances having a hormonal or thyrostatic action and of β-agonists, and repealing Directives 81/602/EEC, 88/146/EEC and 88/299/EEC. OJ L 125, 23 May 1996, pp. 3-9.

Council Directive 96/29/Euratom of 13 May 1996 laying down basic safety standards for the protection of the health of workers and the general public against the dangers arising from ionizing radiation. OJ L 159, 29 June 1996, pp. 1-114.

Council Directive 96/43/EC of 26 June 1996 amending and consolidating Directive 85/73/EEC in order to ensure financing of veterinary inspections and controls on live animals and certain animal products and amending Directives 90/675/EEC and 91/496/EEC. OJ L 162, 1 July 1996, pp. 1-13.

Council Directive 96/29/Euratom of 13 May 1996 laying down basic safety standards for the protection of the health of workers and the general public against the dangers arising from ionizing radiation. OJ L 159, 29 June 1996, pp. 1-114.

Council Directive 97/78/EC of 18 December 1997 laying down the principles governing the organisation of veterinary checks on products entering the Community from third countries. OJ L 24, 30 January 1998, pp. 9-30.

Directive 98/6/EC of the European Parliament and of the Council of 16 February 1998 on consumer protection in the indication of the prices of products offered to consumers. OJ L 80, 18 March 1998, pp. 27-31.

Commission Directive 1999/21/EC of 25 March 1999 on dietary foods for special medical purposes. OJ L 91, 7 April 1999, pp. 29-36.

Directive 98/44/EC of the European Parliament and of the Council of 6 July 1998 on the legal protection of biotechnological inventions. OJ L 213, 30 July 1998, pp. 13-21.

Council Directive 98/81/EC of 26 October 1998 amending Directive 90/219/EEC on the contained use of genetically modified micro-organisms. OJ J L 330, 5 December 1998, pp. 13-31. corrigenda in OJ L 3, 7 January 1999, p. 23 and OJ L 93, 8 April 1999, p. 27.

Council Directive 98/83/EC of 3 November 1998 on the quality of water intended for human consumption. OJ L 330, 5 December 1998, pp. 32-54, corrigendum in OJ L 111, 20 April 2001, p. 31.

Directive 1999/2/EC of the European Parliament and of the Council of 22 February 1999 on the approximation of the laws of the Member States concerning foods and food ingredients treated with ionising radiation. OJ L 66, 13 March 1999, pp. 16-23.

Directive 1999/34/EC of the European Parliament and of the Council of 10 May 1999 amending Council Directive 85/374/EEC on the approximation of the laws, regulations and administrative provisions of the Member States concerning liability for defective products. OJ L 141, 4 June 1999, pp. 20-21, corrigendum in OJ L 283, 6 November 1999, p. 20.

Directive 2000/13/EC of the European Parliament and of the Council of 20 March 2000 on the approximation of the laws of the Member States relating to the labelling, presentation and advertising of foodstuffs. OJ L 109, 6 May 2000, pp. 29-42.

Council Directive 2000/29 of 8 May 2000 on protective measures against the introduction into the Community of organisms harmful to plants or plant products and against their spread within the Community. OJ L 169, 10 July 2000, pp. 1-112.

Directive 2000/36/EC of the European Parliament and of the Council of 23 June 2000 relating to cocoa and chocolate products intended for human consumption. OJ L 197, 3 August 2000, pp. 19-25.

Commission Directive 2001/15/EC of 15 February 2001 on substances that may be added for specific nutritional purposes in foods for particular nutritional uses. OJ L 52, 22 February 2001, pp. 19-25, corrigendum in OJ L 253, 21 September 2001, pp. 34-34.

Directive 2001/18/EC of the European Parliament and of the Council of 12 March 2001 on the deliberate release into the environment of genetically modified organisms and repealing Council Directive 90/220/EEC – Commission Declaration. OJ L 106, 17 April 2001, pp. 1-39.

Directive 2001/82/EC of the European Parliament and of the Council of 6 November 2001 on the Community code relating to veterinary medicinal products. OJ L 311, 28 November 2001, pp. 1-66.

Directive 2001/83/EC of the European Parliament and of the Council of 6 November 2001 on the Community code relating to medicinal products for human use. OJ L 311, 28 November 2001, pp. 67-128.

Council Directive 2001/110/EC of 20 December 2001 relating to honey. OJ L 10, 12 January 2002, pp. 47-52.

Council Directive 2001/112/EC of 20 December 2001 relating to fruit juices and certain similar products intended for human consumption. OJ L 10, 12 January 2002, pp. 58-66.

Council Directive 2001/113/EC of 20 December 2001 relating to fruit jams, jellies and marmalades and sweetened chestnut purée intended for human consumption. OJ L 10, 12 January 2002, pp. 67-72.

Directive 2002/32/EC of the European Parliament and of the Council of 7 May 2002 on undesirable substances in animal feed. OJ L 140, 30.5.2002, pp. 10-22.

Directive 2002/46/EC of the European Parliament and of the Council of 10 June 2002 on the approximation of the laws of the Member States relating to food supplements. OJ L 183, 12 July 2002, pp. 51-57.

Council Directive 2002/99/EC of 16 December 2002 laying down the animal health rules governing the production, processing, distribution and introduction of products of animal origin for human consumption. OJ L 18, 23.1.2003, pp. 11-20.

Directive 2003/99/EC of the European Parliament and of the Council of 17 November 2003 on the monitoring of zoonoses and zoonotic agents, amending Council Decision 90/424/EEC and repealing Council Directive 92/117/EEC. OJ L 325, 12 December 2003, pp. 31-40.

Directive 2004/41/EC of the European Parliament and of the Council of 21 April 2004 repealing certain Directives concerning food hygiene and health conditions for the production and placing on the market of certain products of animal origin intended for human consumption and amending Council Directives 89/662/EEC and 92/118/EEC and Council Decision 95/408/EC. OJ L 157, 30 April 2004. Corrected version in OJ L 195, 2 June 2004, pp. 12-15.

Commission Directive 2006/125/EC of 5 December 2006 on processed cereal-based foods and baby foods for infants and young children. OJ L 339, 6 December 2006, pp. 16-35.

Commission Directive 2006/141/EC of 22 December 2006 on infant formulae and follow-on formulae and amending Directive 1999/21/EC. OJ L 401, 30 December 2006, pp. 1-33.
Commission Directive 2007/42/EC of 29 June 2007 relating to materials and articles made of regenerated cellulose film intended to come into contact with foodstuffs. OJ L 172, 30 June 2007, pp. 71-82.
Directive 2007/45/EC of the European Parliament and of the Council of 5 September 2007 laying down rules on nominal quantities for prepacked products, repealing Council Directives 75/106/EEC and 80/232/EEC, and amending Council Directive 76/211/EEC. OJ L 247, 21 September 2007, pp. 17-20.
Directive 2009/39/EC of the European Parliament and of the Council of 6 May 2009 on foodstuffs intended for particular nutritional uses. OJ L 124, 20 May 2009, pp. 21-29.
Directive 2009/54/EC of the European Parliament and of the Council of 18 June 2009 on the exploitation and marketing of natural mineral waters. OJ L 164, 26 June 2009, pp. 45–58.
Directive 2010/13/EU of the European Parliament and of the Council of 10 March 2010 on the coordination of certain provisions laid down by law, regulation or administrative action in Member States concerning the provision of audiovisual media services. OJ L 95, 15 April 2010, pp. 1-24.
Directive 2011/91/EU of the European Parliament and of the Council of 13 December 2011 on indications or marks identifying the lot to which a foodstuff belongs. OJ L 334, 16 December 20, pp. 1-5.

Regulations

Council Regulation (Euratom) No 3954/87 of 22 December 1987 laying down maximum permitted levels of radioactive contamination of foodstuffs and of feedingstuffs following a nuclear accident or any other case of radiological emergency. OJ L 371, 30 December 1987, pp. 11-13.
Commission Regulation (Euratom) No 944/89 of 12 April 1989 laying down maximum permitted levels of radioactive contamination in minor foodstuffs following a nuclear accident or any other case of radiological emergency. OJ L 101, 13 April 1989, pp. 17-18.
Council Regulation (Euratom) No 2218/89 of 18 July 1989 amending Regulation (Euratom) No 3954/87 laying down maximum permitted levels of radioactive contamination of foodstuffs and of feedingstuffs following a nuclear accident or any other case of radiological emergency. OJ L 211, 22 July 1989, pp. 1-3.
Commission Regulation (Euratom) No 770/90 of 29 March 1990 laying down maximum permitted levels of radioactive contamination of feedingstuffs following a nuclear accident or any other case of radiological emergency. OJ L 83, 30 March 1990, pp. 78-79.
Council Regulation (EEC) No 2377/90 of 26 June 1990 laying down a Community procedure for the establishment of maximum residue limits of veterinary medicinal products in foodstuffs of animal origin. OJ L 224, 18 August 1990, pp. 1-8.
Council Regulation (EEC) No 2913/92 of 12 October 1992 establishing the Community Customs Code. OJ L 302, 19 October 1992, pp. 1-50.
Council Regulation (EEC) No 315/93 of 8 February 1993 laying down Community procedures for contaminants in food. OJ L 37, 13 February 1993, pp. 1-3.

Council Regulation (EC) No 2991/94 of 5 December 1994 laying down standards for spreadable fats. OJ L 316, 9 December 1994, pp. 2–7.

Council Regulation (EC) No 3286/94 of 22 December 1994 laying down Community procedures in the field of the common commercial policy in order to ensure the exercise of the Community's rights under international trade rules, in particular those established under the auspices of the World Trade Organization. OJ L 349, 31 December 1994, pp. 71-78.

Regulation (EC) No 258/97 of the European Parliament and of the Council of 27 January 1997 concerning novel foods and novel food ingredients. OJ L 43, 14 February 1997, pp. 1-6.

Council Regulation (EC) No 820/97 of 21 April 1997 establishing a system for the identification and registration of bovine animals and regarding the labelling of beef and beef products. OJ L 117, 7 May 1997, pp. 1-8.

Regulation (EC) No 1760/2000 of the European Parliament and of the Council of 17 July 2000 establishing a system for the identification and registration of bovine animals and regarding the labelling of beef and beef products and repealing Council Regulation (EC) No 820/97. OJ L 204, 11 August 2000, pp. 1-10.

Commission Regulation (EC) No. 1825/2000 of 25 August 2000 laying down detailed rules for the application of Regulation (EC) No. 1760/2000 of the European Parliament and of the Council as regards the labelling of beef and beef products. OJ L 216, 26 August 2000, pp. 8-12.

Regulation (EC) No 999/2001 of the European Parliament and of the Council of 22 May 2001 laying down rules for the prevention, control and eradication of certain transmissible spongiform encephalopathies. OJ L 147, 31 May 2001, pp. 1-40.

Council Regulation (EC) No 2157/2001 of 8 October 2001 on the Statute for a European company (SE). OJ L 294, 10 November 2001, pp. 1-21.

Commission Regulation (EC) No 2065/2001 of 22 October 2001 laying down detailed rules for the application of Council Regulation (EC) No 104/2000 as regards informing consumers about fishery and aquaculture products. OJ L 278, 23 October 2001, pp. 6-8.

Regulation (EC) No 178/2002 of the European Parliament and of the Council of 28 January 2002 laying down the general principles and requirements of food law, establishing the European Food Safety Authority and laying down procedures in matters of food safety. OJ L 31, 1 February 2002, pp. 1-24.

Regulation (EC) No 1774/2002 of the European Parliament and of the Council of 3 October 2002 laying down health rules concerning animal by-products not intended for human consumption. OJ L 273, 10 October 2002, pp. 1-95.

Commission Regulation (EC) No 1304/2003 of 23 July 2003 on the procedure applied by the European Food Safety Authority to requests for scientific opinions referred to it. OJ L 185, 24 September 2003, p. 6, as corrected by Corrigendum, OJ L 186, 25 July 2003, p. 46.

Regulation (EC) No 1829/2003 of the European Parliament and of the Council of 22 September 2003 on genetically modified food and feed. OJ L 268, 18 October 2003, pp. 1-23.

Regulation (EC) No 1830/2003 of the European Parliament and of the Council of 22 September 2003 concerning the traceability and labelling of genetically modified organisms and the traceability of food and feed products produced from genetically modified organisms and amending Directive 2001/18/EC. OJ L 268, 18 October 2003, pp. 24-28.

Regulation (EC) No 1831/2003 of the European Parliament and of the Council of 22 September 2003 on additives for use in animal nutrition. OJ L 268, 18 October 2003, pp. 29-43.

Regulation (EC) No 1946/2003 of the European Parliament and of the Council of 15 July 2003 on transboundary movements of genetically modified organisms. OJ L 287, 5 November 2003, pp. 1-10.

Regulation (EC) No 2160/2003 of the European Parliament and of the Council of 17 November 2003 on the control of salmonella and other specified food-borne zoonotic agents. OJ L 325, 12 December 2003, pp. 1-15

Regulation (EC) No 2065/2003 of the European Parliament and of the Council of 10 November 2003 on smoke flavourings used or intended for use in or on foods. OJ L 309, 26 November 2003, pp. 1-8.

Council Regulation (EC) No 21/2004 of 17 December 2003 establishing a system for the identification and registration of ovine and caprine animals and amending Regulation (EC) No 1782/2003 and Directives 92/102/EEC and 64/432/EEC. OJ L 5, 9 January 2004, pp. 8-17.

Commission Regulation (EC) No 65/2004 of 14 January 2004 establishing a system for the development and assignment of unique identifiers for genetically modified organisms. OJ L 10, 16 January 2004, pp. 5-10.

Commission Regulation (EC) No 136/2004 of 22 January 2004 laying down procedures for veterinary checks at Community border inspection posts on products imported from third countries. OJ L 21, 28 January 2004, pp. 11-23.

Commission Regulation (EC) No 641/2004 of 6 April 2004 on detailed rules for the implementation of Regulation (EC) No 1829/2003 of the European Parliament and of the Council as regards the application for the authorisation of new genetically modified food and feed, the notification of existing products and adventitious or technically unavoidable presence of genetically modified material which has benefited from a favourable risk evaluation. OJ L 102, 7 April 2004, pp. 14-25.

Regulation (EC) No 852/2004 of the European Parliament and of the Council of 29 April 2004 on the hygiene of foodstuffs. OJ L 139, 30 April 4 2004, pp.1-54. Corrected version in OJ L 226, 25 June 6.2004.

Regulation (EC) No 853/2004 of the European Parliament and of the Council of 29 April 2004 laying down specific hygiene rules for food of animal origin. OJ L 139, 30 April 2004, pp. 55-205.

Regulation (EC) No 854/2004 of the European Parliament and of the Council of 29 April 2004 laying down specific rules for the organisation of official controls on products of animal origin intended for human consumption. OJ L 139, 30 April 2004, pp. 206-320. Corrected version in OJ L 226, 25 June 2004.

Regulation (EC) No 882/2004 of the European Parliament and of the Council of 29 April 2004 on official controls performed to ensure the verification of compliance with feed and food law, animal health and animal welfare rules. OJ L 191, 28 May 2004, pp. 1-141.

Commission Regulation (EC) No 887/2004 of 29 April 2004 establishing the standard import values for determining the entry price of certain fruit and vegetables. OJ L 163, 30 April 2004, pp. 1-2.

Commission Regulation (EC) No 911/2004 of 29 April 2004 implementing Regulation (EC) No 1760/2000 of the European Parliament and of the Council as regards eartags, passports and holding registers. OJ L 163, 30 April 2004, pp. 65-70.

Commission Regulation (EC) No 1756/2004 of 11 October 2004 specifying the detailed conditions for the evidence required and the criteria for the type and level of the reduction of the plant health checks of certain plants, plant products or other objects listed in Part B of Annex V to Council Directive 2000/29/EC. OJ L 313, 12 October 2004, pp. 6-9.

Regulation (EC) No 1935/2004 of the European Parliament and of the Council of 27 October 2004 on materials and articles intended to come into contact with food and repealing Directives 80/590/EEC and 89/109/EEC. OJ L 338, 13 November 2004, pp. 4-17.

Commission Directive 2005/31/EC of 29 April 2005 amending Council Directive 84/500/EEC as regards a declaration of compliance and performance criteria of the analytical method for ceramic articles intended to come into contact with foodstuffs. OJ L 110, 30 April 2005, pp. 36-39.

Regulation (EC) No 183/2005 of the European Parliament and of the Council of 12 January 2005 laying down requirements for feed hygiene. OJ L 35, 8 February 2005, pp. 1-22.

Regulation (EC) No 396/2005 of the European Parliament and of the Council of 23 February 2005 on maximum residue levels of pesticides in or on food and feed of plant and animal origin and amending Council Directive 91/414/EEC. OJ L 70, 16 March 2005, pp. 1-16.

Commission Regulation (EC) No 1895/2005 of 18 November 2005 on the restriction of use of certain epoxy derivatives in materials and articles intended to come into contact with food. OJ L 302, 19 November 2005, pp. 28-32.

Commission Regulation (EC) No 2073/2005 of 15 November 2005 on microbiological criteria for foodstuffs. OJ L 338, 22 December 2005, pp. 1-26.

Commission Regulation (EC) No 2074/2005 of 5 December 2005 laying down implementing measures for certain products under Regulation (EC) No 853/2004 of the European Parliament and of the Council and for the organisation of official controls under Regulation (EC) No 854/2004 of the European Parliament and of the Council and Regulation (EC) No 882/2004 of the European Parliament and of the Council, derogating from Regulation (EC) No 852/2004 of the European Parliament and of the Council and amending Regulations (EC) No 853/2004 and (EC) No 854/2004. OJ L 338, 22 December 2005, pp. 27-59.

Commission Regulation (EC) No 2075/2005 of 5 December 2005 laying down specific rules on official controls for *Trichinella* in meat. OJ L 338, 22 December 2005, pp. 60-82.

Commission Regulation (EC) No 2076/2005 of 5 December 2005 laying down transitional arrangements for the implementation of Regulations (EC) No 853/2004, (EC) No 854/2004 and (EC) No 882/2004 of the European Parliament and of the Council and amending Regulations (EC) No 853/2004 and (EC) No 854/2004. OJ L 338, 22 December 2005, pp. 83-88.

Commission Regulation (EC) No 178/2006 of 1 February 2006 amending Regulation (EC) No 396/2005 of the European Parliament and of the Council to establish Annex I listing the food and feed products to which maximum levels for pesticide residues apply. OJ L 29, 2 February 2006, pp. 3-25.

Commission Regulation (EC) No 401/2006 of 23 February 2006 laying down the methods of sampling and analysis for the official control of the levels of mycotoxins in foodstuffs. OJ L 70, 9 March 2006, pp. 12-34.

Council Regulation (EC) No 510/2006 of 20 March 2006 on the protection of geographical indications and designations of origin for agricultural products and foodstuffs. OJ L 93, 31 March 2006, pp. 12-25.

Commission Regulation (EC) No 575/2006 of 7 April 2006 amending Regulation (EC) No 178/2002 of the European Parliament and of the Council as regards the number and names of the permanent Scientific Panels of the European Food Safety Authority. OJ L 100, 08 April 2006, p. 3.

Commission Regulation (EC) No 1881/2006 of 19 December 2006 setting maximum levels for certain contaminants in foodstuffs. OJ L 364, 20 December 2006, pp. 5-24.

Commission Regulation (EC) No 1883/2006 of 19 December 2006 laying down methods of sampling and analysis for the official control of levels of dioxins and dioxin-like PCBs in certain foodstuffs. OJ L 364, 20 December 2006, pp. 32-43.

Regulation (EC) No 1907/2006 of the European Parliament and of the Council of 18 December 2006 concerning the Registration, Evaluation, Authorisation and Restriction of Chemicals (REACH), establishing a European Chemicals Agency, amending Directive 1999/45/EC and repealing Council Regulation (EEC) No 793/93 and Commission Regulation (EC) No 1488/94 as well as Council Directive 76/769/EEC and Commission Directives 91/155/EEC, 93/67/EEC, 93/105/EC and 2000/21/EC. OJ L 396, 30 December 2006, pp. 1-849.

Regulation (EC) No 1924/2006 of the European Parliament and of the Council of 20 December 2006 on nutrition and health claims made on foods. OJ L 404, 30 December 2006, pp. 9-25. Corrigendum in OJ L 12, 18 January 2007, pp. 3-18.

Commission Regulation (EC) No 2023/2006 of 22 December 2006 on good manufacturing practice for materials and articles intended to come into contact with food. OJ L 384, 29 December 2006, pp. 75-78.

Commission Regulation (EC) No 333/2007 of 28 March 2007 laying down the methods of sampling and analysis for the official control of the levels of lead, cadmium, mercury, inorganic tin, 3-MCPD and benzo(a)pyrene in foodstuffs. OJ L 88, 29 March 2007, pp. 29-38.

Council Regulation (EC) No 834/2007 of 28 June 2007 on organic production and labelling of organic products and repealing Regulation (EEC) No 2092/91. OJ L 189, 20 July 2007, pp. 1-23.

Regulation (EC) No 110/2008 of the European Parliament and of the Council of 15 January 2008 on the definition, description, presentation, labelling and the protection of geographical indications of spirit drinks and repealing Council Regulation (EEC) No 1576/89. OJ L 39, 13 February 2008, pp. 16-54.

Commission Regulation (EC) No 149/2008 of 29 January 2008 amending Regulation (EC) No 396/2005 of the European Parliament and of the Council by establishing Annexes II, III and IV setting maximum residue levels for products covered by Annex I thereto. OJ L 58, 1 March 2008, pp. 1-398.

Commission Regulation (EC) No 202/2008 of 4 March 2008 amending Regulation (EC) No 178/2002 of the European Parliament and of the Council as regards the number and names of the Scientific Panels of the European Food Safety Authority. OJ L 60, 5 March 2008, pp. 17-17.

Commission Regulation (EC) No 282/2008 of 27 March 2008 on recycled plastic materials and articles intended to come into contact with foods and amending Regulation (EC) No 2023/2006. OJ L 86, 28 March 2008, pp. 9-18.

Council Regulation (EC) No 479/2008 of 29 April 2008 on the common organisation of the market in wine, amending Regulations (EC) No 1493/1999, (EC) No 1782/2003, (EC) No 1290/2005, (EC) No 3/2008 and repealing Regulations (EEC) No 2392/86 and (EC) No 1493/1999. OJ L 148, 6 June 2008, pp. 1-61.

Commission Regulation (EC) No 543/2008 of 16 June 2008 laying down detailed rules for the application of Council Regulation (EC) No 1234/2007 as regards the marketing standards for poultrymeat. OJ L 157, 17 June 2008, pp. 46-87.

Commission Regulation (EC) No 133/2008 of 14 February 2008 on imports of pure-bred breeding animals of the bovine species from the third countries and the granting of export refunds thereon (Codified version). OJ L 41, 15 February 2008, pp. 11-14.

Regulation (EC) No 764/2008 of the European Parliament and of the Council laying down procedures relating to the application of certain national technical rules to products lawfully marketed in another Member State and repealing Decision No 3052/95/EC. OJ L 218, 13 August 2008, pp. 21-29.

Regulation (EC) No 765/2008 of the European Parliament and of the Council of 9 July 2008 setting out the requirements for accreditation and market surveillance relating to the marketing of products. OJ 13 August 2008, L 218, pp. 30-47.

Commission Regulation (EC) No 889/2008 of 5 September 2008 laying down detailed rules for the implementation of Council Regulation (EC) No 834/2007 on organic production and labelling of organic products with regard to organic production, labelling and control. OJ L 250, 18 September 2008, pp. 1-84.

Regulation (EC) No 1331/2008 of the European Parliament and of the Council of 16 December 2008 establishing a common authorisation procedure for food additives, food enzymes and food flavourings. OJ L 354, 31 December 2008, pp. 1-6.

Regulation (EC) No 1332/2008 of the European Parliament and of the Council of 16 December 2008 on food enzymes and amending Council Directive 83/417/EEC, Council Regulation (EC) No 1493/1999, Directive 2000/13/EC, Council Directive 2001/112/EC and Regulation (EC) No 258/97. OJ L 354, 31 December 2008, pp. 7-15.

Regulation (EC) No 1333/2008 of the European Parliament and of the Council of 16 December 2008 on food additives. OJ L 354, 31 December 2008, pp. 16-33.

Regulation (EC) No 1334/2008 of the European Parliament and of the Council of 16 December 2008 on flavourings and certain food ingredients with flavouring properties for use in and on foods and amending Council Regulation (EEC) No 1601/91, Regulations (EC) No 2232/96 and (EC) No 110/2008 and Directive 2000/13/EC. OJ L 354, 31 December 2008, pp. 34-50.

Commission Regulation (EC) No 41/2009 of 20 January 2009 concerning the composition and labelling of foodstuffs suitable for people intolerant to gluten. OJ L 16, 21 January 2009, pp. 3-5.

Commission Regulation (EC) No 206/2009 of 5 March 2009 on the introduction into the Community of personal consignments of products of animal origin and amending Regulation (EC) No 136/2004. OJ L 77, 24 March 2009, pp. 1-19.

Commission Regulation (EC) No 450/2009 of 29 May 2009 on active and intelligent materials and articles intended to come into contact with food. OJ L 135, 30 May 2009, pp. 3-11.

Regulation (EC) No 470/2009 of the European Parliament and of the Council of 6 May 2009 laying down Community procedures for the establishment of residue limits of pharmacologically active substances in foodstuffs of animal origin, repealing Council Regulation (EEC) No 2377/90 and amending Directive 2001/82/EC of the European Parliament and of the Council and Regulation (EC) No 726/2004 of the European Parliament and of the Council (Text with EEA relevance). OJ L 152, 16 June 2009, pp. 11-22.

Commission Regulation (EC) No 669/2009 of 24 July 2009 implementing Regulation (EC) No 882/2004 of the European Parliament and of the Council as regards the increased level of official controls on imports of certain feed and food of non-animal origin and amending Decision 2006/504/EC. OJ L 194, 25 July 2009, pp. 11-21.

Regulation (EC) No 767/2009 of the European Parliament and of the Council of 13 July 2009 on the placing on the market and use of feed, amending European Parliament and Council Regulation (EC) No 1831/2003 and repealing Council Directive 79/373/EEC, Commission Directive 80/511/EEC, Council Directives 82/471/EEC, 83/228/EEC, 93/74/EEC, 93/113/EC and 96/25/EC and Commission Decision 2004/217/EC. OJ L 229, 1 September 2009, pp. 1-28.

Commission Regulation (EC) No 953/2009 of 13 October 2009 on substances that may be added for specific nutritional purposes in foods for particular nutritional uses. OJ L 269, 14.10.2009, pp. 9-19.

Council Regulation (EC) No 1048/2009 of 23 October 2009 amending Regulation (EC) No 733/2008 on the conditions governing imports of agricultural products originating in third countries following the accident at the Chernobyl nuclear power station. OJ L 290, 6 November 2009, pp. 4-4.

Regulation (EC) No 1069/2009 of the European Parliament and of the Council of 21 October 2009 laying down health rules as regards animal by-products and derived products not intended for human consumption and repealing Regulation (EC) No 1774/2002 (Animal by-products Regulation). OJ L 300, 14 November 2009, pp. 1-33.

Regulation (EC) No 1107/2009 of the European Parliament and of the Council of 21 October 2009 concerning the placing of plant protection products on the market and repealing Council Directives 79/117/EEC and 91/414/EEC. OJ L 309, 24 November 2009, pp. 1-50.

Commission Regulation (EC) No 1135/2009 of 25 November 2009 imposing special conditions governing the import of certain products originating in or consigned from China, and repealing Commission Decision 2008/798/EC. OJ L 311, 26 November 2009, pp. 3-5.

Commission Regulation (EC) No 1152/2009 of 27 November 2009 imposing special conditions governing the import of certain foodstuffs from certain third countries due to contamination risk by aflatoxins and repealing Decision 2006/504/EC. OJ L 313, 28 November 2009, pp. 40-49.

Regulation (EC) No 1185/2009 of the European Parliament and of the Council of 25 November 2009 concerning statistics on pesticides. OJ L 324, 10 December 2009, pp. 1-22.

Commission Regulation (EU) No 945/2010 of 21 October 2010 adopting the plan allocating to the Member States resources to be charged to the 2011 budget year for the supply of food from intervention stocks for the benefit of the most deprived persons in the EU and derogating from certain provisions of Regulation (EU) No 807/2010. OJ L 278, 22 October 2010, pp. 1-8.

Commission Regulation (EU) No 10/2011 of 14 January 2011 on plastic materials and articles intended to come into contact with food. OJ L 12, 15 January 2011, pp. 1-89.

Commission Regulation (EU) No 142/2011 of 25 February 2011 implementing Regulation (EC) No 1069/2009 of the European Parliament and of the Council laying down health rules as regards animal by-products and derived products not intended for human consumption and implementing Council Directive 97/78/EC as regards certain samples and items exempt from veterinary checks at the border under that Directive Text with EEA relevance. OJ L 54, 26 February 2011, pp. 1-254.

Regulation 182/2011 laying down the rules and general principles concerning mechanisms for control by Member States of the Commission's exercise of implementing powers. OJ L 55, 28 February 2011, pp. 13-18.

Commission Decision 2011/284/EU of 12 May 2011 on the procedure for attesting the conformity of construction products pursuant to Article 20(2) of Council Directive 89/106/EEC as regards power, control and communication cables (notified under document C(2011) 3107). OJ L 131, 18 May 2011, pp. 22-25.

Commission Implementing Regulation (EU) No 297/2011 of 25 March 2011 imposing special conditions governing the import of feed and food originating in or consigned from Japan following the accident at the Fukushima nuclear power station)Text with EEA relevance). OJ L 80, 26 March 2011, pp. 5-8.

Commission Implementing Regulation (EU) No 321/2011 of 1 April 2011 amending Regulation (EU) No 10/2011 as regards the restriction of use of Bisphenol A in plastic infant feeding bottles Text with EEA relevance. OJ L 87, 2 April 2011, pp. 1-2.

Commission Implementing Regulation (EU) No 543/2011 of 7 June 2011 laying down detailed rules for the application of Council Regulation (EC) No 1234/2007 in respect of the fruit and vegetables and processed fruit and vegetables sectors. OJ L 157, 15 June 2011, pp. 1-163.

Commission Implementing Regulation (EU) No 844/2011 of 23 August 2011 approving the pre-export checks carried out by Canada on wheat and wheat flour as regards the presence of ochratoxin A Text with EEA relevance. OJ L 218, 24 August 2011, pp. 4-7.

Commission Regulation (EU) No 1086/2011 of 27 October 2011 amending Annex II to Regulation (EC) No 2160/2003 of the European Parliament and of the Council and Annex I to Commission Regulation (EC) No 2073/2005 as regards salmonella in fresh poultry meat. OJ L 281, 28 October 2011, pp. 7-11.

Regulation 1169/2011 of the European Parliament and of the Council of 25 October 2011 on the provision of food information to consumers, amending Regulations 1924/2006 and 1925/2006 of the European Parliament and of the Council, and repealing Commission Directive 87/250, Council Directive 90/496, Commission Directive 1999/10, Directive 2000/13 of the European Parliament and of the Council, Commission Directives 2002/67 and 2008/5 and Commission Regulation 608/2004. OJ L 304, 15 October 2011. pp. 18-63.

Commission Regulation (EU) No 1282/2011 of 28 November 2011 amending and correcting Commission Regulation (EU) No 10/2011 on plastic materials and articles intended to come into contact with food. OJ L 328, 10 December 2011, pp. 22-29.

Commission Regulation (EU) No 28/2012 of 11 January 2012 laying down requirements for the certification for imports into and transit through the Union of certain composite products and amending Decision 2007/275/EC and Regulation (EC) No 1162/2009. OJ L 12, 14 January 2012, pp. 1-13.

Commission Implementing Regulation (EU) No 284/2012 of 29 March 2012 imposing special conditions governing the import of feed and food originating in or consigned from Japan following the accident at the Fukushima nuclear power station and repealing Implementing Regulation (EU) No 961/2011. OJ L 92, 30 March 2012, pp. 16-23.

Commission Regulation (EU) No 432/2012 of 16 May 2012 establishing a list of permitted health claims made on foods, other than those referring to the reduction of disease risk and to children's development and health. OJ L 136, 25 May 2012, pp. 1-40.

Commission Implementing Regulation (EU) No 996/2012 of 26 October 2012 imposing special conditions governing the import of feed and food originating in or consigned from Japan following the accident at the Fukushima nuclear power station and repealing Implementing Regulation (EU) No 284/2012. OJ L 299, 27 October 2012, pp. 31-41.

Regulation (EU) No 1151/2012 of the European Parliament and of the Council of 21 November 2012 on quality schemes for agricultural products and foodstuffs. OJ L 343, 14 December 2012, pp. 1-29.

Commission Regulation (EU) No 68/2013 of 16 January 2013 on the Catalogue of feed materials. OJ L 29, 30 January 2013, pp. 1-64.

Commission Implementing Regulation (EU) No 208/2013 of 11 March 2013 on traceability requirements for sprouts and seeds intended for the production of sprouts. OJ L 68, 12 March 2013, pp. 16-18.

Commission Regulation (EU) No 209/2013 of 11 March 2013 amending Regulation (EC) No 2073/2005 as regards microbiological criteria for sprouts and the sampling rules for poultry carcases and fresh poultry meat. OJ L 68, 12 March 2013, pp. 19-23.

Commission Regulation (EU) No 210/2013 of 11 March 2013 on the approval of establishments producing sprouts pursuant to Regulation (EC) No 852/2004 of the European Parliament and of the Council. OJ L 68, 12 March 2013, pp. 24-25.

Commission Regulation (EU) No 211/2013 of 11 March 2013 on certification requirements for imports into the Union of sprouts and seeds intended for the production of sprouts. OJ L 68, 12 March 2013, pp. 26-29.

Regulation (EU) No 609/2013 of the European Parliament and of the Council of 12 June 2013 on food intended for infants and young children, food for special medical purposes, and total diet replacement for weight control and repealing Council Directive 92/52/EEC, Commission Directives 96/8/EC, 1999/21/EC, 2006/125/EC and 2006/141/EC, Directive 2009/39/EC of the European Parliament and of the Council and Commission Regulations (EC) No 41/2009 and (EC) No 953/2009. OJ L 181, 29 June 2013, pp. 35-56.

Regulation (EU) No 952/2013 of the European Parliament and of the Council of 9 October 2013 laying down the Union Customs Code. OJ L 269, 10 October 2013, pp. 1-101.

Commission Delegated Regulation (EU) No 1155/2013 of 21 August 2013 amending Regulation (EU) No 1169/2011 of the European Parliament and of the Council on the provision of food information to consumers as regards information on the absence or reduced presence of gluten in food. OJ L 306, 16 November 2013, pp. 7-7.

Regulation (EU) No 1308/2013 of the European Parliament and of the Council of 17 December 2013 establishing a common organisation of the markets in agricultural products and repealing Council Regulations (EEC) No 922/72, (EEC) No 234/79, (EC) No 1037/2001 and (EC) No 1234/2007. OJ L 347, 20 December 2013, pp. 671-854.

Commission Implementing Regulation (EU) No 1321/2013 of 10 December 2013 establishing the Union list of authorised smoke flavouring primary products for use as such in or on foods and/or for the production of derived smoke flavourings. OJ L 333, 12 December 2013, pp. 54-67.

Regulation (EU) No 1379/2013 of the European Parliament and of the Council of 11 December 2013 on the common organisation of the markets in fishery and aquaculture products, amending Council Regulations (EC) No 1184/2006 and (EC) No 1224/2009 and repealing Council Regulation (EC) No 104/2000. OJ L 354, 28 December 2013, pp. 1-21.

Commission Regulation (EU) No 202/2014 of 3 March 2014 amending Regulation (EU) No 10/2011 on plastic materials and articles intended to come into contact with food. OJ L 62, 4 March 2014, pp. 13-15.

Commission Implementing Regulation (EU) No 322/2014 of 28 March 2014 imposing special conditions governing the import of feed and food originating in or consigned from Japan following the accident at the Fukushima nuclear power station. OJ L 95, 29 March 2014, pp. 1-11.

Decisions

Commission Decision 76/642/EEC of 9 June 1976 relating to a proceeding under Article 86 of the Treaty establishing the European Economic Community (IV/29.020 – Vitamins). OJ L 223, 16 August 1976, pp. 27-38.

Decision 292/97/EC of the European Parliament and of the Council of 19 December 1996 on the maintenance of national laws prohibiting the use of certain additives in the production of certain specific foodstuffs. OJ L 48, 19 February1997, pp. 13-15.

Council Decision 1999/468/EC of 28 June 1999 laying down the procedures for the exercise of implementing powers conferred on the Commission. OJ L 184, 17 July 1999, pp. 23-26.

Commission Decision 2000/500/EC of 24 July 2000 on authorising the placing on the market of 'yellow fat spreads with added phytosterol esters' as a novel food or novel food ingredient under Regulation (EC) No 258/97 of the European Parliament and of the Council (notified under document number C(2000) 2121). OJ L 200, 8 August 2000, pp. 59-60.

Commission Decision 2000/678/EC of 23 October 2000 laying down detailed rules for registration of holdings in national databases for porcine animals as foreseen by Council Directive 64/432/EEC (notified under document number C(2000) 3075). OJ L 281, 7 November 2000, pp. 16-17.

Commission Decision 2002/623/EC of 24 July 2002 establishing guidance notes supplementing Annex II to Directive 2001/18/EC of the European Parliament and of the Council on the deliberate release into the environment of genetically modified organisms and repealing Council Directive 90/220/EEC (notified under document number C(2002) 2715). OJ L 200, 30 July 2002, pp. 22-33.

Decision No 1786/2002/EC of the European Parliament and of the Council of 23 September 2002 adopting a programme of Community action in the field of public health (2003-2008) – Commission Statements. OJ L 271, 9 October 2002, pp. 1-12.

Council Commission Decision 2002/994/EC of 20 December 2002 concerning certain protective measures with regard to the products of animal origin imported from China (notified under document number C(2002) 5377). OJ L 348, 21 December 2002, pp. 154-156.

Decision 2003/822/EC of 17 November 2003 on the accession of the European Community to the Codex Alimentarius Commission. OJ L 309, 26 November 2003, pp. 14-21.

Commission Decision 2004/478/EC of 29 April 2004 concerning the adoption of a general plan for food/feed crisis management. OJ L 160, 30 April 2004, pp. 98-110.

Commission Decision 2004/613/EC of 6 August 2004 concerning the creation of an advisory group on the food chain and animal and plant health. OJ L 275, 25.8.2004, pp. 17-19.

Commission Decision 2004/858/EC of 15 December 2004 setting up an executive agency, the 'Executive Agency for the Public Health Programme', for the management of Community action in the field of public health – pursuant to Council Regulation (EC) No 58/2003. OJ L 369, 16 December 2004, pp. 73-75.

Commission Decision 2006/255/EC of 14 March 2006 concerning national provisions imposing on supermarkets an obligation to place genetically modified foods on separate shelves from non-genetically modified foods, notified by Cyprus pursuant to Article 95(5) of the EC Treaty (notified under document number C(2006) 797). OJ L 92, 30 March 2006, pp. 12-14.

Council Decision 2007/5/EC,Euratom of 1 January 2007 determining the order in which the office of President of the Council shall be held. OJ L 1, 4 January 2007, pp. 11-12.

Commission Decision 2007/275/EC of 17 April 2007 concerning lists of animals and products to be subject to controls at border inspection posts under Council Directives 91/496/EEC and 97/78/EC (notified under document number C(2007) 1547). OJ L 116, 4 May 2007, pp. 9-33.

Commission Decision 2007/308/EC of 25 April 2007 on the withdrawal from the market of products derived from GA21xMON810 (MON-ØØØ21-9xMON-ØØ81Ø-6) maize (notified under document number C(2007) 1810). OJ L 117, 5 May 2007, pp. 25-26.

Commission Decision 2008/47/EC of 20 December 2007 approving the pre-export checks carried out by the United States of America on peanuts and derived products thereof as regards the presence of aflatoxins (notified under document number C(2007) 6451). OJ L 11, 15 January 2008, pp. 12-16.

Council Decision 2008/486/EC of 23 June 2008 appointing half of the members of the Management Board of the European Food Safety Authority. OJ L 165, 26 June 2008, pp. 8-9.

Commission Decision 2011/242/EU of 14 April 2011 on the members of the advisory group on the food chain and animal and plant health established by Decision 2004/613/EC. OJ L 101, 15 April 2011, pp. 126-128.

Commission Implementing Decision 2011/402/EU of 6 July 2011 on emergency measures applicable to fenugreek seeds and certain seeds and beans imported from Egypt (notified under document C(2011) 5000). OJ L 179, 7 July 2011, pp. 10-12.

Commission Implementing Decision 2011/718/EU of 28 October 2011 amending Implementing Decision 2011/402/EU on emergency measures applicable to fenugreek seeds and certain seeds and beans imported from Egypt (notified under document C(2011) 7744). OJ L 285, 1 November 2011, pp. 53-55.

Commission Implementing Decision 2012/31/EU of 21 December 2011 amending Annex I to Decision 2007/275/EC concerning the lists of animals and products to be subject to controls at border inspection posts under Council Directives 91/496/EEC and 97/78/EC (notified under document C(2011) 9517). OJ L 21, 24 January 2012, pp. 1-29.

Commission Implementing Decision 2012/482/EU of 20 August 2012 amending Decision 2002/994/EC concerning certain protective measures with regard to the products of animal origin imported from China (notified under document C(2012) 5753). OJ L 226, 22 August 2012, pp. 5-5.

Registers, lists and databases

EFSA Register of questions. http://tinyurl.com/q62snj2.

EU feed materials register. http://tinyurl.com/qbenbh4.

EU List of approved Border Inspection Posts. http://tinyurl.com/nsvthaw.

EU List of authorised countries and establishements for import. http://tinyurl.com/l7lg4od.

EU List of Designated Points of Entry. http://tinyurl.com/poqjomu.

EU Pesticide Database. http://tinyurl.com/qhpcd49.

EU platform for diet, physical activity and health: Database. http://tinyurl.com/om5folr.

EU Register for National Guides to Good Practice. http://tinyurl.com/azgbbn2.

EU Register of approved food establishments. http://tinyurl.com/686drh.

EU Register of authorised GMOs. http://tinyurl.com/3xnedx.

EU TRACES database on monitoring of import and intra-EU trade of live animals and other products of animal origin http://tinyurl.com/q7l5yjr.

EFSA decisions and other documents

Alexander, J., Benford, D., Boobis, A., Eskola, M., Fink-Gremmels, J., Fürst, P., Heppner, C., Schlatter, J. and van Leeuwen, R., 2012. Risk assessment of contaminants in food and feed. EFSA Journal 10(10), s1004, pp. 1-12.

Bureau van Dijk Ingénieurs Conseils with Arcadia International EEIG, Evaluation of EFSA, Final Report, Contract FIN0105, Brussels, 5 December 2005. http://tinyurl.com/q986ocl.

EFSA Management Board Decision MB 16 September 2003-11-Adopted, EFSA Code of good administrative behaviour, Brussels, Belgium. http://tinyurl.com/nu7prpy.

EFSA, Openness, Transparency and confidentiality, MB 16 September 03-13. http://tinyurl.com/opadzmq.

EFSA, Decision concerning access to documents. MB 16 September 2003. http://tinyurl.com/pwgmr88

EFSA Management Board Decision MB 10 March 2005 – 10, concerning implementing measures of transparency and confidentiality requirements. http://tinyurl.com/oh22mof.

EFSA Management Board Decision MB 24 January 2006- 4 on implementing rules concerning the tasks, duties and powers of the Data Protection Officer, Parma. http://tinyurl.com/p34kz3y.

EFSA, Risk Communication Strategy and Plans, MB 12 September 2006. http://tinyurl.com/nd4m5gn.

EFSA, Strategy for cooperation and networking between the EU Member States and EFSA, MB 19 December 2006. http://www.efsa.europa.eu/en/keydocs/docs/msstrategy.pdf.

EFSA Management Board, MB 20 June 2006-4, Proposal for Management Board Conclusions of the External Evaluation of EFSA and Recommendations Arising from The Report. http://tinyurl.com/pfv8njh.

EFSA, Opinion of the scientific panel on biological hazards on microbiological criteria and targets based on risk analysis. EFSA Journal 462, 2007, pp. 1-29. http://tinyurl.com/nnrf8gq.

EFSA Management Board Decision MB 11 September 2007–4.2, Proposal to divide the tasks of the AFC panel, 11 September 2007, Bucharest. http://tinyurl.com/opb9ozr.

EFSA Management Board Decision MB 27 March 2008, Decision concerning the operation of the Advisory Forum, Pafos. http://tinyurl.com/q4mb7y3.

EFSA, 2009. Guidelines on submission of a dossier for safety evaluation by the EFSA of active or intelligent substances present in active and intelligent materials and articles intended to come into contact with food. EFSA Journal 1208.

EFSA, Guidance Document on Declarations of Interests, Annex 2, Annual Declaration of Interests, 08 September 2009, Parma. http://tinyurl.com/ohw4olg.

EFSA, Guidance Document on Declarations of Interests, Annex 3, Specific Declaration of Interests, 08 September 2009, Parma. http://tinyurl.com/ohw4olg.

EFSA, Policy on independence and scientific decision making process, MB 15 December 2011. http://tinyurl.com/q5dlk7e.

EFSA Management Board Decision MB 15 March 2012 concerning the establishment and operations of the Scientific Committee, Scientific Panels and of their Working Groups, Parma. http://tinyurl.com/qj7tz52.

EFSA Management Board Decision MB 15 March 2012. Management plan of the European Food Safety Authority for 2013. http://tinyurl.com/ne987tg.

EFSA Panel on Contaminants in the Food Chain (CONTAM), 2012. Scientific Opinion on Mineral Oil Hydrocarbons in Food. EFSA Journal 10(6), 2704, pp. 1-185.

EFSA Management Board Decision MB 27 June 2013. Rules of Procedure of the Management Board of the European Food Safety Authority, Parma. http://tinyurl.com/nqwbmss.

EFSA Management Board Decision MB 19 December 2013, Financial Regulation of the European Food Safety Authority, Parma. http://tinyurl.com/q9lbp77.

EFSA Panel on Contaminants in the Food Chain (CONTAM), 2013. Guidance on methodological principles and scientific methods to be taken into account when establishing Reference Points for Action (RPAs) for non-allowed pharmacologically active substances present in food of animal origin. EFSA Journal 11(4), 3195, pp. 1-24.

EFSA Annual Report 2013, 20 March 2014, Parma. http://tinyurl.com/ontoz9j.

EFSA Management Board Decision MB 26 June 2014, Implementing Rules for the Financial Regulation, Parma. http://tinyurl.com/p96x7fc.

Proposals for new legislation and preparatory documents

Commission proposal for a Council Decision on general conditions to be followed for establishing microbiological criteria for foodstuffs and feedingstuffs, including the conditions for their preparation, in the veterinary, foodstuffs and animal nutrition sectors. OJ C 252/7, 02 October 1981, pp. 7-10.

COM(2000) 716 final. Proposal for a Regulation of the European Parliament and of the Council laying down the general principles and requirements of food law, establishing the European Food Authority, and laying down procedures in matters of food. OJ C 096 E, 27 March 2001, pp. 247-268.

COM(2003) 52 final. Proposal for a Regulation of the European Parliament and of the Council on official feed and food controls. 5 February 2003, pp. 1-115.

COM(2003) 424 final. Proposal for a Regulation of the European Parliament and of the Council on nutrition and health claims made on foods. 16 July 2003.

COM(2005) 115 final. Proposal for a Decision of the European Parliament and of the Council establishing a Programme of Community action in the field of Health and Consumer protection 2007-2013. 06 April 2005.

COM(2007) 872. Proposal for a Regulation of the European Parliament and of the Council on novel foods and amending Regulation (EC) No XXX/XXXX [common procedure]. 14 January 2008, pp. 1-23.

Proposed legislation on pre-market approval schemes of food enzymes, flavourings and additives. http://tinyurl.com/mov5p.

COM(2008)40 final. Proposal for a Regulation of the European Parliament and of the Council on the provision of food information to consumers. 30 January 2008, pp. 1-84.

COM(2010) 184. Proposal for a Council Regulation (EURATOM) laying down maximum permitted levels of radioactive contamination of foodstuffs and of feedingstuffs following a nuclear accident or any other case of radiological emergency (Recast). 27 April 2010, pp. 1-17.

COM(2010) 375 final. Proposal for a Regulation of the European Parliament and of the Council amending Directive 2001/18/EC as regards the possibility for the Member States to restrict or prohibit the cultivation of GMOs in their territory. 13 July 2010, pp. 1-14. (European Parliament 2010/0208 (COD)).

COM(2013) 265 final. Proposal for a Regulation of the European Parliement and of the Council on official controls and other official activities performed to ensure the application of food and feed law, rules on animal health and welfare, plant health, plant reproductive material, plant protection products and amending Regulations (EC) No 999/2001, 1829/2003, 1831/2003, 1/2005, 396/2005, 834/2007, 1099/2009, 1069/2009, 1107/2009, Regulations (EU) No 1151/2012, [....]/2013 [Office of Publications, please insert number of Regulation laying down provisions for the management of expenditure relating to the food chain, animal health and animal welfare, and relating to plant health and plant reproductive material], and Directives 98/58/EC, 1999/74/EC, 2007/43/EC, 2008/119/EC, 2008/120/EC and 2009/128/EC (Official controls Regulation).

COM(2013) 894 final. Proposal for a Regulation of the European Parliament and of the Council on novel foods. 18 December 2013, pp. 1-51.

COM(2013) 813 final. Proposal for a Rirective of the European Parliament and of the Council on the protection of undisclosed know-how and business information (trade secrets) against their unlawful acquisition, use and disclosure. 28 November 2013, pp. 1-26.

EU Case law

ECJ 5 February 1963. Case 26/62, NV Algemene Transport en Expeditie Onderneming Van Gend en Loos v. Nederlandse Administratie der belastingen. European Court Reports (ECR) 1.

ECJ 15 July 1964. Case 6/64, Flaminio Costa v. ENEL, ECR 585.

ECJ 16 June 1966. Case 54/65, Compagnie des forges de Châtillon, Commentry & Neuves-Maisons v High Authority of the ECSC. ECR 185, p. 195.

ECJ 3 April 1968. Case 28/67, Firma Molkerei-Zentrale Westfalen Lippe GmbH v. Hauptzollamt Paderborn. ECR 143, p. 152.

ECJ 7 February 1972. Case 39/72, Commission v. Italy, Premiums for slaughtering cows, ECR 1973, p. 101.

ECJ 11 July 1974. Case 8/74, Procureur du Roi v Benoît and Gustave Dassonville, ECR 1974, p. 837.

ECJ 8 April 1976. Case 43/75, Gabrielle Defrenne v Société anonyme belge de navigation aérienne Sabena, ECR 1976, p. 455.

ECJ 14 July 1976. Combined cases 3, 4 and 6/76, Cornelis Kramer and others. References for a preliminary ruling: Arrondissementsrechtbank Zwolle et Arrondissementsrechtbank Alkmaar – Netherlands. Biological resources of the sea. ECR 1976, p. 1279.

ECJ 1 February 1977. Case 51/76, Verbond van Nederlandse Ondernemingen v Inspecteur der Invoerrechten en Accijnzen. ECR 1977, p. 113.

ECJ 13 February 1979. Case 85/76, Hoffmann – La Roche & Co. AG v. Commission. ECR 1979, p. 461.

ECJ 20 February 1979, Case 120/78, Rewe-Zentral AG v Bundesmonopolverwaltung für Branntwein (Cassis de Dijon), ECR 1979, p. 649.

ECJ 5 April 1979, Case 148/78, Criminal proceedings against Tullio Ratti. ECR 1979, p. 1629.

ECJ 10 November 1982, Case 261/81, Walter Rau Lebensmittelwerke v De Smedt PVBA. ECR 1982, p. 3961.

ECJ 25 April 1985. Case 207/83, Commission of the European Communities v United Kingdom of Great Britain and Northern Ireland. ECR 1985, p. 1201.

ECJ 12 March 1987. Case 178/84 Commission v Germany ('Reinheitsgebot' or 'German beer'). ECR 1987, p. 1227.

ECJ 8 October 1987. Case 80/86 Criminal proceedings against Kolpinghuis Nijmegen BV. ECR 1987, p. 3969.

ECJ 19 November 1991. Joined Cases C-6/90 and C-9/90, Andrea Francovich and Danila Bonifaci and others v Italian Republic. ECR 1991, I-5357.

ECJ 14 July 1994. Case 17/93, Criminal proceedings against J.J.J. Van der Veldt. ECR 1994, I-03537

ECJ 6 July 1995. Case 470/93, Verein gegen Unwesen in Handel und Gewerbe Köln e.V. v Mars GmbH. ECR 1995 I-01923.

ECJ 5 March 1996. Joined Cases C-46/93 and C-48/93, Brasserie du Pêcheur SA v Bundesrepublik Deutschland and The Queen v Secretary of State for Transport, ex parte: Factortame Ltd and others, ECR 1996, I-01029.

ECJ 5 May 1998. Case C-157/96, The Queen v Ministry of Agriculture, Fisheries and Food, Commissioners of Customs & Excise, ex parte National Farmers' Union, David Burnett and Sons Ltd, R. S. and E. Wright Ltd, Anglo Beef Processors Ltd, United Kingdom Genetics, Wyjac Calves Ltd, International Traders Ferry Ltd, MFP International Ltd, Interstate Truck Rental Ltd and Vian Exports Ltd. ECR 1998, I-02211.

ECJ 5 May 1998. Case C-180/96, U.K. Great Britain and Northern Ireland v. Commission of European Communities, ECR 1996, I-03903.

ECJ 16 July 1998. Case 210/96, Gut Springenheide GmbH and Rudolf Tusky v Oberkreisdirektor des Kreises Steinfurt – Amt für Lebensmittelüberwachung. ECR 1998, I-04657.

ECJ 1 October 1998. Case C-209/96, United Kingdom of Great Britain and Northern Ireland v. Commission of the European Communities. EAGGF – Clearance of accounts – 1992 and 1993 – Beef and veal. ECR 1998, I-05655.

ECJ 13 January 2000. Estée Lauder Cosmetics GmbH & Co. OHG v Lancaster Group GmbH. ECR 2000 I-00117.

ECJ 5 December 2000. Case C-448/98, Criminal proceedings against Jean-Pierre Guimont (Emmenthal cheese case), ECR 2000, I-10663.

ECJ 5 November 2002. Case C-325/00, Commission v. Germany, ECR 2002, I-09977.

ECJ 20 May 2003. Case C-108/01, Consortio del Prociutto di Parma, Salumificio S. Rita SpA v Asda Stores Ltd, Hygrade Foods Ltd, ECR 2003 I-05121.

ECJ 20 May 2003. Cases-469/00 and -108/01, Ravil SARL v Bellon Import SARL and Biraghi SpA, , ECR 2003 I-05053.

ECJ 25 April 2002. Case C-154/00, Commission of the European Communities v Hellenic Republic. ECR 2002, I-03879.

ECJ 10 March 2004. Case T-177/02, Malagutti-Vezinhet SA v.Commission, ECR 2004, II-00827.

ECJ 9 June 2005. Joined Cases C-211/03, C-299/03, C-316/03, C-317/03 to C-318/03. HLH Warenvertriebs GmbH and Orthica BV v Bundesrepublik Deutschland. ECR 2005, I-05141.

CFI 14 December 2005. Case T-383/00, Beamglow Ltd v European Parliament, Council of the European Union and Commission of the European Communities, ECR 2005, II-05459.

ECJ 23 November 2006. Case C-315/05, Lidl Italia Srl v. Comune di Arcole (VR).

ECJ 13 September 2007 Cases C-439/05 and C-454/05 Land Oberösterreich and Republic of Austria v Commission of the European Communities. 2007 I-07141.
ECJ 15 November 2007. Case C-319/05, Commission v. Federal Republic of Germany. ECR 2007, pp. 00000.
CJEU 13 March 2008. Cases 383/06 to 385/06, Vereniging Nationaal Overlegorgaan Sociale Werkvoorziening (C-383/06) and Gemeente Rotterdam (C-384/06) v Minister van Sociale Zaken en Werkgelegenheid and Sociaal Economische Samenwerking West-Brabant (C-385/06) v Algemene Directie voor de Arbeidsvoorziening. ECR 2008, p. 0000.
CJEU 9 December 2008. Case C-121/07 Commission of the European Communities v French Republic. ECR 2008 I-09159.
CJEU 2 April 2009, Case C-421/07, Criminal proceedings against Frede Damgaard. ECR 2009, I-02629.
CJEU 16 July 2009. Case C-165/08 Commission of the European Communities v Republic of Poland, ECR 2009 I-06843.
CJEU 6 September 2010. Case 544/10, Deutsches Weintor eG v. Land Rheinland-Pfalz.
CJEU 22 December 2010. Bavaria NV v Bayerischer Brauerbund eV, 2010 I-13393.
CJEU 11 April 2013. Case C-636/11, Karl Berger v. Freistaat Bayern.
CJEU 18 July 2013. Case C-313/11 European Commission v Republic of Poland.

Soft law, policy documents, reports and general information

Inter-institutional

Spaak Report: Rapport des Chefs de Délégations aux Ministres des Affaires Etrangères. Secretariat of the Intergovernmental Conference, 21 April 1956, Brussels. http://tinyurl.com/lhybf27.
The interinstitutional agreement on better law-making of the European Parliament, Council and Commission. OJ C 321, 31 December 2003, pp. 1-5.

Council and European Parliament

Council of Europe. Convention for the Protection of Human Rights and Fundamental Freedoms, Rome, 4 November 1950. http://tinyurl.com/48lhefh.
Council of Europe. European Social Charter. 1961. http://www.coe.int/T/DGHL/Monitoring/SocialCharter/.
Council Convention 80/934/EEC on the law applicable to contractual obligations opened for signature in Rome on 19 June 1980. OJ L 266, 9 October 1980, pp. 1-19.
Council of the European Communities, Resolution of the Council and of the representatives of the Governments of the Member States, meeting within the Council of 3 December 1990 concerning an action programme on nutrition and health. OJ C 329, 31 December 1990, pp. 1-3.
Council of the European Communities, Conclusions of the Council and the Ministers for Health of the Member States, meeting within the Council on 15 May 1992 on nutrition and health. OJ C 148, 12 June 1992, p. 2.

Council of Europe, Parliamentary Assembly, Report on food and health, 12 April 1994, Doc. 7083. http://tinyurl.com/l9xte7q.

Council of Europe, Parliamentary Assembly. Recommendation 1244 (1994) on food and health. 28 June 1994. http://tinyurl.com/lgu8cwj.

Temporary Committee of Inquiry into BSE, Minutes of proceedings of the sitting of Thursday. 18 July 1996. OJ C 261.

European Council, Presidency Conclusions Luxembourg Summit 1997, 13 December 1997, Nr. 57. http://tinyurl.com/lgokgct.

Manuel Medina Ortega, Report of the Temporary Committee of Inquiry into BSE, set up by the Parliament in July 1996, on the alleged contraventions or maladministration in the implementation of Community law in relation to BSE, without prejudice to the jurisdiction of the Community and the national courts, 7 February 1997, A4-0020/97/A, PE 220.544/fin/A. http://www.mad-cow.org/final_EU.html.

European Parliament, Fact Sheets 4.10.1. Consumer Policy: principles and instruments, Chapter 3 Reform in the wake of the BSE crisis. http://tinyurl.com/nxj6tl8.

Council of the European Union. Council Resolution of 14 December 2000 on health and nutrition. OJ C 20, 23 January 2001, pp. 1-2.

Council of the European Union. Council Conclusions of 2 December 2002 on obesity. OJ C 11, 17 January 2003, p. 3.

Council of the European Union, Council conclusions of 14 May 2004 on promoting heart health. Council document 9627/04, 18 May 2004. http://tinyurl.com/mf33vlb.

European Parliament. Written question E- 2704/04 by Willi Piecyk (PSE) to the Commission. 19 October 2004.

European Parliament. Antwort E-2704/04DE von Herrn Kyprianou im Namen der Kommission. 13 December 2004.

Council of the European Union, Council conclusions of 3 June 2005 on obesity, nutrition and physical activity. Council document 9803/05, 06 June 2005. http://tinyurl.com/m7qkgc3.

Council of the European Union, Council conclusions on the Commission White Paper on a strategy for Europe on nutrition, overweight and obesity-related issues. Council document 15612/07, 5 December 2007, p. 2, 3 et seq. http://tinyurl.com/mtpljnv.

European Parliament, Resolution of 25 September 2008 on the White Paper on nutrition, overweight and obesity-related health issues (2007/2285(INI)). OJ C 8E, 14 January 2010, p. 97.

European Parliament, Position of the European Parliament COD (2010) 208 adopted at first reading on 5 July 2011 with a view to the adoption of Regulation (EU) No .../2011 of the European Parliament and of the Council amending Directive 2001/18/EC as regards the possibility for the Member States to restrict or prohibit the cultivation of GMOs in their territory. 15 July 2011.

Council of the European Union, Council conclusions of 2 December 2003 on healthy lifestyles: education, information and communication. OJ C 22, 27 January 2004, p. 2.

European Commission, agencies and comitology

AFC Management Consulting and CO Concept Marketing Consulting, Evaluation of the School Fruit Scheme Final Report. October 2012, p. 104, 105, 122..http://tinyurl.com/m62xq8a.

COM(85) 310 Final. White Paper Completing the Internal Market. 14 June 1985. http://tinyurl.com/kkflba9

COM(1993) 559 final, Commission communication on the framework for action in the field of public health. 24 November 1993, p. 2, 10, 15, 16.

COM(1997) 176. Commission Green Paper- The General Principles of Food Law in the European Union. 30 June 1997.

COM(97) 183 final. Communication of the European Commission- Consumer Health and Food Safety. 30 April 1997. http://tinyurl.com/owmb8tc.

COM(1998) 230 final. Communication from the Commission to the Council, the European Parliament, the Economic and Social Committee and the Committee of the Regions on the development of public health policy in the European Community. 15 April 1998.

COM(1999) 719 final. White Paper on Food Safety. 12 January 2000. http://tinyurl.com/68mlwn.

COM(2000) 1 final. Communication on the use of the precautionary principle. 2 February 2000. http://tinyurl.com/6du3tm.

COM(2000) 285 final. Communication from the Commission to the Council, the European Parliament, the Economic and Social Committee and the Committee of the Regions on the health strategy of the European Community. 15 June 2000.

COM(2001) 428 final. European Governance. A White Paper. 12 October 2001.

COM(2001) 593. Communication from the commission to the council, the European Parliament and the Economic and Social Committee. Community Strategy for Dioxins, Furans and Polychlorinated Biphenyls. 17 November 2001. http://tinyurl.com/pt422mm.

COM(2005) 9 final. Report from the European Commission to the Council and the European Parliament on the Possibility of Introduction of Electronic Identification for Bovine Animals. 25 January 2005.

COM(2005) 637 final. Green Paper- Promoting healthy diets and physical activity: a European dimension for the prevention of overweight, obesity and chronic diseases. 8 December 2005.

COM(2007) 279 final. White Paper on a Strategy for Europe on Nutrition, Overweight and Obesity related health issues. 30 May 2007, p. 3, 4, 5, 6, 7, 9.

COM(2007) 391 final. White Paper on Sport. 17 July 2007.

COM(2007) 630 final. White Paper – Together for Health: A Strategic Approach for the EU 2008-2013. 23 October 2007.

COM(2007) 640 final. Communication from the Commission to the European Parliament, the Council, the European Economic and Social Committee and the Committee of the Regions, Commission Legislative and Work Programme 2008, Brussels, 23.10.2007.

COM(2009) 215 final. Communication from the Commission to the European Parliament, The Council, The European Economic and Social Committee and the Committee of the Regions. Contributing to Sustainable Development: The role of Fair Trade and non-governmental trade-related sustainability schemes. 5 May 2009.

COM(2009) 567 final. Communication from the Commission to the European Parliament, The Council, The European Economic and Social Committee and the Committee of the Regions. Solidarity in health: Reducing health inequalities in the EU. 20 October 2009.

COM(2010) 785. Report from the Commission to the European Parliament and the Council on the effectiveness and consistency of sanitary and phytosanitary controls on imports of food, feed, animals and plants, 21 December 2010.

COM(2012) 203 final. First Report from the Commission to the European Parliament, the Council, the European Economic and Social Committee and the Committee of the Regions on the application of Directive 2010/13/EU 'Audiovisual Media Service Directive', 04 May 2012, p. 9.

Commission of the European Communities, Bulletin of the European Union. http://tinyurl.com/lezzfmu.

Commission of the European Communities. Communication from the Commission concerning the consequences of the judgment given by the Court of Justice on 20 February 1979 in case 120/78 ('Cassis de Dijon'). OJ C 256, 03 October 1980, pp. 2-3. http://tinyurl.com/nwc9qdn.

Commission of the European Communities. Draft: Communication from the Commission, concerning the creation of an advisory group on the food chain and animal and plant health and the establishment of a consultation procedure on the food chain and animal and plant health through representative European bodies, no date, but prepared Commission Decision (2004/613/EC). OJ L 275, 25 August 2004, pp. 17-19.

Commission Recommendation 97/618/EC of 29 July 1997 concerning the scientific aspects and the presentation of information necessary to support applications for the placing on the market of novel foods and novel food ingredients and the preparation of initial assessment reports under Regulation (EC) No 258/97 of the European Parliament and of the Council. OJ L 253, 16 September 1997, pp. 1-36.

Commission Recommendation 2003/556/EC of 23 July 2003 on guidelines for the development of national strategies and best practices to ensure the coexistence of genetically modified crops with conventional and organic farming (notified under document number C(2003) 2624). OJ L 189, 29 July 2003, pp. 36-47.

Commission Recommendation 2005/108/EC of 4 February 2005 on the further investigation into the levels of polycyclic aromatic hydrocarbons in certain foods (notified under document number C(2005) 256). OJ L 34, 8 February 2005, pp. 43-45.

Commission Recommendation 2006/88/EC of 6 February 2006 on the reduction of the presence of dioxins, furans and PCBs in feedingstuffs and foodstuffs (notified under document number C(2006) 235). OJ L 42, 14 February 2006, pp. 26-28.

Commission Recommendation 2010/307/EU of 2 June 2010 on the monitoring of acrylamide levels in food. OJ L 137, 3 June 2010, pp. 4-10.

Commission Communication — EU best practice guidelines for voluntary certification schemes for agricultural products and foodstuffs, OJ C 341, 16 December 2010.

Commission Recommendation 2013/647/EU of 8 November 2013 on investigations into the levels of acrylamide in food. OJ L 301, 12 November 2013, pp. 15-17.

DG SANCO. Scientific committee on veterinary measures relating to public health. Opinion on the Evaluation of microbiological criteria for food products of animal origin for human consumption. 23 September 1999. http://tinyurl.com/pttbdxv.

DG SANCO. Discussion Paper on Nutrition Claims and Functional Claims. SANCO/1341/2001. http://tinyurl.com/qh3k43y.

DG SANCO. FVO at Home and Away. Consumer Voice 7/02, September 2002. http://tinyurl.com/p72xjh4

DG SANCO. Status report on the European Commission's work in the field of nutrition in Europe October 2002. 2003, p. 5, 7, 23 et seq., 24, Annexes III and IV. http://tinyurl.com/lxf3tqc.

DG SANCO, Opinion of the scientific committee on veterinary measures related to public health on verotoxicogenic *E. coli* (VTEC) in foodstuff. 21-22 January 2003. http://tinyurl.com/m4seksm.

DG SANCO. Opinion of the scientific committee on veterinary measures relating to public health on staphylococcal enterotoxins in milk products, particularly cheeses. 26-27.03.2003. http://tinyurl.com/mzaa2ft.

DG SANCO. Self-Regulation in the EU Advertising Sector: A report of some discussion among interested parties. July 2006. http://tinyurl.com/q3u52fp.

DG SANCO, Healthy democracy. Conclusions and Actions following the DG SANCO 2006 Peer Review Group on Stakeholder Involvement. February 2007. http://tinyurl.com/mgn8cn2.

DG SANCO. Guidance document on certain key questions related to import requirements and the new rules on food hygiene and on official food controls. 2005, revision 2014. http://tinyurl.com/mo6r9xz.

DG SANCO. Discussion paper on strategy for setting microbiological criteria for foodstuffs in Community legislation. 8 March 2005. http://tinyurl.com/m79mz6m.

DG SANCO. Guidance document on the implementation of procedures based on the HACCP principles, and on the facilitation of the implementation of the HACCP principles in certain food businesses. 16 November 2005. http://tinyurl.com/pomg2.

DG SANCO. The Rapid Alert System for Food and Feed (RASFF) Annual report 2006. http://tinyurl.com/ngx5jkn.

DG SANCO. Stakeholder Involvement Event, The Report. 23 May 2007. http://tinyurl.com/m4ygkfs.

DG SANCO. Safer Food for Europe – over a decade of achievement for the FVO. Health & Consumer Voice, November 2007. http://tinyurl.com/putw2tn.

DG SANCO. 03 - Science and stakeholder relations, Joint meeting of the stakeholder dialogue group and the advisory group on the food chain and animal and plant health, summary record. 30 November 2007. http://tinyurl.com/qea3498.

DG SANCO. Guidance document on the implementation of certain provisions of Regulation (EC) No 853/2004 on the hygiene of food of animal origin. 16 February 2009. http://tinyurl.com/ntw9xdx.

DG SANCO. Strategy for Europe on nutrition, overweight and obesity related health issues. Implementation Progress Report. December 2010, p.8, 17, 28, 34, 35, Annex VI. http://tinyurl.com/3xs3cwt.

DG SANCO. EU guidelines on conditions and procedures for the import of polyamide and melamine kitchenware originating in or consigned from People's Republic of China and Hong Kong Special Administrative Region, China. 10 June 2011. http://tinyurl.com/jwysyn2.

DG SANCO. EU Guidance to the Commission Regulation (EC) No 450/2009 of 29 May 2009 on active and intelligent materials and articles intended to come into contact with food. 23 November 2011. http://tinyurl.com/cv7kv47.

DG SANCO. Survey on Members States' Implementation of the EU Salt Reduction Framework. 2012, p. 5. http://tinyurl.com/bxj59qn.

DG SANCO. Guidance document on the implementation of certain provisions of Regulation (EC) No 852/2004on the hygiene of foodstuffs. 18 June 2012. http://tinyurl.com/qrzus.

DG SANCO. Reformulation Frameworks Progress in the Member States, Presentation of the European Commission to the Platform on Diet, Physical Activity and Health. 28 February 2013. http://tinyurl.com/kdxyqlp.

DG SANCO. The Rapid Alert System for Food and Feed (RASFF) Annual report 2013. http://tinyurl.com/p2hfzaj.

DG SANCO. Union Guidelines on Regulation (EU) No 10/2011 on plastic materials and articles intended to come into contact with food. 21 February 2014. http://tinyurl.com/m26nkgh.

DG SANCO. Technical Specifications in Relation to the Master List and the Lists of EU Approved Food Establishments (SANCO/2179/2005 Revision 2014). 10 April 2014. http://tinyurl.com/kkgsfkj.

European Commission. Principles for the development of microbiological criteria for animal products and products for animal origin for human consumption. September 1997. http://tinyurl.com/m2422se.

European Commission. Registration of Prosciutto di Parma. 12 May 1999. http://tinyurl.com/p9nbsaw.

European Commission. Summary of the national legislation on food contact materials. http://tinyurl.com/ps6z29q.

European Commission. Guidelines for the development of community guides to good practice for hygiene and for the application of the HACCP principles, in accordance with Article 9 of Regulation (EC) No 852/2004 on the hygiene of foodstuffs and Article 22 of Regulation (EC) No 183/2005 laying down requirements for feed hygiene. http://tinyurl.com/kke6uwg.

European Commission. Enterprise Directorate-General. Notice to applicants F2/AW D(2003). Guideline on the categorisation of extenson applications (ea) versus variations applications (v) medicinal products for human and veterinary use. October 2003. http://tinyurl.com/mju5cwz.

European Commission. The fight against obesity. Examples of EU projects in the field of nutrition and obesity. 2005. http://tinyurl.com/pnub3sz.

European Commission. EU Platform on Diet, Physical Activity and Health. First Monitoring Progress Report. 2006, pp. 3. http://tinyurl.com/nz2wesl.

European Commission. Impact assessment accompanying the document Proposal for a Regulation of the European Parliament and of the Council on food intended for infants and young children and on food for special medical purpose. http://tinyurl.com/lkvoaam.

European Commission. High Level Group on Nutrition and Physical Activity. EU Framework for national salt initiatives. 2008. http://tinyurl.com/kt2fw7y.

European Commission. Guidance document for Competent Authorities for the control of compliance with EU legislation on aflatoxins, 2010. http://tinyurl.com/q7utxkq.

European Commission. EU Platform on Diet, Physical Activity & Health and High Level Group on Nutrition and Physical Activity. Flash Report. 28 November 2011. http://tinyurl.com/pxkndxp.

European Commission. EU Platform on Diet, Physical Activity and Health. 2012 Annual Report. 2012., p. VII. http://tinyurl.com/p26dkua.

European Commission. High Level Group on Nutrition and Physical Activity. Flash Report. 2 February 2012. http://tinyurl.com/ort74sj.

European Commission. Community Guide to the Principles of Good Practice for the Microbiological Classification and Monitoring of Bivalve Mollusc Production and Relaying Areas with regard to Regulation 854/2004. June 2012. http://tinyurl.com/poze8kt.

European Commission. High Level Group on Nutrition and Physical Activity, Flash Report, 15 November 2012. http://tinyurl.com/muzt43w.

European Commission. High Level Group on Nutrition and Physical Activity, Flash Report, 7 February 2013, http://tinyurl.com/b6anjmx.

European Commission. Commission allocates € 88 million for 2013/2014 School Fruit Scheme. 16 March 2013. http://tinyurl.com/cgudj8l.

European Commission, New strengthened rules for food for infants, young children and food for specific medical purpose. Brussels, 11 June 2013. http://tinyurl.com/q3xnbvc.

Evaluation Partnership, DG SANCO. Evaluation of the European Platform for Action on Diet, Physical Activity and Health, Case Study Report: Advertising and marketing to children. July 2010, p. 4, 49, 50. http://tinyurl.com/3abvqvz.

Standing Committee on the Food Chain and Animal Health. Guidance on the Implementation of Articles 11, 12, 16, 17, 18, 19 and 20 of Regulation (EC) no. 178/2002 on General Food Law. 2004. http://tinyurl.com/m41266d.

Standing Committee on the Food Chain and Animal health (section Toxicological Safety). Modus Operandi for the management of new food safety incidents with a potential for extension involving a chemical substance. January 2006. http://tinyurl.com/lz4cfpb.

Study Group on a European Civil Code, Principles, Definitions and Model Rules of European Private Law. Draft Common Frame of Reference (DCFR). Sellier European Law Publishers, Munich, 2009. http://tinyurl.com/3kldjxf.

European Economic and Social Committee

Opinion of the European Economic and Social Committee on 'Hygiene rules and artisanal food processors' (2006/C 65/25). OJ C 65, 17 March 2006, pp. 141-148.

Other sources of law

International documents

Biosafety Clearing House (OECD). http://tinyurl.com/n5xajsl.

Charter of the United Nations (UN). http://tinyurl.com/oqygw.

Codex Alimentarius Commission. General Principles of Food Hygiene, CAC/RCP 1-1969. http://tinyurl.com/o2wezqm.

Codex Alimentarius Commission. Code of Ethics for International Trade in Food including Concessional and Food Aid Transactions. CAC/RCP 20-1979. http://tinyurl.com/o2wezqm.

Codex Alimentarius Commission. General Standard for the Labelling of Prepackaged Foods. CODEX STAN 1-1985. http://tinyurl.com/o2wezqm.

Codex Alimentarius Commission. General Standard for the Labelling of and Claims for Prepackaged Foods for Special Dietary Uses. Codex STAN 146-1985. http://tinyurl.com/o2wezqm.

Codex Alimentarius Commission. General Standard for Contaminants and Toxins in Food and Feed, CODEX STAN 193-1995. http://tinyurl.com/o2wezqm.

Codex Alimentarius Commission. Principles and Guidelines for the establishment and application of microbiological criteria related to foods, CAC/GL 21 - 1997. http://tinyurl.com/o2wezqm.

Codex Alimentarius Commission. Guidelines for Use of Nutrition and Health Claims. CAC/GL 23-1997. http://tinyurl.com/o2wezqm.

Codex Alimentarius Commission. Codex Alimentarius Commission, General Guidelines for Use of the Term Halal. CAC/GL 24-1997. http://tinyurl.com/o2wezqm.

Codex Alimentarius Commission. Principles and Guidelines for the Conduct of Microbiological Risk Assessment, CAC/GL-30- 1999. http://tinyurl.com/o2wezqm.

Codex Alimentarius Commission. Food Hygiene, Basic Texts, Third Edition, 2003. http://tinyurl.com/mu7mly3.

Codex Alimentarius. Codex Alimentarius Commission, Procedural Manual, 22nd edition, Rome 2014. http://tinyurl.com/mems9qh.

Convention establishing an International Organisation of Legal Metrology. Paris – France, 12 October 1955. Official translation in English made by the British Government and published under Treaty series No 60 of 1962. http://tinyurl.com/kslp9dm.

European Patent Office. Agreement on a Unified Patent Court. http://www.epo.org/law-practice/unitary/patent-court.html.

European Patent Office and the Office for Harmonization in the Internal Market, 2013. Intellectual Property Rights intensive industries: contribution to economic performance and employment in Europe. http://tinyurl.com/ktwno64.

International Trademark Association. Factsheet Introduction to Trademarks. http://tinyurl.com/nlzsep9.

International Union for the Protection of New Varieties of Plants, 1961. International Convention for the Protection of New Varieties of Plants (UPOV). Paris, 2 December 1961. http://tinyurl.com/q29fwe8.

FAO. Committee on Agriculture, Development of a Framework for Good Agricultural Practices. Seventeenth Session, Rome, 31 March-4 April 2003. http://tinyurl.com/lw5u4ew.

FAO. Secretariat of the International Plant Protection Convention. International Plant Protection Convention 1997. 2011. http://tinyurl.com/ktedeuz.

FAO. Voluntary Guidelines to support the progressive realization of the right to adequate food in the context of national food security. Rome, 2005. http://tinyurl.com/cmlspb.

FAO/WHO. Joint FAO/WHO Expert Committee on Nutrition. Report on the second session (WHO Technical Report Series, No. 44). Rome, 10-17 April 1951, p. 43. http://tinyurl.com/m6myeav.

FAO/WHO. International Conference on Nutrition, Final Report of the Conference. Rome, December 1992, p. 7, 13, 24, 45, 46 and seq. http://tinyurl.com/k462xqg.
FAO/WHO. Joint WHO/FAO Expert Consultation on Diet, Nutrition and the Prevention of Chronic Diseases. Geneva, 28 January-1 February 2002. http://tinyurl.com/mbx99rs.
FAO/WHO. Report of the Evaluation of the Codex Alimentarius and other FAO And WHO Food Standards Work. 15 November 2002. http://tinyurl.com/k2h7w8k.
United Nations. Committee on Economic, Social and Cultural Rights, General Comment nr. 12. http://tinyurl.com/qxqfpj5.
United Nations. Charter. San Francisco, 26 June 1945. http://tinyurl.com/oqygw.
United Nations. Universal Declaration of Human Rights, 1948. http://tinyurl.com/cgnkmq.
United Nations. Convention on the Prevention and Punishment of the Crime of Genocide. 12 January 1951. http://tinyurl.com/looz6b2.
United Nations. United Nations Single Convention on Narcotic Drugs, 1961. https://www.unodc.org/pdf/convention_1961_en.pdf.
United Nations. International Covenant on Civil and Political Rights. 16 December 1966. http://tinyurl.com/lh96aub.
United Nations. General Assembly Resolution 2205 (XXI) establishing the United Nations Commission on International Trade Law, 17 December 1966. http://tinyurl.com/q8dxmv6.
United Nations. International Covenant on Economic, Social and Cultural Rights. 16 December 1966. http://tinyurl.com/mrlemdj.
United Nations. Convention on the Elimination of All Forms of Discrimination against Women, 18 December 1979. http://tinyurl.com/bre7p.
United Nations, Convention on the Rights of the Child. 20 November 1989. http://tinyurl.com/nvdsrdy.
United Nations. Convention on Biological Diversity. 1992. http://tinyurl.com/6h3g26u.
United Nations. Cartagena Protocol on Biosafety to the Convention on Biological Diversity. OJ L 201, 31 July 2002, pp. 50-55. http://tinyurl.com/jwuze6j.
United Nations. The Millennium Development Goals Report 2006. New York, 2006. http://tinyurl.com/mfm7y6y.
United Nations. Optional Protocol to the International Covenant on Economic, Social and Cultural Rights. New York, 10 December 2008. http://tinyurl.com/k32x944.
United Nations. Convention on Contracts for the International Sale of Good. New York, 2010. http://tinyurl.com/d84dwxh.
United Nations. General Assembly Resolution on the Political Declaration of the High-level Meeting of the General Assembly on the Prevention and Control of Non-communicable Diseases, A/RES/66/2. 19 September 2011. http://tinyurl.com/mn4xzof.
WHO. Report of an Expert Committee. Arterial hypertension and ischaemic heart disease: preventive aspects, Report of an Expert Committee (WHO Technical Report Series, No. 231). Geneva, 1962. http://tinyurl.com/jwbz39e.
WHO. Arterial hypertension, Report of a WHO Expert Committee (WHO Technical Report Series, 1978 No. 628). Geneva, 1978, p. 40 and 46. http://tinyurl.com/n25f55j.
WHO. Prevention of coronary heart disease, Report of a WHO Expert Committee (WHO Technical Report Series, 1982 No. 678). Geneva, 1982, pp. 5, 12 et seq.

WHO. Diet, nutrition and the prevention of chronic diseases, Report of a WHO Study Group (WHO Technical Report Series, No. 797), Geneva, 1990. http://tinyurl.com/ktdkxae.

WHO, Regional Office for Europe. European food and nutrition policies in action (WHO Regional Publications, European Series,1998 No. 73), 1998, p. 3. http://tinyurl.com/mzggawo.

WHO. Fifty-third World Health Assembly, WHA53.17, Agenda item 12.11, Prevention and Control of Non- Communicable diseases. 20 May 2000. http://tinyurl.com/kbgvwm6.

WHO, Regional Office for Europe. The First Action Plan for Food and Nutrition Policy, WHO European Region, 2000-2005. 2001, p. 1, 41. http://tinyurl.com/l4esf7h.

WHO, Regional Office for Europe. Progress Report on the first action plan for food and nutrition policy in WHO European region 2000-2005. 2002. http://tinyurl.com/knp5sy4.

WHO, Regional Office for Europe.Comparative analysis of food and nutrition policies in WHO European Member States. 2003. http://tinyurl.com/nuoqde8.

WHO. Diet, nutrition and the prevention of chronic diseases, Report of a WHO Study Group (WHO Technical Report Series, No. 916), Geneva, 2003. http://tinyurl.com/m8klmtg.

WHO. Global Strategy on Diet, Physical Activity and Health. 2004. http://tinyurl.com/mj7c4oz.

WHO. World International health regulations 2005, 2nd edition. 2008. http://tinyurl.com/lsx22g5.

WHO. Constitution of the World Health Organization, Basic Documents, 45th edition, Supplement. October 2006. http://tinyurl.com/2yqzar.

WHO. Ministerial Conference on Counteracting Obesity, European Charter on Counteracting Obesity. Istanbul, 15- 17 November 2006, p. 1, 2, 4 and seq. http://tinyurl.com/qht4rt6.

WHO, Regional Office for Europe. Action Plan for Food and Nutrition Policy 2007-2012. 2008, p. 4, 6, 17. http://tinyurl.com/kvugf9h.

WHO. Action Plan for the Global Strategy for the Prevention and Control of Non- Communicable Diseases 2008-2013. 2008, p.2. http://tinyurl.com/lfsa28m.

WHO. Set of recommendations on the marketing of foods and non-alcoholic beverages to children. 2010, p. 4. http://tinyurl.com/cewsq3k.

WHO, Regional Committee for Europe. Health 2020 policy framework and strategy. Malta, 10-13 September 2012, p. 22 et seq. http://tinyurl.com/kzop4xf.

WHO. Global Action Plan for the Prevention and Control of Non-communicable Diseases 2013-2020. 2013, p. 25. http://tinyurl.com/ptbqa4j.

WHO/FAO. The International Food Safety Authorities Network (INFOSAN) Users Guide, October 2006.

WHO/FAO. Understanding the Codex Alimentarius, Rome, 2005. http://tinyurl.com/6rfda3.

World Intelectual Property Organisation. Paris Convention for the Protection of Industrial Property, 20 March 1883. http://tinyurl.com/owy2rkw.

World Intelectual Property Organisation. Berne Convention for the Protection of Literary and Artistic Works. 9 September 1886. http://tinyurl.com/p2tlsro.

World Intelectual Property Organisation. Lisbon Agreement for the Protection of Appellations of Origin and their International Registration. Stockholm, 31 October 1958. http://tinyurl.com/oomnkl4.

World Intelectual Property Organisation. Patent Cooperation Treaty. Washington, 19 June 1970. http://tinyurl.com/kvwu45.

World Intelectual Property Organisation. Strasbourg Agreement Concerning the International Patent Classification. 24 March 1971. http://tinyurl.com/nkchjey.
World Intelectual Property Organisation. Convention Establishing the World Intellectual Property Organization, Stockholm, 14 July 1967, amended 28 September 1979. http://tinyurl.com/obdsn7w.
WTO. Agreement Establishing the World Trade Organization. http://www.wto.org/english/docs_e/legal_e/final_e.htm. Annex II, Dispute Settlement Understanding. http://tinyurl.com/n3cfffn.
WTO. Agreement on Agriculture. http://tinyurl.com/q2wjbk6.
WTO. Agreement on the Application of Sanitary and Phytosanitary Measures (WTO). http://tinyurl.com/lt7vzma.
WTO. Agreement on Technical Barriers to Trade. http://tinyurl.com/llwtb2u.
WTO. Agreement on Trade Related Aspects of Intellectual Property Rights. http://tinyurl.com/azwjj.
WTO. General Agreement on trade in Services. http://tinyurl.com/l8ta35u.
WTO. General Agreement on Tariffs and Trade. http://tinyurl.com/k7qkf6c.

International case law

EC hormones case WT/DS48/AB/R. WTO appellate body. 16 January 1998. http://tinyurl.com/bzjkepx.
Second EC hormones case WT/DS320/R. WTO dispute settlement body. 31 March 2008. http://tinyurl.com/kz4u7p2.
Tuna case WT/DS381/AB/R (US-Tuna II). Measures concerning the importation, marketing and sale of tuna and tuna products. WTO Appellate Body. 16 May 2012. http://tinyurl.com/lskvdh8.

Private standards

BRC. http://tinyurl.com/2vo93lr.
Dutch HACCP. http://tinyurl.com/l5vum9k.
Fair trade. http://tinyurl.com/kfjua66.
GFSI. http://tinyurl.com/6m8x9dx.
GlobalGAP (EurepGAP). http://tinyurl.com/kfvs723.
ICC Official Rules for the Interpretation of Trade Terms, ICC Publication No. 560. 2000. http://tinyurl.com/kg393f8.
IFS. http://tinyurl.com/monoael.
ISO 22.000. http://tinyurl.com/k3c5sh7.
SQF. http://tinyurl.com/lpyxv6m.

EU Member States' and third countries' legislation

Austrian Regulation of 20 August 2009 on the Content of Trans Fatty Acids in Foods.

Constitution of the Republic of South Africa, No. 108 of 1996, 18 December 1996. http://tinyurl.com/kroxe8d.

Danish Executive Order No. 160 of 11 March 2003 on the Content of Trans Fatty Acids in Oils and Fats.

Lebensmittel-, Bedarfsgegenstände- und Futtermittelgesetzbuch (German Code on food, food contact materials and feed). http://tinyurl.com/kzpmue9.

Statutory Instruments 2005 No. 2059 Food, United Kingdom. The Food Hygiene (England) Regulations 2005, Made 21 July 2005.

EU Member States' and third countries' case law

Bundesgerichtsentscheid (Swiss Federal Court). 27 October 1995, 121 I 367, Urteil der II. öffentlichrechtlichen Abteilung vom, i.S. V. gegen Einwohnergemeinde X. und Regierungsrat des Kantons Bern (staatsrechtliche Beschwerde),. http://tinyurl.com/p8znj49.

Constitutional Court of South Africa 4 October 2000. Case CCT 11/00, Government *et al.* v. Grootboom *et al.* http://tinyurl.com/n9cum82.

Hoge Raad (Dutch Supreme Court) 2 February 1990. Case NJ 1990/794 (Bekkers/Staat).

Hoge Raad (Dutch Supreme Court) 23 November 1990. Case NJ 1991/92 (Joemman/Staat).

United States Court. Case US 303 (1980). Diamond v. Chakrabarty 447.

Literature

Agra CEAS Consulting, 2009. Study report of 29 April 2009 by Agra CEAS Consulting concerning the revision of Directive 2009/39/EC.

Ahluwalia, P., 2004. The implementation of the right to food at the national level: a critical examination of the Indian Campaign on the right to food as an effective operationalization of Article 11 of ICESCR. Center for Human Rights and Global Justice Working Paper, Economic, Social And Cultural Rights Series, No. 8.

Alemanno, A., 2010. L'approche européenne de la sécurité des importations- Concilier protection des consommateurs et accès au marché après l'affaire du lait chinois frelaté. Revue du Droit de l'Union Européenne 3, pp. 527-548.

Ammendrup, S. and Füssell, A.E., 2001. Legislative requirements for the identification and traceability of farm animals within the European Union. Revue Scientifique et Technique (International Office of Epizootics) 20, pp. 437-444.

Beekman, V., Kornelis, M., Pronk, A., Teeuw, J. and Smelt, A.J., 2006. Stimulering eigen verantwoordelijkheid. Zorgen dat producenten en consumenten zorgen voor voedselveiligheid. Rapport 5 June 2005, LEI, The Hague, the Netherlands.

Bergsma, N., 2010. Voedselveiligheid: certificatie en overheidstoezicht, Praktijkgids. Warenwet Sdu, The Hague, the Netherlands.

Berends, G. and Carreno, I., 2005. Safeguards in food law – ensuring food scares are scarce. European Law Review 30, pp. 386-405.

Bevilacqua, D., 2006. The Codex Alimentarius Commission and its influence on European and National Food Policy. EFFL 1, pp. 3-16.

Boin, G., 2013. Horsemeat crisis about to tighten French Food Law. EFFL 4, pp. 247-249.

Bondt, N., Deneux, S.D.C., Van Dijke, I., Jong, O. de, Smelt, A., Splinter, G.M., Tromp, S.O. and De Vlieger, J.J, 2006. Voedselveiligheid, ketens en toezicht op controle. Rapport 5 June 01, LEI, The Hague, the Netherlands.

Borda, I.A., Philen, R.M., Posada de la Paz, M., De la Cámara, A.G., Ruiz-Navarro, M.D., Giménez Ribota, O., Alvargonzález Soldevilla, J., Terracini, B., Peña, S.S., Fuentes Leal, C. and Kilbourne, E.M., 1998. Toxic oil syndrome mortality: the first thirteen years. International Journal of Epidemiology 27, pp. 1057-1063.

BRC, 2008. BRC Global Standard for Food Safety, Issue 5.

Bremmers, H.J., Van der Meulen, B.M.J. and Purnhagen, K.P., 2013. Multistakeholder responses to the European health claims requirements. Journal on Chain and Network Science 13, pp.161-172.

Broberg, M.P., 2008. Transforming the European Community's Regulation of Food Safety. Sieps 2008 (e-book). http://tinyurl.com/p8xoc5w.

Buchanan, R.L., 1995. The role of microbiological criteria and risk assessment in HACCP. Food Microbiology 12, pp. 421-424.

Bunte, F., Van Galen, M., de Winter, M., Dobson, P., Bergès-Sennou, F., Monier-Dilhan, S., Juhász, A., Moro, D., Sckokai, P., Soregaroli, C., Van der Meulen, B.J.M. and Szajkowska, A., 2011. The impact of private labels on the competitiveness of the European food supply chain, European Union 2011. http://tinyurl.com/o6dc5jr.

Capelli, F.and Klaus, B., 2009. Is garlic a food or a drug? EFFL 6, pp. 390-399.

Cardenas D., 2013. Let not they food be confused with thy medicine: Hippocratic misquotation. e-SPEN Journal 8, pp. 260-262.

Carlson, S., Carvajal, R., Coutrelis, N., Desjeux, J.F., Morelli, L., Van Dael, P, Van der Meulen, B.J.M., Tops, A. and Weirl, C., 2010. Publish and perish: a disturbing trend in the European Union's Regulation of nutrition health claims made on foods. FDLI Update Magazine, September/October 2010. http://tinyurl.com/ot2xq8h.

Carpenter, K., 2003. A short history of nutritional science: part 4 (1945-1985). Journal of Nutrition 133, pp. 3331-3342.

Cheftel, J.C., 2005. Food and nutrition labelling in the European Union. Food Chemistry 93, pp. 531-550.

Chia-Hui Lee, G., 2006. Private food standards and their impacts on developing countries. European Commission, DG Trade Unit G2 (internship report). http://trade.ec.europa.eu/doclib/docs/2006/november/tradoc_127969.pdf.

Coffee Jr., J.C., 2001. The rise of dispersed ownership: the roles of law & the state in the separation of ownership and control. Yale Law Journal 111, pp. 62-63.

Commission on Intellectual Property Rights, 2000. Integrating intellectual property rights and development policy. http://tinyurl.com/mz4tc6f.

Coppens, P., 2008. The use of botanicals in food supplements and medicinal products: the co-existance of two legal frameworks. EFFL 2, pp. 93-100

Coutrelis, N., 2000. Product liability in the food sector. International Business Lawyer 28(5).

Coutrelis, N., 2011. EU 'new approach' also for food law? In: Van der Meulen, B.J.M. (ed.), Private Food Law. Governing food chains through contract law, self-regulation, private standards, audits and certification schemes. Wageningen Academic Publishers, Wageningen, the Netherlands, pp. 381-390.

Dannecker G. (ed.), 1994. Criminal Law on food processing and distribution, and administrative sanctions in the European Union. Trier Academy of European Law series, issue 10. Bundesanzeiger Verlag, Cologne, Germany.

De Bakker, E., Backus, G., Selnes, T., Meeusen, M., Ingenbleek, P. and Van Wagenberg, C., 2007. Nieuwe rollen, nieuwe kansen? Een programmeringsstudie voor toezicht op controle in het agro-foodcomplex. Rapport 6 July 08, LEI, The Hague, the Netherlands. http://tinyurl.com/lu6fy8p.

De Jonge B., 2014. Plant variety protection in Sub-Saharan Africa. Balancing commercial and smallholder farmers' interests. Journal of Politics and Law 7(3), pp. 100-111.

De Swarte, C.S. and Donker, R.A., 2005. Towards an FSO/ALOP based food safety policy. Food Control 16, pp. 825-830.

Eide, W.B. and Kracht, U., 2005. Food and human rights in development. Vol. I. Legal and Institutional dimensions and selected topics. Itersentia, Antwerpen, Belgium.

Eide, W.B., and Kracht, U., 2007. Food and human rights in development. Vol. II. Evolving issues and emerging applications. Itersentia, Antwerpen, Belgium.

EU Pledge, 2012 Monitoring Report. November 2012, p. 22 *et seq.* http://www.eu-pledge.eu/content/annual-reports.

Eurodiet, 2000. Nutrition & Diet for Healthy Lifestyles in Europe, Science & Policy Implications. http://nutrition.med.uoc.gr/eurodiet/EurodietCoreReport.pdf.

Fogden, M., 2001. Hygiene. In: K. Goodburn (ed.), EU food law. A practical guide. Woodhead Publishing. Cambridge, UK.

Food Safety Authority of Ireland, 2010. Guidance on Food Additives. http://tinyurl.com/k3o89km.

Food Standards Agency, 2005. General guidance for food business operators EC Regulation No. 2073/2005 on microbiological criteria for foodstuffs. http://tinyurl.com/m5x6fn3.

Gallagher, M., and Thomas, I., 2010. Food fraud: the deliberate adulteration and misdescription of foodstuffs. EFFL 6, pp. 347-353.

Garcia-Cela, E.G., Ramos, A.J., Sanchis, V.S. and Marin, S.M., 2012. Emerging risk management metrics in food safety: FSO, PO. How do they apply to the mycotoxin hazard? Food Control 25, pp. 797-808.

Gelpí, E., Posada de la Paz, M., Terracini, B., Abaitua, I., de la Cámara, A.G., Kilbourne, E.M., Lahoz, C., Nemery, B., Philen, R.M., Soldevilla, L., Tarkowski, S. and WHO/CISAT Scientific Committee for the Toxic Oil Syndrome, 2002. The Spanish toxic oil syndrome twenty years after its onset: a multidisciplinary review of scientific knowledge. Environmental Health Perspectives 110, pp. 457-464.

GlobalGAP, 2008. Control Points and Compliance Criteria Plant Propagation Material. http://tinyurl.com/me4sxlh.

Gollin, M., 2008. Driving innovation. intellectual property strategies for a dynamic world, Cambridge University Press, New York, USA.

Hager, B.M., 2000. The rule of law: a lexicon for policy makers. 2nd edition, Mansfield Center for Pacific Affairs. http://tinyurl.com/pdr5ozq.

Hasler, C. M., 1998. Functional foods: their role in disease prevention and health promotion. Food Technology 52, pp. 57-62.

Havinga, T., 2006. Private regulation of food safety by supermarkets. Law & Policy 28, pp. 515-533.

Holland, D. and Pope, H., 2004. EU food law and policy. Kluwer Law International, 2004, pp. 110-120.

Hospes, O., 2009. Regulating biofuels in the name of sustainability or the right to food? In: Hospes, O. and Van der Meulen, B.J.M. (eds.) Fed up with the right to food? The Netherlands' policies and practices regarding the human right to adequate food. Wageningen Academic Publishers, Wageningen, the Netherlands, pp. 121-135.

Human Rights Law Network, 2009. Right to Food, Human Rights Network 4th edition. New Delhi, India.

ICTSD Bridges weekly. http://tinyurl.com/n5e4vtw.

International Association for the Study of Obesity, 2002. International Obesity Task Force and European Association for the Study of Obesity. Obesity in Europe – The Case for Action. 2002, p. 4 et seq. http://tinyurl.com/m2rkrog.

James, W.P.T., in collaboration with Feno-Luzzi, A., Isaksson, B., and Szostak, W.B., 1989. Healthy nutrition, preventing nutrition-related diseases in Europe. Nutrition Bulletin 14, pp. 134-135.

James, W.P.T., 1998. Food and nutrition policy in this century. In: WHO, Regional Office for Europe, European food and nutrition policies in action (WHO Regional Publications, European Series, No. 73), pp. 21-22.

James, P., Kemper, F. and Pascal, G., 1999. A European Food and Public Health Authority. The future of scientific advice in the EU. http://tinyurl.com/ma73bj2.

Josling, T., 2006. The war on terroir: geographical indications as a transatlantic trade conflict. Journal of Agricultural Economics 57, pp. 337-363.

Kemp, J.A., 2013. Exclusion from patentability of plant varieties and essentially biological processes for the production of plants. http://tinyurl.com/nboo5g3.

Kent, G., 2005. Freedom from want. The human right to adequate food. Georgetown University Press, Washington, DC, USA. http://tinyurl.com/lcre9wn.

Kimbrell, E., 2000. What is Codex Alimentarius. AgBioForum 3, pp. 197-202.

Kireeva, I., and Black, R., 2011. Chemical safety of food: setting of MRLs for plant protection products ('pesticides') in the European Union and in the Russian Federation. EFFL 3, pp. 174-186.

Klaus, B., 2011. 'Distinction between Feed Materials and Feed Additives in Consideration of the General Principles of Law'. EFFL 1, pp. 6-7.

Klaus, B., 2005. Der gemeinschaftsrechtliche Lebensmittelbegriff. Inhalt und Konsequenzen für die Praxis insbesondere im Hinblick auf die Abgrenzung von Lebensmitteln und Arzneimitteln, Verlag P.C.O. Bayreath (diss.), Bayreath, Germany.

Knipschild, K., 2003. Lebensmittelsicherheit als Aufgabe des Veterinär- und Lebensmittelrechts (diss.). Nomos Verlagsgesellschaft, Baden-Baden, Germany.

Knudsen, G. and Matikainen-Kallström, M., 1999. Joint Parliamentary Committee Report on food safety in the EEA, Brussels, 6 April 1999. http://tinyurl.com/md4ubtt.

Korzycka- Iwanow, M. and Zboralska, M., 2010. Never-ending debate on food supplements: harmonisation or disharmonsation of the Law? EFFL 3, pp. 124-135.

Kotler, Ph., and Keller, K., 2012. Marketing management. 14th Edition, Prentice Hall, New York, NY, USA.

Kozah, S., 2002. Realising the Right to Food in South Africa: not by policy alone – a need for framework legislation. South African Journal on Human Rights 20, pp. 664-683.

Lawrence, D., Kennedy, J. and Hattan, E., 2002. New controls on the deliberate release of GMOs. European Environmental Law Review 11, pp. 51-56.

Lelieveld, H. and Keener, L., 2007. Global harmonization of food regulations and legislation - the Global Harmonization Initiative. Trends in Food Science & Technology 18, pp. S15-S19.

Louwaars, N., Dons, H,. Van Overwalle, G., Raven, H., Arundel, A., Eaton, D. and Nelis, A., 2009. Breeding business: the future of plant breeding in the light of developments in patent rights and plant breeder's rights. CGN, Wageningen, the Netherlands. http://tinyurl.com/ld2svbw.

Luning, P.A., Marcelis, W.J. and Jongen, W.M.F., 2002. Food quality management. a techno-managerial approach. Wageningen Academic Publishers, Wageningen, the Netherlands.

MacMaolain, C., 2007. EU food law: protecting consumers and health in a common market. Hart Publishing, Oxford, UK, pp. 241-263.

MacRae, R., 1990. A history of sustainable agriculture. Ecological Agriculture Projects. Excerpt from thesis, 'Strategies for overcoming the barriers to the transition to sustainable agriculture'. http://tinyurl.com/oprt9vp.

Malanczuk, P., 1997. Akehurst's Modern Introduction to International Law, 7th ed. Akehurst, London, UK.

Masson-Matthee, M.D., 2007. The Codex Alimentarius Commission and its standards. An examination of the legal aspects of the Codex Alimentarius Commission. Asser Press, The Hague, the Netherlands, 2007.

McMullin, M.S. and Bell, R.P., 2003. Beerbrella. US Patent 6 637 447. http://tinyurl.com/pszgzkd.

Meisterernst. A., 2011. Foods for particular nutritional uses – death sentence passed for sound reasons? EFFL 6, pp. 315-327.

Navarro, S. and Wood, R., 2001. Codex Deciphered. Leatherhead, Surry UK.

Nöhle, U., 2005. Risikokommunication und Risikomanagement in der erweiterten EU. Zeitung fur Lebensmittelrecht 3, pp. 297-305.

O'Rourke, R., 205. European Food Law. 3rd Edition, Sweet & Maxwell, London, UK.

O'Rourke, R., 2012. EU Measures on the safety of food imports from Japan following the nuclear accident at Fukushima. European Journal of Risk Regulation 3, pp. 82-86.

Paganizza, V., 2011. Fukushima, RASFF and ECURIE – Condizioni speciali per l'importazione di alimenti e mangimi provenienti dal Giappone dopo l'11 marzo 2011. Rivista di diritto alimentare V, pp. 1-13.

Pagnattaro, M.A. and Peirce, E.R., 2010. From China to your plate – An analysis of new regulatory efforts and stakeholder responsibility to ensure food safety. The George Washington International Law Review 42, pp. 1-55.

Portman, O.W. and Hegsted, D.M., 1957. Nutrition. Annual Review of Biochemistry 26, pp. 307-326.

Riedl, R. and Riedl, C., 2008. Shortcomings of the New European Food Hygiene Legislation from the viewpoint of a competent authority. EFFL 2, pp. 64-83.

Robertson, A., Lobstein, T. and Knai C., 2007. Obesity and socio-economic groups in Europe: Evidence review and implications for action. http://tinyurl.com/mqok3sf.

Rodriguez Font, M., 2012. The 'Cucumber Crisis': legal gaps and lack of precision in the risk analysis system in food safety. Rivista di diritto alimentare VI, pp. 1-15.

Romero Melchor, S. and Timmermans, L., 2009. 'It's the dosage, stupid': The ECJ clarifies the border between medicines and botanical food supplements. EFFL 3, pp. 185-191.

Roosevelt, F.D., 1941. Four freedoms address to the USA Congress. 6 January 1941. http://tinyurl.com/aqsso4.

OECD, 2012. Working Party on Agricultural Policies and Markets. Final Report on Private Standards and the Shaping of the Agro-Food System. 31 July 2006. OECD, Health at a Glance: Europe 2012. OECD Publishing, p.8 and seq. http://tinyurl.com/l2wmnas.

Rosso Grossman, M., 2005. traceability and labeling of genetically modified crops, food and feed in the European Union. Journal of Food Law & Policy 1, pp. 43-85.

Rosso Grossman, M., 2006. Animal identification and traceability under the US National Animal Identification System. Journal of Food Law & Policy 2, pp. 231-315.

Scholten-Verheijen, I., Appelhof, T., Van den Heuvel, R. and Van der Meulen, B.J.M., 2012. Roadmap to EU Food Law. Eleven international publishing, The Hague, the Netherlands.

Scholten-Verheijen, I. and Tychon, H., 2010. The marketing and use of feed – a new Regulation. EFFL 3, pp. 160.

Senden, L.A.J., 2004. Soft Law in European Community Law. Hart Publishing, Oxford, UK.

Shaw, D.J., 2009. Global food and agricultural institutions. Routledge, London, UK.

Sirsi, E., 2012. GM food and feed. In: Costate, L. and Albisinni, F. (eds.), European Food Law, CEDAM Wolters Kluwer, Taranto, Italy, pp. 337-350.

Somsen, H. (ed.), 2007. The regulatory challenge of biotechnology. Human Genetics, Food and Patents, Biotechnology Regulation Series. Edward Elgar, Cheltenham UK, pp. 139-155.

Sprong, C., Van den Bosch, R., Iburg, S., de Moes, K., Paans, E., Sutherland Borja, S., Van der Velde, H., Van Kranen, H., Van Loveren, H., Van der Meulen, B.J.M. and Verhagen, H., 2014. Grey area novel foods: an investigation into criteria with clear boundaries. European Journal of Nutrition & Food Safety 4, pp. 342-363.

Szajkowska , A., 2010. The impact of the definition of the precautionary principle in EU food law. Common Market Law Review 47, pp. 173-196.

Szajkowska, A., 212. Regulating food law: risk analysis and the precautionary principle as general principles of EU food law. Wageningen Academic Publishers, Wageningen, the Netherlands.

Titz, A., 2005. The borderline between medicinal products and food supplements. Pharmaceuticals Policy and Law 8, pp. 37-49.

University of Reading EATWELL Project, 2013. Effectiveness of policy interventions to promote healthy eating and recommendations for future action. http://tinyurl.com/myflkp2.

Van der Meulen, B.M.J., 2004. The right to adequate food. Food Law between the market and human rights, Elsevier, The Hague, the Netherlands.

Van der Meulen, B.M.J., and Van der Velde, M., 2004. Food Safety Law in the European Union. Wageningen Academic Publishers, Wageningen, the Netherlands.

Van der Meulen, B.M.J. and Freriks, A.A., 2006. 'Beastly bureaucracy' animal traceability, identification and labeling in EU Law. Journal of Food Law & Policy 2, pp. 317-359.

Van der Meulen, B.M.J. and Freriks, A.A., 2006. Millefeuille. The emergence of a multi-layered controls system in the European food sector. Utrecht Law Review 2, pp. 156-176.

Van der Meulen, B.M.J. and Van der Velde, M., 2008. European Food Law Handbook, Wageningen Academic Publishers, Wageningen, the Netherlands.

Van der Meulen, B.M.J. and Hospes, O. (eds)., 2009. Fed up with the right to food? The Netherlands' policies and practices regarding the human right to adequate food, Wageningen Academic Publishers, Wageningen, the Netherlands.

Van der Meulen, B.M.J., 2009. The system of food law in the European Union. Deakin Law Review 14, pp. 305-339.

Van der Meulen, B.M.J., 2009. Reconciling food law to competitiveness. Wageningen Academic Publishers, Wageningen, the Netherlands.

Van der Meulen, B.M.J., 2010. The global arena of food law: emerging contours of a meta-framework. Erasmus Law Review 3, pp. 217-240.

Van der Meulen, B.M.J., 2010. The freedom to feed oneself: food in the struggle for paradigms in Human Rights Law. In: Hospes, O. and Hadiprayitno, I. (eds.), Governing Food Security. Wageningen Academic Publishers. Wageningen, the Netherlands, pp. 81-104.

Van der Meulen, B.M.J., 2010. The function of food law. On objectives of food law, legitimate factors and interests taken into account. EFFL 2, pp. 83-90.

Van der Meulen, B.M.J. (ed.), 2011. Private Food Law. Governing food chains through contract law, self-regulation, private standards, audits and certification schemes. Wageningen Academic Publishers, Wageningen, the Netherlands.

Van der Meulen, B.M.J., Bremmers, H.J., Wijnands, J.H.M. and Poppe, J., 2012. Structural precaution: the application of premarket approval schemes in EU food legislation. Food and Drug Law Journal 67, pp. 453-473.

Van der Meulen B.M.J., 2012. International Food Law, Part I. In: Scholten-Verheijen, I., Appelhof, T., Van den Heuvel, R. and Van der Meulen, B.J.M. (eds.) Roadmap to EU Food Law. Eleven international publishing, The Hague, the Netherlands, pp. 3-48.

Van der Meulen B.M.J., 2012. The Core of Food Law: A Critical Reflection on the Single Most Important Provision in All of EU Food Law, EFFL, pp. 117-125.

Van der Meulen, B.M.J., 2013. Governance in Law. Charting legal intuition. In: Colombi Ciacchi, A.L.B. (ed.). Law & Governance. Beyond the Public-Private Law Divide? Eleven international publishing, The Hague, the Netherlands, pp. 275-309.

Van der Meulen B.M.J., 2013. The structure of European Food Law. Laws 2, pp. 69-98.

Van der Meulen, B.M.J. and Van der Zee, E., 2013. 'Through the wine gate.' First steps towards human rights awareness in EU food (labelling) law. EFFL 8, pp. 41-52.

Van der Spiegel, M., 2004. Measuring effectiveness of food quality management (diss.). Wageningen University, Wageningen, The Netherlands.

Van Genugten, W., Meijknecht, A., Maister, B., Van Woensel, C., De Jonge, B., Tumushabe, G., Barungi, J., Louwaars, N., Napier, N., Gumbi, S. and de Wit, T.R., 2011. Harnessing intellectual property rights for development objectives. Wolf Legal Publishers, Nijmegen, the Netherlands.

Van Plaggenhoef, W., Batterink, M. and Trienekens, J.H., 2003. International trade and food safety. Overview of legislation and standards. Wageningen University, Wageningen, the Netherlands.

Van Schothorst, M., Zwietering, M.H., Ross, T., Buchanan, R.L., Cole, M.B., 2009. Relating microbiological criteria to food safety objectives and performance objectives. Food Control 20, pp. 967-979.

Verhagen, H., te Boekhorst, J., Kamps, L., van Lieshout, M.J., Ploeger, H., Verreth, D., Salminen, S. and van Loveren, H., 2009. Novel foods: an explorative study into their grey area. British Journal of Nutrition 101, pp. 1270-1277

Vidar, M., 2006. State recognition of the right to food at the national level. Unu-Wider Research Paper No. 2006/61, June 2006. http://tinyurl.com/qxma5pz.

Wahlgren, P., 2009. Legal Reasoning. A Jurisprudential Model. Stockholm Institute for Scandianvian Law. http://tinyurl.com/o3x6l9b

Watal, J. and Kampf, R., 2007. The TRIPs Agreement and Intellectual Property in Health and Agriculture. In: Krattiger, A., Mahoney, R.T., Nelsen, L., Thomson, J.A., Bennett, A.B., Satyanarayana, K., Graff, G.D., Fernandez, C. and Kowalski, S.P., (eds.) Intellectual Property Management in Health and Agricultural Innovation: A Handbook of Best Practices. MIHR Oxford, UK, p. 253.

Wernaart, B., 2014. The enforceability of the human right to adequate food. A comparative study. Wageningen Academic Publishers, Wageningen, the Netherlands.

Wijnands, J.H.M., Van der Meulen, B.M.J. and Poppe, K.J., 2007. Competitiveness of the European Food Industry. An economic and legal assessment, Office for Official Publications of the European Communities. http://tinyurl.com/p5xk7b6.

Will, M. and Guenther, D., 2007. Food quality and safety standards, as required by EU Law and the private industry with special reference to the MEDA countries' exports of fresh and processed fruits & vegetables, herbs & spices a practitioners' reference book. 2nd edition. GTZ. http://tinyurl.com/7xn9xq9.

Yeats, S., 2011. Latest developments in patenting plant inventions in Europe. EPO presentation, Brussels, 11 October 2011. http://tinyurl.com/l6q32ks.

Press

BBC News 4 July 1954: Housewives celebrate end of rationing. http://tinyurl.com/dl4vk.

BBC 16 may 1990 on BSE. http://tinyurl.com/2bezj.

BBC News, 5 June 2002. Belgian Hormone Killers Jailed, http://tinyurl.com/o5md5hk.

Butler, K., Four Men Found Guilty of Contract Hit on Vet. Independent (London), 5 June 2002, at 11.

Butler, K., Why the Mafia is into Your Beef: The EU Ban on Growth Hormones for Cows has Created a Lucrative Black Market, Independent (London), 19 March 1996, at 13.

European Voice 13 March 2013. Parliament brings CAP reform closer to Commission proposal. http://tinyurl.com/necozon.

HLN.BE 30 September 2009. http://tinyurl.com/np656re.

Graff, J., One Sweet Mess, Time, July 21, 2002. http://tinyurl.com/ngcab5t.

Whitney, C., Brussels Journal, Food Scandal Adds to Belgium's Image of Disarray, New York Times, 9 June 1999. http://tinyurl.com/ojwwtoa.

Websites

Civitas Iniative. http://tinyurl.com/n8fu3n4.
Codex Alimentarius. http://tinyurl.com/yk2hrov.
DG SANCO. Questions and Answers on Food Additives. http://tinyurl.com/o3ttkgh.
DG SANCO. Legislation on dietetic foods/ Foods for special groups. http://tinyurl.com/mhbwhrc.
DG SANCO. Animal and Plant Health Package: Smarter rules for safer food. http://tinyurl.com/bwqnyv3.
DG SANCO. Food. http://tinyurl.com/mzg8bn8.
DG SANCO. Food Contact Materials. http://tinyurl.com/7sygkka.
EFSA Application Helpdesk. http://tinyurl.com/ojxewc5.
EFSA. Food Additives. http://tinyurl.com/loshy87.
EFSA. Food Contact Materials Applications. http://tinyurl.com/phjq6hs.
EFSA. Food Contact Materials News. http://tinyurl.com/5vl4g3.
EPO Member States. http://tinyurl.com/75dml5n.
EPO. Unitary Patent. http://tinyurl.com/pgk895n.
EU Pledge. http://tinyurl.com/q9wx98m.
European Commission. Article 31 Euratom Treaty. http://tinyurl.com/lyh3o44.
European Commission. Joint research Centre. European Union Reference Laboratory for Food Contact Materials. http://tinyurl.com/cqg249c.
European Commission. School Fruit Scheme. http://tinyurl.com/qgohsde.
European Commission. EU agricultural product quality policy. http://tinyurl.com/3n9xcj8.
European Parliament. Legislative Observatory. http://tinyurl.com/l5679ue.
Fediaf. http://tinyurl.com/p89le6n.
Foodinsight. Background on Functional Foods. http://tinyurl.com/mfqojxx.
FAO. http://tinyurl.com/yu2wt.
IFS. Packaging Guideline. http://tinyurl.com/pjcftmg.
Madrid – The International Trademark System. http://tinyurl.com/49jq3xf.
No Patent on seeds. http://tinyurl.com/q997v3l.
Novel foods (approved and other). http://tinyurl.com/2cba37t.
Proposed legislation for market regulation. http://tinyurl.com/2capk2.
The United States Patent and Trademark Office. http://tinyurl.com/2fh8.
USDA. Honey Bees and Colony Collapse Disorder. http://tinyurl.com/533udq.
WHO. http://tinyurl.com/6dsta7.
WHO. Obesity. http://tinyurl.com/plhhezq.
WIPO. The International Design System. http://tinyurl.com/4sf8haq.
WTO. http://tinyurl.com/2tv8ls.
WTO. International Trade Statistics 2007. http://tinyurl.com/5efnl8.
WTO. Organization chart. http://tinyurl.com/8a76o5z.

About the authors

Frank Andriessen studied veterinary medicine at Utrecht University, the Netherlands. He joined the European Commission in 1994 as inspector/auditor dealing mainly with import controls, third country audits, and BSE-related issues. He managed the Quality and Development Unit within FVO before taking up his current position as Head of Unit SANCO.F1 'Country Profiles, Co-ordination of Follow-up'. E-mail: frank.andriessen@ec.europa.eu

Dr. **Harry J. Bremmers** is Associate Professor Food Law and Economics at the Law and Governance Goup of Wageningen University, Netherlands.
E-mail: harry.bremmers@wur.nl

Dr. **Morten P. Broberg** is Professor of Law and Jean Monnet Chair-holder at the Faculty of Law at the University of Copenhagen, Denmark.
E-mail: morten.broberg@jur.ku.dk

David Byrne is Chancellor *emeritus* of Dublin City University, Ireland. At the time of the recast of EU food law he was EU Commissioner Health and Consumer Affairs.

Dionne C. Chan graduated from the Master International and European law with specialisation in International Trade and Investment Law and the Master (Dutch) Private Law at the University of Amsterdam in the Netherlands. Her contribution results from her internship at the Law and Governance Group at Wageningen University in the Netherlands. E-mail: dionne.c.chan@gmail.com

Dr. **Bram De Jonge** is Researcher at Law & Governance Group, Wageningen University, the Netherlands, where he works in the field of Intellectual Property Rights, genetic resources policies, and international development. He is currently a visiting researcher at the Intellectual Property Law and Policy Research Unit of the University of Cape Town, South Africa. E-mail: bram.dejonge@wur.nl

Martin Holle is Professor of Food Law and Administrative Law at the Hamburg University of Applied Sciences, Germany. Until June 2013 he was General Counsel, Foods at Unilever PLC, London, UK, advising on all matters of Food and Advertising Law. He is member of the advisory board of the German Scientific Association for Food Law (WGfL) and author of a guide to the EU Regulation on Nutrition and Health Claims. E-mail: martin.holle@haw-hamburg.de

Jaap D. Kluifhooft is Head of Regulatory and Scientific Affairs at Stepan Lipid Nutrition, Amsterdam, the Netherlands. He was educated as a food technologist in Wageningen University, the Netherlands where he build a profound practical experience and knowledge in obtaining and maintaining the safety and regulatory compliance of a food ingredient portfolio in a B2B-environment. He successfully managed Novel Food submissions in EU, Canada, Australia and New Zealand and FDA GRAS notifications. E-mail: jaap.kluifhooft@lipidnutrition.com

Dr. **Karola Krell Zbinden** is Attorney-at-law at Markwalder Emmenegger in Berne, Switzerland. E-mail: karola.krell@mepartners.ch

Dr. **Martha Cecilia Kühn** focuses on technical and regulatory compliance for EU market entry as Project Director at EU Food Comply (www.foodcomply.eu; the Netherlands). She is also Adjunct food law researcher and lecturer at Wageningen University, the Netherlands and was formerly Project Group Manager and Product Development Manager at Unilever. She is author of 'The Functional Foods Dossier: building solid health claims', a practical industrial guide to prepare a scientific dossier for health claims of European functional foods. E-mail: c.kuhn@foodcomply.eu

Bernard Maister M.D., J.D., LL.M. practiced law in the USA before joining the IP Unit at the University of Cape Town, South Africa where he researches, writes and teaches in the area of IP law. E-mail: maisterb@gmail.com

Francesco Montanari holds a PhD in EU law from the University of Bologna, Italy. He is a former official of the European Commission, and currently runs his own food law practice in Portugal. He is also a member of the European Food Law Association, and Senior Associate at Arcadia International and Food Agriculture Requirements (FARE). E-mail: francesco.montanari@arcadia-international.net

Dr. Ir. **Ans Punt** is Assistant Professor at the division of Toxicology of Wageningen University, the Netherlands. E-mail: ans.punt@wur.nl

Dr. **Hanna Schebesta** is Lecturer at the Law and Governance Group at Wageningen University, the Netherlands and Academic Assistant of the Alias-Project at the European University Institute in Florence, Italy. E-mail: hanna.schebesta@wur.nl

Irene E.M. Scholten-Verheijen is Attorney-at-law and partner at Legaltree, the Netherlands. She specialises in regulated markets, with a focus on the pharma and food sectors. Irene is a member of the Dutch Food Law Association (NVLR) and the European Food Law Association (EFLA) and a board member of the European Institute for Food Law. She has several publications to her name (among others: 'Landkaart levensmiddelenrecht', Praktijkgidsen Warenwet, and Private Food Law). E-mail: irene.scholten@legaltree.nl

Rozita Spirovska Vaskoska works as an independent consultant and is also active as external teaching staff at the Faculty of Technology and Technical Sciences – Veles, University St. Kliment Ohridski-Bitola, Macedonia.
E-mail: rozitavaskoska@gmail.com

Anna Szajkowska is Case Handler – Infringements at the European Commission, Directorate-General for the Environment. She studied law at Warsaw University in Poland. Before joining the European Commission in 2012, she was a researcher at the Law and Governance Group at Wageningen University, the Netherlands, where she also obtained her PhD with a thesis entitled 'Regulating food law. Risk analysis and the precautionary principle as general principles of EU food law'. In 2005 she was awarded an academic scholarship from the Italian National Research Council. E-mail: anna.szajkowska@ec.europa.eu

Bernd M.J. van der Meulen is Professor of Law and Governance at Wageningen University, the Netherlands. The Law and Governance Group focus their research and teaching on the legal and legal-economic aspects of agro-food-chain, including regulatory affairs, intellectual property, comparative food law and human rights. E-mail: bernd.vandermeulen@wur.nl

Sofie M.M. van der Meulen is Attorney-at-law at Axon Lawyers, Amsterdam, the Netherlands. She graduated at Leiden University, the Netherlands with a degree in administrative and constitutional law. She is involved in the food sector through the organization of the Food Law Academy and is co-author of the EU food law roadmap. She specializes in legal and regulatory aspects of medical technology. E-mail: sofie.vandermeulen@axonlawyers.com.

Menno van der Velde was until his retirement in 2013 Associate Professor of Law and Governance with a specialisation in European Union law at Wageningen University, the Netherlands.

Dasep Wahidin is a PhD researcher in food law at Wageningen University, the Netherlands. E-mail: dasep.wahidin@wur.nl

Index

A

B

C

D

E

G

H

I

J

K

L

M

N

S

T

U

V

W

Y

Z